Our users say it best...

"The major enhancement in the fourth edition involves the addition of substantial information and detailed examples on mixed models using PROC MIXED as well as thorough comparisons of PROC MIXED and PROC GLM analyses. It serves as an excellent introduction to PROC MIXED for the most common mixed-models situations (nested, two-way cross-classifications, split-plots, models with mixtures of crossed and nested effects and repeated measures), using classical random-effects assumptions. There are good discussions about random versus fixed effects, problems with unbalanced data or missing cells, and techniques of analysis."

Leigh W. Murray
Professor of Experimental Statistics and Director
University Statistics Center
New Mexico State University

"The authors aim to write a book that offers a broad coverage of regression and ANOVA models. They have completed the mission. The first chapter, 'Introduction,' is clear and should be read first to get a sense of the road map to the linear models. Because of the breadth rather than depth of its content, it fits intermediate users; however, advanced users may use it for quick reference. That is, this book is good for an overview as well as a reference. The whole book is user friendly, and it is easy to follow the content. Its special feature is the comparison of current advancements in selecting methods (such as PROC ANOVA and PROC GLM) for analyzing linear models. I highly recommend this book for ANOVA/SAS courses."

Mayling M. Chu, Ph.D.
California State University, Stanislaus

SAS® FOR LINEAR MODELS

Fourth Edition

Ramon C. Littell

Walter W. Stroup

Rudolf J. Freund

The correct bibliographic citation for this manual is as follows: Littell, Ramon C., Walter W. Stroup, Rudolf J. Freund. 2002. *SAS® for Linear Models, Fourth Edition*. Cary, NC: SAS Institute Inc.

SAS® for Linear Models, Fourth Edition

SAS Institute Inc. ISBN 1-59047-023-0
John Wiley & Sons, Inc. ISBN 0-471-22174-0

SAS Institute Inc., SAS Campus Drive, Cary, North Carolina 27513.

1st printing, March 2002
2nd printing, April 2005

SAS Publishing provides a complete selection of books and electronic products to help customers use SAS software to its fullest potential. For more information about our e-books, e-learning products, CDs, and hard-copy books, visit the SAS Publishing Web site at **support.sas.com/pubs** or call 1-800-727-3228.

Contents

Chapter 5   Unbalanced Data Analysis: Basic Methods

Chapter 7   Analysis of Covariance

Chapter 8   Repeated-Measures Analysis

Chapter 11   Examples of Special Applications

x

Acknowledgments

We would like to acknowledge several people at SAS whose efforts have contributed to the completion of this book. First of all, we are grateful to Jim Goodnight, who originally encouraged us to write the book. We also thank Jim Ashton, Donna Fulenwider, and Jenny Kendall, who reviewed the original chapters and contributed many useful comments.

The work of several colleagues has influenced our writing. In particular, we acknowledge Professors Walt Harvey of Ohio State University; Ron Hocking, formerly of Texas A&M University; George Milliken and Dallas Johnson of Kansas State University; Bill Sanders of the University of Tennessee; Shayle Searle of Cornell University; and David Weeks of Oklahoma State University.

Walter Stroup would like to thank colleague Erin Blankenship and graduate students in the UNL biometry department, especially Kurt Brumbaugh, Prabhakar Dhungana, LeAnna Guerin, Haifeng Guo, Rebecca Hensberry, Lana Olson, Yaobing Sui, and Yuli Xie, for their valuable feedback during the writing of this book.

For the fourth edition we gratefully acknowledge the SAS reviewers, Brent Cohen, Mike Patetta, Kathleen Kiernan, and Randy Tobias. We are also grateful to the outside reviewers, Stephen Kachman of the University of Nebraska, Mark Payton of Oklahoma State University, and William Bridges Jr. of Clemson University.

Also important to this effort is the SAS production team of copyeditor Tate Renner, technical publishing specialists Candy Farrell and Janet Howard, cover designer Cate Parrish, and marketing analysts Patricia Spain and Ericka Wilcher. We express a special thank-you to acquisitions editor Judy Whatley for her many efforts keeping us on track and keeping us organized, and for her attention to the many details of this project. We literally could not have done it without her.

Finally, we thank our wives and families for their support and for their patience during the many hours spent away from them working on this book.

Chapter 1 Introduction

1.1 About This Book

The fourth edition of *SAS® for Linear Models*, like earlier editions, plays a role somewhere between a textbook on applied linear models and a manual for using certain procedures in SAS. It can serve as a companion to various courses offered at colleges and universities. Most universities offer a course on statistical methods directed at graduate students in scientific fields that utilize analysis of variance and regression analysis. These topics are covered at an appropriate level for such a course. In addition, this book makes a useful companion to a graduate course in linear models for statistics students to bridge the gap between theory and application. It also is useful to persons engaged in data analysis as a reference for statistical topics and SAS programming to implement a multitude of methods that fall under the general classification of "linear models."

The list of topics covered in *SAS® for Linear Models* is intentionally very broad to at least touch the bases mentioned in the previous paragraph. Other books offer more detailed information of a similar nature on regression analysis, mixed models, and other topics. These are *SAS® System for Regression, Third Edition, SAS® System for Mixed Models,* and *Categorical Data Analysis Using the SAS® System, Second Edition,* all published by SAS.

1.2 Statistical Topics and SAS Procedures

The broad range of topics are itemized and detailed below, according to chapter. Previous editions of *SAS® System for Linear Models* used the REG, MEANS, TTEST, ANOVA, and GLM procedures in SAS. In the decade since publication of the third edition, there have been enormous advances in computer technology and statistical methodology, as well as enhancements to SAS. These have resulted in changes in the most efficient and appropriate ways of analyzing data, and standards for data analysis have been raised by scientific journals and regulatory agencies. Accordingly, we have shifted emphasis of the procedures and topics in the fourth edition. Most of the applications of PROC ANOVA in the first three editions now use PROC GLM, and PROC MIXED is used for most mixed-model applications. A brief treatment of a new topic, *generalized* linear models (as opposed to *general* linear models), is introduced, illustrated primarily with PROC GENMOD.

Chapter 2 describes linear regression analysis for single and multiple independent variables. However, the main purpose is not to provide a comprehensive illustration of regressions methods. Rather, the intent is to lay the groundwork for linear models and the inferential techniques for estimating and testing hypotheses regarding linear combinations of parameters by using the REG and GLM procedures.

Chapter 3 introduces the basic methods for estimating and comparing means by using *t*-tests and analysis of variance. The MEANS, TTEST, and GLM procedures are demonstrated. The ANOVA procedure is only briefly mentioned because PROC GLM has all the capabilities of PROC ANOVA.

Chapter 4 describes ANOVA methods for data with random effects. Analyses for nested sampling classifications, randomized-blocks designs, and two-way mixed-models are illustrated using the GLM and MIXED procedures. Results from GLM and MIXED are compared for the two-way mixed model.

Chapter 5 presents analysis of unbalanced data. The fixed-effects situation is discussed first, covering the four types of sums of squares in PROC GLM. Least-squares means and the concept of estimable functions are covered next, and then analysis of unbalanced mixed-model data is discussed, using both PROC GLM and PROC MIXED.

Chapter 6 discusses details of the linear model. This chapter presents the principles and theory needed to understand the methods used by PROC GLM and PROC MIXED.

Chapter 7 describes analysis of covariance—that is, methods for comparing means in the presence of a continuous concomitant variable. Also, general issues are described when both qualitative and quantitative variables are included in a model.

Chapter 8 focuses on repeated-measures analysis. Analysis methods using PROC GLM are presented, similar to the third edition. We also present an introduction to the topic using PROC MIXED, which models the covariance structure of the repeated measures.

Chapter 9 discusses multivariate analysis, and presents essentially the same material as in the third edition.

Chapter 10 introduces the topic of *generalized* linear models. These are models with a non-normally distributed response variable. A common subtopic is *logistic regression* for the case of a binary response variable. This is a new topic for the fourth edition, which introduces PROC GENMOD. Since the third edition, this methodology has become accepted in many subject matter areas, and expected in some.

Finally, Chapter 11 presents examples of special analyses. These are examples that extend and combine methods that have been described in previous chapters.

Chapter 2 Regression

2.1 Introduction

Regression analysis refers to statistical techniques that are used to relate a variable to one or more other variables. The relationship is expressed as an equation that is estimated from a set of data that contains observations on the variables. When the equation is linear, we have a **linear regression analysis.** An example of a linear regression equation is

$$y = \beta_0 + \beta_1 x_1 + \beta_2 x_2 + \varepsilon \tag{2.1}$$

With this equation, we would be interested in relating y to the variables x_1 and x_2. The variable y is called the **dependent** variable, and x_1 and x_2 are called **independent** variables. The manner in which y depends on x_1 and x_2 is determined by the values of the **regression coefficients**, β_0, β_1, and β_2. The coefficient β_0 is called the **intercept**, and is the expected value of y when both x_1 and x_2 are equal to zero. The coefficient β_1 is called the **slope** in the x_1 direction. It measures the expected amount of change in y for each unit change in x_1, holding x_2 fixed. Likewise, β_2 measures the expected amount of change in y for each unit change in x_2, holding x_1 fixed. Finally, ε is an unobservable random variable that accounts for the fact that the relation between the dependent variable y and the independent variables x_1 and x_2 is not deterministic. In other words, the value of y is not strictly determined by values of x_1 and x_2. It also depends on *random variation* through the value of ε. In most situations, ε is assumed to be a random variable with an expected value equal to 0, which we write as $E(\varepsilon)=0$. With this assumption, $E(y)= \beta_0 + \beta_1 x_1 + \beta_2 x_2$. If there is only one independent variable, the regression equation is called a **simple linear regression** equation. If there are two or more independent variables, then the regression equation is called a **multiple linear regression** equation.

When we use equation (2.1) to represent the relation between y and x_1 and x_2 for different values of the variables in a data set, we write

$$y_j = \beta_0 + \beta_1 x_{1j} + \beta_2 x_{2j} + \varepsilon_j \qquad\qquad (2.2)$$

where the value of j indicates the observation number. Equation (2.2), along with assumptions about the probability distributions of the ε_j, is called the statistical *model*. Frequently, we assume the ε_j are distributed normally and with equal variances, and independently of each other. In applications, we estimate the parameters β_0, β_1, and β_2, and denote their estimates $\hat{\beta}_0$, $\hat{\beta}_1$, and $\hat{\beta}_2$. We then obtain the prediction equation

$$\hat{y} = \hat{\beta}_0 + \hat{\beta}_1 x_1 + \hat{\beta}_2 x_2$$

We assume you are familiar with the meaning of linear equations, and that you have a basic knowledge of regression analysis. However, a short review of regression methods and computations is presented in Section 2.4. SAS contains several procedures that are capable of doing the computations for linear regression analysis. The main regression procedure is PROC REG. It can perform computations for numerous linear regression methods and has extensive graphics capabilities. You can read about using PROC REG in Section 2.2. You can learn about many other applications of PROC REG in *SAS®* *System for Regression, Third Edition* (Freund and Littell, 2000). In the present book, you will learn about regression methods applied to other situations, such as analysis of variance and models with random effects. Other SAS procedures, PROC GLM and PROC MIXED, are better suited for these types of applications. Following the introduction to PROC REG, we switch to PROC GLM and PROC MIXED for most of the remainder of the book. The main purpose of showing the regression applications of PROC GLM in this chapter is to give you the basis for general linear model applications in terms of regression techniques.

2.2 The REG Procedure

The basic use of PROC REG is to obtain an analysis of data based on a linear regression model, as specified in Section 2.4, "Statistical Background." The following SAS statements invoke PROC REG:

```
proc reg;
   model list-of-dependent-variables=list-of-independent-
   variables;
```

PROC REG can perform a regression of a single dependent variable y on a single independent variable x using the following SAS statements:

```
proc reg;
   model y=x;
```

The results of a simple linear regression of this type are shown in Output 2.2.

Many options are available in PROC REG. Some of these are specified following a slash (/) at the end of the MODEL statement (see Section 2.2.2, "The P, CLM, and CLI Options: Predicted Values and Confidence Limits," for examples). Other options are specified as separate SAS statements (see Section 2.2.5, "Tests of Subsets and Linear Combinations of Coefficients" and Section 2.2.6, "Fitting Restricted Models: RESTRICT Statement and NOINT Option," for examples).

2.2.1 Using the REG Procedure to Fit a Model with One Independent Variable

This section illustrates a regression model with one independent variable. A SAS data set named MARKET contains a variable named CATTLE whose values are numbers of head of cattle (in thousands) that were sold in 19 livestock auction markets. The data set also contains a variable named COST whose values are costs of operation of the auction market in thousands of dollars. Of course, auction markets that sell larger numbers of head of cattle also have higher operating costs. We want to use simple linear regression analysis to relate cost of operation to number of head of cattle sold. There are three primary objectives in the regression analysis:

❑   to estimate the COST of operation of the auction market per head of CATTLE

❑   to estimate the average cost of operating a market that sells a specified number of head of cattle

❑   to determine how much of the variation in cost of operation is attributable to variation in the number of head of cattle.

The SAS data set named MARKET with variables MKT, CATTLE, and COST appears in Output 2.1.

Output 2.1
Data Set
MARKET

```
                         The SAS System

              OBS    MKT    CATTLE     COST
               1      1      3.437    27.698
               2      2     12.801    57.634
               3      3      6.136    47.172
               4      4     11.685    49.295
               5      5      5.733    24.115
               6      6      3.021    33.612
               7      7      1.689     9.512
               8      8      2.339    14.755
               9      9      1.025    10.570
              10     10      2.936    15.394
              11     11      5.049    27.843
              12     12      1.693    17.717
              13     13      1.187    20.253
              14     14      9.730    37.465
              15     15     14.325   101.334
              16     16      7.737    47.427
              17     17      7.538    35.944
              18     18     10.211    45.945
              19     19      8.697    46.890
```

A simple linear regression model is used in the initial attempt to analyze the data. The equation for the simple linear regression model is

$$COST = \beta_0 + \beta_1\, CATTLE + \varepsilon \tag{2.3}$$

This is called the **population** model, and it describes the relation between COST and CATTLE in a population of markets. The 19 markets in the data set represent a random sample from this population. The population model equation (2.3) uses a straight line with slope β_1 and intercept

β_0 to represent the relationship between *expected* COST of operation and number of head of CATTLE sold. If the true relationship between expected COST and CATTLE is essentially linear over the range of values for CATTLE in the data set, then the parameter β_1 measures the expected cost of operation per head of cattle sold. The parameter β_0, in theory, would be the expected cost for a market that sold no CATTLE, but it has questionable practical value. The straight line might provide a good approximation over the range from, say, CATTLE=3.0 through CATTLE=15, but it might not be of much use outside this range. Finally, the ε term in the model is a random variable, commonly called an error term, that accounts for variation of COST among markets that sell the same number of CATTLE. We make the following assumptions about the ε's:

❑ The ε's are independent.

❑ The ε's have expected value 0, $E(\varepsilon) = 0$.

❑ The ε's have variance σ^2, $V(\varepsilon) = \sigma^2$.

Use the following SAS statements to perform a simple linear regression with PROC REG:

```
proc reg data=market;
   model cost=cattle;
```

The PROC statement invokes PROC REG, and DATA=MARKET tells SAS to apply PROC REG to the data set named MARKET. In the remainder of this book, you may assume that the data set used is the last data set created unless DATA= *data-set-name* is specified.

The MODEL statement contains the equation COST=CATTLE, which corresponds to the statistical model of equation (2.3). The left side specifies the dependent variable (in this case COST), and the right side specifies the independent variable (in this case CATTLE). Note that no term is specified in the MODEL statement corresponding to the intercept (β_0) because an intercept term is automatically assumed by PROC REG unless indicated otherwise with the NOINT option (see Section 2.2.6). In addition, no term is indicated in the MODEL statement corresponding to the error term (ε). PROC REG produces ordinary least-squares (OLS) estimates of the parameters. The OLS estimates are optimal (unbiased with minimum variance) if the errors are independent and have equal variances.

Results from PROC REG appear in Output 2.2. Technical details are given in Section 2.4.

Output 2.2
Results of
Regression
with One
Independent
Variable

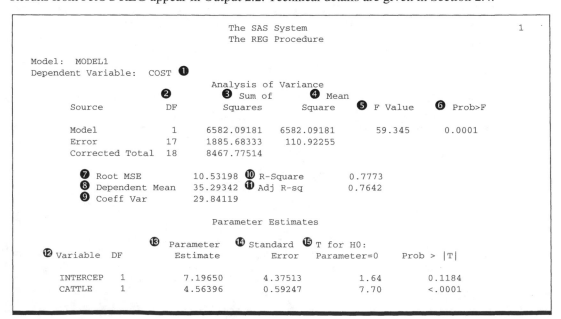

```
                                     The SAS System                              1
                                    The REG Procedure

Model:  MODEL1
Dependent Variable:  COST ❶
                                   Analysis of Variance
                         ❷           ❸ Sum of      ❹ Mean
        Source          DF           Squares        Square    ❺ F Value   ❻ Prob>F

        Model            1         6582.09181     6582.09181     59.345      0.0001
        Error           17         1885.68333      110.92255
        Corrected Total 18         8467.77514


            ❼ Root MSE          10.53198 ❿ R-Square        0.7773
            ❽ Dependent Mean    35.29342 ⓫ Adj R-sq        0.7642
            ❾ Coeff Var         29.84119


                              Parameter Estimates

                   ⓭ Parameter    ⓮ Standard  ⓯ T for H0:
    ⓬ Variable  DF    Estimate        Error     Parameter=0      Prob > |T|

      INTERCEP    1     7.19650       4.37513        1.64           0.1184
      CATTLE      1     4.56396       0.59247        7.70          <.0001
```

The callout numbers have been added to the output to key the descriptions that follow:

❶ The name of the Dependent Variable (COST).

❷ The degrees of freedom (DF) associated with the sums of squares (SS).

❸ The Regression SS (called MODEL SS) is 6582.09181, and the Residual SS (called ERROR SS) is 1885.68333. The sum of these two sums of squares is the CORRECTED TOTAL SS=8467.77514. This illustrates the basic identity in regression analysis that TOTAL SS=MODEL SS + ERROR SS. Usually, good models result in the MODEL SS being a large fraction of the CORRECTED TOTAL SS.

❹ The Mean Squares (MS) are the Sum of Squares divided by the respective DF. The MS for ERROR (MSE) is an unbiased estimate of σ^2, provided the model is correctly specified.

❺ The value of the *F*-statistic, 59.345, is the ratio of the MODEL Mean Square divided by the ERROR Mean Square. It is used to test the hypothesis that all coefficients in the model, except the intercept, are equal to 0. In the present case, this hypothesis is $H_0 : \beta_1 = 0$.

❻ The *p*-value (Prob>F) of 0.0001 indicates that there is less than one chance in 10,000 of obtaining an *F* this large or larger by chance if, in fact, $\beta_1 = 0$. Thus, you can conclude that β_1 is not equal to 0.

❼ Root MSE=10.53198 is the square root of the ERROR MS and estimates the error standard deviation, σ (if the model is adequate).

❽ Dependent Mean=35.29342 is simply the average of the values of the variable COST over all observations in the data set.

❾ Coeff Var=29.84119 is the coefficient of variation expressed as a percentage. This measure of relative variation is the ratio of Root MSE to Dependent Mean, multiplied by 100. In this example, the error standard deviation is 29.84119% of the overall average value of COST. The C.V. is sometimes used as a standard to gauge the relative magnitude of error variation compared with that from similar studies.

❿ R-Square=0.7773 is the square of the multiple correlation coefficient. It is also the ratio of MODEL SS divided by TOTAL SS and, thereby, represents the fraction of the total variation in the values of COST due to the linear relationship to CATTLE.

⓫ Adj R-sq is an alternative to R-Square and is discussed in Section 2.2.3, "A Model with Several Independent Variables."

⓬ The labels INTERCEP and CATTLE identify the coefficient estimates.

⓭ The number 7.19650 to the right of the label INTERCEP is the estimate of β_0, and the number 4.56396 to the right of the label CATTLE is the estimate of β_1. These give the fitted model

$$\widehat{COST} = 7.19650 + 4.56396(CATTLE)$$

These parameter estimates and this equation provide some of the information sought from the data. The quantity 4.564 is the *estimate* of the COST of operation per head of CATTLE. The equation can be used to estimate the cost of operating a market that sells any prescribed number of cattle. For example, the cost of operating a market selling ten thousand head of cattle is estimated to be $7.196 + 4.564(10.0) = 52.836$, that is, \$52,836.

Section 2.2.2 shows how PROC REG can provide this estimate directly, compute its standard error, and construct confidence limits.

⑭ The (estimated) Standard Errors of the estimates of β_0 and β_1 are 4.375 and 0.592. These can be used to construct confidence intervals for the model parameters. For example, a 95% confidence interval for β_1 is

$$\left(4.564 - 2.110(0.592),\ 4.564 + 2.110(0.592)\right) = (3.315,\ 5.813)$$

where 2.110 is the .05 level tabulated *t*-value with 17 degrees of freedom. Thus, you can infer, with 95% confidence, that the mean cost of market operation per head of cattle is between 3.315 and 5.813 dollars.

⑮ The *t*-statistic for testing the null hypothesis that $\beta_0 = 0$ is *t*=1.64 with *p*-value (Prob > |T|) of 0.1184. The statistic for testing the null hypothesis that $\beta_1 = 0$ is *t*=7.70 with a *p*-value of 0.0001. For this one-variable model, the *t*-test is equivalent to the *F*-test for the model because $F=t^2$.

In this example, a simple linear regression model is fitted using ordinary least squares. There are two assumptions that are necessary (but not sufficient) for inferences to be valid. These assumptions are listed below:

❑   The true relationship between COST of operation and number of CATTLE is linear.

❑   The variance of COST is the same for all values of CATTLE.

These assumptions are typically not completely met in real applications. If they are only slightly violated, then inferences are valid for most practical purposes.

Output 2.3 is produced by the following statements:

```
proc reg data=auction;
   model cost=cattle;
   plot cost*cattle;
run;
```

The graph shows the values of COST plotted versus CATTLE with the regression plotted through the data. The plot indicates that both assumptions are reasonable, except for the exceptionally large value of COST for CATTLE=14.325.

Output 2.3
Plot of
CATTLE Data

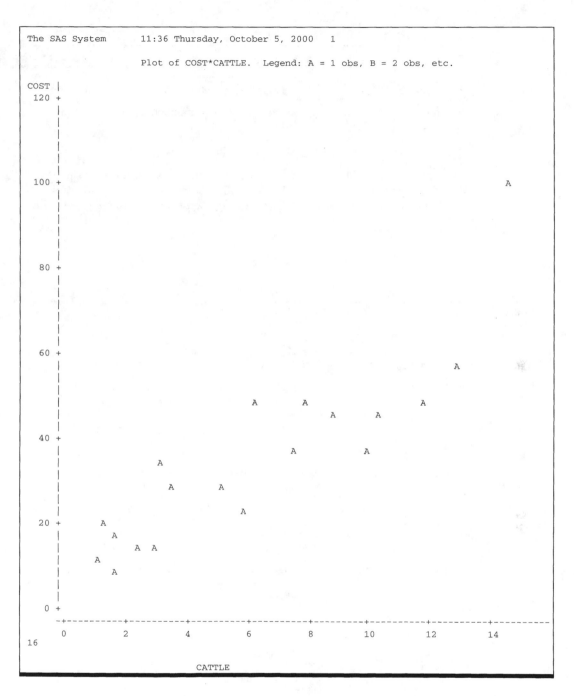

2.2.2 The P, CLM, and CLI Options: Predicted Values and Confidence Limits

A common objective of regression analysis is to compute the predicted value

$$\hat{y} = \hat{\beta}_0 + \hat{\beta}_1 x_1 + \ldots + \hat{\beta}_k x_k$$

for some selected values of $x_1, \ldots, x_k$. You can do this in several ways using PROC REG or other regression procedures in SAS. In PROC REG, the most direct way is to use the P (for predicted) option in the MODEL statement. Output 2.4 shows the results from the following SAS statements:

```
proc reg data-auction;
   id cattle;
   model cost=cattle / p clm cli;
```

Output 2.4
PROC REG
with the P,
CLM, and
CLI Options

```
                              The SAS System                              1


Dep  Var    Predicted    Std Err
     Obs     CATTLE        COST      Value  Mean Predict      95%   CL Mean

      1        3.437     27.6980    22.8828      2.9041   16.7558   29.0099
      2       12.801     57.6340    65.6197      4.6192   55.8741   75.3653
      3        6.136     47.1720    35.2009      2.4162   30.1031   40.2987
      4       11.685     49.2950    60.5263      4.0704   51.9386   69.1140
      5        5.733     24.1150    33.3617      2.4292   28.2365   38.4868
      6        3.021     33.6120    20.9842      3.0477   14.5541   27.4143
      7        1.689      9.5120    14.9050      3.5837    7.3440   22.4661
      8        2.339     14.7550    17.8716      3.3095   10.8891   24.8541
      9        1.025     10.5700    11.8746      3.8834    3.6814   20.0677
     10        2.936     15.3940    20.5963      3.0787   14.1009   27.0917
     11        5.049     27.8430    30.2399      2.5037   24.9576   35.5222
     12        1.693     17.7170    14.9233      3.5820    7.3659   22.4806
     13        1.187     20.2530    12.6139      3.8087    4.5783   20.6496
     14        9.730     37.4650    51.6038      3.2127   44.8257   58.3819
     15       14.325    101.3340    72.5752      5.4094   61.1624   83.9880
     16        7.737     47.4270    42.5078      2.5914   37.0405   47.9751
     17        7.538     35.9440    41.5996      2.5511   36.2172   46.9820
     18       10.211     45.9450    53.7991      3.4072   46.6104   60.9877
     19        8.697     46.8900    46.8892      2.8468   40.8831   52.8954


     Obs            CATTLE       95% CL Predict    Residual

      1              3.437     -0.1670   45.9326     4.8152
      2             12.801     41.3560   89.8834    -7.9857
      3              6.136     12.4031   57.9987    11.9711
      4             11.685     36.7041   84.3486   -11.2313
      5              5.733     10.5577   56.1656    -9.2467
      6              3.021     -2.1480   44.1164    12.6278
      7              1.689     -8.5667   38.3767    -5.3930
      8              2.339     -5.4202   41.1634    -3.1166
      9              1.025    -11.8083   35.5575    -1.3046
     10              2.936     -2.5542   43.7467    -5.2023
     11              5.049      7.4002   53.0797    -2.3969
     12              1.693     -8.5472   38.3938     2.7937
     13              1.187    -11.0149   36.2428     7.6391
     14              9.730     28.3725   74.8351   -14.1388
     15             14.325     47.5951   97.5552    28.7588
     16              7.737     19.6246   65.3911     4.9192
     17              7.538     18.7365   64.4627    -5.6556
     18             10.211     30.4447   77.1535    -7.8541
     19              8.697     23.8713   69.9072     0.0008
```

Specifying the P option in the MODEL statement causes PROC REG to compute predicted values corresponding to each observation in the data set. These computations are printed following the basic PROC REG output under the heading Predict Value. The P option also causes the observed *y* values (in this case COST) to be printed under the heading Dep Var COST, along with the residuals, where Residual = Dep Var COST – Predict Value.

The estimated standard errors of the predicted values are also printed as a result of the P option. These are computed according to the formula Std Err Predict = $\sqrt{V(\hat{y})}$, where $V(\hat{y})$ is given in equation (1.1) in Section 2.4.3, "Hypothesis Tests and Confidence Intervals."

The ID statement identifies each observation according to the value of the variable indicated in the statement; in this case CATTLE. This enables you to relate the observed and predicted values to the values of another variable. Usually, one of the independent variables is the most meaningful ID variable.

The CLM option in the MODEL statement gives upper and lower 95% confidence limits for the mean of the subpopulation corresponding to specific values of the independent variables. The CLM option makes the following computation:

$$\left(\hat{y} - t_{\alpha/2}(\text{STD ERR PREDICT}),\ \hat{y} + t_{\alpha/2}(\text{STD ERR PREDICT}) \right)$$

The CLI option in the MODEL statement gives upper and lower 95% prediction intervals for a future observation. The CLI option makes the following computation:

$$\left(\hat{y} - t_{\alpha/2}\sqrt{\hat{V}(y - \hat{y})},\ \hat{y} + t_{\alpha/2}\sqrt{\hat{V}(y - \hat{y})} \right)$$

$\hat{V}(y - \hat{y})$ is given in Section 2.4.3, with σ^2 replaced by MSE.

Consider OBS 4, which has CATTLE=11.685 (see Output 2.5). The observed value of COST for OBS 4 is 49.2950. The predicted value for OBS 4 is 60.5263 (see Output 2.4). This number is used to estimate the expected cost of operation corresponding to CATTLE=11.685, computed as 60.526 = 7.196 + 4.564(11.685). The 95% confidence limits for this mean are (51.9386, 69.1140). The predicted value of 60.5263 is also used to predict COST (not yet observed) of some other randomly drawn auction that has CATTLE=12.685. The 95% prediction limits for this individual sample are (36.7041, 84.3486).

In short, the CLM option yields a confidence interval for the subpopulation mean, and the CLI option yields a prediction interval for a value to be drawn at random from the subpopulation. The CLI limits are always wider than the CLM limits, because the CLM limits accommodate only variability in $\hat{y}$, whereas the CLI limits accommodate variability in $\hat{y}$ and variability in the future value of *y*. This is true even though $\hat{y}$ is used as an estimate of the subpopulation mean as well as a predictor of the future value.

2.2.3   A Model with Several Independent Variables

This section illustrates regression analyses of data with several independent variables. Output 2.5 shows a data set named AUCTION that has several independent variables.

Output 2.5
Data for
Regression
with Several
Independent
Variables

```
                              The SAS System

    OBS   MKT   CATTLE   CALVES    HOGS    SHEEP     COST    VOLUME   TYPE
     1     1     3.437    5.791    3.268   10.649   27.698   23.145    O
     2     2    12.801    4.558    5.751   14.375   57.634   37.485    O
     3     3     6.136    6.223   15.175    2.811   47.172   30.345    O
     4     4    11.685    3.212    0.639    0.694   49.295   16.230    B
     5     5     5.733    3.220    0.534    2.052   24.115   11.539    B
     6     6     3.021    4.348    0.839    2.356   33.612   10.564    B
     7     7     1.689    0.634    0.318    2.209    9.512    4.850    O
     8     8     2.339    1.895    0.610    0.605   14.755    5.449    B
     9     9     1.025    0.834    0.734    2.825   10.570    5.418    O
    10    10     2.936    1.419    0.331    0.231   15.394    4.917    B
    11    11     5.049    4.195    1.589    1.957   27.843   12.790    B
    12    12     1.693    3.602    0.837    1.582   17.717    7.714    B
    13    13     1.187    2.679    0.459   18.837   20.253   23.162    O
    14    14     9.730    3.951    3.780    0.524   37.465   17.985    B
    15    15    14.325    4.300   10.781   36.863  101.334   66.269    O
    16    16     7.737    9.043    1.394    1.524   47.427   19.698    B
    17    17     7.538    4.538    2.565    5.109   35.944   19.750    B
    18    18    10.211    4.994    3.081    3.681   45.945   21.967    B
    19    19     8.697    3.005    1.378    3.338   46.890   16.418    B
```

The AUCTION data set is an expansion of the MARKET data set. It contains data from the 19 livestock auction markets for sales of four classes of livestock—CATTLE, CALVES, HOGS, and SHEEP. The objective is to relate the annual cost (in thousands of dollars) of operating a livestock market (COST) to the number (in thousands) of livestock in various classes (CATTLE, CALVES, HOGS, SHEEP) that were sold in each market. This is done with a multiple regression analysis. The variables VOLUME and TYPE are used later.

The multiple regression model

$$\text{COST} = \beta_0 + \beta_1 \text{ CATTLE} + \beta_2 \text{ CALVES} + \beta_3 \text{ HOGS} + \beta_4 \text{ SHEEP} + \varepsilon \qquad (2.4)$$

relates COST to the four predictor variables. To fit this model, use the following SAS statements:

```
proc reg;
    model cost=cattle calves hogs sheep;
```

The results appear in Output 2.6.

Output 2.6
Results for
Regression
with Several
Independent
Variables

```
                              The SAS System                              1

Model: MODEL1
Dependent variable: COST

                         Analysis of Variance
                            Sum of          Mean
         Source      DF    Squares         Square ❶  F Value ❷  Prob>F

         Model        4   7936.73649    1984.18412      52.610     0.0001
         Error       14    531.03865      37.93133
         C Total     18   8467.77514


            Root MSE         6.15884 ❸ R-square         0.9373

            Dep Mean        35.29342 ❹ Adj R-sq         0.9194
            C.V.            17.45040


                          Parameter Estimates

                  ❺  Parameter   ❻  Standard  ❼  T for H0:
         Variable  DF    Estimate        Error   Parameter=0 ❽ Prob > |T|

         INTERCEP   1    2.288425    3.38737222       0.676       0.5103
         CATTLE     1    3.215525    0.42215239       7.616       0.0001
         CALVES     1    1.613148    0.85167539       1.894       0.0791
         HOGS       1    0.814849    0.47073855       1.737       0.1054
         SHEEP      1    0.802579    0.18981766       4.228       0.0008
```

The upper portion of the output, as in Output 2.2, contains the partitioning of TOTAL SS, labeled C Total, into MODEL SS and ERROR SS, along with the corresponding mean squares. The callout numbers have been added to the output to key the descriptions that follow:

❶ The *F*-value of 52.610 is used to test the null hypothesis

$$H_0 : \beta_1 = \beta_2 = \beta_3 = \beta_4 = 0$$

❷ The associated *p*-value (Prob>F) of 0.0001 leads to a rejection of this hypothesis and to the conclusion that some of the β's are not 0.

❸ R-square=0.9373 shows that a large portion of the variation in COST can be explained by variation in the independent variables in the model.

❹ Adj R-sq=0.9194 is an alternative to R-square that is adjusted for the number of parameters in the model according to the formula

$$\text{ADJ R-SQ} = 1 - \left(1 - \text{R-SQUARE}\right)\left(\left(n - 1\right) / \left(n - m - 1\right)\right)$$

where *n* is the number of observations in the data set and *m* is the number of regression parameters in the model, excluding the intercept. This adjustment is used to overcome an objection to R-square as a measure of goodness of fit of the model. This objection stems from the fact that R-square can be driven to 1 simply by adding superfluous variables to the model with no real improvement in fit. This is not the case with Adj R-sq, which tends to stabilize to a certain value when an adequate set of variables is included in the model.

❺ The Parameter Estimates give the fitted model

$$\widehat{COST} = 2.29 + 3.22(CATTLE) + 1.61(CALVES) + 0.81(HOGS) + 0.80(SHEEP)$$

Thus, for example, one head of CATTLE contributes \$3.22 to the COST of operating a market, if all other numbers of livestock are held fixed. Remember that the COST of operating the markets was given in \$1000 units, whereas the numbers of animals were given in 1000-head units.

❻ These are the (estimated) Standard Errors of the parameter estimates and are useful for constructing confidence intervals for the parameters, as shown in Section 2.4.3.

❼ The *t*-tests (T for H_0 : Parameter=0) are used for testing hypotheses about individual parameters. It is important that you clearly understand the interpretation of these tests. This can be explained in terms of comparing the fits of complete and reduced models. (Review Section 2.4.2, "Partitioning the Sums of Squares.") The complete model for all of these *t*-tests contains all the variables on the right-hand side of the MODEL statement. The reduced model for a particular test contains all these variables except the one being tested. Thus, the *t*-statistic=1.894 for testing the hypothesis $H_0 : \beta_2 = 0$ is actually testing whether the complete model containing CATTLE, CALVES, HOGS, and SHEEP fits better than the reduced model containing only CATTLE, HOGS, and SHEEP.

❽ The *p*-value (Prob > | T |) for this test is *p*=0.0791.

2.2.4 The SS1 and SS2 Options: Two Types of Sums of Squares

PROC REG can compute two types of sums of squares associated with the estimated coefficients in the model. These are referred to as Type I and Type II sums of squares and are computed by specifying SS1 or SS2, or both, as MODEL statement options. For example, the following SAS statements produce Output 2.7:

```
proc reg;
   model cost=cattle calves hogs sheep / ss1 ss2;
run;
```

*Output 2.7
PROC REG
with the SS1
and SS2
MODEL
Statement
Options*

```
                              The SAS System

Model: MODEL1
Dependent variable: COST

                          Analysis of Variance
                              Sum of       Mean
            Source        DF  Squares     Square    F Value   Prob>F

            Model          4  7936.73649  1984318412  52.310   0.0001
            Error         14   531.03865    37.93133
            C Total       18  8467.77514

                Root MSE       6.15884  R-square    0.9373
                Dep Mean      35.29342  Adj R-sq    0.9194
                C.V.          17.45040

                          Parameter Estimates

                       Parameter     Standard    T for H0:
            Variable  DF  Estimate      Error   Parameter=0   Prob > |T|

            INTERCEP   1  2.288425   3.38737222    0.676      0.5103
            CATTLE     1  3.215525   0.42215439    7.617      0.0001
            CALVES     1  1.613148   0.85467539    1.894      0.0791
            HOGS       1  0.814849   0.47073855    1.731      0.1054
            SHEEP      1  0.802579   0.18981766    4.228      0.0008

            Variable  DF  Type I SS    Type II SS
            INTERCEP   1      23667    17.311929
            CATTLE     1  6582.091806  2200.712494
            CALVES     1   186.671101   136.081196
            HOGS       1   489.863790   113.656260
            SHEEP      1   678.109792   678.109792
```

These sums of squares are printed as additional columns and are labeled Type I SS and Type II SS. You may find it helpful at this point to review the material in Section 2.4. In particular, the concepts of partitioning sums of squares, complete and reduced models, and reduction notation are useful in understanding the different types of sums of squares.

The Type I SS are commonly called **sequential** sums of squares. They represent a partitioning of the MODEL SS into component sums of squares due to adding each variable sequentially to the model in the order prescribed by the MODEL statement.

The Type I SS for the INTERCEP is simply $n\bar{y}^2$. It is called the **correction for the mean**. The Type I SS for CATTLE (6582.09181) is the MODEL SS for a regression equation that contains only CATTLE. The Type I SS for CALVES (186.67110) is the increase in MODEL SS due to adding CALVES to the model that already contains CATTLE. In general, the Type I SS for a particular variable is the sum of squares due to adding that variable to a model that already contains all the variables that preceded the particular variable in the MODEL statement. Continuing the pattern, you see that the Type I SS for HOGS (489.86379) is the increase in MODEL SS due to adding HOGS to a model that already contains CATTLE and CALVES. Finally, the Type I SS for SHEEP (678.10979) is the increase in MODEL SS due to adding SHEEP to a model that already contains CATTLE, CALVES, and HOGS. Note that

$$\text{MODEL SS} = 7936.7 = 6582.1 + 186.7 + 489.9 + 678.1$$

illustrates the sequential partitioning of the MODEL SS into the Type I components that correspond to the variables in the model.

The Type II SS are commonly called the **partial** sums of squares. For a given variable, the Type II SS is equivalent to the Type I SS for that variable if it were the last variable in the MODEL statement. (Note that the Type I SS and Type II SS for SHEEP are equal in Output 2.7.) In other words, the Type II SS for a particular variable is the increase in MODEL SS due to adding the variable to a model that already contains all the other variables in the MODEL statement. The Type II SS, therefore, do not depend on the order in which the independent variables are listed in the MODEL statement. Furthermore, they do not yield a partitioning of the MODEL SS unless the independent variables are uncorrelated.

The Type I SS and Type II SS are shown in the table below in reduction notation (see Section 2.4.2):

	Type I **(Sequential)**	**Type II** **(Partial)**
CATTLE	$R(\beta_1 \mid \beta_0)$	$R(\beta_1 \mid \beta_0, \beta_2, \beta_3, \beta_4)$
CALVES	$R(\beta_2 \mid \beta_0, \beta_1)$	$R(\beta_2 \mid \beta_0, \beta_1, \beta_3, \beta_4)$
HOGS	$R(\beta_3 \mid \beta_0, \beta_1, \beta_2)$	$R(\beta_3 \mid \beta_0, \beta_1, \beta_2, \beta_4)$
SHEEP	$R(\beta_4 \mid \beta_0, \beta_1, \beta_2, \beta_3)$	$R(\beta_4 \mid \beta_0, \beta_1, \beta_2, \beta_3)$

The reduction notation provides a convenient device to determine the complete and reduced models that are compared if you construct an F-test using one of these sums of squares. First, note that each sum of squares for a particular variable has one degree of freedom, so that the sums of squares are also mean squares. Thus, for example, a Type I F-test for CALVES is given by

$$F = \frac{\text{Type I SS for CALVES}}{\text{MSE}} = \frac{186.7}{37.9} = 4.92$$

The reduction notation shows that this F-value would be used to test whether the complete model containing CATTLE and CALVES fits the data significantly better than the reduced model containing only CATTLE. Similarly, a Type II F-statistic for CALVES is given by

$$F = \frac{\text{Type II SS for CALVES}}{\text{MSE}} = \frac{136.1}{37.9} = 3.59$$

It would be used to test whether the complete model containing CATTLE, CALVES, HOGS, and SHEEP fits significantly better than the reduced model containing CATTLE, HOGS, and SHEEP. In this example, the difference between these two F-values is not great. The difference between Type I and Type II F-values for HOGS is considerably greater. The variation due to HOGS that is not due to CATTLE and CALVES is 489.86379, but the variation due to HOGS that is not due to CATTLE, CALVES, and SHEEP is only 113.65626. The former is significant at the 0.003 level, whereas the latter is significant only at the 0.105 level. Thus, a model containing CATTLE and CALVES is significantly improved by adding HOGS, but a model containing CATTLE, CALVES, and SHEEP is improved much less by adding HOGS.

PROC REG does not compute F-values for the Type I and Type II sums of squares, nor does it compute the corresponding significance probabilities. However, you can use PROC GLM to make the same computations discussed in this chapter. There are several other distinctions between the capabilities of PROC REG and PROC GLM. Not all analyses using PROC REG can be easily performed by PROC GLM. Some of these distinctions are discussed in subsequent sections.

It is now appropriate to note that the Type II F-tests are exactly equivalent to the t-tests for the parameters because they are comparing the same complete and reduced models. In fact, the Type II F-statistic for a given variable is equal to the square of the t-statistic for the same variable.

For most applications, the desired test for a single parameter is based on the Type II sums of squares, which are equivalent to the *t*-tests for the parameter estimates. Type I sums of squares, however, are useful if there is need for a specific sequencing of tests on individual coefficients as, for example, in polynomial models.

2.2.5   Tests of Subsets and Linear Combinations of Coefficients

Tests of hypotheses that individual coefficients are equal to 0 are given by the *t*-tests on the parameters in the basic PROC REG output, as discussed in Section 2.2.3. In this section, a direct procedure is demonstrated for testing that subsets of coefficients are equal to 0. These tests are specified in the optional TEST statement in PROC REG. The TEST statement can also be used to test that linear functions of parameters are equal to specified constants.

The TEST statement must follow a MODEL statement in PROC REG. Several TEST statements can follow one MODEL statement. The general form of the TEST statement is

label: **TEST** *equation* <, . . . , *equation*>< / *option*>;

The label is optional and serves only to identify results in the output. The equations provide the technical information that PROC REG uses to determine what hypotheses are to be tested. These tests can be interpreted in terms of comparing complete and reduced (or restricted) models in the same manner as discussed in previous sections. The complete model for all tests specified by a TEST statement is the model containing all variables on the right side of the MODEL statement. The reduced model is derived from the complete model by imposing the conditions implied by the equations indicated in the TEST statement.

For illustration, recall the AUCTION data set in Section 2.2.3 and the SAS statement

```
model cost=cattle calves hogs sheep;
```

which fits the complete regression model

$$COST = \beta_0 + \beta_1(CATTLE) + \beta_2(CALVES) + \beta_3(HOGS) + \beta_4(SHEEP) + \varepsilon$$

To test the hypothesis $H_0 : \beta_3 = 0$, use the statement

```
hogs:   test hogs=0;
```

This statement tells PROC REG to construct an *F*-test to compare the complete model with the reduced model

$$COST = \beta_0 + \beta_1(CATTLE) + \beta_2(CALVES) + \beta_4(SHEEP) + \varepsilon$$

Similarly, to test the hypothesis $H_0 : \beta_3 = \beta_4 = 0$, which is equivalent to $H_0 : (\beta_3 = 0 \text{ and } \beta_4 = 0)$, use the statement

```
hogsheep:   test hogs=0, sheep=0;
```

The hypothesis of a 0 intercept, $H_0 : \beta_1 = 0$, is specified with the statement

```
intercep:   test intercept=0;
```

If the right side of an equation in a TEST statement is 0, you don't need to specify it. PROC REG will assume a right-side value of 0 by default.

More general linear functions are tested in a similar fashion. For example, to test that the average cost of selling one hog is one dollar, you test the hypothesis $H_0 : \beta_3 = 1$. This hypothesis is specified in the following TEST statement:

```
hogone:   test hogs=1;
```

Another possible linear function of interest is to test whether the average cost of selling one hog differs from the average cost of selling one sheep. The null hypothesis is $H_0 : \beta_3 = \beta_4$, which is equivalent to $H_0 : \beta_3 - \beta_4 = 0$, and is specified by the following statement:

```
hequals:   test hogs-sheep=0;
```

The results of all five of these TEST statements appear in Output 2.8.

Output 2.8
Results of the
TEST Statement

```
                             The SAS System                              1
Dependent Variable: COST
Test: HOGS        Numerator:    113.6563  DF:   1  F value:   2.9964
                  Denominator:   37.93133 DF:  14  Prob>F:    0.1054


Dependent Variable: COST
Test: HOGSHEEP    Numerator:    583.9868  DF:   1  F value:  15.3959
                  Denominator:   37.93133 DF:  14  Prob>F:    0.0003


Dependent Variable: COST
Test: INTERCEP    Numerator:     17.3119  DF:   1  F value:   0.4564
                  Denominator:   37.93133 DF:  14  Prob>F:    0.5103


Dependent Variable: COST
Test: HOGONE      Numerator:      5.8680  DF:   1  F value:   0.1547
                  Denominator:   37.93133 DF:  14  Prob>F:    0.7000


Dependent Variable: COST
Test: HEQUALS     Numerator:      0.0176  DF:   1  F value:   0.0005
                  Denominator:   37.93133 DF:  14  Prob>F:    0.9831
```

For each TEST statement indicated, a sum of squares is computed with degrees of freedom equal to the number of equations in the TEST statement. From these quantities, a mean square that forms the numerator of an F-statistic is computed. The denominator of the F-ratio is the mean square for error. The value of F is printed, along with its p-value. The test labeled HOGS is, of course, the equivalent of the t-test for HOGS in Output 2.5.

Note: If there are linear dependencies or inconsistencies among the equations in a TEST statement, then PROC REG prints a message that the test failed, and no F-ratio is computed.

2.2.6 Fitting Restricted Models: The RESTRICT Statement and NOINT Option

Subject to linear restrictions on the parameters, models can be fitted by using the RESTRICT statement in PROC REG. The RESTRICT statement follows a MODEL statement and has the general form

RESTRICT *equation* <, . . ., *equation*>;

where each equation is a linear combination of the model parameters set equal to a constant.

Consider again the data set AUCTION and the following MODEL statement:

```
model cost=cattle calves hogs sheep;
```

This model is fitted in Section 2.2.3. Inspection of Output 2.5 shows that the INTERCEP estimate is close to 0 and that the parameter estimates for HOGS and SHEEP are similar in value. Hypotheses pertaining to these conditions are tested in Section 2.2.5. The results suggest a model that has 0 intercept and equal coefficients for HOGS and SHEEP, namely

$$COST = \beta_1(CATTLE) + \beta_2(CALVES) + \beta(HOGS) + \beta(SHEEP) + \varepsilon$$

where β is the common value of β_3 and β_4.

This model can be fitted with the following RESTRICT statement:

```
restrict intercept=0, hogs-sheep=0;
```

The results of these statements appear in Output 2.9.

Output 2.9
Results of the
RESTRICT
Statement

```
Model: MODEL1
NOTE: Restrictions have been applied to parameter estimates.
Dependent variable: COST

                        Analysis of Variance

                         Sum of        Mean
        Source      DF   Squares       Square     F Value   Prob>F

        Model        2   7918.75621   3959.37811  115.388   0.0001
        Error       16    549.01893     34.31368
        C Total     18   8467.77514

             Root MSE       5.85779  R-square      0.9352
             Dep Mean      35.29342  Adj R-sq      0.9271
             C.V.          16.59739

                        Parameter Estimates

                     Parameter     Standard     T for H0:
        Variable  DF   Estimate       Error    Parameter=0   Prob > |T|

        INTERCEP   1  1.110223E-15  0.00000000      .             .
        CATTLE     1      3.300043  0.38314175    8.613        0.0001
        CALVES     1      1.943717  0.59107649    3.328        0.0043
        HOGS       1      0.806825  0.13799841    5.847        0.0001
        SHEEP      1      0.806825  0.13799841    5.847        0.0001
        RESTRICT  -1      7.905632  10..92658296  0.724        0.4798
        RESTRICT  -1     -9.059424  64.91322402  -0.140        0.8907
```

Note that the INTERCEP parameter estimate is 0 (except for round-off error), and the parameter estimates for HOGS and SHEEP have the common value .806825. Note also that there are parameter estimates and associated *t*-tests for the two equations in the RESTRICT statement. These pertain to the Lagrangian parameters that are incorporated in the restricted minimization of the Error SS.

You will find it useful to compare Output 2.9 with results obtained by invoking the restrictions explicitly. The model with the RESTRICT statement is equivalent to the model

$$COST = \beta_1(CATTLE) + \beta_2(CALVES) + \beta(SHEEP + HOGS) + \varepsilon$$

This is a three-variable model with an intercept of 0. The variables are CATTLE, CALVES, and HS, where the variable HS=HOGS + SHEEP. The model is then fitted using PROC REG with the MODEL statement

```
model cost=cattle calves hs / noint;
```

where NOINT is the option that specifies that no intercept be included. In other words, the fitted regression plane is forced to pass through the origin. The results appear in Output 2.10, from which the following fitted equation is obtained:

$$COST = 3.300 \ (CATTLE) + 1.967(CALVES) + 0.807(HS)$$
$$= 3.300 \ (CATTLE) + 1.967(CALVES) + 0.807(HOGS)$$
$$+0.807(SHEEP)$$

Output 2.10
Results of
Regression
with Implicit
Restrictions

```
                      The SAS System
                      The REG Procedure
                       Model: MODEL1
                   Dependent Variable: cost

      NOTE: No intercept in model. R-Square is redefined.

                      Analysis of Variance

                            Sum of         Mean
Source               DF     Squares       Square    F Value    Pr > F

Model                 3       31586        10529     306.83    <.0001
Error                16   549.01893     34.31368
Uncorrected Total    19       32135


          Root MSE              5.85779    R-Square     0.9829
          Dependent Mean       35.29342    Adj R-Sq     0.9797
          Coeff Var            16.59739


                      Parameter Estimates

                      Parameter      Standard
        Variable   DF  Estimate         Error    t Value    Pr > |t|

        cattle      1   3.30004       0.38314       8.61     <.0001
        calves      1   1.96717       0.59108       3.33     0.0043
        hs          1   0.80682       0.13800       5.85     <.0001
```

This is the same fitted model obtained in Output 2.9 by using the RESTRICT statements.

As just pointed out, equivalent restrictions can be imposed in different ways using PROC REG. When the NOINT option is used to restrict β_0 to be 0, however, caution is advised in trying to interpret the sums of squares, *F*-values, and R-squares. Notice, for instance, that the MODEL SS do not agree in Output 2.9 and Output 2.10. The ERROR SS and degrees of freedom do agree. In Output 2.9, the MODEL SS and ERROR SS sum to the Corrected Total SS (C Total), whereas in Output 2.10 they sum to the Uncorrected Total SS (U Total). In Output 2.10, the MODEL *F*-statistic is testing whether the fitted model fits better than a model containing no parameters, a test that has little or no practical value.

Corresponding complications arise regarding the R-square statistic with no-intercept models. Note that R-Square=0.9829 for the no-intercept model in Output 2.10 is greater than R-Square=0.9373 for the model in Output 2.6, although the latter has two more parameters than the former. This seems contrary to the general phenomenon that adding terms to a model causes the R-square to increase. This seeming contradiction occurs because the denominator of the R-square is the Uncorrected Total SS when the NOINT option is used. This is the reason for the message that R-square is redefined at the top of Output 2.10. It is, therefore, not meaningful to compare an R-square for a model that contains an intercept with an R-square for a model that does not contain an intercept.

2.2.7 Exact Linear Dependency

Linear dependency occurs when exact linear relationships exist among the independent variables, that is, when one or more of the columns of the **X** matrix (see Section 2.4) can be expressed as a linear combination of the other columns. In this event, the **X′X** matrix is singular and cannot be inverted in the usual sense to obtain parameter estimates. If this occurs, PROC REG uses a generalized inverse to compute parameter estimates. However, care must be exercised to determine exactly what parameters are being estimated. More technically, the generalized inverse approach yields one particular solution to the normal equations. The PROC REG computations are illustrated with the auction market data.

Recall the variable VOLUME in Output 2.5. VOLUME represents the total of all major livestock sold in each market. This is an example of exact linear dependency: VOLUME is exactly the sum of the variables CATTLE, CALVES, HOGS, and SHEEP.

The following statements produce the results shown in Output 2.11:

```
proc reg;
     model cost=cattle calves hogs sheep volume;
```

Output 2.11
Results of
Exact
Collinearity

```
NOTE: Model is not full rank. Least-squares solutions for the parameters are not unique.
Some statistics will be misleading. A reported DF of 0 or B means that the estimate is
biased.
NOTE: The following parameters have been set to 0, since the variables are a linear
combination of other variables as shown.

                    VOLUME = CATTLE + CALVES + HOGS + SHEEP
                         Parameter Estimates

                              Parameter      Standard
            Variable    DF      Estimate        Error     t Value    Pr > |t|

            Intercept    1       2.28842       3.38737       0.68      0.5103
            CATTLE       B       3.21552       0.42215       7.62      <.0001
            CALVES       B       1.61315       0.85168       1.89      0.0791
            HOGS         B       0.81485       0.47074       1.73      0.1054
            SHEEP        B       0.80258       0.18982       4.23      0.0008
            VOLUME       0             0           .          .          .
```

In Output 2.11, the existence of the collinearity is indicated by the notes in the model summary, followed by the equation describing the collinearity. The parameter estimates that are printed are equivalent to the ones that are obtained if VOLUME were not included in the MODEL statement. A variable that is a linear combination of other variables that precede it in the MODEL statement is indicated by 0 under the DF and Parameter Estimate headings. Other variables involved in the linear dependencies are indicated with a B, for bias, under the DF heading. The estimates are, in fact, unbiased estimates of the parameters in a model that does not include the variables indicated with a 0 but are biased for the parameters in the other models. Thus, the parameter estimates printed in Output 2.11 are unbiased estimates for the coefficients in the model

$$COST = \beta_0 + \beta_1(CATTLE) + \beta_2(CALVES) + \beta_3(HOGS) + \beta_4(SHEEP) + \varepsilon$$

But these estimates are biased for the parameters in a model that included VOLUME but did not include, for example, SHEEP.

2.3 The GLM Procedure

The REG procedure was used in Section 2.2 to introduce simple and multiple linear regression. In this section, PROC GLM will be used to obtain some of the same analyses you saw in Section 2.2. The purpose is to provide a springboard from which to launch the linear model analysis methodology that will be used throughout the remainder of this book. The basic syntax of PROC GLM is very similar to that of PROC REG for fitting linear regression models. Beyond the basic applications the two procedures become more specialized in their capabilities. PROC REG has greater regression diagnostic and graphic features, but PROC GLM has the ability to create "dummy" variables. This makes PROC GLM suited for *analysis of variance, analysis of covariance*, and certain *mixed-model* applications.

The following SAS statements invoke PROC GLM:

```
proc glm;
    model list-of-dependent-variables=list-of-independent-
    variables;
```

Options on the PROC, MODEL, and other statements permit specialized and customized analyses.

2.3.1 Using the GLM Procedure to Fit a Linear Regression Model

PROC GLM can fit linear regression models and compute essentially the same basic statistics as PROC REG. These statements apply PROC GLM to fit a multiple linear regression model to the AUCTION data. Results appear in Output 2.12.

```
proc glm;
    model cost=cattle calves hogs sheep;
run;
```

Output 2.12
Results Using
PROC GLM
for Multiple
Linear
Regression

```
                                    The SAS System

                                   The GLM Procedure

                          Number of observations    19
                                    The SAS System

                                   The GLM Procedure

Dependent Variable: COST

                                        Sum of
        Source                 DF       Squares     Mean Square    F Value    Pr > F
        Model                   4    7936.736489    1984.184122      52.31    <.0001
        Error                  14     531.038650      37.931332
        Corrected Total        18    8467.775139

                    R-Square      Coeff Var      Root MSE      COST Mean

                    0.937287      17.45040       6.158842      35.29342


        Source                 DF      Type I SS     Mean Square    F Value    Pr > F

        CATTLE                  1    6582.091806    6582.091806     173.53    <.0001
        CALVES                  1     186.671101     186.671101       4.92    0.0436
        HOGS                    1     489.863790     489.863790      12.91    0.0029
        SHEEP                   1     678.109792     678.109792      17.88    0.0008


        Source                 DF     Type III SS    Mean Square    F Value    Pr > F

        CATTLE                  1    2200.712494    2200.712494      58.02    <.0001
        CALVES                  1     136.081196     136.081196       3.59    0.0791
        HOGS                    1     113.656260     113.656260       3.00    0.1054
        SHEEP                   1     678.109792     678.109792      17.88    0.0008


                                           Standard
        Parameter          Estimate          Error      t Value    Pr > |t|

        Intercept        2.288424577      3.38737222       0.68    0.5103
        CATTLE           3.215524803      0.42215239       7.62    <.0001
        CALVES           1.613147614      0.85167539       1.89    0.0791
        HOGS             0.814849491      0.47073855       1.73    0.1054
        SHEEP            0.802578622      0.18981766       4.23    0.0008
```

Compare Output 2.12 with Output 2.6. You find the same partition of sums of squares, with slightly different labels. You also find the same R-square, Root MSE, *F*-test for the Model, and other statistics. In addition, you find the same parameter estimates for the fitted model, giving the regression equation

$$COST = 2.288 + 3.215 \ CATTLE + 1.612 \ CALVES + 0.815 \ HOGS + 0.803 \ SHEEP$$

Compare Output 2.12 with Output 2.7. You see that the Type I sums of squares are the same from PROC GLM and PROC REG. You also see that the Type III sums of squares from GLM are the same as the Type II sums of squares from PROC REG. In fact, there are also Types II and IV sums of squares available from PROC GLM whose values would be the same as the Type II sums of squares from PROC REG. In general, Types II, III, and IV sums of squares are identical for regression models, but differ for some ANOVA models, as you will see in subsequent chapters.

2.3.2 Using the CONTRAST Statement to Test Hypotheses about Regression Parameters

Fitting a regression model is just one step in a regression analysis. You usually want to do more analyses to answer specific questions about the variables in the model. In Section 2.2.3 you learned how to test the hypothesis

$$H_0 : \beta_1 = \beta_2 = \beta_3 = \beta_4 = 0$$

using the test statistic F = MS Model / MS Error. You also learned how to test the hypotheses of individual parameters. For example, the hypothesis that the CALVES parameter is zero,

$$H_0 : \beta_2 = 0$$

can be tested using *t*-statistics for the parameter estimates. In Section 2.2.4, you learned how to test the hypothesis

$$H_0 : \beta_2 = 0$$

using an *F*-statistic based on either the Type I or the Type II sum of squares. Output 2.12 shows the *F*-values of 4.92 based on the Type I sum of squares, and 3.59 based on the Type III (= Type II) sum of squares. These *F*-values have *p*-values of 0.0436 and 0.0791. These are different because the reference models are different for the hypotheses. The distinction is even more dramatic for hypotheses about the SHEEP regression parameter, β_3. The Type I *F*-test for $H_0 : \beta_3 = 0$ has *p*-value of 0.0029, and the Type III *F*-test for $H_0 : \beta_3 = 0$ has *p*-value of 0.1054.

You can test a hypothesis of the form

$$H_0 : \beta_1 = \beta_2 = 0$$

in the reference model COST $= \beta_0 + \beta_1$ CATTLE $+ \beta_2$ CALVES $+ \beta_3$ +HOGS $+ \varepsilon$ using the *F*-statistic

$$F = \text{MS}(\beta_1 , \beta_2 | \beta_0 , \beta_3) / \text{MS(ERROR)},$$

where $\text{MS}(\beta_1 , \beta_2 | \beta_0 , \beta_3) = \text{R}(\beta_1 , \beta_2 | \beta_0 , \beta_3) / 2$, and

$$\text{R}(\beta_1 , \beta_2 | \beta_0 , \beta_3) = \text{R}(\beta_1 | \beta_0) + \text{R}(\beta_2 | \beta_0 , \beta_1).$$

These computations can be obtained by using Type I sums of squares:

$$\begin{aligned} \text{MS}(\beta_1 , \beta_2 | \beta_0 , \beta_3) &= (2200.71 + 136.08)/2 \\ &= 1168.40, \end{aligned}$$

and thus $F = 1168.40/37.93 = 30.80$. The *p*-value for $F = 30.80$ is $<.0001$.

Other hypotheses about regression parameters can be tested using a special tool in PROC GLM called **CONTRAST statements**. CONTRAST statements can be used to test hypotheses about any linear combination of parameters in the model.

Next you will see how to test the following hypotheses using the CONTRAST statement:

☐ $H_0 : \beta_3 = 0$ (Cost of HOGS is zero)

☐ $H_0 : \beta_3 = \beta_4$ (Cost of HOGS is equal to cost of SHEEP)

☐ $H_0 : \beta_3 = \beta_4 = 0$ (Costs of HOGS and SHEEP are both equal to zero)

The basic syntax of the CONTRAST statement is

    contrast 'label' effect values;

where **effect values** refers to coefficients of a linear combination of parameters, and **label** simply refers to a character string to identify the results in the PROC GLM output. For our regression example, a hypothesis about a linear combination of parameters would have the form

$$H_0 : a_0 \beta_0 + a_1 \beta_1 + a_2 \beta_2 + a_3 \beta_3 + a_4 \beta_4 = 0$$

where a_0, a_1, a_2, a_3, and a_4 are numbers. This hypothesis would be indicated in a CONTRAST statement as

    contrast 'label' intercept a_0 cattle a_1 calves a_2 hogs a_3 sheep a_4;

The linear combination in the hypothesis $H_0 : \beta_3 = 0$ is simply β_3, so $a_3 = 1$ and a_0, a_1, a_2, and a_4 and are all equal to zero. Therefore, you could use the CONTRAST statement

```
contrast 'hogcost=0' intercept 0 cattle 0 calves 0 hogs 1 sheep 0;
```

Actually, you only need to specify the nonzero constants. Others will be set to zero implicitly. So you could write the CONTRAST statement to test the hypothesis $H_0 : \beta_3 = 0$ as

```
contrast 'hogcost=0' hogs 1;
```

You can follow the same logic and get a test of the hypothesis $H_0 : \beta_3 = \beta_4$. Write this hypothesis as $H_0 : \beta_3 - \beta_4 = 0$. The linear combination you are testing is $\beta_3 - \beta_4$, so $a_3 = 1$ and $a_4 = -1$, and the other constants are zero. This leads to the CONTRAST statement

```
contrast 'hogcost=sheepcost' hogs 1 sheep -1;
```

The third hypothesis $H_0 : \beta_3 = \beta_4 = 0$ is different from the first two because it entails two linear combinations simultaneously. Actually, the hypothesis should be written

$$H_0 : \beta_3 = 0 \text{ and } \beta_4 = 0$$

Then you see that the two linear combinations are β_3 and β_4. You can specify two or more linear combinations in a CONTRAST statement by separating their coefficients with commas. A CONTRAST statement to test this hypothesis is

```
contrast 'hogcost=sheepcost=0' hogs 1, sheep 1;
```

Now run the statements

```
proc glm;
   model cost=cattle calves hogs sheep;
   contrast 'hogcost=0' hogs 1;
   contrast 'hogcost=sheepcost' hogs 1 sheep -1;
   contrast 'hogcost=sheepcost=0' hogs 1, sheep 1;
run;
```

Results of the CONTRAST statements appear in Output 2.13.

Output 2.13
Results from the
CONTRAST
Statements in
Multiple Linear
Regression

Contrast	DF	Contrast SS	Mean Square	F Value	Pr > F
hogcost=0	1	113.656260	113.656260	3.00	0.1054
hogcost=sheepcost	1	0.017568	0.017568	0.00	0.9831
hogcost=sheepcost=0	2	1167.973582	583.986791	15.40	0.0003

2.3.3 Using the ESTIMATE Statement to Estimate Linear Combinations of Parameters

You learned how to test hypotheses about linear combinations of parameters using the CONTRAST statement in Section 2.3.2. Sometimes it is more meaningful to estimate the linear combinations. You can do this with the ESTIMATE statement in PROC GLM. The ESTIMATE statement is used in essentially the same way as the CONTRAST statement. But instead of *F*-tests for linear combinations, you get estimates of them along with standard errors. However, the ESTIMATE statement can estimate only one linear combination at a time, whereas the CONTRAST statement could be used to test two or more linear combinations simultaneously.

Consider the first two linear combinations you tested in Section 2.3.2, namely β_3 and $\beta_3 - \beta_4$. You can estimate these linear combinations by using the ESTIMATE statements

```
estimate 'hogcost=0' hogs 1;
estimate 'hogcost=sheepcost' hogs 1 sheep -1;
```

Results are shown in Output 2.14.

Output 2.14
Results from the
ESTIMATE
Statements in
Multiple Linear
Regression

Parameter	Estimate	Standard Error	t Value	Pr > |t|
hogcost-sheepcost	0.0122709	0.57018280	0.02	0.9831
cost predict	24.7405214	1.75311927	14.11	<.0001

Compare Output 2.14 with Output 2.13. You see that the *F*-values in Output 2.13 are equal to the squares of the *t*-values in Output 2.14. Therefore, they are equivalent test statistics. You can confirm this by noting that the *F*-tests in Output 2.13 have the same *p*-values as the *t*-tests in Output 2.14. The advantage of the ESTIMATE statement is that it gives you estimates of the linear combinations and standard errors in addition to the tests of significance.

2.4 Statistical Background

The REG and GLM procedures implement a multiple linear regression analysis according to the model

$$y = \beta_0 + \beta_1 x_1 + \beta_2 x_2 + \ldots + \beta_m x_m + \varepsilon$$

which relates the behavior of a dependent variable y to a linear function of the set of independent variables $x_1, x_2, \ldots, x_m$. The β_i's are the parameters that specify the nature of the relationship, and ε is the random error term. Although it is assumed that you have a basic understanding of regression analysis, it may be helpful to review regression principles, terminology, notation, and procedures.

2.4.1 Terminology and Notation

The principle of least squares is applied to a set of n observed values of y and the associated values of x_i to obtain estimates $\hat{\beta}_0, \hat{\beta}_1, \ldots, \hat{\beta}_m$ of the respective parameters $\beta_0, \beta_1, \ldots, \beta_m$. These estimates are then used to construct the fitted model

$$\hat{y} = \hat{\beta}_0 + \hat{\beta}_1 x_1 + \ldots + \hat{\beta}_m x_m$$

Many regression computations are illustrated conveniently in matrix notation. Letting y_j, x_{ij}, and ε_j denote the values of y, x_i, and ε, respectively, in the jth observation, the **Y** vector, the **X** matrix, and the ε vector can be defined as

$$\mathbf{Y} = \begin{bmatrix} y_1 \\ \vdots \\ y_n \end{bmatrix} \quad \mathbf{X} = \begin{bmatrix} 1 & x_{11} & \cdots & x_{m1} \\ \vdots & \vdots & & \vdots \\ 1 & x_{1n} & \cdots & x_{mn} \end{bmatrix} \quad \varepsilon = \begin{bmatrix} \varepsilon_1 \\ \vdots \\ \varepsilon_n \end{bmatrix}$$

Then the model in matrix notation is

$$\mathbf{Y} = \mathbf{X}\boldsymbol{\beta} + \varepsilon$$

where $\boldsymbol{\beta}' = (\beta_0, \beta_1, \ldots, \beta_m)$ is the parameter vector.

The vector of least-squares estimates, $\hat{\boldsymbol{\beta}}' = (\hat{\beta}_0, \hat{\beta}_1, \ldots, \hat{\beta}_m)$, is obtained by solving the set of normal equations

$$\mathbf{X}'\mathbf{X}\hat{\boldsymbol{\beta}} = \mathbf{X}'\mathbf{Y}$$

Assuming that $\mathbf{X}'\mathbf{X}$ has full rank, there is a unique solution to the normal equations given by

$$\hat{\boldsymbol{\beta}} = (\mathbf{X}'\mathbf{X})^{-1}\mathbf{X}'\mathbf{Y}$$

The matrix $(\mathbf{X}'\mathbf{X})^{-1}$ is useful in regression analysis and is often denoted by

$$
\left(\mathbf{X}'\mathbf{X}\right)^{-1} = \mathbf{C} = \begin{bmatrix}
c_{00} & c_{01} & \cdots & c_{0m} \\
c_{10} & c_{11} & \cdots & c_{1m} \\
\cdot & \cdot & & \cdot \\
\cdot & \cdot & & \cdot \\
\cdot & \cdot & & \cdot \\
c_{m0} & c_{m1} & \cdots & c_{mm}
\end{bmatrix}
$$

The **C** matrix provides the variances and covariances of the regression parameter estimates $V(\hat{\beta}) = \sigma^2 \mathbf{C}$, where $\sigma^2 = V(\varepsilon_i)$.

2.4.2 Partitioning the Sums of Squares

A basic identity results from least squares, specifically,

$$
\Sigma \left(y - \bar{y}\right)^2 = \Sigma \left(\hat{y} - \bar{y}\right)^2 + \Sigma \left(y - \hat{y}\right)^2
$$

This identity shows that the total sum of squared deviations from the mean, $\Sigma(y - \bar{y})^2$, can be partitioned into two parts: the sum of squared deviations from the regression line to the overall mean, $\Sigma(\hat{y} - \bar{y})^2$, and the sum of squared deviations from the observed y values to the regression line, $\Sigma(y - \hat{y})^2$. These two parts are called the sum of squares due to *regression* (or model) and the *residual* (or error) sum of squares, respectively. In most SAS procedures, the total sum of squared deviations from the mean is labeled *TOTAL SS*. The regression sum of squares is labeled *MODEL SS*, and the residual sum of squares is labeled *ERROR SS*. Thus,

TOTAL SS = MODEL SS + ERROR SS

The TOTAL SS always has the same value for a given set of data, regardless of the model that is fitted. However, partitioning into MODEL SS and ERROR SS depends on the model. Generally, the addition of a new x variable to a model will increase the MODEL SS and, correspondingly, reduce the RESIDUAL SS. In matrix notation, the residual or error sum of squares is

$$
\begin{aligned}
\text{ERROR SS} &= \mathbf{Y}'(\mathbf{I} - \mathbf{X}(\mathbf{X}'\mathbf{X})^{-1}\mathbf{X}')\mathbf{Y} \\
&= \mathbf{Y}'\mathbf{Y} - \mathbf{Y}'\mathbf{X}(\mathbf{X}'\mathbf{X})^{-1}\mathbf{X}'\mathbf{Y} \\
&= \mathbf{Y}'\mathbf{Y} - \hat{\beta}\mathbf{X}'\mathbf{Y}
\end{aligned}
$$

The error mean square

$$
s^2 = \text{MSE} = \text{ERROR SS} / \left(n - m - 1\right)
$$

is an unbiased estimate of σ^2, the variance of ε_i.

PROC REG and PROC GLM compute several sums of squares. Each sum of squares can be expressed as the difference between the regression sums of squares for two models, which are called **complete** and **reduced** models. This approach relates a given sum of squares to the comparison of two regression models.

Denote by MODEL SS1 the MODEL SS for a regression with $m = 5$ x variables:

$$
y = \beta_0 + \beta_1 x_1 + \beta_2 x_2 + \beta_3 x_3 + \beta_4 x_4 + \beta_5 x_5 + \varepsilon
$$

and by MODEL SS$_2$ the MODEL SS for a reduced model not containing x_4 and x_5:

$$y = \beta_0 + \beta_1 x_1 + \beta_2 x_2 + \beta_3 x_3 + \varepsilon$$

Reduction in sum of squares notation can be used to represent the difference between regression sums of squares for the two models. For example,

$$R(\beta_4, \beta_5 \mid \beta_0, \beta_1, \beta_2, \beta_3) = \text{MODEL SS}_1 - \text{MODEL SS}_2$$

The difference, or reduction in error, $R(\beta_4, \beta_5 \mid \beta_0, \beta_1, \beta_2, \beta_3)$ indicates the increase in regression sums of squares due to the addition of β_4 and β_5 to the reduced model. It follows that

$$R(\beta_4, \beta_5 \mid \beta_0, \beta_1, \beta_2, \beta_3) = \text{MODEL SS}_1 - \text{MODEL SS}_2$$

Since TOTAL SS=MODEL SS+ERROR SS, it follows that

$$R(\beta_4, \beta_5 \mid \beta_0, \beta_1, \beta_2, \beta_3) = \text{ERROR SS}_2 - \text{ERROR SS}_1$$

The expression $R(\beta_4, \beta_5 \mid \beta_0, \beta_1, \beta_2, \beta_3)$ is also commonly referred to as

❑   the sum of squares due to β_4 and β_5 (or x_4 and x_5) adjusted for $\beta_0, \beta_1, \beta_2, \beta_3$ (or the intercept and x_1, x_2, x_3)

❑   the sum of squares due to fitting x_4 and x_5 after fitting the intercept and x_1, x_2, x_3

❑   the effects of x_4 and x_5 above and beyond, or partial of, the effects of the intercept and x_1, x_2, x_3.

2.4.3 Hypothesis Tests and Confidence Intervals

Inferences about model parameters are highly dependent on the other parameters in the model under consideration. Therefore, in hypothesis testing it is important to emphasize the parameters for which inferences have been adjusted. For example, tests based on $R(\beta_3 \mid \beta_0, \beta_1, \beta_2)$ and $R(\beta_3 \mid \beta_0, \beta_1)$ may measure entirely different concepts. Consequently, a test of $H_0 : \beta_3 = 0$ versus $H_0 : \beta_3 \neq 0$ may have one result for the model $y = \beta_0 + \beta_1 x_1 + \beta_2 x_2 + \beta_3 x_3 + \varepsilon$ and another result for the model $y = \beta_0 + \beta_1 x_1 + \beta_3 x_3 + \varepsilon$. Differences reflect relationships among independent variables rather than inconsistencies in statistical methodology.

Statistical inferences can also be made in terms of linear functions of the parameters in the form

$$H_0 : \quad \ell_0 \beta_0 + \ell_1 \beta_1 + \ldots + \ell_m \beta_m = 0$$

where the ℓ_i are constants chosen to correspond to a specified hypothesis. Such functions are estimated by the corresponding linear function

$$\mathbf{L}\hat{\boldsymbol{\beta}} = \ell_0 \hat{\beta}_0 + \ell_1 \hat{\beta}_1 + \ldots + \ell_m \hat{\beta}_m$$

of the least-squares estimates $\hat{\boldsymbol{\beta}}$. The variance of $\mathbf{L}\hat{\boldsymbol{\beta}}$ is

$$V\left(\mathbf{L}\hat{\beta}\right) = \left(\mathbf{L}\left(\mathbf{X'X}\right)^{-1}\mathbf{L'}\right)\sigma^2$$

A *t*-test or *F*-test is then used to test $H_0 : (\mathbf{L}\boldsymbol{\beta}) = 0$. The denominator usually uses the error mean square MSE as the estimate of σ^2. Because the variance of the estimated function is based on statistics computed for the entire model, the test of the hypothesis is made in the presence of all model parameters. These tests can be generalized to simultaneous tests of several linear functions. Confidence intervals can be constructed to correspond to the tests.

Three common types of statistical inference are

❑ a test that all slope parameters $(\beta_1, \ldots, \beta_m)$ are 0.

This test compares the fit of the complete model to the model containing only the mean, using the statistic

$$F = (\text{MODEL SS} / m) / \text{MSE}$$

where

$$\text{MODEL SS} = R(\beta_1, \beta_2, \ldots, \beta_m, \,|\beta_0)^1$$

The *F*-statistic has $(m, n\text{-}m\text{-}1)$ degrees of freedom (DF).

❑ a test that the parameters in a subset are 0.

This test compares the fit of the complete model

$$y = \beta_0 + \beta_1 x_1 + \ldots + \beta_g x_g + \beta_{g+1} x_{g+1} + \ldots + \beta_m x_m + \varepsilon$$

with the fit of the reduced model

$$y = \beta_0 + \beta_1 x_1 + \ldots + \beta_g x_g + \varepsilon$$

An *F*-statistic is used to perform the test

$$F = \left(R\left(\beta_{g+1}, \ldots, \beta_m \,|\, \beta_0, \beta_1, \ldots, \beta_g\right) / (m - g)\right) / \text{MSE}$$

Note that reordering the variables produces a test for any desired subset of parameters. If the subset contains only one parameter, β_m, the test is

$$\begin{aligned} F &= \left(R\left(\beta_m \,|\, \beta_0, \beta_1, \ldots, \beta_{m-1}\right) / 1\right) / \text{MSE} \\ &= (\text{partial SS due to } \beta_m) / \text{MSE} \end{aligned}$$

which is equivalent to the *t*-test

$$t = \hat{\beta}_m / \sqrt{c_{mm}\text{MSE}}$$

The corresponding $(1 - \alpha)$ confidence interval about β_m is

$$\hat{\beta}_m \pm t_{\alpha/2} \sqrt{c_{mm}\text{MSE}}$$

❑ an estimate of a subpopulation mean corresponding to a specific **x**. For a given set of *x* values described by a vector **x**, the subpopulation mean is

$$E(y_x) = \beta_0 + \beta_1 x_1 + \ldots + \beta_m x_m = \mathbf{x}'\boldsymbol{\beta}$$

---

[1] $R(\beta_0, \beta_1, \ldots, \beta_m)$ is rarely used. For more information, see Section 2.2.6, "Fitting Restricted Models: The RESTRICT Statement and NOINT Option."

The estimate of $E(y_x)$ is

$$\hat{y}_\mathbf{x} = \hat{\beta}_0 + \hat{\beta}_1 x_1 + \ldots + \hat{\beta}_m x_m = \mathbf{x}'\hat{\boldsymbol{\beta}}$$

The vector $\mathbf{x}$ is constant; hence, the variance of $\hat{y}_x$ is

$$V(\hat{y}_x) = \mathbf{x}'(\mathbf{X}'\mathbf{X})^{-1}x\sigma^2$$

This is useful for computing the confidence intervals.

A related inference is to predict a future single value of y corresponding to a specified $\mathbf{x}$. The predicted value is $\hat{y}$, the same as the estimate of the subpopulation mean corresponding to $\mathbf{x}$. But the relevant variance is

$$V(y - \hat{y}_x) = \left(1 + \mathbf{x}'(\mathbf{X}'\mathbf{X})^{-1}x\right)\sigma^2$$

2.4.4 Using the Generalized Inverse

Many applications, especially those involving PROC GLM, involve an $\mathbf{X}'\mathbf{X}$ matrix that is not of full rank and, therefore, has no unique inverse. For such situations, both PROC GLM and PROC REG compute a generalized inverse, $(\mathbf{X}'\mathbf{X})^-$, and use it to compute a regression estimate, $\mathbf{b} = (\mathbf{X}'\mathbf{X})^- \mathbf{X}'\mathbf{Y}.$

A generalized inverse of a matrix $\mathbf{A}$ is any matrix $\mathbf{G}$ such that $\mathbf{AGA=A}$. Note that this also identifies the inverse of a full-rank matrix.

If $\mathbf{X}'\mathbf{X}$ is not of full rank, then there is an infinite number of generalized inverses. Different generalized inverses lead to different solutions to the normal equations that will have different expected values. That is, $E(\mathbf{b}) = (\mathbf{X}'\mathbf{X})^- \mathbf{X}'\mathbf{X}\boldsymbol{\beta}$ depends on the particular generalized inverse used to obtain $\mathbf{b}$. Thus, it is important to understand what is being estimated by a particular solution.

Fortunately, not all computations in regression analysis depend on the particular solution obtained. For example, the error sum of squares has the same value for all choices of $(\mathbf{X}'\mathbf{X})^-$ and is given by

$$\text{SSE} = \mathbf{Y}'\left(\mathbf{I} - \mathbf{X}(\mathbf{X}'\mathbf{X})^- \mathbf{X}'\right)\mathbf{Y}$$

Hence, the model sum of squares also does not depend on the particular generalized inverse obtained.

The generalized inverse has played a major role in the presentation of the theory of linear statistical models, such as in the books of Graybill (1976) and Searle (1971). In a theoretical setting, it is often possible, and even desirable, to avoid specifying a particular generalized inverse. To apply the generalized inverse to statistical data with computer programs, however, a generalized inverse must actually be calculated. Therefore, it is necessary to declare the specific generalized inverse being computed. Consider, for example, an $\mathbf{X}'\mathbf{X}$ matrix of rank k that can be partitioned as

$$\mathbf{X}'\mathbf{X} = \begin{bmatrix} \mathbf{A}_{11} & \mathbf{A}_{12} \\ \mathbf{A}_{21} & \mathbf{A}_{22} \end{bmatrix}$$

where $\mathbf{A}_{11}$ is $k \times k$ and of rank k. Then $\mathbf{A}_{11}^{-1}$ exists, and a generalized inverse of $\mathbf{X}'\mathbf{X}$ is

$$\left(\mathbf{X}'\mathbf{X}\right)^{-} = \begin{matrix} \mathbf{A}_{11}^{-1} & \phi_{12} \\ \phi_{21} & \phi_{22} \end{matrix}$$

where each ϕ_{ij} is a matrix of zeros of the same dimensions as $\mathbf{A}_{ij}$.

This approach to obtaining a generalized inverse can be extended indefinitely by partitioning a singular matrix into several sets of matrices, as shown above. Note that the resulting solution to the normal equations, $\mathbf{b} = \left(\mathbf{X}'\mathbf{X}\right)^{-} \mathbf{XY},$ has zeros in the positions corresponding to the rows filled with zeros in $\left(\mathbf{X}'\mathbf{X}\right)^{-}$. This is the solution printed by PROC GLM and PROC REG and is regarded as a biased estimate of β.

Because $\mathbf{b}$ is not unique, a linear function $\mathbf{Lb}$, and its variance, are generally not unique either.

However, there is a class of linear functions called **estimable functions**, and they have the following properties:

❑ $\mathbf{Lb}$ and its variance are invariant through all possible generalized inverses. In other words, $\mathbf{Lb}$ and $V(\mathbf{Lb})$ are unique.

❑ $\mathbf{Lb}$ is an unbiased estimate of $\mathbf{L}\beta$.

❑   The vector $\mathbf{L}$ is a linear combination of rows of $\mathbf{X}$.

Analogous to the full-rank case, the variance of $\mathbf{Lb}$ is given by

$$V\left(\mathbf{Lb}\right) = \left(\mathbf{L}\left(\mathbf{X}'\mathbf{X}\right)^{-}\mathbf{L}'\right)\sigma^2$$

This expression is used for statistical inference. For example, a test of $H_0 : \mathbf{L}\beta = \mathbf{0}$ is given by the t-test

$$t = \mathbf{Lb} / \sqrt{\left(\mathbf{L}\left(\mathbf{X}'\mathbf{X}\right)^{-}\mathbf{L}'\right)\text{MSE}}$$

Chapter 3 Analysis of Variance for Balanced Data

3.1 Introduction

The arithmetic mean is the basic descriptive statistic associated with the linear model. In some studies, you only want to estimate a single mean. More commonly, you want to compare the means of two or more treatments. For one- or two-sample (that is, one- or two-treatment) analyses, *t*-tests, or confidence intervals based on the *t*-distribution, are often used. The MEANS procedure and TTEST procedures can perform one- and two-sample *t*-tests. In most cases, either you want to compare more than two treatments, or you must use a more complex design in order to adequately control extraneous variation. For these situations, you need to use analysis of variance. In fact the two-sample tests are merely special cases of analysis of variance, so the analysis of variance is actually a general tool applicable to a wide variety of applications, for two or more treatments.

This chapter begins by presenting one- and two-sample analyses of means using the MEANS and TTEST procedures. Then, more complex analyses using the ANOVA and GLM procedures are discussed. Most of the focus is on analysis of variance and related methods using PROC GLM. [1]

3.2 One- and Two-Sample Tests and Statistics

In addition to a wide selection of descriptive statistics, SAS can provide *t*-tests for a single sample, for paired samples, and for two independent samples.

3.2.1 One-Sample Statistics

The following single-sample statistics are available with SAS:

mean:
$$\bar{x} = \frac{\sum_i x_i}{n}$$

standard deviation:
$$s = \sqrt{\frac{\sum_i (x_i - \bar{x})^2}{n-1}}$$

standard error of the mean:
$$s_{\bar{x}} = \frac{s}{\sqrt{n}}$$

student's *t*:
$$\frac{\bar{x}}{s_{\bar{x}}}$$

The statistics $\bar{x}$, s, and $s_{\bar{x}}$ estimate the population parameters μ, σ, and $\sigma_{\bar{x}} = \sigma / \sqrt{n}$, respectively. Student's *t* is used to test the null hypothesis H_0: $\mu=0$.

PROC MEANS can compute most common descriptive statistics and calculate *t*-tests and the associated significance probability (*p*-value) for a single sample. The basic syntax of the MEANS procedure is as follows:

PROC MEANS options;
VAR variables;
BY variables;
CLASS variables;
WHERE variables;
FREQ variables;
WEIGHT variable;
ID variables;
OUTPUT options;

---

[1] SAS can provide other descriptive statistics with the UNIVARIATE, MEANS, and SUMMARY procedures. PROC SUMMARY is useful for creating data sets of descriptive statistics.

The VAR statement is optional. If this statement is not included, PROC MEANS computes statistics for all numeric variables in the data set. The BY, CLASS, and WHERE statements enable you to obtain separate computations for subgroups of observations in the data set. The FREQ, WEIGHT, ID, and OUTPUT statements can be used with PROC MEANS to perform functions such as weighting or creating an output data set. For more information about PROC MEANS, consult the *SAS/STAT User's Guide* in SAS OnlineDoc, Version 8.

The following example shows a single-sample analysis. In order to design a mechanical harvester for bell peppers, an engineer determined the angle (from a vertical reference) at which 28 peppers hang on the plant (ANGLE). The following statistics are needed:

❏ the sample mean $\bar{x}$, an estimate of the population mean, μ

❏ the sample standard deviation s, an estimate of the population standard deviation, σ.

❏ the standard error of the mean, $s_{\bar{x}}$, a measure of the precision of the sample mean.

Using these computations, the engineer can construct a 95% confidence interval for the mean, the endpoints of which are $\bar{x} - t_{.05}s_{\bar{x}}$ and $\bar{x} + t_{.05}s_{\bar{x}}$ where $t_{.05}$ is obtained from a table of *t*-values. The engineer can also use the statistic $t = \bar{x}/s_{\bar{x}}$ to test the hypothesis that the population mean is equal to 0.

The following SAS statements print the data and perform these computations:

```
data peppers;
    input angle @@;
datalines;
3 11 -7 2 3 8 -3 -2 13 4 7
-1 4 7 -1 4 12 -3 7 5 3 -1
9 -7 2 4 8 -2
;
proc print;
proc means mean std stderr t prt;
run;
```

This PROC MEANS statement specifically calls for the mean (MEAN), the standard deviation (STD), the standard error of the mean (STDERR), the *t*-statistic for testing the hypothesis that the population mean is 0 (T), and the *p*-value (significance probability) of the *t*-test (PRT). These represent only a few of the descriptive statistics that can be requested in a PROC MEANS statement. The data, listed by PROC PRINT, and output from PROC MEANS, appear in Output 3.1.

Output 3.1
PROC
MEANS for
Single-
Sample
Analysis

```
                              Obs     angle

                               1        3
                               2       11
                               3       -7
                               4        2
                               5        3
                               6        8
                               7       -3
                               8       -2
                               9       13
                              10        4
                              11        7
                              12       -1
                              13        4
                              14        7
                              15       -1
                              16        4
                              17       12
                              18       -3
                              19        7
                              20        5
                              21        3
                              22       -1
                              23        9
                              24       -7
                              25        2
                              26        4
                              27        8
                              28       -2


                         The MEANS Procedure

                      Analysis Variable : angle
```

Mean	Std Dev	Std Error	t Value	Pr > |t|
3.1785714	5.2988718	1.0013926	3.17	0.0037

A *t*-table shows $t_{.05}=2.052$ with 27 degrees of freedom (DF). The confidence interval for the mean ANGLE is, therefore, $3.179 \pm 2.052(1.0014)$, which yields the interval (1.123, 5.333). The value of $t=3.17$ has a significance probability of $p=0.0037$, indicating that the engineer can reject the null hypothesis that the mean ANGLE in the population, μ, is 0.

You can compute the confidence interval by adding the option CLM to the PROC MEANS statement. The default is a 95% confidence interval. You can add the ALPHA option to change the level of confidence. For example, ALPHA=0.1 gives you a 90% confidence interval. Alternatively, you can use the OUTPUT statement, along with additional programming statements, to compute the confidence interval. First insert the following statements immediately before the RUN statement in the above program:

```
output out=stats
mean=xbar stderr=sxbar;
```

Then use the following program statements:

```
data stats; set stats;
   t=tinv(27,.05);
   bound=t*sxbar;
   lower=xbar-bound;
   upper=xbar+bound;
proc print;
run;
```

This might seem a little complicated just to get a confidence interval. However, it illustrates the use of the OUTPUT statement to obtain computations from a procedure and the use of a DATA step to make additional computations. Similar methods can be used with other procedures such as the REG procedure, the GLM procedure discussed later in this chapter, the MIXED procedure introduced in Chapter 4, and the GENMOD procedure introduced in Chapter 10.

You should note that a test of H_0: μ=C, where C≠0, can be obtained by subtracting C from each observation. You can do this in the DATA step by adding a command after the INPUT statement, and then applying the single-sample analysis to the revised response variable. For example, you could test H_0: μ=5 with the following statements:

```
data peppers;
    set peppers;
    diff5=angle-5;
proc means t;
run;
```

3.2.2 Two Related Samples

You can apply a single-sample analysis to the difference between paired measurements to make inferences about means from paired samples. This type of analysis is appropriate for randomized-blocks experiments with two treatments. It is also appropriate in many experiments that use before-treatment and after-treatment responses on the same experimental unit, as shown in the example below.

A combination stimulant-relaxant drug is administered to 15 animals whose pulse rates are measured before (PRE) and after (POST) administration of the drug. The purpose of the experiment is to determine if there is a change in the pulse rate as a result of the drug.

The appropriate *t*-statistic is $t = \overline{d} / s_{\overline{d}}$ where $\overline{d} = \sum_i d_i / n$, d_i = the difference between the PRE and POST measurement for the *i*th animal, for example, PRE-POST, $s_{\overline{d}} = s_d / \sqrt{n}$, and

$$s_d = \sqrt{\frac{\sum_i (d_i - \overline{d})^2}{n-1}}.$$

The *t* for the paired differences tests the null hypothesis of no change in pulse rate. You can compute the differences, D=PRE-POST, for each subject and the one-sample *t*-test based on the differences with the following SAS statements:

```
data pulse;
    input pre post;
    d=pre-post;
datalines;
    62 61
    63 62
    58 59
    64 61
    64 63
    61 58
    68 61
    66 64
    65 62
    67 68
    69 65
```

```
        61 60
        64 65
        61 63
        63 62
        ;
        proc print;
        proc means mean std stderr t prt;
           var d;
        run;
```

In this example, the following SAS statement creates the variable *D* (the difference in rates):

```
        d=pre-post;
```

Remember that a SAS statement that generates a new variable is part of a DATA step.

The PROC MEANS statements here and in the preceding example are identical. The statement

```
        var d;
```

following the PROC MEANS statement restricts the PROC MEANS analysis to the variable *D*. Otherwise, computations would also be performed on PRE and POST. The data listed by PROC PRINT and output from PROC MEANS appear in Output 3.2.

Output 3.2
*Paired-
Difference
Analysis*

Obs	pre	post	d
1	62	61	1
2	63	62	1
3	58	59	-1
4	64	61	3
5	64	63	1
6	61	58	3
7	68	61	7
8	66	64	2
9	65	62	3
10	67	68	-1
11	69	65	4
12	61	60	1
13	64	65	-1
14	61	63	-2
15	63	62	1

The MEANS Procedure

Analysis Variable : d

Mean	Std Dev	Std Error	t Value	Pr > |t|
1.4666667	2.3258383	0.6005289	2.44	0.0285

The *t*-value of 2.44 with *p*=0.0285 indicates a statistically significant change in mean pulse rate. Because the mean of D (1.46) is positive, the drug evidently decreases pulse rate.

You can also compute the paired test more simply by using PROC TTEST. The TTEST procedure computes two-sample paired *t*-tests for both the paired and independent case. The latter is shown in Section 3.2.3., "Two Independent Samples." For the paired test, use the following SAS statements:

```
proc ttest;
  paired pre*post;
run;
```

The statement PAIRED PRE*POST causes the test to be computed for the paired difference PRE-POST. The results appear in Output 3.3. The estimated mean difference of PRE-POST, 1.4667, appears in the column labeled MEAN. The lower and upper 95% confidence limits appear in the columns labeled Lower CL Mean and Upper CL Mean, respectively.

Output 3.3
Paired-
Difference
Analysis
Using PROC
TTEST with
the PAIRED
Option

```
                              The TTEST Procedure

                                 Statistics

                     Lower CL           Upper CL  Lower CL           Upper CL
     Difference      N    Mean     Mean     Mean   Std Dev  Std Dev  Std Dev

     pre - post     15   0.1787  1.4667   2.7547   1.7028   2.3258   3.6681

                                 Statistics

          Difference      Std Err    Minimum      Maximum

          pre - post       0.6005        -2            7


                                  T-Tests

          Difference        DF     t Value     Pr > |t|

          pre - post        14       2.44       0.0285
```

You can also use the single mean capability of PROC TTEST with the *D* variable:

```
proc ttest;
   var d;
run;
```

As mentioned at the beginning of this section, the paired two-sample test is a special case of the test for treatment effects in a randomized-blocks design, pairs being a special case of blocks. Section 3.5, "Randomized-Blocks Designs," presents the analysis of blocked designs.

3.2.3 Two Independent Samples

You can test the significance of the difference between means from two independent samples with the *t*-statistic

$$t = (\overline{x}_1 - \overline{x}_2) \bigg/ \sqrt{s^2 \left(\frac{1}{n_1} + \frac{1}{n_2} \right)}$$

where $\overline{x}_1$, $\overline{x}_2$, and n_1, n_2 refer to the means and sample sizes of the two groups, respectively, and s^2 refers to the *pooled variance* estimate,

$$s^2 = \frac{(n_1 - 1)s_1^2 + (n_2 - 1)s_2^2}{n_1 + n_2 - 2}$$

Note that s_1^2 and s_2^2 are the sample variances for the two groups, respectively. The pooled variance estimate should be used if it is reasonable to assume that the population variances of the two groups, σ_1^2 and σ_2^2 are equal. If this assumption cannot be justified, then you should use an approximate *t*-statistic given by

$$t = (\bar{x}_1 - \bar{x}_2) \Big/ \sqrt{\left(\frac{s_1^2}{n_1} + \frac{s_2^2}{n_2} \right)}$$

You can use PROC TTEST to compute both of these *t*'s along with the (folded) *F*-statistic

$$F' = (\text{larger of } s_1^2, s_2^2) \big/ (\text{smaller of } s_1^2, s_2^2)$$

to test the assumption $\sigma_1^2 = \sigma_2^2$. Analysis-of-variance procedures, for example, PROC ANOVA and PROC GLM, give equivalent results but do not test equality of the variances and perform the approximate *t*-test.

An example of this test is the comparison of muzzle velocities of cartridges made from two types of gunpowder (POWDER). The muzzle velocity (VELOCITY) was measured for eight cartridges made from powder type 1 and ten cartridges from powder type 2. The data appear in Output 3.4.

Output 3.4
PROC PRINT
of BULLET
Data for Two
Independent
Samples

Obs	powder	velocity
1	1	27.3
2	1	28.1
3	1	27.4
4	1	27.7
5	1	28.0
6	1	28.1
7	1	27.4
8	1	27.1
9	2	28.3
10	2	27.9
11	2	28.1
12	2	28.3
13	2	27.9
14	2	27.6
15	2	28.5
16	2	27.9
17	2	28.4
18	2	27.7

The two-sample *t*-test is appropriate for testing the null hypothesis that the muzzle velocities are equal. You can obtain such a *t*-test with these SAS statements:

```
proc ttest data=bullets;
   var velocity;
   class powder;
run;
```

PROC TTEST performs the two-sample analysis. The variable POWDER in the CLASS statement identifies the groups (or treatments) whose means are to be compared. CLASS variables may be numeric or character variables. This CLASS statement serves the same purpose as it does in all other procedures that require identification of groups of treatments. In PROC TTEST, the CLASS variable must have exactly two values. Otherwise, the procedure issues an error message and stops processing. The VAR statement identifies the variable whose means you want to compare. Note that PROC TTEST is limited to comparing two groups. To compare more than two groups, you use analysis-of-variance procedures, discussed in Section 3.3, "The Comparison of Several Means: Analysis of Variance."

Output 3.5 shows the data from PROC PRINT and the results of PROC TTEST.

*Output 3.5
PROC TTEST
for Two
Independent
Samples*

```
                         The TTEST Procedure

                             Statistics

                             Lower CL          Upper CL  Lower CL
   Variable  Class       N     Mean    Mean      Mean    Std Dev  Std Dev

   velocity            1  8   27.309  27.638   27.966    0.2596   0.3926
   velocity            2 10   27.841  28.06    28.279    0.2106   0.3062
   velocity  Diff (1-2)       -0.771  -0.422   -0.074    0.2582   0.3467

                             Statistics

                               Upper CL
       Variable  Class         Std Dev   Std Err   Minimum   Maximum

       velocity            1    0.799    0.1388     27.1      28.1
       velocity            2    0.5591   0.0968     27.6      28.5
       velocity  Diff (1-2)     0.5276   0.1644


                               T-Tests

       Variable   Method          Variances    DF   t Value   Pr > |t|

       velocity   Pooled          Equal        16    -2.57     0.0206
       velocity   Satterthwaite   Unequal    13.1    -2.50     0.0267


                        Equality of Variances

         Variable    Method     Num DF   Den DF   F Value   Pr > F

         velocity    Folded F      7        9      1.64     0.4782
```

The first part of PROC TTEST output gives you the number of observations, mean, standard deviation, standard error of the mean, the minimum and maximum observations of VELOCITY for the two levels of POWDER, and the upper and lower 95% confidence limits. The second part gives you the *t*-test results, the *t*-statistic (T), the degrees of freedom (DF), and the *p*-value (Pr> |t|). You can see that there are two sets of statistics. These correspond to two types of assumptions: the usual two-sample *t*-test that assumes equal variances (Equal) or an approximate *t*-test that does not assume equal variances (Unequal). The approximate *t*-test uses Satterthwaite's approximation for the sum of two mean squares (Satterthwaite 1946) to calculate the significance probability Pr> |t|. Section 4.5.3, "Satterthwaite's Formula for Approximate Degrees of Freedom," presents the approximation in some detail.

The *F*-test at the bottom of Output 3.5 is used to test the hypothesis of equal variances. An $F=1.64$ with a significance probability of $p=0.4782$ provides insufficient evidence to conclude that the variances are unequal. Therefore, use the test that assumes equal variances. For this test $t=2.5694$ with a *p*-value of 0.0206. This is strong evidence of a difference between the mean velocities for the two powder types, with the mean velocity for powder type 2 greater than that for powder type 1.

The two-sample independent test of the difference between treatment means is a special case of one-way analysis of variance. Thus, using analysis of variance for the BULLET data, shown in Section 3.3 is equivalent to the *t*-test procedures shown above, assuming equal variances for the two samples. This point is developed in the next section.

3.3 The Comparison of Several Means: Analysis of Variance

Analysis of variance and related mean comparison procedures are the primary tools for making statistical inferences about a set of two or more means. SAS offers several procedures. Two of them, PROC ANOVA and PROC GLM, are specifically intended to compute analysis of variance. Other procedures, such as PROC TTEST, PROC NESTED, and PROC VARCOMP, are available for specialized types of analyses.

PROC ANOVA is limited to balanced or orthogonal data sets. PROC GLM is more general—it can be used for both balanced and unbalanced data sets. While the syntax is very similar, PROC ANOVA is simpler computationally than PROC GLM. At one time, this was an issue, because large models using the GLM procedure often exceeded the computer's capacity. With contemporary computers, GLM's capacity demands are rarely an issue, and so PROC GLM has largely superseded PROC ANOVA.

PROC MIXED can compute all of the essential analysis-of-variance statistics. In addition, MIXED can compute statistics specifically appropriate for models with random effects that are not available with any other SAS procedure. For this reason, MIXED is beginning to supplant GLM for data analysis, much as GLM previously replaced ANOVA. However, GLM has many features not available in MIXED that are useful for understanding underlying analysis-of-variance concepts, so it is unlikely that GLM will ever be completely replaced.

The rest of this chapter focuses on basic analysis of variance with the main focus on PROC GLM. Random effects and PROC MIXED are introduced in Chapter 4.

3.3.1 Terminology and Notation

Analysis of variance partitions the variation among observations into portions associated with certain factors that are defined by the classification scheme of the data. These factors are called sources of variation. For example, variation in prices of houses can be partitioned into portions associated with region differences, house-type differences, and other differences. Partitioning is done in terms of sums of squares (SS) with a corresponding partitioning of the associated degrees of freedom (DF). For three sources of variation (A, B, C),

TOTAL SS = SS(A) + SS(B) + SS(C) + RESIDUAL SS

The term TOTAL SS is normally the sum of the squared deviations of the data values from the overall mean, $\sum_i (y_i - \overline{y})^2$, where y_i represents the observed response for the ith observation.

The formula for computing SS(A), SS(B), and SS(C) depends on the situation. Typically, these terms are sums of squared differences between means. The term RESIDUAL SS is simply what is left after subtracting SS(A), SS(B), and SS(C) from TOTAL SS.

Degrees of freedom are numbers associated with sums of squares. They represent the number of independent differences used to compute the sum of squares. For example, $\sum_i (y_i - \overline{y})^2$ is a sum of squares based upon the differences between each of the n observations and the mean, that is, $y_1 - \overline{y}$, $y_2 - \overline{y}$, ... , $y_n - \overline{y}$. There are only n-1 linearly independent differences, because any one of these differences is equal to the negative of the sum of the others. For example, consider the following:

$$y_n - \overline{y} = \sum_{i=1}^{n-1} (y_i - \overline{y})$$

Total degrees of freedom are partitioned into degrees of freedom associated with each source of variation and the residual:

TOTAL DF = DF(A) + DF(B) + DF(C) + RESIDUAL DF

Mean squares (MS) are computed by dividing each SS by its corresponding DF. Ratios of mean squares, called *F*-ratios, are then used to compare the amount of variability associated with each source of variation. Tests of hypotheses about group means can be based on *F*-ratios. The computations are usually displayed in the familiar tabular form shown below:

Source of Variation	DF	SS	MS	F	*p*-value
A	DF(A)	SS(A)	MS(A)	F(A)	*p* for A
B	DF(B)	SS(B)	MS(B)	F(B)	*p* for B
C	DF(C)	SS(C)	MS(C)	F(C)	*p* for C
Residual	Residual DF	SS(Residual)	Residual MS		
Total	Total DF	SS(Total)			

Sources of variation in analysis of variance typically measure treatment factor effects. Three kinds of effects are considered in this chapter: main effects, interaction effects, and nested effects. Each is discussed in terms of its SS computation. Effects can be either fixed or random, a distinction that is developed in Chapter 4, "Analyzing Data with Random Effects." All examples in this chapter assume fixed effects.

A main effect sum of squares for a factor A, often called the sum of squares for treatment A, is given by

$$SS(A) = \sum_i n_i (\bar{y}_i - \bar{y}_.)^2 \tag{3.1}$$

or alternatively by

$$SS(A) = \sum_i \frac{y_i^2}{n_i} - \frac{y_.^2}{n_.} \tag{3.2}$$

where

n_i equals the number of observations in level *i* of factor A.

y_i equals the total of observations in level *i* of factor A.

$\bar{y}_{i.}$ equals the mean of observations in level *i* of factor A.

$n_.$ equals the total number of observations ($\sum_i n_i$)

$y_.$ equals the total of all observations ($\sum_i y_i$)

$\bar{y}_.$ equals the mean of all observations ($y_./n_.$).

As equation (3.1) implies, the SS for a main effect measures variability among the means corresponding to the levels of the factor. If A has *a* levels, then SS(A) has (*a* − 1) degrees of freedom.

For data with a single factor, the main effect and treatment SS are one and the same. For data with two or more factors, treatment variation must be partitioned into additional components. The structure of these multiple factors determines what SS besides main effects are appropriate. The two basic structures are **crossed** and **nested** classifications. In a crossed classification, every level of each factor occurs with each level of the other factors. In a nested classification, each level of one factor occurs with different levels of the other factor. See also Figures 4.1 and 4.2 in Chapter 4 for an illustration.

3.3.1.1 Crossed Classification and Interaction Sum of Squares

In crossed classifications, you partition the SS for treatments into main effect and **interaction** components. To understand an interaction, you must first understand **simple effects**. It is easiest to start with a two-factor crossed classification. Denote $\bar{y}_{ij}$ the mean of the observations on the *ij*th factor combination, that is, the treatment receiving level *i* of factor A and level *j* of factor B. The *ij*th factor combination is also defined as the *ij*th **cell**. A simple effect is defined as

$$A \mid B_j = \bar{y}_{ij} - \bar{y}_{i'j}, \text{ for differences between two levels of } i \neq i' \text{ of factor A at level } j \text{ of factor B}$$

or alternatively

$$B \mid A_i = \bar{y}_{ij} - \bar{y}_{ij'}, \text{ for differences between two levels of } j \neq j' \text{ of factor B at level } i \text{ of factor A}$$

If the simple effects "$A \mid B_j$" are not the same for all levels of factor B, or, equivalently, if the "$B \mid A_i$" are not the same for all levels of factor A, then an interaction is said to occur. If all simple effects are equal, there is no interaction. An interaction effect is thus defined by $\bar{y}_{ij} - \bar{y}_{i'j} - \bar{y}_{ij'} + \bar{y}_{i'j'}$. If it is equal to zero, there is no interaction; otherwise, there is an "A by B" interaction.

It follows that you calculate the sum of squares for the interaction between the factors A and B with the equation

$$SS(A*B) = \sum_{ij} n_{ij} (\bar{y}_{ij} - \bar{y}_{i.} - \bar{y}_{.j} + \bar{y}_{..})^2 \tag{3.3}$$

or alternatively

$$SS(A*B) = \sum_{ij} y_{ij}^2 / n - \sum_i y_{i.}^2 / bn - \sum_i y_{.j}^2 / an + y_{..}^2 / abn \tag{3.4}$$

where

n	equals the number of observations on the *ij*th cell.
a and b	are the number of levels of A and B, respectively.
y_{ij}	equals the total of all observations in the *ij*th cell.
$y_{i.}$	is equal to $\sum_j y_{ij}$, the total of all observations on the *i*th level of A.

| $y_{\cdot j}$ | is equal to $\sum_i y_{ij}$, the total of all observations on the jth level of B. |

| $y_{\cdot\cdot}$ | is equal to $\sum_{ij} y_{ij}$, the grand total of all observations. |

The sum of squares for A*B has

$$(a-1)(b-1) = ab - a - b + 1$$

degrees of freedom.

3.3.1.2 Nested Effects and Nested Sum of Squares

For nested classification, suppose factor B is nested within factor A. That is, a different set of levels of B appears with each level of factor A. For this classification, you partition the treatment sum of squares into the main effect, SS(A) and the SS for nested effect, written B(A). The formula for the sum of squares of B(A) is

$$SS[B(A)] = \sum_{ij} n_{ij} (\bar{y}_{ij} - \bar{y}_{i\cdot})^2 \tag{3.5}$$

or alternatively

$$SS[B(A)] = \sum_{ij} y_{ij}^2 \big/ n_{ij} - \sum_i y_{i\cdot}^2 \big/ n_{i\cdot} \tag{3.6}$$

where

| n_{ij} | equals the number of observations on level j of B and level i of A. |

| y_{ij} | equals the total of observations for level j of B and level i of A. |

| $\bar{y}_{ij}$ | equals the mean of observations for level j of B and level i of A. |

| $n_{i\cdot}$ | is equal to $\sum_j n_{ij}$. |

| $y_{i\cdot}$ | is equal to $\sum_j y_{ij}$. |

| $\bar{y}_{i\cdot}$ | is equal to $y_{i\cdot}/n_{i\cdot}$. |

Looking at equation (3.5) as

$$SS(B(A)) = \sum_i \left(\sum_j n_{ij} (\bar{y}_{ij} - \bar{y}_{i\cdot})^2 \right) \tag{3.7}$$

you see that SS(B(A)) measures the variation among the levels of B within each level of A and then pools, or adds, across the levels of A. If there are b_i levels of B in level i of A, then there are $(b_i - 1)$ DF for B in level i of A, for a total of $\sum_i (b_i - 1)$ DF for the B(A) effect.

3.3.2 Using the ANOVA and GLM Procedures

Because of its generality and versatility, PROC GLM is the preferred SAS procedure for analysis of variance, provided all model effects are fixed effects. For one-way and balanced multiway classifications, PROC ANOVA produces the same results as the GLM procedure. The term **balanced** means that each cell of the multiway classification has the same number of observations.

This chapter begins with a one-way analysis of variance example. Because the computations used by PROC ANOVA are easier to understand without developing matrix algebra concepts used by PROC GLM, the first example begins using PROC ANOVA. Subsequent computations and all remaining examples use PROC GLM, because GLM is the procedure data analysts ordinarily use in practice. These examples are for basic experimental designs (completely random, randomized blocks, Latin square) and factorial treatment designs.

Generally, PROC ANOVA computes the sum of squares for a factor A in the classification according to equation (3.2). Nested effects are computed according to equation (3.6). A two-factor interaction sum of squares computed by PROC ANOVA follows equation (3.4), which can be written more generally as

$$SS(A*B) = \sum_{ij} y_{ij}^2 \big/ n_{ij} \; - \; \sum_{i} y_{i\cdot}^2 \big/ n_{i\cdot} \; - \; \sum_{i} y_{\cdot j}^2 \big/ n_{\cdot j} \; + \; y_{\cdot\cdot}^2 \big/ n_{\cdot\cdot} \qquad (3.8)$$

where n_{ij} is the number of observations and y_{ij} is the observed total for the ijth A×B treatment combination. If n_{ij} has the same value for all ij, then equation (3.8) is the same as equation (3.4). Equation (3.4) is not correct unless all the n_{ij} are equal to the same value, and this formula could even produce a negative value because it would not actually be a sum of squares. If a negative value is obtained, PROC ANOVA prints a value of 0 in its place. Sums of squares for higher-order interactions follow a similar formula.

The ANOVA and GLM procedures share much of the same syntax. The GLM procedure has additional features described later in this section. The shared basic syntax is as follows:

PROC ANOVA (or **GLM**) options;
CLASS variables;
MODEL dependents=effects / options;
MEANS effects / options;
ABSORB variables;
FREQ variable;
TEST H=effects E=effect;
MANOVA H=effects E=effect M=equations / options;
REPEATED factor-name levels / options;
BY variables;

The CLASS and MODEL statements are required to produce the ANOVA table. The other statements are optional. The ANOVA output includes the F-tests of all effects in the MODEL statement. All of these F-tests use residual mean squares as the error term. PROC GLM produces four types of sums of squares. In the examples considered in this chapter, the different types of sums of squares are all the same, and are identical to those computed by PROC ANOVA. Distinctions among the types of SS occur with unbalanced data, and are discussed in detail in Chapters 5 and 6.

The MEANS statement produces tables of the means corresponding to the list of effects. Several multiple comparison procedures are available as options in the MEANS statement. Section 3.3.3, "Multiple Comparisons and Preplanned Comparisons," and Section 3.4.2, "Computing Means, Multiple Comparisons of Means, and Confidence Intervals," illustrate these procedures.

The TEST statement is used for tests where the residual mean square is not the appropriate error term, such as certain effects in mixed models and main-plot effects in split-plot experiments (see Chapter 4). You can use multiple MEANS and TEST statements, but only one MODEL statement. The ABSORB statement implements the technique of absorption, which saves time and reduces storage requirements for certain types of models. This is illustrated in Chapter 11, "Examples of Special Applications."

The MANOVA statement is used for multivariate analysis of variance (see Chapter 9, "Multivariate Linear Models"). The REPEATED statement can be useful for analyzing repeated-measures designs (see Chapter 8, "Repeated-Measures Analysis"), although the more sophisticated repeated-measures analysis available with PROC MIXED is preferable in most situations. The BY statement specifies that separate analyses are performed on observations in groups defined by the BY variables. Use the FREQ statement when you want each observation in a data set to represent *n* observations, where *n* is the value of the FREQ variable.

Most of the analysis-of-variance options in PROC GLM use the same syntax as PROC ANOVA. The same analysis-of-variance program in PROC ANOVA will work for GLM with little modification. GLM has additional statements—CONTRAST, ESTIMATE, and LSMEANS. The CONTRAST and ESTIMATE statements allow you to test or estimate certain functions of means not defined by other multiple comparison procedures. These are introduced in Section 3.4, "Analysis of One-Way Classification of Data." The LSMEANS statement allows you to compute means that are adjusted for the effects of unbalanced data, an extremely important consideration for unbalanced data, which is discussed in Chapter 5. LSMEANS has additional features useful for factorial experiments (see Section 3.7, "Two-Way Factorial Experiment") and analysis of covariance (see Chapter 7).

For more information about PROC ANOVA and PROC GLM, see their respective chapters in the *SAS/STAT User's Guide* in SAS OnlineDoc, Version 8.

As an introductory example, consider the BULLET data from Section 3.2.3. You can compute the one-way analysis of variance with PROC ANOVA using the following statements:

```
proc anova;
  class powder;
  model velocity=powder;
```

The data appear in Output 3.6.

Output 3.6
Analysis-of-Variance Table for BULLET Two-Sample Data

```
                          The ANOVA Procedure

Dependent Variable: velocity

                               Sum of
Source              DF        Squares    Mean Square   F Value   Pr > F

Model                1     0.79336111     0.79336111      6.60   0.0206

Error               16     1.92275000     0.12017188

Corrected Total     17     2.71611111


             R-Square     Coeff Var     Root MSE     velocity Mean

             0.292094     1.243741      0.346658        27.87222


Source              DF       Anova SS    Mean Square   F Value   Pr > F

powder               1     0.79336111     0.79336111      6.60   0.0206
```

The output gives the sum of squares and mean square for the treatment factor, POWDER, and for residual, called ERROR in the output. Note that the MODEL and POWDER sum of squares are identical. Treatment and MODEL statistics are always equal for one-way analysis of variance, but not for the more complicated analysis-of-variance models discussed starting with Section 3.6, "Latin Square Design with Two Response Variables." The *F*-value, 6.60, is the square of the two-sample *t*-value assuming equal variances shown previously in Output 3.5. The *p*-value for the two-sample *t*-test and the ANOVA *F*-test shown above are identical. This equivalence of the two-sample test and one-way ANOVA holds whenever there are two treatments and the samples are independent. However, ANOVA allows you to compare more than two treatments.

Alternatively, you can use PROC GLM to compute the analysis of variance. You can also use the ESTIMATE statement in GLM to compute the estimate and standard error of the difference between the means of the two POWDER levels. The statements and results are not shown here, but you can obtain them by following the examples in Section 3.4, "Analysis of One-Way Classification of Data." The estimate and standard error of the difference for the BULLET data are identical to those given in Output 3.5.

3.3.3 Multiple Comparisons and Preplanned Comparisons

The *F*-test for a factor in an analysis of variance tests the null hypothesis that all the factor means are equal. However, the conclusion of such a test is seldom a satisfactory end to the analysis. You usually want to know more about the differences among the means (for example, which means are different from which other means or if any groups of means have common values).

Multiple comparisons of the means are commonly used to answer these questions. There are numerous methods for making multiple comparisons, most of which are available in PROC ANOVA and PROC GLM. In this chapter, only a few of the methods are illustrated.

One method of multiple comparisons is to conduct a series of *t*-tests between pairs of means; this is essentially the method known as least significant difference (LSD). Refer to Steel and Torrie (1980) for examples.

Another method of multiple comparisons is Duncan's multiple-range test. With this test, the means are first ranked from largest to smallest. Then the equality of two means is tested by referring the difference to tabled critical points, the values of which depend on the range of the ranks of the two means tested. The larger the range of the ranks, the larger the tabled critical point (Duncan 1955).

The LSD method and, to a lesser extent, Duncan's method, are frequently criticized for inflating the Type I error rate. In other words, the overall probability of falsely declaring some pair of means different, when in fact they are equal, is substantially larger than the stated α-level. This overall probability of a Type I error is called the **experimentwise** error rate. The probability of a Type I error for one particular comparison is called the **comparisonwise** error rate. Other methods are available to control the experimentwise error rate, including Tukey's method.

You can request the various multiple comparison tests with options in the MEANS statement in the ANOVA and GLM procedures.

Multiple comparison procedures, as described in the previous paragraphs, are useful when there are no particular comparisons of special interest. But in most situations there is something about the factor that suggests specific comparisons. These are called **preplanned comparisons** because you can decide to make these comparisons prior to collecting data. Specific hypotheses for preplanned comparisons can be tested by using the CONTRAST, ESTIMATE, or LSMEANS statement in PROC GLM, as discussed in Section 3.4.3, "Planned Comparisons for One-Way Classification: The CONTRAST Statement."

3.4 The Analysis of One-Way Classification of Data

One-way classification refers to data that are grouped according to some criterion, such as the values of a classification variable. The gunpowder data presented in Section 3.2.3, "Two Independent Samples," and in Section 3.3.2, "Using the ANOVA and GLM Procedures," are an example of a one-way classification. The values of VELOCITY are classified according to POWDER. In this case, there are two levels of the classification variable—1 and 2. Other examples of one-way classifications might have more than two levels of the classification variable. Populations of U.S. cities could be classified according to the state containing the city, giving a one-way classification with 50 levels (the number of states) of the classification variable. One-way classifications of data can result from sample surveys. For example, wages determined in a survey of migrant farm workers could be classified according to the type of work performed. One-way classifications also result from a completely randomized designed experiment. For example, strengths of monofilament fiber can be classified according to the amount of an experimental chemical used in the manufacturing process, or sales of a new facial soap in a marketing test could be classified according to the color of the soap. The type of statistical analysis that is appropriate for a given one-way classification of data depends on the goals of the investigation that produced the data. However, you can use analysis of variance as a tool for many applications.

The levels of a classification variable are considered to correspond to different populations from which the data were obtained. Let k stand for the number of levels of the classification criterion, so there are data from k populations. Denote the population means as $\mu_1, \ldots, \mu_k$. Assume that all the populations have the same variance, σ^2 and that all the populations are normally distributed. Also, consider now those situations for which there are the same number of observations from each population (denoted n). Denote the jth observation in the ith group of data by y_{ij}. You can summarize this setup as follows:

$$y_{11}, \ldots, y_{1,n} \text{ is a sample from } N(\mu_1, \sigma^2)$$

$$y_{21}, \ldots, y_{2,n} \text{ is a sample from } N(\mu_2, \sigma^2)$$

.

.

.

$$y_{k1}, \ldots, y_{k,n} \text{ is a sample from } N(\mu_k, \sigma^2)$$

$N(\mu_i, \sigma^2)$ refers to a normally distributed population with mean μ_i and variance σ^2. Sometimes it is useful to express the data in terms of linear models. One way of doing this is to write

$$y_{ij} = \mu_i + e_{ij}$$

where μ_i is the mean of the ith population and e_{ij} is the departure of the observed value y_{ij} from population mean. This is called a **means model**. Another model is called an **effects model**, and is denoted by the equation

$$y_{ij} = \mu + \tau_i + e_{ij}$$

The effects model simply expresses the *i*th population mean as the sum of two components, $\mu_i = \mu + \tau_i$. In both models e_{ij} is called the **error** and is normally distributed with mean 0 and variance σ^2. Moreover, both of these models are regression models, as you will see in Chapter 6. Therefore, results from regression analysis can be used for these models, as discussed in subsequent sections.

Notice that the models for one-way analysis of variance assume that the observations within each classification level are normally distributed and that the variances among the observations for each level are equal. The latter assumption was addressed in Section 3.2.3, "Two Independent Samples." The analysis-of-variance procedure is robust, meaning that only severe failures of these assumptions compromise the results. Nonetheless, these assumptions should be checked. You can obtain simple but useful visual tools by sorting the data by classification level and running PROC UNIVARIATE. For example, for the BULLET data, use the following SAS statements:

```
proc sort; by powder;
proc univariate normal plot; by powder;
  var velocity;
run;
```

Output 3.7 shows results selected for relevance.

Output 3.7
PROC
UNIVARIATE
Output for
BULLET
Data to
Check
ANOVA
Assumptions

1. Normal Probability Plots

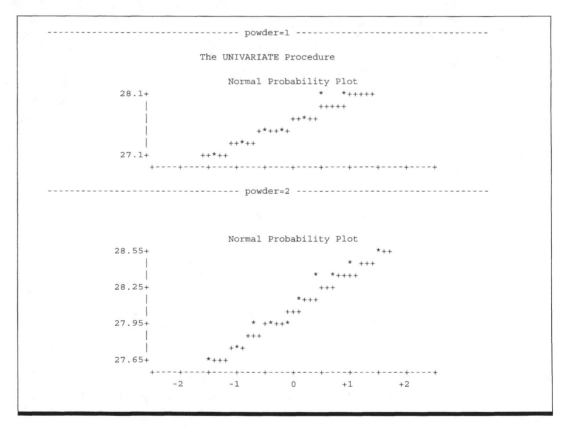

2. Side-by-Side Box-and-Whisker Plots

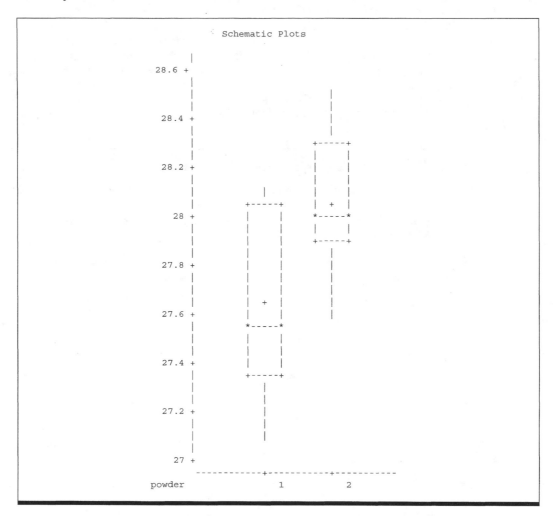

These plots allow you to look for strong visual evidence of failure of assumptions. You can check for non-normality using the normal probability plots. Some departure from normality is common and has no meaningful effect on ANOVA results. In fact, many statisticians argue that true normal distributions are rare in nature, if they exist at all. Highly skewed distributions, however, can seriously affect ANOVA results; strongly asymmetric box-and-whisker plots give you a useful visual cue to detect such situations. The side-by-side box-and-whisker plot also allows you to detect heterogeneous variances. Note that neither plot suggests failure of assumptions. The box-and-whisker plot does suggest that the typical response to POWDER 1 is less than the response to POWDER 2.

The UNIVARIATE output contains many other statistics, such as the variance, skewness, and kurtosis by treatment and formal tests of normality. These can be useful, for example, for testing equal variance. It is beyond the scope of this text to discuss model diagnostics in great detail. You can find such discussions in most introductory statistical methods texts. The MEANS statement of PROC GLM has an option, HOVTEST, that computes statistics to test homogeneity of variance. The GLM procedure and the MEANS statement are discussed in more detail in the remaining sections of this chapter. An example of the HOVTEST output appears in Output 3.9. In any event, you should be aware that many statisticians consider formal tests of assumptions to be of limited usefulness because the number of observations per treatment is often quite small. In most cases, strong visual evidence is the best indicator of trouble.

When analysis-of-variance assumptions fail, a common strategy involves transforming the data and computing the analysis of variance on the transformed data. Section 4.2, "Nested Classifications," contains an example using this approach. Often, assumptions fail because the distribution of the data is known to be something other than normal. Generalized linear models are essentially regression and ANOVA models for data whose distribution is known but not necessarily normal. In such cases, you can use methods illustrated in Chapter 10, "Generalized Linear Models."

Section 3.4.1 presents an example of analysis of variance for a one-way classification.

3.4.1 Computing the ANOVA Table

Four specimens of each of five brands (BRAND) of a synthetic wood veneer material are subjected to a friction test. A measure of wear is determined for each specimen. All tests are made on the same machine in completely random order. Data are stored in a SAS data set named VENEER.

```
data veneer;
    input brand $ wear;
cards;
ACME 2.3
ACME 2.1
ACME 2.4
ACME 2.5
CHAMP 2.2
CHAMP 2.3
CHAMP 2.4
CHAMP 2.6
AJAX 2.2
AJAX 2.0
AJAX 1.9
AJAX 2.1
TUFFY 2.4
TUFFY 2.7
TUFFY 2.6
TUFFY 2.7
XTRA 2.3
XTRA 2.5
XTRA 2.3
XTRA 2.4
;
proc print data=veneer;
run;
```

Output 3.8 shows the data.

Output 3.8
Data for One-Way Classification

Obs	brand	wear
1	ACME	2.3
2	ACME	2.1
3	ACME	2.4
4	ACME	2.5
5	CHAMP	2.2
6	CHAMP	2.3
7	CHAMP	2.4
8	CHAMP	2.6
9	AJAX	2.2
10	AJAX	2.0
11	AJAX	1.9
12	AJAX	2.1
13	TUFFY	2.4
14	TUFFY	2.7
15	TUFFY	2.6
16	TUFFY	2.7
17	XTRA	2.3
18	XTRA	2.5
19	XTRA	2.3
20	XTRA	2.4

An appropriate analysis of variance has the basic form:

Source of Variation	DF
BRAND	4
Error	15
Total	19

The following SAS statements produce the analysis of variance:

```
proc glm data=veneer;
    class brand;
    model wear=brand;
    means brand/hovtest;
run;
```

Since the data are classified only according to the values of BRAND, this is the only variable in the CLASS statement. The variable WEAR is the response variable to be analyzed, so WEAR appears on the left side of the equal sign in the MODEL statement. The only source of variation (effect in the ANOVA table) other than ERROR (residual) and TOTAL is variation due to brands; therefore, BRAND appears on the right side of the equal sign in the MODEL statement. The MEANS statement causes the treatment means to be computed. The HOVTEST option computes statistics to test the homogeneity of variance assumption. The treatment means are not shown here; they are considered in more detail later. Output from the MODEL and HOVTEST statements appear in Output 3.9.

Output 3.9
Analysis of
Variance for
One-Way
Classification
with a
Homogeneity-
of-Variance
Test

```
                              The GLM Procedure

Dependent Variable: wear

                                   Sum of
Source                   DF        Squares     Mean Square   F Value   Pr > F

Model                     4     0.61700000      0.15425000      7.40   0.0017

Error                    15     0.31250000      0.02083333

Corrected Total          19     0.92950000

                R-Square    Coeff Var      Root MSE      wear Mean

                0.663798     6.155120      0.144338       2.345000

Source                   DF      Type I SS    Mean Square   F Value   Pr > F

brand                     4     0.61700000     0.15425000      7.40   0.0017


Source                   DF    Type III SS    Mean Square   F Value   Pr > F

brand                     4     0.61700000     0.15425000      7.40   0.0017


               Levene's Test for Homogeneity of wear Variance
               ANOVA of Squared Deviations from Group Means

                            Sum of        Mean
          Source      DF    Squares      Square    F Value   Pr > F

          brand        4   0.000659     0.000165     0.53    0.7149
          Error       15   0.00466      0.000310
```

The results in Output 3.9 are summarized in the following ANOVA table:

Source	DF	SS	MS	F	p
BRAND	4	0.6170	0.1542	7.40	0.0017
ERROR	15	0.3125	0.0208		
TOTAL	19	0.9295			

Notice that you get the same computations from PROC GLM as from PROC ANOVA for the analysis of variance, although they are labeled somewhat differently. For one thing, in addition to the MODEL sum of squares, PROC GLM computes two sets of sums of squares for BRAND— Type I and Type III sums of squares—rather than the single sum of squares computed by the ANOVA procedure. For the one-way classification, as well as for balanced multiway classifications, the GLM-Type I, GLM-Type III, and PROC ANOVA sums of squares are identical. For unbalanced multiway data and for multiple regression models, the Type I and Type III SS are different. Chapter 5 discusses these differences. For the rest of this chapter, only the Type III SS will be shown in example GLM output.

The HOVTEST output appears as "Levene's Test for Homogeneity of WEAR Variance." The *F*-value, 0.53, tests the null hypothesis that the variances among observations within each treatment are equal. There is clearly no evidence to suggest failure of this assumption for these data.

3.4.2 Computing Means, Multiple Comparisons of Means, and Confidence Intervals

You can easily obtain means and multiple comparisons of means by using a MEANS statement after the MODEL statement. For the VENEER data, you get BRAND means and LSD comparisons of the BRAND means with the statement

```
means brand/lsd;
```

Results appear in Output 3.10.

Output 3.10
Least
Significant
Difference
Comparisons
of BRAND
Means

```
                        t Tests (LSD) for wear

      NOTE: This test controls the Type I comparisonwise error rate, not the
                         experimentwise error rate.


                   Alpha                           0.05
                   Error Degrees of Freedom          15
                   Error Mean Square           0.020833
                   Critical Value of t          2.13145
                   Least Significant Difference  0.2175


          Means with the same letter are not significantly different.


             t Grouping         Mean       N     brand

                      A        2.6000       4     TUFFY

                      B        2.3750       4     XTRA
                      B
                      B        2.3750       4     CHAMP
                      B
                      B        2.3250       4     ACME

                      C        2.0500       4     AJAX
```

Means and the number of observations (*N*) are produced for each BRAND. Because LSD is specified as an option, the means appear in descending order of magnitude. Under the heading "T Grouping" are sequences of A's, B's, and C's. Means are joined by the same letter if they are not significantly different, according to the *t*-test or equivalently if their difference is less than LSD. The BRAND means for XTRA, CHAMP, and ACME are not significantly different and are joined by a sequence of B's. The means for AJAX and TUFFY are found to be significantly different from all other means so they are labeled with a single C and A, respectively, and no other means are labeled with A's or C's.

You can obtain confidence intervals about means instead of comparisons of the means if you specify the CLM option:

```
means brand/lsd clm;
```

Results in Output 3.11 are self-explanatory.

Output 3.11
Confidence
Intervals for
BRAND
Means

```
                    t Confidence Intervals for wear

               Alpha                              0.05
               Error Degrees of Freedom             15
               Error Mean Square             0.020833
               Critical Value of t            2.13145
               Half Width of Confidence Interval 0.153824


          brand        N        Mean      95% Confidence Limits

          TUFFY        4      2.60000      2.44618      2.75382
          XTRA         4      2.37500      2.22118      2.52882
          CHAMP        4      2.37500      2.22118      2.52882
          ACME         4      2.32500      2.17118      2.47882
          AJAX         4      2.05000      1.89618      2.20382
```

You can also obtain confidence limits for differences between means, as discussed in Section 3.5.2., "Additional Multiple Comparison Methods."

3.4.3 Planned Comparisons for One-Way Classification: The CONTRAST Statement

Multiple comparison procedures, as demonstrated in the previous section, are useful when there are no particular comparisons of special interest and you want to make all comparisons among the means. But in most situations there is something about the classification criterion that suggests specific comparisons. For example, suppose you know something about the companies that manufacture the five brands of synthetic wood veneer material. You know that ACME and AJAX are produced by a U.S. company named A-Line, that CHAMP is produced by a U.S. company named C-Line, and that TUFFY and XTRA are produced by a non-U.S. companies.

Then you would probably be interested in comparing certain groups of means with other groups of means. For example, you would want to compare the means for the U.S. companies with the means for the non-U.S. companies; you would want to compare the means for the two U.S. companies with each other; you would want to compare the two A-Line means; and you would want to compare the means for the two non-U.S. brands. These would be called **planned comparisons**, because they are suggested by the structure of the classification criterion (BRAND) rather than the data. You know what comparisons you want to make before you look at the data. When this is the case, you ordinarily obtain a more relevant analysis of the data by making the planned comparisons rather than using a multiple comparison technique, because the planned comparisons are focused on the objectives of the study.

You use **contrasts** to make planned comparisons. In SAS, PROC ANOVA does not have a CONTRAST statement, but the GLM procedure does, so you must use PROC GLM to compute contrasts. You use CONTRAST as an optional statement the same way you use a MEANS statement.

To define contrasts and get them into a form you can use in the GLM procedure, you should first express the comparisons as null hypotheses concerning linear combinations of means to be tested. For the comparisons indicated above, you would have the following null hypotheses:

❑ U.S. versus non-U.S.

H_0: $1/3(\mu_{ACME} + \mu_{AJAX} + \mu_{CHAMP}) = 1/2(\mu_{TUFFY} + \mu_{XTRA})$

❑ A-Line versus C-Line

H_0: $1/2(\mu_{ACME} + \mu_{AJAX}) = \mu_{CHAMP}$

❏   ACME versus AJAX

H_0: $\mu_{ACME} = \mu_{AJAX}$

❏   TUFFY versus XTRA

H_0: $\mu_{TUFFY} = \mu_{XTRA}$

The basic form of the CONTRAST statement is

CONTRAST 'label' *effect-name effect-coefficients*;

where *label* is a character string used for labeling output, *effect-name* is a term on the right-hand side of the MODEL statement, and effect-coefficients is a list of numbers that specifies the linear combination of parameters in the null hypothesis. The ordering of the numbers follows the **alphameric** ordering (numbers first, in ascending order, then alphabetical order) of the levels of the classification variable, unless specified otherwise with an ORDER= option in the PROC GLM statement.

Starting with one of the simpler comparisons, ACME versus AJAX, you want to test H_0:$\mu_{ACME}=\mu_{AJAX}$. This hypothesis must be expressed as a linear combination of the means equal to 0, that is, H_0: $\mu_{ACME} - \mu_{AJAX}=0$. In terms of all the means, the null hypothesis is

$$H_0: 1 * \mu_{ACME} - 1 * \mu_{AJAX} + 0 * \mu_{CHAMP} + 0 * \mu_{TUFFY} + 0 * \mu_{XTRA} = 0 .$$

Notice that the BRAND means are listed in alphabetical order. All you have to do is insert the coefficients on the BRAND means in the list of effect coefficients in the CONTRAST statement. The coefficients for the levels of BRAND follow the alphabetical ordering.

```
proc glm; class brand;
   model wear = brand;
      contrast 'ACME vs AJAX'  brand  1 -1  0  0  0;
```

Results appear in Output 3.12.

*Output 3.12
Analysis of
Variance and
Contrast with
PROC GLM*

Contrast	DF	Contrast SS	Mean Square	F Value	Pr > F
ACME vs AJAX	1	0.15125000	0.15125000	7.26	0.0166

Output from the CONTRAST statement, labeled ACME vs AJAX, shows a sum of squares for the contrast, and an *F*-value for testing H_0: $\mu_{ACME}=\mu_{AJAX}$. The *p*-value tells you the means are significantly different at the 0.0166 level.

Actually, you don't have to include the trailing zeros in the CONTRAST statement. You can simply use

```
contrast 'ACME vs AJAX' brand 1 -1;
```

By default, if you omit the trailing coefficients they are assumed to be zeros.

Following the same procedure, to test H_0: $\mu_{TUFFY}=\mu_{XTRA}$, use the statement

```
contrast 'TUFFY vs XTRA' brand 0 0 0 1 -1;
```

The contrast U.S. versus non-U.S. is a little more complicated because it involves fractions. You can use the statement

```
contrast 'US vs NON-U.S.' brand .33333 .33333 .33333 -.5 -.5;
```

Although the continued fraction for 1/3 is easily written, it is tedious. Other fractions, such as 1/7, are even more difficult to write in decimal form. It is usually easier to multiply all coefficients by the least common denominator to get rid of the fractions. This is legitimate because the hypothesis you are testing with a CONTRAST statement is that a linear combination is equal to 0, and multiplication by a constant does not change whether the hypothesis is true or false. (Something is equal to 0 if and only if a constant times the something is equal to 0.) In the case of U.S. versus non-U.S., the assertion is that

$$H_0: \ 1/3(\mu_{ACME} + \mu_{AJAX} + \mu_{CHAMP}) = 1/2(\mu_{TUFFY} + \mu_{XTRA})$$

is equivalent to

$$H_0: \ 2(\mu_{ACME} + \mu_{AJAX} + \mu_{CHAMP}) - 3(\mu_{TUFFY} + \mu_{XTRA}) = 0$$

This tells you the appropriate CONTRAST statement is

```
contrast 'US vs NON-U.S.' brand  2 2 2 -3 -3;
```

The GLM procedure enables you to run as many CONTRAST statements as you want, but good statistical practice ordinarily indicates that this number should not exceed the number of degrees of freedom for the effect (in this case 4). Moreover, you should be aware of the inflation of the overall (experimentwise) Type I error rate when you run several CONTRAST statements.

To see how CONTRAST statements for all four comparisons are used, run the following program:

```
proc glm; class brand;
   model wear = brand;
        contrast 'US vs NON-U.S.' brand  2  2  2 -3 -3;
        contrast 'A-L vs C-L'     brand  1  1 -2  0  0;
        contrast 'ACME vs AJAX'   brand  1 -1  0  0  0;
        contrast 'TUFFY vs XTRA'  brand  0  0  0  1 -1;
   run;
```

Output 3.13
Contrasts
among
BRAND
Means

Contrast	DF	Contrast SS	Mean Square	F Value	Pr > F
US vs NON-U.S.	1	0.27075000	0.27075000	13.00	0.0026
A-L vs C-L	1	0.09375000	0.09375000	4.50	0.0510
ACME vs AJAX	1	0.15125000	0.15125000	7.26	0.0166
TUFFY vs XTRA	1	0.10125000	0.10125000	4.86	0.0435

Results in Output 3.13 indicate statistical significance for each of the contrasts. Notice that the *p*-value for ACME vs AJAX is the same in the presence of other CONTRAST statements as it was when run as a single contrast in Output 3.12. Computations for one CONTRAST statement are unaffected by the presence of other CONTRAST statements. The contrasts in Output 3.13 have a special property called orthogonality, which is discussed in Section 3.4.6, "Orthogonal Contrasts."

3.4.4 Linear Combinations of Model Parameters

Thus far, the coefficients in a CONTRAST statement have been discussed as coefficients in a linear combination of means. In fact, these are coefficients on the effect parameters in the MODEL statement. It is easier to think in terms of means, but PROC GLM works in terms of model parameters. Therefore, you must be able to translate between the two sets of parameters.

Models are discussed in more depth in Chapter 4. For now, all you need to understand is the relationship between coefficients on a linear combination of means and the corresponding coefficients on linear combinations of model effect parameters. For the linear combinations representing comparisons of means (that is, with coefficients summing to 0), this relationship is very simple for the one-way classification. The coefficient of an effect parameter in a linear combination of effect parameters is equal to the coefficient on the corresponding mean in the linear combination of means. This is because of the fundamental relationship between means and effect parameters, that is, $\mu_i = \mu + \tau_i$. For example, take the contrast A-Line versus C-Line. The linear combination in terms of means is

$$2\mu_{CHAMP} - \mu_{ACME} - \mu_{AJAX}$$
$$= 2(\mu + \tau_{CHAMP}) - (\mu + \tau_{ACME}) - (\mu + \tau_{AJAX})$$
$$= 2\tau_{CHAMP} - \tau_{ACME} - \tau_{AJAX}$$

You see that the coefficient on τ_{CHAMP} is the same as the coefficient on μ_{CHAMP}; the coefficient on τ_{ACME} is equal to the coefficient on μ_{ACME}, and so on. Moreover, the parameter disappears when you convert from means to effect parameters, because the coefficients on the means sum to 0.

It follows that, for comparisons in the one-way classification, you may derive coefficients in terms of means and simply insert them as coefficients on model effect parameters in a CONTRAST statement. For more complicated applications, such as two-way classifications, the task is not so straightforward. You'll see this in Section 3.7, "A Two-Way Factorial Experiment," and subsequent sections in this chapter.

3.4.5 Testing Several Contrasts Simultaneously

Sometimes you need to test several contrasts simultaneously. For example, you might want to test for differences among the three means for U.S. BRANDs. The null hypothesis is

$$H_0: \mu_{ACME} = \mu_{AJAX} = \mu_{CHAMP}$$

This hypothesis equation actually embodies two equations that can be expressed in several ways. One way to express the hypothesis in terms of two equations is

$$H_0: \mu_{ACME} = \mu_{AJAX} \quad \text{and} \quad H_0: \mu_{ACME} = \mu_{CHAMP}$$

Why are the two hypotheses equivalent? Because the three means are all equal if and only if the first is equal to the second and the first is equal to the third.

You can test this hypothesis by writing a CONTRAST statement that expresses sets of coefficients for the two equations, separated by a comma. An appropriate CONTRAST statement is

```
contrast 'US BRANDS' brand 1 -1 0 0 0,  brand 1 0 -1 0 0;
```

Results appear in Output 3.14.

Output 3.14
Simultaneous
Contrasts
among U.S.
BRAND
Means

Contrast	DF	Contrast SS	Mean Square	F Value	Pr > F
US BRANDS	2	0.24500000	0.12250000	5.88	0.0130

Notice that the sum of squares for the contrast has 2 degrees of freedom. This is because you are testing two equations simultaneously. The *F*-statistic of 5.88 and associated *p*-value tell you the means are different at the 0.0130 level of significance.

Another way to express the hypothesis in terms of two equations is

$$H_0: \mu_{ACME} = \mu_{AJAX} \quad \text{and} \quad H_0: 2\,\mu_{CHAMP} = \mu_{ACME} + \mu_{AJAX}$$

A contrast for this version of the hypothesis is

```
contrast 'US BRANDS'    brand  1 -1  0  0  0,
                        brand  1  1 -2  0  0;
```

Results from this CONTRAST statement, not included here, are identical to Output 3.10.

3.4.6 Orthogonal Contrasts

Notice that the sum of squares, 0.245, in Output 3.14 is equal to the sum of the sums of squares for the two contrasts ACME vs AJAX (0.15125) and A-L vs C-L (0.09375) in Output 3.13. That occurs because the two sets of coefficients in this CONTRAST statement are orthogonal. Arithmetically, this means the sum of products of coefficients for the respective means is 0—that is, $(1 \times 1) + [1 \times (-1)] + (0 \times 2) = 0$. Moreover, all four of the contrasts in Output 3.13 form an orthogonal set. You can verify this by multiplying pairs of coefficients and adding the products. Therefore, the sum of the four contrast sums of squares in Output 3.9 is equal to the overall BRAND SS (0.617) in Output 3.9.

Statistically, orthogonal means that the sums of squares for the two contrasts are independent. The outcome of one of them in no way influences the outcome of any other. Sets of orthogonal comparisons are commonly considered desirable, because the result of any one of them tells you (essentially) nothing about what to expect from any other comparison. However, desirable as it is to have independent tests, it is more important to construct sets of contrasts to address the objectives of the investigation. Practically meaningful contrasts are more desirable than simply orthogonal ones.

3.4.7 Estimating Linear Combinations of Parameters: The ESTIMATE Statement

The CONTRAST statement is used to construct an *F*-test for a hypothesis that a linear combination of parameters is equal to 0. In many applications, you want to obtain an estimate of the linear combination of parameters, along with the standard error of the estimate. You can do this with an ESTIMATE statement. The ESTIMATE statement is used in much the same way as a CONTRAST statement. You could estimate the difference $\mu_{ACME} - \mu_{AJAX}$ with the following statement:

```
estimate 'ACME vs AJAX'   brand  1 -1  0  0  0;
```

This statement is exactly like the CONTRAST statement for ACME vs AJAX, with the keyword CONTRAST replaced by the keyword ESTIMATE.

Output 3.15
Estimating
the Difference
between
BRAND
Means

Parameter	Estimate	Standard Error	t Value	Pr > |t|
ACME vs AJAX	0.27500000	0.10206207	2.69	0.0166

Results shown in Output 3.15 include the value of the estimate, a standard error, a *t*-statistic for testing whether the difference is significantly different from 0, and a *p*-value for the *t*-statistic. Note the *p*-value (0.0166) for the *t*-test is the same as for the *F*-test for the contrast in Output 3.12. This is because the two tests are equivalent; the *F* is equal to the square of the *t*.

For the present application, the estimate of $\mu_{ACME} - \mu_{AJAX}$ can be computed as

$$\bar{y}_{ACME} - \bar{y}_{AJAX}$$

The standard error is

$$\sqrt{MS(ERROR) \times \left(\frac{1}{n_1} + \frac{1}{n_2}\right)}$$

In more complicated examples, such as two-way classification with unbalanced data, more complicated computations for means are required.

Suppose you want to estimate $\mu_{CHAMP} - 1/2(\mu_{ACME} + \mu_{AJAX})$. You can use the following statement:

```
estimate 'AL vs CL' brand -.5 -.5 1 0 0;
```

The coefficients in the above ESTIMATE statement are *not* equivalent to the coefficients (–1 –1 2 0 0) as they would be in a CONTRAST statement. The latter set of coefficients would actually estimate *twice* the mean difference of interest. You can avoid the fractions by using the DIVISOR option:

```
estimate 'AL vs CL' brand -1 -1 2 0 0 / divisor=2;
```

Now suppose you want to estimate a linear combination of means that does not represent a comparison of two groups of means. For example, maybe you want to estimate the average of the three U.S. means, $1/3(\mu_{ACME} + \mu_{AJAX} + \mu_{CHAMP})$. The coefficients do not sum to 0, so you can't simply take coefficients of the means and use them in the ESTIMATE statement as coefficients on model effect parameters. The μ parameter does not disappear when you convert from means to effect parameters:

$$1/3(\mu_{ACME} + \mu_{AJAX} + \mu_{CHAMP})$$

$$= 1/3(\mu + \tau_{ACME} + \mu + \tau_{AJAX} + \mu + \tau_{CHAMP})$$

$$= \mu + 1/3(\tau_{ACME} + \tau_{AJAX} + \tau_{CHAMP})$$

You see that the parameter μ remains in the linear combination of model effect parameters. This parameter is called INTERCEPT in CONTRAST and ESTIMATE statements. This is because μ shows up as the intercept in a regression model, as discussed in Chapter 4 where the connection between analysis-of-variance models and regression models is explained. An appropriate ESTIMATE statement is

```
estimate 'US MEAN' intercept 1 brand .33333 .33333 .33333 0 0;
```

or equivalently

```
estimate 'US MEAN' intercept 3 brand 1 1 1 0 0 / divisor=3;
```

Results from this ESTIMATE statement appear in Output 3.16.

Output 3.16
Estimating
the Mean of
U.S. BRANDS

Parameter	Estimate	Standard Error	t Value	Pr > |t|
US MEAN	2.25000000	0.04166667	54.00	<.0001

In this application the estimate and its standard error are useful. For example, you can construct a 95% confidence interval:

$$2.25 \pm 2.13(0.0417)$$

Again, the estimate is 2.25 = 1/3(2.325 + 2.050 + 3.375), and the standard error is $[(1/4 + 1/4 + 1/4)MS(ERROR)]^{1/2}$. Since MS(ERROR) is the basic variance estimate in this formula, the degrees of freedom for the *t*-statistic are there for MS(ERROR). The *t*-statistic is computed to test the null hypothesis

$$H_0: \mu_{ACME} + \mu_{AJAX} + \mu_{CHAMP} = 0$$

Of course, this hypothesis is not of practical interest.

3.5 Randomized-Blocks Designs

The randomized-blocks design assumes that a population of experimental units can be divided into a relatively homogeneous subpopulations that are called blocks. The treatments are then randomly assigned to experimental units within the blocks. If all treatments are assigned in each block, the design is called a randomized-**complete-blocks** design. Blocks usually represent naturally occurring differences not related to the treatments. In analysis of variance, the extraneous variation among blocks can be partitioned out, usually reducing the error mean square. Also, differences between treatment means do not contain block variation. In this sense, the randomized-blocks design controls block variation.

A classic example of blocks is an agricultural field that is divided into smaller, more homogeneous subfields. Other examples of blocks include days of the week, measuring or recording devices, and operators of a machine. The paired two-sample design, such as the PULSE data in Section 3.2.2, "Two Related Samples," is a special case of the randomized-complete-blocks design with the two samples as treatments and pairs as the blocks.

In the following example, five methods of applying irrigation (IRRIG) are applied to a Valencia orange tree grove. The trees in the grove are arranged in eight blocks (BLOC) to account for local variation. That is, variation among trees within a block is minimized. Assignment of the irrigation method to trees within each block is random, giving a randomized-blocks design. Each of the five irrigation methods appears in all eight blocks and there are no missing data, making this a randomized-complete-blocks design. At harvest, for each plot the fruit is weighed in pounds. The objective is to determine if method of irrigation affects fruit weight (FRUITWT) and to rank the irrigation treatments accordingly.

The data appear in Output 3.17. The following SAS DATA step often provides a convenient shortcut for data entry, because it allows you to put the data for all eight blocks on a single line for each treatment. You can modify these statements to put the data for all treatments on a single line for each block.

```
data methods;
    input irrig $ @@;
        do bloc=1 to 8;
            input fruitwt @@;
            output;
        end;
datalines;
trickle         450 469 249 125 280 352 221 251
basin           358 512 281  58 352 293 283 186
spray           331 402 183  70 258 281 219  46
sprnkler        317 423 379  63 289 239 269 357
flood           245 380 263  62 336 282 171  98
;

proc sort;
    by bloc irrig;

proc print;
    var bloc irrig fruitwt;
run;
```

Output 3.17
Data for the Randomized-Complete-Blocks Design

Obs	bloc	irrig	fruitwt
1	1	basin	358
2	1	flood	245
3	1	spray	331
4	1	sprnkler	317
5	1	trickle	450
6	2	basin	512
7	2	flood	380
8	2	spray	402
9	2	sprnkler	423
10	2	trickle	469
11	3	basin	281
12	3	flood	263
13	3	spray	183
14	3	sprnkler	379
15	3	trickle	249
16	4	basin	58
17	4	flood	62
18	4	spray	70
19	4	sprnkler	63
20	4	trickle	125
21	5	basin	352
22	5	flood	336
23	5	spray	258
24	5	sprnkler	289
25	5	trickle	280
26	6	basin	293
27	6	flood	282
28	6	spray	281
29	6	sprnkler	239
30	6	trickle	352
31	7	basin	283
32	7	flood	171
33	7	spray	219
34	7	sprnkler	269
35	7	trickle	221
36	8	basin	186
37	8	flood	98
38	8	spray	46
39	8	sprnkler	357
40	8	trickle	251

3.5.1 Analysis of Variance for Randomized-Blocks Design

The following analysis of variance for the randomized-complete-blocks design provides a test for the differences among irrigation methods:

Source	DF
BLOC	7
IRRIG	4
ERROR	28
TOTAL	39

Use the following SAS statements to compute the basic analysis of variance:

```
proc glm;
    class bloc irrig;
    model fruitwt=bloc irrig;
```

BLOC and IRRIG appear in the CLASS statement because the data are classified according to these variables. The MODEL statement specifies that the response variable to be analyzed is FRUITWT. The two sources of variation in the analysis-of-variance table (other than ERROR and TOTAL) are BLOC and IRRIG, so these variables appear on the right side of the MODEL statement. The analysis appears in Output 3.18.

Output 3.18
Analysis of
Variance for the
Randomized-
Complete-
Blocks Design

```
Dependent Variable: fruitwt

                                  Sum of
Source                DF          Squares      Mean Square    F Value    Pr > F

Model                 11      445334.0250      40484.9114      12.04     <.0001

Error                 28       94146.7500       3362.3839

Corrected Total       39      539480.7750


          R-Square     Coeff Var      Root MSE      fruitwt Mean

          0.825486     21.71153       57.98607        267.0750


Source                DF        Type I SS      Mean Square    F Value    Pr > F

bloc                   7      401308.3750      57329.7679      17.05     <.0001
irrig                  4       44025.6500      11006.4125       3.27     0.0254


Source                DF      Type III SS      Mean Square    F Value    Pr > F

bloc                   7      401308.3750      57329.7679      17.05     <.0001
irrig                  4       44025.6500      11006.4125       3.27     0.0254
```

The top section contains lines labeled MODEL, ERROR, and CORRECTED TOTAL. The total variation, as measured by the total sum of squares, is partitioned into two components: variation due to the effects in the model (MODEL) and variation not due to effects in the model (ERROR). The bottom section of the output contains lines labeled BLOC and IRRIG. These partition the MODEL sum of squares into two components: sum of squares due to the effects of blocks (BLOC) and sum of squares due to the effects of treatment (IRRIG). In most cases when MODEL is partitioned into two or more sources of variation, the *F*-test for MODEL has no useful interpretation; you want to interpret the BLOC and IRRIG sources of variation separately.

In the GLM output, there are two sets of sums of squares, TYPE I and TYPE III. For balanced data such as the randomized-complete-blocks design with no missing data, these sums of square types are identical. In analysis of variance, the types of sum of squares matter when the data are unbalanced. An example is when you have missing data or **incomplete-blocks** designs. Chapter 5 discusses the different types of sums of squares.

You can summarize the key features of Output 3.18 in the following ANOVA table:

Source	DF	SS	MS	F	*p*-value
BLOC	7	401308.375			
IRRIG	4	44025.650	11006.4125	3.27	0.0254
ERROR	28	94146.750	3362.3839		
TOTAL	39	539480.775			

3.5.2 Additional Multiple Comparison Methods

In Section 3.4, "The Analysis of One-Way Classification of Data," you saw how to compare treatment means using least significant difference tests, basically two-sample *t*-tests in the context of analysis of variance, and contrasts. In addition, there are many multiple comparison tests available in PROC ANOVA and PROC GLM. It bears repeating that you should use contrasts whenever the structure of the treatment design permits, and it is usually advisable to structure the treatment design to facilitate using contrasts tailored to the specific objectives of the study. However, there are many situations where no obvious structure exists, and imposing structure would be artificial and inappropriate. These are the cases for which you should use multiple comparison tests.

In this section the randomized-blocks example shown above is used to illustrate some of these tests. The tests illustrated in this example are summarized below, including information pertaining to their error rates and option keywords.

❑ Least Significant Difference (LSD)

   comparisonwise error rate (ALPHA=probability of Type I error for any one particular comparison)

❑ Duncan's New Multiple Range (DUNCAN)

   error rate comparable to $k-1$ orthogonal comparisons tested simultaneously

❑ Waller-Duncan (WALLER)

   error rate dependent on value of analysis-of-variance *F*-test

❑ Tukey's Honest Significant Difference (TUKEY)

   experimentwise error rate (ALPHA=probability of one or more Type I errors altogether).

You may wonder why there are so many different tests. In mean comparisons, two types of error are possible. The test may incorrectly declare treatment means that are actually equal to be different; this is called a Type I error. Or the test may fail to declare a difference between treatment means that are not equal; this is called a Type II error. Multiple comparison tests differ in their Type I error rate, that is, the probability of incorrectly declaring treatment means to be different. The LSD test has the highest Type I error rate and Tukey's the lowest, with Duncan and Waller-Duncan in the middle. When Type I error rate is reduced, all other things being equal, Type II error rate increases. The reason for so many tests is that different situations call for different priorities. For example, in the early stages of research, when you may be trying to identify new treatments that show any evidence of promise, Type I error may not be serious because follow-up research will reveal spurious differences. However, Type II error *is* serious because a potentially valuable treatment will go unnoticed. On the other hand, in later stages of research, Type I error may be much more serious, because it may mean allowing an ineffective product to be recommended as if it were effective, possibly with tragic consequences. As you can imagine, there is no one "correct" test for all situations—you must evaluate each case based on the relative consequences of Type I and Type II error.

For the LSD, DUNCAN, and TUKEY options, ALPHA=.05 unless the ALPHA= option is specified. Only ALPHA= values of .01, .05, or .1 are allowed with the Duncan's test. The Waller test is based on Bayesian principles and utilizes the Type I/Type II error seriousness ratio, called the *k*-ratio, instead of an ALPHA=value. In practice, ALPHA=.05 for the DUNCAN option and KRATIO=100 for the WALLER option produce similar results.

The following SAS statements illustrate the options:

```
proc glm;
    class bloc irrig;
    model fruitwt=bloc irrig;
    means irrig/duncan lsd tukey waller;
    means irrig/duncan tukey alpha=0.1;
```

Note that you can ask for more than one multiple comparison option for a given ALPHA level in the same MEANS statement. The results in Output 3.19 reveal that, among the methods illustrated, the LSD option tends to produce the most significant differences, the TUKEY option tends to produce the least, and the DUNCAN tends to be somewhere in between.

Output 3.19
Several Types
of Multiple
Comparison
Procedures

1. LSD

```
                    t Tests (LSD) for fruitwt

        NOTE: This test controls the Type I comparisonwise error rate, not the
                            experimentwise error rate.


                    Alpha                           0.05
                    Error Degrees of Freedom          28
                    Error Mean Square           3362.384
                    Critical Value of t          2.04841
                    Least Significant Difference    59.39


            Means with the same letter are not significantly different.


                t Grouping         Mean       N    irrig

                        A         299.63       8    trickle
                        A
                        A         292.00       8    sprnkler
                        A
                        A         290.38       8    basin

                        B         229.63       8    flood
                        B
                        B         223.75       8    spray
```

2. DUNCAN

```
                        Duncan's Multiple Range Test for fruitwt

        NOTE: This test controls the Type I comparisonwise error rate, not the
                              experimentwise error rate.

                        Alpha                            0.05
                        Error Degrees of Freedom           28
                        Error Mean Square           3362.384

            Number of Means          2          3          4          5
            Critical Range       59.39      62.40      64.35      65.74

        Means with the same letter are not significantly different.


            Duncan Grouping           Mean        N     irrig

                             A        299.63       8     trickle
                             A
                        B    A        292.00       8     sprnkler
                        B    A
                        B    A        290.38       8     basin
                        B
                        B    C        229.63       8     flood
                             C
                             C        223.75       8     spray
```

3. WALLER

```
                        Waller-Duncan K-ratio t Test for fruitwt

        NOTE: This test minimizes the Bayes risk under additive loss and certain
                                 other assumptions.

                        Kratio                            100
                        Error Degrees of Freedom           28
                        Error Mean Square           3362.384
                        F Value                          3.27
                        Critical Value of t           2.23982
                        Minimum Significant Difference  64.939

        Means with the same letter are not significantly different.


            Waller Grouping           Mean        N     irrig

                             A        299.63       8     trickle
                             A
                        B    A        292.00       8     sprnkler
                        B    A
                        B    A        290.38       8     basin
                        B
                        B    C        229.63       8     flood
                             C
                             C        223.75       8     spray
```

4. TUKEY

```
            Tukey's Studentized Range (HSD) Test for fruitwt

NOTE: This test controls the Type I experimentwise error rate, but it
      generally has a higher Type II error rate than REGWQ.

              Alpha                                   0.05
              Error Degrees of Freedom                  28
              Error Mean Square                   3362.384
              Critical Value of Studentized Range  4.12030
              Minimum Significant Difference        84.471


          Means with the same letter are not significantly different.

          Tukey Grouping          Mean      N    irrig

                       A        299.63      8    trickle
                       A
                       A        292.00      8    sprnkler
                       A
                       A        290.38      8    basin
                       A
                       A        229.63      8    flood
                       A
                       A        223.75      8    spray
```

5. DUNCAN with Type I error level set to ALPHA=0.10

```
                Duncan's Multiple Range Test for fruitwt

NOTE: This test controls the Type I comparisonwise error rate, not the
                      experimentwise error rate.

                   Alpha                          0.1
                   Error Degrees of Freedom        28
                   Error Mean Square          3362.384

         Number of Means         2        3        4        5
         Critical Range      49.32    52.01    53.71    54.90


          Means with the same letter are not significantly different.

          Duncan Grouping         Mean      N    irrig

                       A        299.63      8    trickle
                       A
                       A        292.00      8    sprnkler
                       A
                       A        290.38      8    basin

                       B        229.63      8    flood
                       B
                       B        223.75      8    spray
```

6. TUKEY with Type I error level set to ALPHA=0.10

```
              Tukey's Studentized Range (HSD) Test for fruitwt

    NOTE: This test controls the Type I experimentwise error rate, but it
          generally has a higher Type II error rate than REGWQ.

              Alpha                                    0.1
              Error Degrees of Freedom                  28
              Error Mean Square                   3362.384
              Critical Value of Studentized Range  3.66039
              Minimum Significant Difference        75.042

       Means with the same letter are not significantly different.

          Tukey Grouping           Mean      N    irrig

                         A        299.63      8    trickle
                         A
                    B    A        292.00      8    sprnkler
                    B    A
                    B    A        290.38      8    basin
                    B    A
                    B    A        229.63      8    flood
                    B
                    B             223.75      8    spray
```

Some multiple comparison results can be expressed as confidence intervals for differences between pairs of means. This provides more information regarding the differences than simply joining nonsignificantly different means with a common letter, but more space is required to print the results. Specifying the CLDIFF option selects the confidence interval option. For example, the following SAS statement produces Output 3.20:

```
means irrig/tukey alpha=0.1 cldiff;
```

Output 3.20
Simultaneous
Confidence
Intervals for
Differences

```
              Tukey's Studentized Range (HSD) Test for fruitwt

       NOTE: This test controls the Type I experimentwise error rate.


              Alpha                                    0.1
              Error Degrees of Freedom                  28
              Error Mean Square                   3362.384
              Critical Value of Studentized Range  3.66039
              Minimum Significant Difference        75.042
```

*Output 3.20
(Continued)
Simultaneous
Confidence
Intervals for
Differences*

```
             Comparisons significant at the 0.1 level are indicated by ***

                                   Difference      Simultaneous
                     irrig          Between        90% Confidence
                  Comparison          Means           Limits

               trickle  - sprnkler     7.63       -67.42    82.67
               trickle  - basin        9.25       -65.79    84.29
               trickle  - flood       70.00        -5.04   145.04
               trickle  - spray       75.88         0.83   150.92   ***
               sprnkler - trickle     -7.63       -82.67    67.42
               sprnkler - basin        1.63       -73.42    76.67
               sprnkler - flood       62.38       -12.67   137.42
               sprnkler - spray       68.25        -6.79   143.29
               basin    - trickle     -9.25       -84.29    65.79
               basin    - sprnkler    -1.63       -76.67    73.42
               basin    - flood       60.75       -14.29   135.79
               basin    - spray       66.63        -8.42   141.67
               flood    - trickle    -70.00      -145.04     5.04
               flood    - sprnkler   -62.38      -137.42    12.67
               flood    - basin      -60.75      -135.79    14.29
               flood    - spray        5.88       -69.17    80.92
               spray    - trickle    -75.88      -150.92    -0.83   ***
               spray    - sprnkler   -68.25      -143.29     6.79
               spray    - basin      -66.63      -141.67     8.42
               spray    - flood       -5.88       -80.92    69.17
```

The three asterisks (***) appear to the right of each difference whose confidence interval does not include 0. Such confidence intervals, for instance TRICKLE-SPRAY and SPRAY-TRICKLE in Output 3.20, indicate the difference is significant at the ALPHA rate. The confidence interval method of presentation is the default for some methods when the means are based on different numbers of observations because the required difference for significance depends on the numbers of observations in the means.

3.5.3 Dunnett's Test to Compare Each Treatment to a Control

In some experiments, the primary objective is to screen treatments by making pairwise comparisons of each treatment to a "control" or reference treatment. For example, Keuhl (2000) notes that for the Valencia orange irrigation data, FLOOD is the standard method and hence the reference treatment against which the others are to be evaluated.

Dunnett's procedure is a specialized procedure intended to control experimentwise error rate when mean comparisons are limited to pairwise tests between the reference treatment and each other treatment. The MEANS statement provides a DUNNETT option, that is, you use the SAS statement

```
means irrig/dunnett alpha=0.1;
```

The ALPHA=0.1 is optional. If you omit it, the default α – level is 0.05. Output 3.21 shows the result.

Output 3.21
Dunnett's Test
for Differences
between
Reference and
Other
Treatments

```
                      Dunnett's t Tests for fruitwt

   NOTE: This test controls the Type I experimentwise error for comparisons of
                     all treatments against a control.

                  Alpha                            0.1
                  Error Degrees of Freedom          28
                  Error Mean Square            3362.384
                  Critical Value of Dunnett's t 2.26128
                  Minimum Significant Difference  65.561


   Comparisons significant at the 0.1 level are indicated by ***.

                               Difference      Simultaneous
                   irrig        Between        90% Confidence
                Comparison       Means            Limits

             trickle  - basin      9.25      -56.31    74.81
             sprnkler - basin      1.63      -63.94    67.19
             flood    - basin    -60.75     -126.31     4.81
             spray    - basin    -66.63     -132.19    -1.06   ***
```

The style of Output 3.21 is similar to the confidence interval presentation for the Tukey procedure shown in Output 3.20. Three asterisks indicate significant treatment differences, or equivalently, confidence intervals that do not include 0. In this case, the SPRAY-BASIN difference is statistically significant at $\alpha = 0.10$.

Note, however, that BASIN was used as the reference, not FLOOD, as required by the objectives. The default for the Dunnett procedure is to use the first treatment in alphameric order as the reference. If you want another treatment to be used as the control, you need to modify the MEANS statement by including an option that names the reference treatment:

```
means irrig/dunnett ('flood') alpha=0.1;
```

Output 3.22 shows the modified Dunnett procedure.

Output 3.22
The Dunnett
Procedure
with FLOOD
Specified as
the Control

```
                      Dunnett's t Tests for fruitwt

   NOTE: This test controls the Type I experimentwise error for comparisons of
                     all treatments against a control.

                  Alpha                            0.1
                  Error Degrees of Freedom          28
                  Error Mean Square            3362.384
                  Critical Value of Dunnett's t 2.26128
                  Minimum Significant Difference  65.561


   Comparisons significant at the 0.1 level are indicated by ***.


                               Difference      Simultaneous
                   irrig        Between        90% Confidence
                Comparison       Means            Limits

             trickle  - flood     70.00        4.44   135.56   ***
             sprnkler - flood     62.38       -3.19   127.94
             basin    - flood     60.75       -4.81   126.31
             spray    - flood     -5.88      -71.44    59.69
```

You can see from this output that only the TRICKLE treatment yields a greater mean fruit weight by a margin that is statistically significant at $\alpha = 0.10$.

3.6 A Latin Square Design with Two Response Variables

As described in Section 3.5, the randomized-blocks design controls one source of extraneous variation. It often happens, however, that there are two or more identifiable sources of variation. Such a situation may call for a Latin square design. The Latin square design is a special case of the more general row-column design, which controls two sources of extraneous variation, usually referred to as rows and columns. The Latin square is an orthogonal design, so that PROC ANOVA, and GLM Type I and Type III sums of square yield equivalent results. Treatments are randomly assigned to experimental units with the restriction that each treatment occurs once in each row and once in each column.

Consider the following example of a Latin square: Four materials (A, B, C, and D) used in permanent-press garments are subjected to a test for weight loss and shrinkage. The four materials (MAT) are placed in a heat chamber that has four control settings or positions (POS). The test is conducted in four runs (RUN), with each material assigned to each of the four positions in one execution of the experiment:

Run	Position			
	1	2	3	4
1	B	D	A	C
2	D	B	C	A
3	A	C	B	D
4	C	A	D	B

The weight loss (WTLOSS) and shrinkage (SHRINK) are measured on each sample following each test. The data appear in Output 3.23.

Output 3.23
Data for the
Latin Square
Design

Obs	run	pos	mat	wtloss	shrink
1	2	4	A	251	50
2	2	2	B	241	48
3	2	1	D	227	45
4	2	3	C	229	45
5	3	4	D	234	46
6	3	2	C	273	54
7	3	1	A	274	55
8	3	3	B	226	43
9	1	4	C	235	45
10	1	2	D	236	46
11	1	1	B	218	43
12	1	3	A	268	51
13	4	4	B	195	39
14	4	2	A	270	52
15	4	1	C	230	48
16	4	3	D	225	44

The following table shows the sources of variation and degrees of freedom for an analysis of variance for the Latin square design:

Source	DF
RUN	3
POS	3
MAT	3
ERROR	6

Use the following SAS statements to obtain the analysis of variance:

```
proc glm data=garments;
   class run pos mat;
   model wtloss shrink = run pos mat;
run;
```

The data are classified according to RUN, POS, and MAT, so these variables appear in the CLASS statement. The response variables to be analyzed are WTLOSS and SHRINK, and the sources of variation in the ANOVA table are RUN, POS, and MAT. Note that one MODEL statement handles both response variables simultaneously. Output 3.24 shows the results.

Output 3.24
Analysis of
Variance for
the Latin
Square Design

```
Dependent Variable: wtloss

                              Sum of
Source               DF       Squares      Mean Square   F Value   Pr > F

Model                 9    7076.500000     786.277778     12.84    0.0028

Error                 6     367.500000      61.250000

Corrected Total      15    7444.000000

          R-Square     Coeff Var      Root MSE     wtloss Mean

          0.950631     3.267740       7.826238      239.5000

Source               DF     Type III SS    Mean Square   F Value   Pr > F

run                   3     986.500000     328.833333      5.37    0.0390
pos                   3    1468.500000     489.500000      7.99    0.0162
mat                   3    4621.500000    1540.500000     25.15    0.0008
Dependent Variable: shrink

                              Sum of
Source               DF       Squares      Mean Square   F Value   Pr > F

Model                 9     265.7500000    29.5277778      9.84    0.0058

Error                 6      18.0000000     3.0000000

Corrected Total      15     283.7500000

          R-Square     Coeff Var      Root MSE     shrink Mean

          0.936564     3.675439       1.732051      47.12500

Source               DF     Type III SS    Mean Square   F Value   Pr > F

run                   3      33.2500000    11.0833333      3.69    0.0813
pos                   3      60.2500000    20.0833333      6.69    0.0242
mat                   3     172.2500000    57.4166667     19.14    0.0018
```

The following table is a summary of the results.

| WTLOSS | | | | | |
Source	DF	SS	MS	F	p
RUN	3	986.5			
POS	3	1468.5			
MAT	3	4621.5	1540.5	25.15	0.0008
ERROR	6	367.5	61.25		
TOTAL	15	7444.0			

| SHRINK | | | | | |
Source	DF	SS	MS	F	p
RUN	3	33.25			
POS	3	60.25			
MAT	3	172.25	57.42	19.14	0.0018
ERROR	6	18.00	3.00		
TOTAL	15	283.75			

The *F*-tests for MAT indicate differences between materials in both WTLOSS and SHRINK. For a more detailed discussion of Latin square designs, see Steel and Torrie (1980).

3.7 A Two-Way Factorial Experiment

Two of the basic aspects of the design of experiments are treatment structure and error control. Choosing between randomization schemes, such as completely randomized, randomized blocks, and so on, is part of error control. This aspect is sometimes called the **experiment design**. On the other hand the structure of the treatments, what factor(s) and factor levels are to be observed is called the **treatment design**. The **factorial** treatment design is one of the most important and widely used treatment structures. The factorial treatment design can be used with any randomization scheme, or experiment design. This section introduces the analysis of variance and mean comparison procedures used with factorial experiments.

A complete factorial experiment consists of all possible combinations of levels of two or more variables. Levels can refer to numeric quantities of variables, such as pounds of fertilizer ingredients or degrees of temperature, as well as qualitative categories, such as names of breeds or drugs. Variables, which are called factors, can be different fertilizer ingredients (N, P, K), operating conditions (temperature, pressure), biological factors (breeds, varieties), or any combination of these. An example of a factorial experiment is a study using nitrogen, phosphorus, and potassium, each at three levels. Such an experiment has $3^3 = 27$ treatment combinations.

Factorial experiments can be used to investigate several types of treatment effects. Following from the discussion of sums of squares and related terminology in Section 3.3.1, these are

❑ **simple effects**, that is, how levels of one factor affect the response variable holding the other factor constant at a given level

❑ **interactions**, that is, how levels of one factor affect the response variable across levels of another factor—do the simple effects remain constant (no interaction) or do they change (interaction)

❑ **main effects**, that is, overall differences between levels of each factor averaged over all the levels of the other factor.

For example, suppose three seed growth-promoting methods (METHOD) are applied to seed from each of five varieties (VARIETY) of turf grass. Six pots are planted with seed from each METHOD×VARIETY combination. The resulting 90 pots are randomly placed in a uniform growth chamber and the dry matter yields (YIELD) are measured after clipping at the end of four weeks. In this experiment, the concern is only about these five varieties and three growth methods. VARIETY and METHOD are regarded as fixed effects. A complete description of the experiment, for example, for a scientific article, includes the treatment design, a 3×5 factorial, and the randomization scheme, a completely randomized design. The two factors are METHOD and VARIETY.

Data are recorded in a SAS data set called GRASSES, which appears in Output 3.25. For convenience, the six replicate measurements are recorded as Y1-Y6 in the same data line.

Output 3.25
Data for the
Factorial
Experiment

Obs	method	variety	y1	y2	y3	y4	y5	y6	trt
1	a	1	22.1	24.1	19.1	22.1	25.1	18.1	a1
2	a	2	27.1	15.1	20.6	28.6	15.1	24.6	a2
3	a	3	22.3	25.8	22.8	28.3	21.3	18.3	a3
4	a	4	19.8	28.3	26.8	27.3	26.8	26.8	a4
5	a	5	20.0	17.0	24.0	22.5	28.0	22.5	a5
6	b	1	13.5	14.5	11.5	6.0	27.0	18.0	b1
7	b	2	16.9	17.4	10.4	19.4	11.9	15.4	b2
8	b	3	15.7	10.2	16.7	19.7	18.2	12.2	b3
9	b	4	15.1	6.5	17.1	7.6	13.6	21.1	b4
10	b	5	21.8	22.8	18.8	21.3	16.3	14.3	b5
11	c	1	19.0	22.0	20.0	14.5	19.0	16.0	c1
12	c	2	20.0	22.0	25.5	16.5	18.0	17.5	c2
13	c	3	16.4	14.4	21.4	19.9	10.4	21.4	c3
14	c	4	24.5	16.0	11.0	7.5	14.5	15.5	c4
15	c	5	11.8	14.3	21.3	6.3	7.8	13.8	c5

3.7.1 ANOVA for a Two-Way Factorial Experiment

An analysis of variance for the experiment has the following form:

Source	DF
METHOD	2
VARIETY	4
METHOD×VARIETY	8
ERROR	75

The METHOD×VARIETY interaction is a measure of whether differences among METHOD means depend on the VARIETY being used. If the interaction is present, it may be necessary to compare METHOD means separately for each VARIETY, that is, evaluate the simple effects of METHOD|VARIETY. If the interaction is not present, a comparison of METHOD averaged over all levels of VARIETY, that is, the main effect of METHOD, is appropriate.

Because a single YIELD value is needed for each observation instead of six values, the data set GRASSES shown in Output 3.25 must be rearranged to permit analysis. This data manipulation would not be necessary if the values of YIELD had originally been recorded using one data line per replication. Use the following SAS statements to rearrange the data:

```
data fctorial;  set grasses;  drop y1-y6;
    yield=y1; output;
    yield=y2; output;
    yield=y3; output;
    yield=y4; output;
    yield=y5; output;
    yield=y6; output;
run;
```

This creates a new data set, named FCTORIAL, containing the rearranged data. The following SAS statements sort the data by METHOD and VARIETY, and then compute and plot means for visual inspection:

```
proc sort;
    by method variety;
proc means data=fctorial noprint;
    by method variety;
    output out=factmean mean=yldmean;
    proc print data=factmean;
run;
```

The PROC MEANS statement instructs SAS to compute means and standard errors of the means of each METHOD×VARIETY combination. Note that you must first use PROC SORT to sort the data in the same order as the BY statement used with PROC MEANS. The NOPRINT option suppresses PROC MEANS from printing its computations. The OUTPUT statement creates a new SAS data set named FACTMEAN. The MEAN= option creates a new variable named YLDMEAN, whose values are the means of the variable YIELD for each combination of the values of the variables METHOD and VARIETY. The data set FACTMEAN appears in Output 3.26.

Output 3.26
Cell Means
for the
Factorial
Experiment

Obs	method	variety	_TYPE_	_FREQ_	yldmean
1	a	1	0	6	21.7667
2	a	2	0	6	21.8500
3	a	3	0	6	23.1333
4	a	4	0	6	25.9667
5	a	5	0	6	22.3333
6	b	1	0	6	15.0833
7	b	2	0	6	15.2333
8	b	3	0	6	15.4500
9	b	4	0	6	13.5000
10	b	5	0	6	19.2167
11	c	1	0	6	18.4167
12	c	2	0	6	19.9167
13	c	3	0	6	17.3167
14	c	4	0	6	14.8333
15	c	5	0	6	12.5500

You can use PROC PLOT or PROC GPLOT to plot YLDMEAN in order to visually show METHOD and VARIETY effects. The following statements cause PROC PLOT to make a low-resolution plot of the mean yields for each variety:

```
proc plot data=factmean;
    plot yldmean*variety=method;
```

The PLOT statement plots the values of YLDMEAN on the vertical axis versus the VARIETY values on the horizontal axis and labels the points according to METHOD names A, B, or C. You can use PROC GPLOT to construct a higher resolution version of this interaction plot, which appears in Output 3.27. Use the statements

```
axis1 value=(font=swiss2 h=2) label=(f=swiss h=2 'Mean
    Yield');
axis2 value=(font=swiss h=2 )label=(f=swiss h=2 'Variety');
legend1 value=(font=swiss h=2  ) label=(f=swiss h=2 'Method');
symbol1 color=black interpol=join
      line=1  value='A' font=swiss;
symbol2 color=black interpol=join
      line=2 value='B' font=swiss;
symbol3 color=black interpol=join
      line=20 value='C' font=swiss;

proc gplot data=factmean;
    plot yldmean*variety=method/caxis=black ctext=black
        vaxis=axis1 haxis=axis2 legend=legend1;
```

Output 3.27
Plots of Cell
Mean for the
Factorial
Experiment

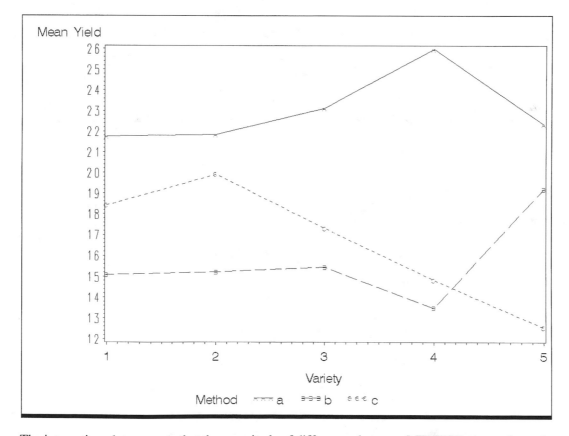

The interaction plot suggests that the magnitude of differences between METHOD means depends on which VARIETY is used. This should be formally tested, however, since the graph only shows treatment means, not their underlying variation.

Run the following SAS statements to compute the analysis:

```
proc glm data=fctorial;
    class method variety;
    model yield=method variety method*variety;
run;
```

Note that both treatment factors, METHOD and VARIETY, are classification variables and thus appear in the CLASS statement. The MODEL statement specifies that the analysis of YIELD contain sources of variation METHOD, VARIETY, and METHOD*VARIETY. You can see that the syntax for interaction is *factor A*factor B*.

Output 3.28 contains the results.

Output 3.28
Analysis of
Variance for
the Factorial
Experiment

```
Dependent Variable: yield


                                   Sum of
Source                    DF       Squares    Mean Square    F Value    Pr > F

Model                     14    1339.024889      95.644635       4.87    <.0001

Error                     75    1473.766667      19.650222

Corrected Total           89    2812.791556


             R-Square    Coeff Var     Root MSE     yield Mean

             0.476048     24.04225     4.432857       18.43778

Source                    DF    Type III SS    Mean Square    F Value    Pr > F

method                     2    953.1562222     476.5781111      24.25    <.0001
variety                    4     11.3804444       2.8451111       0.14    0.9648
method*variety             8    374.4882222      46.8110278       2.38    0.0241
```

Note that the METHOD*VARIETY effect is significant at the $p=0.0241$ level, confirming the apparent interaction observed by visual inspection of Output 3.27.

This example contains balanced data because every METHOD×VARIETY combination contains six observations. You could, therefore, obtain a valid analysis of variance using PROC ANOVA. However, if the number of observations had not been equal in all METHOD×VARIETY combinations, PROC ANOVA would not necessarily provide valid computations of sums of squares. Moreover, the GLM Type I and Type III sums of square would no longer be equal and you would need to make appropriate decisions regarding interpretation. Chapter 5 discusses these issues in detail.

3.7.2 Multiple Comparisons for a Factorial Experiment

If the interaction is *not significant*, you can perform multiple comparisons on the *main effect means* by adding the following SAS statement to PROC GLM (or ANOVA). This statement produces the main effect means for METHOD and VARIETY and for the METHOD*VARIETY treatment combination means as well.

```
means method variety method*variety;
```

As you saw in Section 3.5, "Randomized-Blocks Designs," the MEANS statement has several options for multiple comparison tests. However, these options will only compute multiple comparisons for the METHOD and VARIETY means, not for the METHOD*VARIETY means. Alternatively, you can use the LSMEANS statement with PROC GLM, which is described below. LSMEANS computes both main effect means and factorial treatment combination means such as METHOD*VARIETY. It will also compute multiple comparison tests for these means, but with the following caveat: Many statisticians do not consider multiple comparisons appropriate for testing differences among treatment combination means in a factorial experiment. Several authors have written articles critical of the frequent misuse of such procedures. See, for example, Chew (1976) and Little (1978). The main point of these objections is that with factorial treatment designs, the main focus should be on interactions first, then simple effects or main effects (but not both) depending on whether the interaction is negligible or not. Multiple comparisons tend to obscure the essential information contained in the data and make interpretation needlessly complicated and confusing. Instead, you should proceed as follows.

Because the METHOD*VARIETY interaction is significant in the GRASSES example, it is appropriate to compare simple effects. This example shows how to compare the METHOD means separately for each VARIETY. You can easily adapt this example to compare VARIETY means for each METHOD if that is more consistent with the research objectives.

In the past, it was common practice to rerun the analysis using PROC GLM (or ANOVA) with a BY statement, resulting in one analysis-of-variance table per level of the BY variable. However, this is very inefficient, because the error DF for each analysis can be quite small. In essence, you are throwing out most of the data for each analysis. For example, if you do a separate analysis BY VARIETY you get ANOVA's with 2 DF for METHOD and 10 DF for error. Unless you have lots of data to waste, this seriously reduces the power of the resulting tests. New features in PROC GLM and PROC MIXED allow you to avoid this problem.

The GLM and MIXED procedures have options in the LSMEANS statement that allow you to test each factor at each level of the other factor. The LSMEANS statement computes an estimate of the treatment mean called a **least-squares mean**, or **LS mean** as it is hereafter referred to in this text. For analysis of variance with balanced data, the sample treatment mean, $\bar{y}_i$, and the LS mean are the same. For other analyses, for example, ANOVA with unbalanced data or analysis of covariance, LS means use a definition of treatment means that avoids serious problems associated with sample means. These issues are explained in subsequent chapters.

For now, all you need to know is that the LSMEANS statement is just another way to obtain the treatment means and it has some useful features for factorial experiments. One of them is the SLICE option. Include the following statement after the MODEL statement in the GLM program given earlier:

```
lsmeans method*variety/slice=variety;
```

The SLICE statements obtain *F*-tests for simple effects. For example, SLICE=VARIETY causes a separate *F*-statistic to be computed for the METHOD effect at each VARIETY. Formally, the null hypotheses are H_0: $\mu_{1j} = \mu_{2j} = \mu_{3j}$ for each VARIETY $j = 1, 2, 3, 4, 5$, where μ_{ij} denotes the mean of METHOD i and VARIETY j. Note that you can have multiple slices in the LSMEANS statement. For example, the following two statements are equivalent ways of obtaining both sets of simple effect tests:

```
lsmeans method*variety/slice=variety slice=method;
lsmeans method*variety/slice=(variety method);
```

Only the results for SLICE=VARIETY are shown here. They appear in Output 3.29.

Output 3.29
SLICE Option
to Test Simple
Effects of
METHOD at
Each
VARIETY

		method*variety Effect Sliced by variety for yield			
variety	DF	Sum of Squares	Mean Square	F Value	Pr > F
1	2	134.001111	67.000556	3.41	0.0383
2	2	138.903333	69.451667	3.53	0.0341
3	2	192.703333	96.351667	4.90	0.0100
4	2	562.293333	281.146667	14.31	<.0001
5	2	299.743333	149.871667	7.63	0.0010

You can see that magnitudes of the METHOD effects vary among the VARIETYs. You can also see that there is a statistically significant METHOD effect for every VARIETY. Unfortunately, the SLICE option does not reveal any further detail about the simple effects. To do this, additional mean comparisons are required.

3.7.3 Multiple Comparisons of METHOD Means by VARIETY

In order to compare the simple effects of METHOD within each VARIETY, you can compute multiple comparison statistics among the METHOD*VARIETY LS means, and then use the subset of those statistics that specifically pertain to the simple effect comparisons. This section shows you how to use either the GLM or MIXED procedure to do this. The MIXED procedure allows you to get what you need more easily, but for continuity, GLM is shown first. Use the following statements:

```
proc glm; class method variety;
    model yield=method|variety;
    lsmeans method*variety/cl pdiff adjust=tukey;
```

In the MODEL statement, METHOD|VARIETY is programming shorthand for METHOD VARIETY and METHOD*VARIETY. The PDIFF option computes *p*-values for all possible treatment differences. The default *p*-values use the LSD test. The ADJUST=TUKEY option modifies the *p*-values according to Tukey's test. There are several options to use different tests, such as TUKEY and DUNNETT. The CL option computes confidence limits. If you use it without the PDIFF option, confidence limits for the treatment combination means are computed. When you use CL and PDIFF together, the confidence limits are for differences. You can use the options and the SLICE option in the same statement. The default α – level is 0.05, hence 95% confidence, but you can use the ALPHA= option shown previously to change it.

There are $3 \times 5 = 15$ treatments combinations and thus $15 \times 14 = 210$ mean comparisons. The full output gives you all possible treatment combination comparisons. It is not shown here because it is so lengthy and using all of these comparisons is controversial. Some statisticians do not object as long as their use can be justified by the objectives, but most statisticians discourage this practice. Because the number of comparisons far exceeds the degrees of freedom for treatment, the experimentwise error rate is extremely high. A better practice is to be selective about the comparisons you use—for example, only use the simple effect comparisons. Output 3.30 shows the output from the LSMEANS TRT statement edited so that only the simple effects appear.

Output 3.30
Confidence
Limits for
Simple Effect
Differences
between
METHOD by
VARIETY

```
                          Least Squares Means
                 Adjustment for Multiple Comparisons: Tukey


                                                 LSMEAN
              method     variety    yield LSMEAN  Number

                a          1       21.7666667       1
                a          2       21.8500000       2
                a          3       23.1333333       3
                a          4       25.9666667       4
                a          5       22.3333333       5
                b          1       15.0833333       6
                b          2       15.2333333       7
                b          3       15.4500000       8
                b          4       13.5000000       9
                b          5       19.2166667      10
                c          1       18.4166667      11
                c          2       19.9166667      12
                c          3       17.3166667      13
                c          4       14.8333333      14
                c          5       12.5500000      15

                        Difference      Simultaneous 95%
                          Between      Confidence Limits for
         i      j          Means        LSMean(i)-LSMean(j)

         1      6        6.683333      -2.292241    15.658908
         1     11        3.350000      -5.625575    12.325575
         6     11       -3.333333     -12.308908     5.642241

         2      7        6.616667      -2.358908    15.592241
         2     12        1.933333      -7.042241    10.908908
         7     12       -4.683333     -13.658908     4.292241

         3      8        7.683333      -1.292241    16.658908
         3     13        5.816667      -3.158908    14.792241
         8     13       -1.866667     -10.842241     7.108908

         4      9       12.466667       3.491092    21.442241
         4     14       11.133333       2.157759    20.108908
         9     14       -1.333333     -10.308908     7.642241

         5     10        3.116667      -5.858908    12.092241
         5     15        9.783333       0.807759    18.758908
        10     15        6.666667      -2.308908    15.642241
```

Output 3.30 takes some orientation to read. The first table of LS means shows the treatment combinations, their LS means, and an "LSMEAN Number" assigned to each treatment combination. The combination METHOD=a, VARIETY=1 is LSMEAN Number=1. Method=a, variety=2 is LS Mean number 2, and so forth. Thus, the simple effects for Variety 1 are LS Mean 1 vs. 6 (variety 1, method a vs. b) 1 vs. 11 (variety 1, method a vs. c) and 6 vs. 11 (variety 1, method b vs. c). The sets of three differences in Output 3.30 are arranged by variety.

Editing the output from the LSMEANS METHOD*VARIETY/CL PDIFF statement is awkward and time-consuming. Also, the LSMEANS statement does not have an option to compute the standard error of a treatment difference, which, with considerable prodding from the statistics community, many journals now require or at least strongly encourage. The MIXED procedure addresses both of these problems. Chapter 4 introduces the MIXED procedure. For now, we show the following SAS program and the results because it is a far more convenient way to create the table of simple effects. Use the statements

```
proc mixed data=fctorial;
    class variety method;
    model yield=method variety method*variety;
    lsmeans method*variety/diff;
```

```
        ods output diffs=cld;
run;

data smpleff;
    set cld;
    if variety=_variety;
proc print data=smpleff;
var variety _variety method _method estimate stderr df tvalue
        probt;
```

Note that the CLASS, MODEL, and LSMEANS statements for PROC MIXED are identical to PROC GLM. In the CLASS statement, placing VARIETY before METHOD causes the levels of METHOD to be listed within each VARIETY. The ODS OUTPUT statement uses the SAS Output Delivery System to create a new data set (CLD) containing information about the differences among the METHOD*VARIETY means. The DATA step creates a new data set that only uses elements of the data set of differences (CLD) if the treatment combinations have the same VARIETY. Thus, data set SMPLEFF contains only the desired simple effects. The variables METHOD, VARIETY and _METHOD, _VARIETY identify the two treatment combination means whose difference is estimated. Output 3.31 shows the final result from PROC PRINT. The VAR statement, which restricts the output to these variables, is of interest.

Output 3.31
Simple Effects
of METHOD
by VARIETY
from the ODS
MIXED Output

Obs	variety	_variety	method	_method	Estimate	StdErr	DF	tValue	Probt
1	1	1	a	b	6.6833	2.5593	75	2.61	0.0109
2	1	1	a	c	3.3500	2.5593	75	1.31	0.1945
3	1	1	b	c	-3.3333	2.5593	75	-1.30	0.1968
4	2	2	a	b	6.6167	2.5593	75	2.59	0.0117
5	2	2	a	c	1.9333	2.5593	75	0.76	0.4524
6	2	2	b	c	-4.6833	2.5593	75	-1.83	0.0712
7	3	3	a	b	7.6833	2.5593	75	3.00	0.0036
8	3	3	a	c	5.8167	2.5593	75	2.27	0.0259
9	3	3	b	c	-1.8667	2.5593	75	-0.73	0.4681
10	4	4	a	b	12.4667	2.5593	75	4.87	<.0001
11	4	4	a	c	11.1333	2.5593	75	4.35	<.0001
12	4	4	b	c	-1.3333	2.5593	75	-0.52	0.6039
13	5	5	a	b	3.1167	2.5593	75	1.22	0.2271
14	5	5	a	c	9.7833	2.5593	75	3.82	0.0003
15	5	5	b	c	6.6667	2.5593	75	2.60	0.0111

Output 3.31 gives the estimated difference between the treatment combinations (Estimate), the standard error of the difference, and the DF, *t*-statistic, and *p*-value (Probt) for the comparison. By definition, the latter are the test statistics for the LSD mean comparison test. The MIXED LSMEANS statement has an ADJUST= option similar to the GLM LSMEANS statement to adjust the *p*-value for procedures other than the LSD. For example, you could specify ADJUST=TUKEY. There is also a CL option to compute confidence limits. If you use it, Output 3.31 would also include the lower and upper limits of the confidence interval.

3.7.4 Planned Comparisons in a Two-Way Factorial Experiment

You can use CONTRAST and ESTIMATE statements to make planned comparisons among means in a two-way classification just like you did in the one-way classification. Recall that these statements can be used with PROC GLM (or MIXED) but not PROC ANOVA.

In Section 3.7.3, METHODs were compared separately for each VARIETY using a multiple comparison procedure. The comparisons were made separately for each variety because of the significant METHOD*VARIETY interaction. The multiple comparison procedure was used because no knowledge of the METHODs was assumed that might suggest specific comparisons among the METHOD means. Now assume that you know something about the METHODs that might suggest a specific comparison. Assume that METHOD A is a new technique that is being evaluated in relation to the industry standard techniques, METHODs B and C. So you might want to compare a mean for METHOD A with the average of means for METHODs B and C, referred to here as A vs B,C. In general terms, assume you want to estimate the difference

$$\mu_A - \frac{1}{2}(\mu_B + \mu_C)$$

There are several ways to make this comparison:

- ❑ compare A vs B,C separately for each VARIETY (simple effect)
- ❑ compare A vs B,C averaged across all VARIETY levels (main effect)
- ❑ compare A vs B,C averaged across subsets of VARIETY (compromise)

Which way is appropriate depends on how the comparison interacts with VARIETY. The first comparison (simple effect) would be appropriate if the comparisons were generally different from one VARIETY to the next, that is, if the comparison interacts with VARIETY. The second comparison (main effect) would be appropriate if the comparison did not interact with VARIETY, that is, if the comparison had essentially the same value (within the range of random error) for all the varieties. The third way is a compromise between simple effect and main effect comparisons. It would be appropriate if there were subsets of varieties so that the comparison did not interact with VARIETY within the subsets. Each way of making the comparison can be done with CONTRAST or ESTIMATE statements. This illustrates the tremendous flexibility of the CONTRAST and ESTIMATE statements as tools for statistical analysis.

Once again, it is easier to think in terms of means, but PROC GLM works in terms of model parameters. For this reason some notation is needed to relate means to model parameters. Denote by μ_{ij} the (population) mean for METHOD i with VARIETY j. This is called a **cell mean** for the ijth **cell**, or METHOD×VARIETY combination. For example, μ_{B3} is the cell mean for METHOD B with VARIETY 3. A GLM model for this two-way classification specifies that

$$\mu_{ij} = \mu + \alpha_i + \beta_j + (\alpha\beta)_{ij}$$

This equation is the basic relationship between the means and model parameters. In words, the mean for METHOD i with VARIETY j is equal to a constant (or **intercept**) plus an effect of METHOD i plus an effect of VARIETY j plus an effect of the interaction for METHOD i and VARIETY j. In terms of the data,

$y_{ijk} = k$th observed value in METHOD i with VARIETY j

$= \mu_{ij} + e_{ijk}$

$= \mu + \alpha_i + \beta_j + (\alpha\beta)_{ij} + e_{ijk}$

where e_{ijk} is the random error representing the difference between the observed value and the mean of the population from which the observation was obtained.

Writing CONTRAST and ESTIMATE statements can be a little tricky, especially in multiway classifications. You can use the basic relationship between the means and model parameters to construct CONTRAST and ESTIMATE statements. Following is a three-step process that always works. In Section 3.7.9, "An Easier Way to Set Up CONTRAST and ESTIMATE Statements," a simpler way of accomplishing the same task is presented. First, however, it is instructive to go through this three-step approach to demonstrate how the process works:

1. Write the linear combination you want to test or estimate in terms of means.

2. Convert means into model parameters.

3. Gather like terms.

The resulting expression will have coefficients for model parameters that you can directly insert into a CONTRAST or an ESTIMATE statement.

3.7.5 Simple Effect Comparisons

To set up a comparison of the first type (a comparison of A vs B,C in VARIETY 1) use the basic relationship between means and model parameters. This is a simple effect comparison because you are comparing METHOD means within a particular VARIETY. Use an ESTIMATE statement to estimate A vs B,C in VARIETY 1.

1. Writing the linear combination in terms of cell means gives

$$\mu_{A1} - 0.5(\mu_{B1} - \mu_{C1})$$

2. Converting to model parameters gives

$$\mu + \alpha_A + \beta_1 + (\alpha\beta)_{A1} - 0.5[\mu + \alpha_B + (\alpha\beta)_{B1} + \mu + \alpha_c + \beta_1 + (\alpha\beta)_{C1}]$$

3. Gathering like terms gives

$$(1 - .5 - .5)\mu + \alpha_A - 0.5\,\alpha_B - 0.5\,\alpha_C + (1 - .5 - .5)\beta_1 + (\alpha\beta)_{A1} - 0.5(\alpha\,\beta)_{B1} - 0.5(\alpha\,\beta)_{C1}$$

$$= \alpha_A - 0.5\alpha_B - 0.5\alpha_C + (\alpha\,\beta)_{A1} - 0.5(\alpha\,\beta)_{B1} - 0.5(\alpha\,\beta)_{C1}$$

Now you have the information you need to set up the ESTIMATE statement to go with the PROC GLM model. The required statements are

```
proc glm; class method variety;
   model yield = method variety method*variety
   estimate 'A vs B,C in V1' method 1 -.5 -.5
      method*variety 1 0 0 0 0 -.5 0 0 0 0 -.5 0 0 0 0;
```

Note the following:

❑ The μ and β parameters disappeared from the expression, so you don't need INTERCEPT or VARIETY terms in the ESTIMATE statement. Leaving them out is equivalent to setting their coefficients equal to 0.

❑ The ordering of the METHOD*VARIETY coefficients is determined by the CLASS statement. In this CLASS statement, METHOD comes before VARIETY. For this reason, VARIETY levels change within METHOD levels.

If you only wanted a test of the hypothesis H_0: $\mu_{A1} - 0.5(\mu_{B1} + \mu_{C1}) = 0$, you could replace the ESTIMATE statement with a CONTRAST statement containing the same coefficients:

```
contrast 'A vs B,C in V1' method 1 -.5 -.5
   method*variety 1 0 0 0 0 -.5 0 0 0 0 -.5 0 0 0 0;
```

Rather than examine output for the single ESTIMATE statement make the comparison for all five varieties. You would probably want to estimate the comparison A vs B,C separately for each VARIETY if the comparison interacts with VARIETY, that is, if the value of the comparison differs from one VARIETY to the next.

As an exercise, see if you can go through the three-step process to get the coefficients for estimates of A vs B,C in each of VARIETY 2, 3, 4, and 5. Here is a complete PROC GLM step with the correct ESTIMATE statements for A vs B,C in each of the five varieties:

```
proc glm; class method variety;
    model yield=method variety method*variety / ss1;
run;
    estimate 'A vs B,C in V1' method 1 -.5 -.5
        method*variety 1 0 0 0 0  -.5 0 0 0 0  -.5 0 0 0 0;
    estimate 'A vs B,C in V2' method 1 -.5 -.5
        method*variety 0 1 0 0 0  0 -.5 0 0 0  0 -.5 0 0 0;
    estimate 'A vs B,C in V3' method 1 -.5 -.5
        method*variety 0 0 1 0 0  0 0 -.5 0 0  0 0 -.5 0 0;
    estimate 'A vs B,C in V4' method 1 -.5 -.5
        method*variety 0 0 0 1 0  0 0 0 -.5 0  0 0 0 -.5 0;
    estimate 'A vs B,C in V5' method 1 -.5 -.5
        method*variety 0 0 0 0 1  0 0 0 0 -.5  0 0 0 0 -.5;
run;
```

The results appear in Output 3.32.

Output 3.32 Estimates of Method Differences by Variety

Parameter	Estimate	Standard Error	t Value	Pr > |t|
A vs B,C in V1	5.0166667	2.21642856	2.26	0.0265
A vs B,C in V2	4.2750000	2.21642856	1.93	0.0575
A vs B,C in V3	6.7500000	2.21642856	3.05	0.0032
A vs B,C in V4	11.8000000	2.21642856	5.32	<.0001
A vs B,C in V5	6.4500000	2.21642856	2.91	0.0048

Notice that the estimates differ considerably between VARIETY, an indication of interaction between the comparison A vs B,C and VARIETY. This is no surprise, because there was interaction between METHOD and VARIETY in the analysis-of-variance table in Section 3.7.1, "ANOVA for a Two-Way Factorial Experiment." It is possible that VARIETY could interact with METHOD in general, but not interact with the comparison A vs B,C. In Section 3.7.7, "Simultaneous Contrasts in Two-Way Classifications," you see how to set up a test for the statistical significance of the interaction between the comparison A vs B,C and VARIETYs.

3.7.6 Main Effect Comparisons

If the comparison A vs B,C did not interact with VARIETY (that is, if the comparison had essentially the same value across all VARIETYs), then you would want to average all the simple effect estimates to get a better estimate of the common value of the comparison. This is called a main effect comparison. In terms of means, the main effect of A vs B,C is

$$0.2[\mu_{A1} - 0.5(\mu_{B1} + \mu_{C1})] + \dots + 0.2[\mu_{A5} - 0.5(\mu_{B5} + \mu_{C5})]$$

To estimate this main effect with an ESTIMATE statement, convert to model parameters and simplify. You will obtain

$$\alpha_A - 0.5(\alpha_B + \alpha_C) + 0.2(\alpha\beta)_{A1} + ... + 0.2(\alpha\beta)_{A5}$$
$$- 0.1(\alpha\beta)_{B1} - ... - 0.1(\alpha\beta)_{B5}$$
$$- 0.1(\alpha\beta)_{C1} - ... - 0.1(\alpha\beta)_{C5}$$

So an appropriate ESTIMATE statement is

```
estimate 'A vs B,C Overall' method 1 -.5 -.5
    method*variety .2 .2 .2 .2 .2  -.1 -.1 -.1 -.1 -.1 -.1 -.1
        -.1 -.1 -.1;
```

Results from this statement appear in Output 3.33. You can verify by hand that, in fact, this estimate is the average of all the estimates in Output 3.32. Moreover, the standard error in Output 3.33 is only 1/5 times as large as the standard errors in Output 3.32, so you can see the benefit of averaging the estimates if they are all estimates of the same quantity.

Output 3.33
Estimate of A
vs B,C
Averaged
over All
Varieties

Parameter	Estimate	Standard Error	t Value	Pr > |t|
A vs B,C Overall	6.85833333	0.99121698	6.92	<.0001

3.7.7 Simultaneous Contrasts in Two-Way Classifications

This section illustrates setting up simultaneous contrasts in a two-way classification by constructing a test for significance of interaction between the comparison A vs B,C and VARIETY. The hypothesis of no interaction between A vs B,C and VARIETY is

$$H_0: [\mu_{A1} - 0.5(\mu_{B1} + \mu_{C1})] = ... = [\mu_{A5} - 0.5(\mu_{B5} + \mu_{C5})]$$

This hypothesis is actually a set of four equations, which can be written in different but equivalent ways. One way to express the equality of all the comparisons is to specify that each is equal to the last. This gives the hypothesis in the equations

$$H_0: [\mu_{A1} - 0.5(\mu_{B1} + \mu_{C1})] = [\mu_{A5} - 0.5(\mu_{B5} + \mu_{C5})] \text{ and}$$
$$[\mu_{A2} - 0.5(\mu_{B2} + \mu_{C2})] = [\mu_{A5} - 0.5(\mu_{B5} + \mu_{C5})] \text{ and}$$
$$[\mu_{A3} - 0.5(\mu_{B3} + \mu_{C3})] = [\mu_{A5} - 0.5(\mu_{B5} + \mu_{C5})] \text{ and}$$
$$[\mu_{A4} - 0.5(\mu_{B4} + \mu_{C4})] = [\mu_{A5} - 0.5(\mu_{B5} + \mu_{C5})] \text{ and}$$

going through the three-step process for each of these equations results in the following CONTRAST statement:

```
contrast 'A vs BC * Varieties'
    method * variety 1 0 0 0 -1  -.5 0 0 0 .5  -.5 0 0 0 .5,
    method * variety 0 1 0 0 -1  0 -.5 0 0 .5  0 -.5 0 0 .5,
    method * variety 0 0 1 0 -1  0 0 -.5 0 .5  0 0 -.5 0 .5,
    method * variety 0 0 0 1 -1  0 0 0 -.5 .5  0 0 0 -.5 .5;
```

As mentioned in Section 3.4.3, concerning the CONTRAST statement for simultaneous comparisons in the one-way classification, there are several ways to specify a set of four equations that would be equivalent to the null hypothesis that the comparison A vs B,C is the same in all five VARIETYs. No matter how you set up the four equations, a CONTRAST statement derived from those equations would produce the results in Output 3.34.

Output 3.34
Test for A vs
BC * Varieties
Interaction

Contrast	DF	Contrast SS	Mean Square	F Value	Pr > F
A vs BC * Varieties	4	138.6555556	34.6638889	1.76	0.1450

The *F*-test for the A vs B,C *Varieties interaction in Output 3.34 is significant at the 0=0.145 level. In many hypothesis-testing situations, you might not consider this significant. However, the *F*-test for the interaction is a preliminary test in the model-building phase to decide whether simple effects or main effects should be reported for the contrast. The decision should be based on a rather liberal cutoff level of significance, such as .2 or .25. You want to relax the Type I error rate in order to decrease the Type II error rate. It might be a serious mistake to declare there is no interaction when in fact there is interaction (a Type II error); you would then report main effects when you should report simple effects. The estimated main effect might not be a good representation of any of the simple effects. It is usually a less serious mistake to declare there is interaction when in fact there is not (a Type I error); you would then report simple effects when you should report main effects. In this event, you still have unbiased estimates, but you lose precision.

3.7.8 Comparing Levels of One Factor within Subgroups of Levels of Another Factor

There are sometimes good reasons to report simple effects averaged across subgroups of levels of another factor (or factors). This is especially desirable when there are a large number of levels of the second factor. For example, if there were twenty varieties in the example instead of five, it would not be feasible to report a separate comparison of methods for each of the twenty varieties. You might want to consider trying to find subgroups of varieties such that the method comparison does not interact with the varieties within the subgroups. It would be legitimate to report the method comparison averaged across the varieties within the subgroups. You should search for the subgroups with caution, however. Identification of potential subgroups should be on the basis of some prior knowledge of the varieties, such as subgroups that have some property in common.

In our example, suppose VARIETY 1 and VARIETY 2 have a similar genetic background, and VARIETY 3 and VARIETY 4 have a similar genetic background (but different from varieties 1 and 2). This presents a natural basis for forming subgroups. You might want to group VARIETY 1 and VARIETY 2 together and report a single result for the comparison A vs B,C averaged across these two varieties, and do the same thing for VARIETY 3 and VARIETY 4. The validity of these groupings, however, is contingent upon there being no interaction between the comparison A vs B,C and VARIETY within the groups.

A test for the significance of interaction between the comparison and the varieties within the respective subgroups is presented here. If the *p*-value for a test is less than .2, then assume interaction to be sufficiently large to suggest separate comparisons for the two varieties within a group. Otherwise, assume that interaction is negligible, and average the comparison across the varieties within a group.

The null hypothesis of no interaction between the comparison A vs B,C and VARIETY 1 and VARIETY 2 is

$$H_0: [\mu_{A1} - 0.5(\mu_{B1} + \mu_{C1})] = [\mu_{A2} - 0.5(\mu_{B2} + \mu_{C2})]$$

You have probably become familiar with the three-step process of converting null hypothesis equations into CONTRAST statements. You can determine that the CONTRAST statement to test this hypothesis is

```
contrast 'A vs B,C * V1,V2'
    method*variety 1 -1 0 0 0  -.5 .5 0 0 0  -.5 .5 0 0 0;
```

Likewise, the null hypothesis of no interaction between A vs B,C and VARIETY 3 and VARIETY 4 is

$$H_0: [\mu_{A3} - 0.5(\mu_{B3}+\mu_{C3})] = [\mu_{A4} - 0.5(\mu_{B4}+\mu_{C4})]$$

and the associated CONTRAST statement is

```
contrast 'A vs B,C * V3,V4'
    method*variety 0 0 1 -1 0  0 0 -.5 .5 0  0 0 -.5 .5 0;
```

Results of these CONTRAST statements appear in Output 3.35.

Output 3.35
Interaction between A vs B,C and VARIETY Subsets

Contrast	DF	Contrast SS	Mean Square	F Value	Pr > F
A vs B,C * V1,V2	1	1.1001389	1.1001389	0.06	0.8136
A vs B,C * V3,V4	1	51.0050000	51.0050000	2.60	0.1114

You can see that the *F*-test for the interaction between A vs B,C and VARIETY 1 and VARIETY 2 has a *p*-value of only 0.8136, which is about as nonsignificant as you can hope to get. Assume that this interaction is negligible, and average the comparison across VARIETY 1 and VARIETY 2. On the other hand, the *F*-test for interaction between A vs B,C and VARIETY 3 and VARIETY 4 has a *p*-value of .1114, which can be considered sufficiently significant to require separate estimates of A vs B,C in each to VARIETY 3 and VARIETY 4. Estimates of A vs B,C obtained separately for VARIETY 3 and VARIETY 4 were given in Section 3.7.5, "Simple Effect Comparisons." Additionally, you need the comparison A vs B,C averaged across VARIETY 1 and VARIETY 2.

You want an estimate of

$$0.5\{[\mu_{A1} - 0.5(\mu_{B1}+\mu_{C1})] + [\mu_{A2} - 0.5(\mu_{B2}+\mu_{C2})]]\}$$

The three-step process yields the following ESTIMATE statement:

```
estimate 'A vs B,C in V1,V2' method 1 -.5 -.5
    method*variety .5 .5 0 0 0  -.25 -.25 0 0 0  -.25 -.25 0 0
        0;
```

Output 3.36 shows the results.

Output 3.36
Estimate of A vs B,C Averaged over VARIETY 1 and VARIETY 2

Parameter	Estimate	Standard Error	t Value	Pr > |t|
A vs B,C in V1,V2	4.64583333	1.56725166	2.96	0.0041

Note that the estimate 4.64 is the average of the two estimates 5.02 for VARIETY 1 and 4.27 for VARIETY 2 in Output 3.32. The advantage of averaging is the smaller standard error of 1.57 for the combined estimate compared with 2.21 (see Output 3.32) for the individual estimates.

3.7.9 An Easier Way to Set Up CONTRAST and ESTIMATE Statements

You have used the three-step process given in Section 3.7.5, "Simple Effect Comparisons," to obtain coefficients for a CONTRAST or ESTIMATE statement. This process always works, but it can be tedious. Now that you understand the process, here is a simpler diagrammatic method. This method works because of two basic principles that are easy to understand in terms of a two-way classification with factors A and B having a and b levels, respectively. Recall the relation between the cell means and model parameters $\mu_{ij} = \mu + \alpha_i + \beta_j + (\alpha\beta)_{ij}$.

When you convert a linear combination of cell means to a linear combination of model parameters, the coefficients on the interaction parameters are equal to the coefficients on the cell means. Certain conditions must hold regarding coefficients of model parameters:

❑ Coefficients on the $(\alpha\beta)_{ij}$ terms for a fixed i must add up to the coefficient on α_i.

❑ Coefficients on the $(\alpha\beta)_{ij}$ terms for a fixed j must add up to the coefficient on β_j.

❑ Coefficients on the α_i's and coefficients on the β_j's must both sum to the coefficient on μ.

Let c_{ij} stand for the coefficient on μ_{ij}. Put the coefficients in a diagram as follows:

		Factor B				
		1	2	...	b	subtotals
	1	c_{11}	c_{12}	...	c_{1b}	$c_{1.}$
	2	c_{21}	c_{22}	...	c_{2b}	$c_{2.}$
Factor A	.	.	.	.	.	
	.	.	.	.	.	
	.	.	.	.	.	
	a	c_{a1}	c_{a2}	...	c_{ab}	$c_{a.}$
subtotals		$c_{.1}$	$c_{.2}$	...	$c_{.b}$	$c_{..}$

Then c_{ij} will also be the coefficient on $(\alpha\beta)_{ij}$, $c_{i.}$ will be the coefficient on α_i, $c_{.j}$ will be the coefficient on β_j, and $c_{..}$ will be the coefficient on μ.

To use this for a particular linear combination, take A vs B,C in VARIETY 1.

The linear combination in terms of cell means is

$$\mu_{A1} - 0.5(\mu_{B1} + \mu_{C1})$$

First put the c_{ij} coefficients into the body of the table, then sum down columns and across rows to get the coefficients on the α's and β's. Finally, sum the coefficients on either the α's or the β's to get the coefficient on μ :

		VARIETY					
		1	2	3	4	5	
	A	1	0	0	0	0	1
METHOD	B	−.5.	0	0	0	0	−.5
	C	−.5.	0	0	0	0	−.5
		0	0	0	0	0	0

You can see that the linear combination in terms of model parameters is

$$\alpha_A - 0.5\,\alpha_B - 0.5\,\alpha_C + (\alpha\beta)_{A1} - 0.5(\alpha\beta)_{B1} - 0.5(\alpha\beta)_{C1}$$

which we derived using the three-step process discussed in Section 3.7.5, "Simple Effect Comparisons."

Chapter 4   Analyzing Data with Random Effects

4.1   Introduction

Chapter 3 looked at factors whose levels were chosen intentionally. Typically, the objective of a particular study dictates that a specific set of treatments, or treatment factor levels, be included. The effects corresponding to factors chosen in this way are called **fixed effects**.

On the other hand, many studies use factors whose levels represent a larger population. Many studies incorporate blocking factors to provide replication over a selection of different conditions. Investigators are not normally interested in the specific performance of individual blocks, but rather in what the average across blocks reveals. For example, you might want to test chemical compounds at several laboratories to compare the compounds averaged across all laboratories. Laboratories are selected to represent some broader possible set of laboratories. Other studies employ experimental factors in which the levels of the factors are a sample of a much larger population of possible levels. If you work in industry, you probably have seen experiments that use a selection of batches of a raw material, or a sample of workers on an assembly line, or a subset of machines out of a much larger set of machines that are used in a production process. Such factors (laboratories, batches, workers, machines, or whatever) are called **random** effects. Theoretically, the levels of a factor that are in the experiment are considered to be a random sample from a broader population of possible levels of the factor.

Traditionally, one criterion used to distinguish **fixed** and **random** effects was the type of inference. If you were interested in specific treatments, or laboratories, or machines—for example, you wanted to estimate means or test treatment differences—then by definition that effect was fixed. If, instead, your interest was in what happened across the broader collection of laboratories or batches or workers or machines, rather than in what happened with a particular laboratory or batch or worker or machine, and the only parameter to be estimated was the variance associated with that factor, then your effect was random.

With contemporary linear model theory for fixed and random effects, the distinction is more subtle. Fixed effects remain defined as they have been. However, for random effects, while interest always focuses on estimating the variance, in some applications you may also be interested in specific levels. For example, your laboratories may be a sample of a larger population, and for certain purposes you want a population-wide average, but you may also want to look at certain laboratories. Workers may represent a population for certain purposes, but the supervisor may also want to use the data for performance evaluations of individual workers. Animal breeding pioneered this approach: randomly sampled sires were used to estimate variance among sires for genetic evaluation, but individual "sire breeding values" were also determined to identify valuable sires. You can do this with random effects as long as you take into account the distribution among the random effect levels, a method called **best linear unbiased prediction**.

In contemporary linear model theory, there is only one truly meaningful distinction between fixed and random effects. If the effect level can reasonably be assumed to represent a probability distribution, then the effect is random. If it does not represent a probability distribution, then the effect is fixed. Period. Treatments are almost invariably fixed effects, because interest focuses almost exclusively on mean differences and, most importantly, the treatment levels result from deliberate choice, not sampling a distribution. On the other hand, blocks, laboratories, workers, sires, machines, and so forth, typically (but not always!) represent a sample (although often an imperfect sample) of a population with a probability distribution. Random effects raise two issues in analyzing linear models. First is how you construct test statistics and confidence intervals. Second, if you are interested in specific levels, you need to use best linear unbiased prediction rather than simply calculating sample means.

With balanced data, random factors do not present a major issue for the estimation of treatment means or differences between treatment means. You simply compute means or differences between means, averaged across the levels of random factors in the experiment. However, the presence of random effects has a major impact on the construction of test statistics and standard errors of estimates, and hence on appropriate methods for testing hypotheses and constructing confidence intervals. It is safe to say that improper attention to the presence of random effects is one of the most common and serious mistakes in statistical analysis of data.

Random effects probably occur in one form or another in the majority of statistical studies. The RANDOM statement in the GLM procedure can help you determine correct methods in many common applications. The MIXED procedure provides an even more comprehensive set of tools for working with random effects. In many common applications, methods that are essential are available in MIXED but not in GLM.

4.2 Nested Classifications

Data may be organized into two types of classification patterns, crossed (Figure 4.1) or nested (Figure 4.2).

Figure 4.1 *Crossed Classification*

A ⇓ B ⇒	1	2
1	A_1B_1	A_1B_2
2	A_2B_1	A_2B_2

Figure 4.2 *Nested Classification*

A ⇒	1		2	
B ⇒	1	2	3	4
unit ⇒	A_1B_1	A_1B_2	A_2B_3	A_2B_4

Nested classifications of data have sampling units that are classified in a hierarchical manner. Typically, these samples are taken in several stages:

1. selection of main units (analogous to level A in Figure 4.2)

2. selection of subunits from each main unit (analogous to level B in Figure 4.2)

3. selection of sub-subunits from the subunits, and so on.

Normally, the classification factors at each stage are considered random effects, but in some cases a classification factor may be considered fixed, especially one corresponding to level A in Figure 4.2, that is, the first stage of sampling.

Here is an example of a nested classification. Microbial counts are made on samples of ground beef in a study whose objective is to assess sources of variation in numbers of microbes. Twenty packages of ground beef (PACKAGE) are purchased and taken to a laboratory. Three samples (SAMPLE) are drawn from each package, and two replicate counts are made on each sample. Output 4.1 shows the raw data.

Output 4.1
Microbial
Counts in
Ground Beef

Obs	package	ct11	ct12	ct21	ct22	ct31	ct32
1	1	527	821	107	299	1382	3524
2	2	2813	2322	3901	4422	383	479
3	3	703	652	745	995	2202	1298
4	4	1617	2629	103	96	2103	8814
5	5	4169	2907	4018	882	768	271
6	6	67	28	68	111	277	199
7	7	1612	1680	6619	4028	5625	6507
8	8	195	127	591	399	275	152
9	9	619	520	813	956	1219	923
10	10	436	555	58	54	236	188
11	11	1682	3235	2963	2249	457	2950
12	12	6050	3956	2782	7501	1952	1299
13	13	1330	758	132	93	1116	3186
14	14	1834	1200	18248	9496	252	433
15	15	2339	4057	106	146	430	442
16	16	31229	84451	6806	9156	12715	12011
17	17	1147	3437	132	175	719	1243
18	18	3440	3185	712	467	680	205
19	19	8196	4565	1459	1292	9707	8138
20	20	1090	1037	4188	1859	8464	14073

The data are plotted in Output 4.2, with points identified according to their SAMPLE number.

Output 4.2
Plots of Count
versus Package
Number

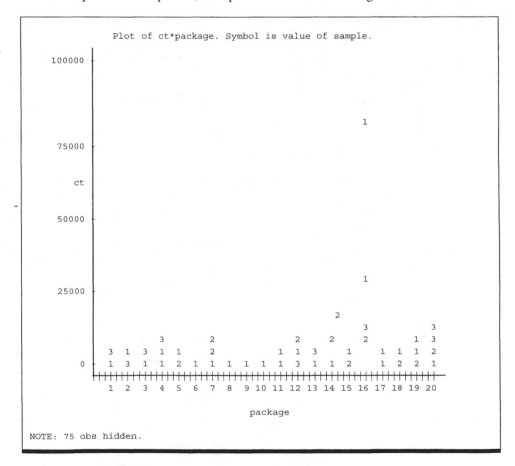

You can see the larger variation among larger counts. In order to stabilize the variance, the logarithm (base 10) of the counts (LOGCT) was computed and serves as the response variable to be analyzed. The plot of LOGCT, which appears in Output 4.3, indicates the transformation was successful in stabilizing the variance.

Output 4.3
Plot of Log Count versus Package Number

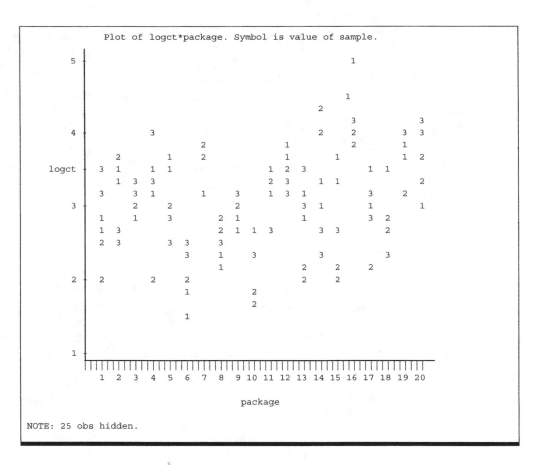

NOTE: 25 obs hidden.

Logarithms are commonly computed for microbial data for the additional reason that interest is in differences in the order of magnitude rather than in interval differences.

A model for the data is

$$y_{ijk} = \mu + a_i + b(a)_{ij} + e_{ijk} \qquad (4.1)$$

where

y_{ijk} is the $\log_{10}$ count for the kth replicate of the jth sample from the ith package.

μ is the overall mean of the sampled population.

a_i are the effects of packages, that is, random variables representing differences between packages, with variance σ_p^2, $i = 1, \ldots, 20$.

$b(a)_{ij}$ are random variables representing differences between samples in the same package, with variance σ_s^2, $i = 1, \ldots, 20, j = 1, 2, 3$.

e_{ijk} are random variables representing differences between replicate counts in the same sample, with variance σ^2, $i = 1, \ldots, 20, j = 1, 2, 3$ and $k = 1, 2$.

The random variables a_i, $b(a)_{ij}$, and e_{ijk} are assumed to be normal distributed and independent with means equal to 0. Note several conventions used in this text for denoting fixed versus random effects. Greek symbols denote fixed effects, as they have for all models in previous chapters, and for μ in this model. Latin symbols denote random effects. If you consider packages to be fixed, instead of random, you would denote the package effects as α_i instead of a_i. The notation $b(a)$ is used for nested factors, in this case factor B (samples) nested within factor A (packages). You could denote the effects of replicates within samples as $c(ab)_{ijk}$, but by convention the smallest subunit in the model is generally denoted as e_{ijk}.

The variance (V) of the log counts can be expressed as

$$V(y_{ijk}) = \sigma_y^2$$
$$= \sigma_p^2 \ + \ \sigma_s^2 \ + \ \sigma^2$$

Expressing the equation with words, the variance of the logarithms of microbial count is equal to the sum of the variances due to differences among packages, among samples in the same package, and between replicates in the same sample. These individual variances are therefore called **components of variance**. The first objective is to estimate the variance components, and there are several statistical techniques for doing so, including analysis of variance (ANOVA) and maximum likelihood (or, more commonly, *restricted* maximum likelihood, or REML). In this chapter, both ANOVA and REML methods are used. For balanced data, ANOVA and REML produce identical results. The first examples in this chapter use ANOVA because it is easier to see how the method works. PROC MIXED, introduced later in this chapter, uses REML because it is easier to generalize to more complex models.

4.2.1 Analysis of Variance for Nested Classifications

An analysis-of-variance table for the ground beef microbial counts has the following form:

Source of Variation	DF
packages	19
samples in packages	40
replicates in samples	60

You can produce this table by using the GLM procedure (see Chapter 3, "Analysis of Variance for Balanced Data"). You can also use the ANOVA, NESTED, and VARCOMP procedures to produce this table. The MIXED procedure does not compute the analysis of variance table per se, but it computes statistics that are typical end points for the analysis of data with random effects. Which procedure is best to use depends on your objectives.

As noted in Chapter 3, PROC ANOVA computes analysis of variance for balanced data only. PROC GLM computes the same analysis of variance but can be used for unbalanced data as well (see Chapter 5). In the early days of computing, limited capacity often forced users to use PROC ANOVA for large data sets. However, with contemporary computers this is rarely an issue and hence there is rarely any reason to use PROC ANOVA instead of PROC GLM. PROC NESTED is a specialized procedure that is useful only for nested classifications. It provides estimates of the components of variance using the analysis-of-variance method of estimation. Because PROC NESTED is so specialized, it is easy to use. However, PROC GLM can compute the same analysis of variance as PROC NESTED, but it does so within the framework of a much broader range of applications. Finally, PROC MIXED and PROC VARCOMP compute the variance component estimates. The MIXED procedure can also compute a variety of statistics not available with any other procedure. Many of these statistics have become increasingly important in the analysis of data with random effects. For these reasons, this chapter focuses on using PROC GLM to compute the analysis of variance, and later sections introduce PROC MIXED to compute additional statistics typically of interest.

The program statements for PROC GLM are similar to those introduced in Chapter 3. You add a RANDOM statement to compute the expected values of the mean squares—that is, what is being estimated by the individual mean squares. Here are the proper SAS statements:

```
proc glm;
  class package sample;
  model logct=package sample(package);
  random package sample(package);
```

You can see that the syntax for a nested effect, in this case SAMPLE nested within PACKAGE, follows from the notation used for nested effects in model (4.1). The RANDOM statement is simply a list of effects in the model to be considered random. In most practical situations, you add a TEST option to the RANDOM statement in order to compute the proper test statistics. Section 4.2.3 illustrates the TEST option. However, you should first understand the analysis-of-variance statistics that PROC GLM computes by default.

The analysis-of-variance results appear in Output 4.4, and the expected mean square coefficients are given in Output 4.5.

Output 4.4
GLM Analysis of Variance of Log Count

Source	DF	Sum of Squares	Mean Square	F Value	Pr > F
Model	59	50.46346700	0.85531300	22.23	<.0001
Error	60	2.30863144	0.03847719		
Corrected Total	119	52.77209844			

R-Square	Coeff Var	Root MSE	logct Mean
0.956253	6.432487	0.196156	3.049459

Source	DF	Type I SS	Mean Square	F Value	Pr > F
package	19	30.52915506	1.60679763	41.76	<.0001
sample(package)	40	19.93431194	0.49835780	12.95	<.0001

Source	DF	Type III SS	Mean Square	F Value	Pr > F
package	19	30.52915506	1.60679763	41.76	<.0001
sample (package)	40	19.93431194	0.49835780	12.95	<.0001

Note: The *F*-statistics computed by PROC GLM for the basic analysis of variance of models with random effects are not necessarily correct. For the basic *F*-statistics shown above, GLM always uses MS(ERROR) for the denominator. For example, the *F*-statistic for PACKAGE is incorrect because MS(ERROR) is not the correct denominator mean square. Section 4.2.3, in conjunction with Table 4.1, shows you how to use the expected mean squares to determine the correct *F*-statistics.

Output 4.5
Expected Mean
Squares for Log
Count Data

```
                              Type III Expected Mean Square
Source

package              Var(Error) + 2 Var(sample(package)) + 6 Var(package)

sample(package)      Var(Error) + 2 Var(sample(package))
```

Now consider the output labeled "Type III Expected Mean Square." This part of the output gives you the expressions for the expected values of the mean squares. Table 4.1 shows you how to interpret the coefficients of expected mean squares.

Table 4.1 *Coefficients of Expected Mean Squares*

Variance Source	Source of Variation	DF	Expected Mean Squares	This Tells You:
PACKAGE	packages	19	$\sigma^2 + 2\sigma_s^2 + 6\sigma_p^2$	MS(PACKAGE) estimates $\sigma^2 + 2\sigma_s^2 + 6\sigma_p^2$
SAMPLE	samples in packages	40	$\sigma^2 + 2\sigma_s^2$	MS(SAMPLE) estimates $\sigma^2 + 2\sigma_s^2$
ERROR	replicates in samples	60	σ^2	MS(ERROR) estimates σ^2

From the table of coefficients of expected mean squares you get the estimates of variance components. These estimates are

❑ $\hat{\sigma}^2 = 0.0385 = $ MS(ERROR)

❑ $\hat{\sigma}_s^2 = 0.2299 = $ [MS(SAMPLE) − MS(ERROR)]/2

❑ $\hat{\sigma}_p^2 = 0.1847 = $ [MS(PACKAGE) − MS(SAMPLE)]/6

The variance of a single microbial count is

$$\hat{\sigma}_y^2 = \text{TOTAL Variance Estimate}$$

$$= \hat{\sigma}^2 + \hat{\sigma}_s^2 + \hat{\sigma}_p^2$$

$$= 0.0385 + 0.2299 + 0.1847$$

$$= 0.4532$$

Note: The expression TOTAL Variance Estimate does not refer to MS(TOTAL) = 0.4435,

  although the values are similar. From these estimates, you see that

❑ 8.49% of TOTAL variance is attributable to ERROR variance

❑ 50.74% of TOTAL variance is attributable to SAMPLE variance

❑ 40.77% of TOTAL variance is attributable to PACKAGE variance.

4.2.2 Computing Variances of Means from Nested Classifications and Deriving Optimum Sampling Plans

The variance of a mean can also be partitioned into portions attributable to individual sources of variation. The variance of a mean $\overline{Y}_{..}$ computed from a sample of n_p packages, n_s samples per package, and n replicates per sample, is estimated to be

$$\hat{\sigma}_{y_{..}}^2 = \hat{\sigma}_p^2 / n_p + \hat{\sigma}_s^2 / n_p n_s + \hat{\sigma}^2 / n_p n_s n$$

Output 4.4 showed that the overall mean is 3.0494. Its standard error can be determined from the square root of the formula for the variance of the mean. For these data, the standard error is

$$\hat{\sigma}_{y_{..}}^2 = [(0.1847)/20 + (0.2299) / 20 * 3 + (0.0385) / 20 * 3250 * 2]^{(1/2)} = 0.1157$$

The formula for the variance of a mean can also be used to derive an optimum sampling plan, subject to certain cost constraints. Suppose you are planning a study, for which you have a budget of $500. Each package costs $5, each sample costs $3, and each replicate count costs $1. The total cost is

$$\text{cost} = \$5 * n_p + \$3 * n_p * n_s + \$1 * n_p * n_s * n$$

You can create a SAS data set by taking various combinations of n_p, n_s, and n for which the cost is $500, and compute the variance estimate for the mean. Then you can choose the combination of n_p, n_s, and n that minimizes $\hat{\sigma}_{y_{..}}^2$

4.2.3 Analysis of Variance for Nested Classifications: Using Expected Mean Squares to Obtain Valid Tests of Hypotheses

Expected mean squares tell you how to set up appropriate tests of a hypothesis regarding the variance components. Suppose you want to test the null hypothesis H_0: $\sigma_p^2 = 0$. If this null hypothesis is true, then the expected mean square for PACKAGE and the expected mean square for SAMPLE (samples in packages) are both equal to $\sigma^2 + 2\sigma_s^2$. Therefore, MS(PACKAGE) and

MS(SAMPLE) should have approximately the same value if H_0: $\sigma_p^2 = 0$ is true. On the other hand, if H_0: $\sigma_p^2 = 0$ is false, then the MS(PACKAGE) should be larger than the MS(SAMPLE). It follows that you can compare the value of MS(PACKAGE) with the value of MS(SAMPLE) to get an indication of whether the null hypothesis is true or false.

Formally, you do this with an F-statistic: divide MS(PACKAGE) by MS(SAMPLE). The result has an F-distribution with $n_p - 1$ DF in the numerator and $n_p(n_s - 1)$ DF in the denominator. For the microbial count data, $F = 1.607/0.498 = 3.224$, with numerator DF=19 and denominator DF=40, which is significant at the $p = 0.0009$ level. Therefore, you reject H_0: $\sigma_p^2 = 0$, and conclude $\sigma_p^2 > 0$.

You can go through the same process of using the table of expected mean squares to set up a test of the null hypothesis H_0: $\sigma_s^2 = 0$. You see that the appropriate test statistic is $F = $ MS(SAMPLE)/MS(REPLICATE), with numerator DF=40 and denominator DF=60. This $F = 12.952$ is significant at the $p < 0.0001$ level. Again, you conclude $\sigma_s^2 > 0$.

You can compute the test statistics for H_0: $\sigma_p^2 = 0$ and H_0: $\sigma_s^2 = 0$ using PROC GLM. Either you can use the TEST option with the RANDOM statement, or you can use a TEST statement to define the F-statistic you want to use. The needed SAS statements are

```
proc glm;
  class package sample;
  model logct=package sample(package);
  random package sample(package)/test;
  test h=package e=sample(package);
```

Notice that you only need to use either the TEST option in the RANDOM statement or the TEST statement, but not both. The former uses the expected mean squares determined by the RANDOM statement to define the appropriate F-statistics. The latter requires you to know what ratio needs to be computed. In the TEST statement, H= refers to the numerator MS, and E= specifies the denominator MS to be used for the F-statistic. For the balanced data sets presented in this chapter, the F-statistics computed by the RANDOM statement's TEST option and by the TEST statement are the same. This is not always true for unbalanced data, which is discussed in Chapters 5 and 6. Note that you do not need a TEST statement for H_0: $\sigma_s^2 = 0$ because the default F-ratio MS[SAMPLE(PACKAGE)] / MS(ERROR) is correct. Output 4.6 gives the results for the RANDOM statement TEST option and the TEST statement.

Output 4.6
RANDOM TEST
Option and TEST
Statement Results in
PROC GLM for
Log Count Data

```
Output from RANDOM statement TEST option:


    Tests of Hypotheses for Random Model Analysis of Variance

Dependent Variable: logct

Source            DF   Type III SS    Mean Square    F Value    Pr > F

package           19     30.529155       1.606798       3.22     0.0009

Error             40     19.934312       0.498358
Error: MS(sample(package))



Source            DF    Type III SS    Mean Square    F Value    Pr > F

sample(package)   40     19.934312        0.498358      12.95    <.0001

Error: MS(Error)  60      2.308631        0.038477



Output from TEST H=PACKAGE E=SAMPLE(PACKAGE) statement:

Tests of Hypotheses Using the Type III MS for sample(package) as an Error Term

Source            DF    Type III SS    Mean Square    F Value    Pr > F

package           19    30.52915506    1.60679763       3.22     0.0009
```

4.2.4 Variance Component Estimation for Nested Classifications: Analysis Using PROC MIXED

PROC MIXED is SAS' most sophisticated and versatile procedure for working with models with random effects. You can duplicate the tests and estimates discussed in previous sections. In addition, there are several linear models statistics that only PROC MIXED can compute. This section introduces the basic features of PROC MIXED for random-effects models. The required statements for model (4.1) are

```
proc mixed;
  class package sample;
  model logct= ;
  random package sample(package);
```

In some respects, the statements for PROC GLM and PROC MIXED are the same—the CLASS statement, the MODEL statement up to the left-hand side, and the RANDOM statement are all identical to the statements used above for PROC GLM. The MODEL statement contains the big difference. In PROC GLM, you list all the effects in the model other than the intercept and error, regardless of whether they are fixed or random. In PROC MIXED, you list *ONLY the fixed effects*, if there are any, other than the intercept. For this model, the only fixed effect is μ, so nothing appears to the right of the equal sign except a space and a semicolon. Output 4.7 shows the results for these statements.

Output 4.7
Basic PROC
MIXED Results
for Log Count
Data

```
                              Iteration History

      Iteration      Evaluations      -2 Res Log Like        Criterion

          0               1            245.73109785
          1               1            128.18662316         0.00000000

                      Convergence criteria met.

                       Covariance Parameter
                            Estimates

                    Cov Parm              Estimate

                    package                0.1847
                    sample(package)        0.2299
                    Residual               0.03848

                         Fit Statistics

              -2 Res Log Likelihood           128.2
              AIC (smaller is better)         134.2
              AICC (smaller is better)        134.4
              BIC (smaller is better)         137.2
```

The essential part of this output appears under "Covariance Parameter Estimates." These are the estimates of σ_p^2, σ_s^2, and σ^2, the PACKAGE, SAMPLE(PACKAGE), and RESIDUAL (or error) variances. You can see that they are the same as the estimates obtained using the analysis of variance and expected mean squares in Section 4.2.1. This will always be true for balanced data, but not necessarily for unbalanced data. Chapter 5 presents examples with unbalanced data. PROC MIXED uses **restricted maximum likelihood**, usually referred to by its acronym, REML, to estimate variance components. Chapter 6 contains a brief explanation of REML and gives references for additional information.

The REML procedure requires numerical iteration. The "Iteration History" appears immediately before the variance component estimates. You should look for the expression "Convergence Criteria Met." If it appears, fine. For the types of data sets in this book, convergence problems are extremely rare. If you do get a failure to converge, it is probably because the data are being misread, or the CLASS, MODEL, or RANDOM statements are mistyped.

The basic MIXED output does not contain an analysis-of-variance table. If you want to test hypotheses about the variance components, you can use PROC GLM to compute the *F*-tests as shown previously. PROC MIXED does allow you two ways to test the variance components—the **Wald** test and the **likelihood ratio** test. Of the two, the likelihood ratio test is preferable for most applications.

The likelihood ratio test uses the **residual log likelihood**—more precisely, –2 times the residual log likelihood, that is, "–2 Res Log Likelihood" in Output 4.7—to construct a test statistic that has an approximate χ^2 distribution. For the model that contains both effects, PACKAGE and SAMPLE(PACKAGE), the –2 residual log likelihood is 128.2. If you drop SAMPLE(PACKAGE) from the RANDOM statement and rerun the analysis, you can test H_0: $\sigma_s^2 = 0$. Run the SAS statements

```
proc mixed;
 class package sample;
 model logct= ;
 random package;
```

You get the "Fit Statistics" shown in Output 4.8.

Output 4.8
Statistics for
Fit of Model
without
SAMPLE
(PACKAGE)
for Log
Count Data

```
                    Fit Statistics

        -2 Res Log Likelihood        201.2
        AIC (smaller is better)      205.2
        AICC (smaller is better)     205.3
        BIC (smaller is better)      207.2
```

The –2 residual log likelihood for this model is 201.2. The difference between this and the value of the full model is 201.2–128.2 = 73.0. This is the **likelihood ratio statistic**. It has an approximate χ^2 distribution with 1 DF. The 1 DF is because one variance component has been removed from the model. The $\alpha=0.05$ critical value for χ^2 is 3.84, the *p*-value for $\chi^2=73.0$ is <0.0001. Therefore, you reject $H_0: \sigma_s^2 = 0$.

You can then drop PACKAGE from the model to test $H_0: \sigma_p^2 = 0$. The statements

```
proc mixed;
  class package sample;
  model logct= ;
```

yield a –2 residual log likelihood of 245.7. The likelihood ratio test statistic for $H_0: \sigma_p^2 = 0$ is 245.7–201.2 = 44.5. Again, comparing it to $\chi^2_{(1)}$ shows that you reject H_0. The *p*-value is <0.0001.

You can also obtain Wald statistics. You use the option COVTEST in the PROC MIXED statement to compute approximate standard errors for the variance components. The Wald statistic is the ratio $Z = \dfrac{\text{variance estimate}}{\text{standard error}}$. *Z* is assumed to have an approximate standard normal distribution. Use the following SAS statements:

```
proc mixed covtest;
  class package sample;
  model logct= ;
  random package sample(package);
```

Output 4.9 shows the results.

Output 4.9
PROC MIXED
Variance
Component
Estimates Using
the COVTEST
Option

```
              Covariance Parameter Estimates
                                  Standard      Z
     Cov Parm           Estimate    Error     Value      Pr Z

     package            0.1847     0.08885     2.08      0.0188
     sample(package)    0.2299     0.05583     4.12      <.0001
     Residual           0.03848    0.007025    5.48      <.0001
```

Using $H_0: \sigma_p^2 = 0$ as an example, the asymptotic standard error of $\hat{\sigma}_p^2$ is 0.08885 and thus the approximate Z-statistic is $\dfrac{0.1847}{0.08885} = 2.08$. You can compare $Z = 2.08$ to a table value from the standard normal distribution, or note that the *p*-value (Pr>Z) is 0.0188.

Important Note: For tests of variance components, the normal approximation is *very poor* unless the sample size (in this case the number of packages) is in the hundreds or, preferably, thousands. For this reason, use of the Wald statistic to test variance components is *strongly discouraged* unless your sample size is very large. The ANOVA *F*-tests using PROC GLM are preferable. Alternatively, you can construct likelihood ratio tests using PROC MIXED, although doing so requires multiple runs and is therefore less convenient. See Littell et al. (1996) for more about likelihood ratio tests.

4.2.5 Additional Analysis of Nested Classifications Using PROC MIXED: Overall Mean and Best Linear Unbiased Prediction

Section 4.2.2 presented the estimate and standard error of the overall mean. The PROC GLM output provides the estimate $\bar{y}_{...}$, but not the variance component estimates. You have to calculate these from the mean squares and then hand-calculate the standard error of the mean. PROC MIXED does not compute these numbers by default, but you can obtain them either with the SOLUTION option in the MODEL statement or with an ESTIMATE statement. The SAS statements are

```
model logct= / solution;
  random package sample(package);
  estimate 'overall mean' intercept 1;
```

Output 4.10 shows the results.

Output 4.10
PROC
MIXED
Estimates of
Overall Mean
for Log Count
Data

```
                      Solution for Fixed Effects

                                Standard
        Effect      Estimate     Error      DF    t Value    Pr > |t|

        Intercept    3.0495      0.1157     19     26.35      <.0001

                             Estimates

                                Standard
        Label       Estimate     Error      DF    t Value    Pr > |t|

        overall mean 3.0495      0.1157     19     26.35      <.0001
```

You can see that the SOLUTION option and the ESTIMATE statement produce identical results. In this case, they are simply different ways of including $\hat{\mu}$ in the output.

In some applications with random-effects models, the equivalent of treatment means may be of interest. Henderson (1963,1975) first developed this procedure in the context of animal breeding. Random sires were used to estimate variance components that had genetic interpretations. At the same time, breeders wanted to assess the "breeding value" of each sire, conceptually similar to the mean sire performance. However, because sire effects are random, and there is information about their probability distribution, this affects how you estimate breeding value. The mixed-model procedure called **best linear unbiased prediction**, or BLUP, was developed for this purpose. In addition to animal breeding, there are many other applications of BLUP. In clinical trials, random samples of patients provide estimates of the mean performance of a treatment for the population of inference, but BLUPs are essential for physicians to monitor individual patients. In quality assurance, a sample of workers can provide estimates of the mean performance of a machine, but BLUPs can help supervisors monitor the performance of individual employees.

To illustrate using the log count data, suppose you want to know the "mean" log count of the first package (PACKAGE=1 in the SAS data set). In terms of model (4.1) you want to estimate $\mu + a_1$, that is, the overall mean plus the effect of package 1. If PACKAGE were a fixed effect, you would simply calculate the sample mean for package 1, $\bar{y}_{1..} = 2.8048$. However, because the package effect a_1 is a random variable, its probability distribution must be taken into account, which the sample mean does not do. Instead, you compute the BLUP.

The best linear predictor of $\mu + a_1$ is equal to the estimate of $\mu + E(a_1|y)$, the expected value of the effects of package 1 given the data. Because μ is a fixed effect, its estimate, as you have seen, is its sample mean. The conditional expectation of a_1 turns out to be $E(a_1) + \text{cov}(\hat{a}_1, \bar{y}_{1..})[\text{var}(\bar{y}_{1..})]^{-1}(\bar{y}_{1..} - \bar{y}_{...})$. From model (4.1), $E(a_1)=0$. The resulting expression is equal to

$$\left(\frac{\hat{\sigma}_p^2}{\hat{\sigma}_p^2 + \dfrac{\hat{\sigma}_s^2}{n_s} + \dfrac{\hat{\sigma}^2}{n_s n}} \right)(\bar{y}_{1..} - \bar{y}_{...}) = \left(\frac{0.1847}{0.1847 + \dfrac{0.2299}{3} + \dfrac{0.03848}{3*2}} \right)(2.8048 - 3.0495) = -0.1688$$

This expression is often called an "EBLUP" because it uses estimated variance components. A "true" BLUP assumes the variance components are known. Note that the "usual" fixed effects estimate of the package effect, $\bar{y}_{1..} - \bar{y}_{...} = -0.2447$, but EBLUP is smaller. In effect, the estimate of a_1 is shrunk to account for its probability distribution. For this reason, the BLUP is often called a **shrinkage estimator**.

You can obtain EBLUPs for the random effects, such as $\hat{a}_1$, using a SOLUTION option with the RANDOM statement, and for linear combinations of fixed or RANDOM effects, such as $\hat{\mu} + \hat{a}_1$, using the ESTIMATE statement. Use the following SAS statements:

```
proc mixed;
 class package sample;
 model logct= / solution;
 random package sample(package)/solution;
 estimate 'pkg 1 blup' intercept 1 | package 1 0;
```

Output 4.11 gives the results for the PACKAGE random effects and PKG 1 BLUP, the EBLUP analog of the PACKAGE 1 mean. The SOLUTION statement also causes the SAMPLE(PACKAGE) EBLUPs to be printed, but these are not shown. **Note:** When random effects are involved, the ESTIMATE statement requires a vertical bar (|). Fixed effects (for example, μ) go before the bar; random effects (for example, a_1) go after the bar. Otherwise, you use the same syntax as in any other ESTIMATE statement.

Output 4.11
EBLUPs of
Package Effects
and Package
"Mean" for Log
Count Data

```
                         Solution for Random Effects

                                            Std Err
Effect          package  sample  Estimate    Pred    DF  t Value  Pr > |t|

package            1             -0.1688    0.2523   60   -0.67    0.5061
package            2              0.1171    0.2523   60    0.46    0.6442
package            3             -0.03558   0.2523   60   -0.14    0.8883
package            4             -0.04658   0.2523   60   -0.18    0.8542
package            5              0.07525   0.2523   60    0.30    0.7666
package            6             -0.7363    0.2523   60   -2.92    0.0049
package            7              0.3593    0.2523   60    1.42    0.1596
package            8             -0.4495    0.2523   60   -1.78    0.0799
package            9             -0.09742   0.2523   60   -0.39    0.7008
package           10             -0.5484    0.2523   60   -2.17    0.0337
package           11              0.1601    0.2523   60    0.63    0.5282
package           12              0.3226    0.2523   60    1.28    0.2060
package           13             -0.1901    0.2523   60   -0.75    0.4542
package           14              0.1521    0.2523   60    0.60    0.5491
package           15             -0.2128    0.2523   60   -0.84    0.4024
package           16              0.8166    0.2523   60    3.24    0.0020
package           17             -0.1594    0.2523   60   -0.63    0.5300
package           18             -0.06795   0.2523   60   -0.27    0.7886
package           19              0.3966    0.2523   60    1.57    0.1213
package           20              0.3132    0.2523   60    1.24    0.2194


                              Estimates

                            Standard
   Label          Estimate    Error     DF    t Value   Pr > |t|

   pkg 1 blup      2.8807     0.2420     19     11.90    <.0001
```

Note that the PKG 1 BLUP is $\hat{\mu} + \hat{a}_1 = 3.0495 - 0.1688$. Consistent with the idea of shrinkage estimation, that is, using what is known about the distribution of the package effects, the estimate is "shrunk" toward the overall mean relative to the package 1 sample mean, 2.8048.

4.3 Blocked Designs with Random Blocks

Chapter 3 presented the analysis of variance for randomized-complete-blocks designs. The analysis implicitly assumed fixed blocks. In many cases, it is more reasonable to assume that blocks are random. Technically, the question of fixed versus random blocks depends on whether the blocks observed constitute the *entire population* of possible blocks—if so, they are fixed. Also considered is whether it is reasonable to assume there is a larger population of blocks with some probability distribution of block effects and the blocks observed are representatives of that population—if so, blocks are random. Seen from this perspective, it is hard to imagine a fixed-block experiment of any inferentially interesting consequence. From a practical viewpoint, however, many experiments do not represent the population of blocks well enough to make use of the additional inference possible with random blocks. Furthermore, with complete-blocks designs, inference for treatment *differences* is identical for fixed-blocks and random-blocks models. As you will see in Chapters 5 and 6, even when you have missing data or use an incomplete blocks design, fixed versus random typically has only a trivial effect on inference about treatment differences. Thus, for experiments whose exclusive goal is to estimate treatment effects, there is often little point in fulfilling the extra design requirements for random-blocks inference.

On the other hand, if estimating treatments *means* is important, especially obtaining confidence intervals for means, then the choice of a fixed- or random-blocks model matters a great deal.

The purpose of this section is to present the **random-blocks** analysis of the randomized-complete-blocks design, and to compare it with fixed-blocks ANOVA. Section 4.3.1 takes the data first presented in Output 3.17 and shows how to use PROC MIXED to do the random-blocks analysis. Section 4.3.2 discusses differences and similarities between the random-blocks analysis and fixed-blocks ANOVA. Section 4.3.2 also presents general guidelines for deciding whether fixed- or random-blocks analysis is more appropriate.

4.3.1 Random-Blocks Analysis Using PROC MIXED

The analysis of variance for the randomized-blocks design uses the model equation

$$y_{ij} = \mu + \tau_i + \beta_j + e_{ij} \tag{4.2}$$

where

y_{ij} is the observation on treatment *i* and block *j*.

μ is the intercept.

τ_I is the *i*th treatment effect.

β_j is the *j*th block effect.

e_{ij} is the residual, or error, for the *ij*th observation, assumed i.i.d. $N(0,\sigma^2)$.

The Valencia orange data set presented in Output 3.17 was an example. The τ_i's were the IRRIG method effects.

The analysis of variance presented in Section 3.5 implicitly assumes that treatments are fixed effects, but it does not depend on any assumptions about the block effects. However, additional features of the analysis of variance using PROC GLM reveals GLM computations assume fixed blocks. Although fixed treatments is usually a reasonable assumption, fixed blocks may not be.

Model (4.2) can be modified for random blocks. Usually, blocks are assumed i.i.d. $N(0, \sigma_b^2)$. In keeping with the Greek-fixed, Latin-random convention, the block effect β_j is changed to b_j, so the model equation becomes $y_{ij} = \mu + \tau_i + b_j + e_{ij}$. Also, the block and error effects are assumed to be mutually independent. Note that the random-blocks model for the randomized-blocks design is a **mixed model** because it contains both fixed- and random-model effects.

You use the following PROC MIXED statements to compute the mixed-model random-blocks analysis:

```
proc mixed;
  class bloc irrig;
  model fruitwt=irrig;
  random bloc;
```

Compared to the GLM statements in Section 3.5, you delete BLOC from the MODEL statement and add a RANDOM statement for the BLOC effect. Output 4.12 shows the results.

Output 4.12
Random-Blocks Analysis of Valencia Orange Data Using PROC MIXED

```
                        Covariance Parameter
                             Estimates

                   Cov Parm        Estimate

                   bloc                10793
                   Residual          3362.38


                        Fit Statistics

             -2 Res Log Likelihood           413.8
             AIC (smaller is better)         417.8
             AICC (smaller is better)        418.2
             BIC (smaller is better)         417.9


                  Type 3 Tests of Fixed Effects

                         Num      Den
             Effect       DF       DF     F Value    Pr > F

             irrig         4       28       3.27     0.0254
```

The output contains the essential information contained in the analysis of variance, but in different form. Instead of sums of squares and mean squares for the various sources of variation, the output provides variance estimates for random-effects sources of variation and *F*-statistics and associated degrees of freedom and *p*-values for fixed-effects sources of variation. Note that $F=3.27$ and $p=0.0254$ are identical to the results for IRRIG in the analysis of variance (Output 3.18).

You can compute least-squares means (LS means) and estimated treatment differences by adding the statement

```
lsmeans irrig / diff;
```

to the above PROC MIXED program. Output 4.13 shows the results.

Output 4.13
Estimated Least-Squares Means and Treatment Differences Using PROC MIXED

```
                             Least Squares Means

                                      Standard
        Effect      irrig      Estimate      Error      DF     t Value     Pr > |t|

        irrig       basin       290.37      42.0652     28       6.90       <.0001
        irrig       flood       229.62      42.0652     28       5.46       <.0001
        irrig       spray       223.75      42.0652     28       5.32       <.0001
        irrig       sprnkler    292.00      42.0652     28       6.94       <.0001
        irrig       trickle     299.62      42.0652     28       7.12       <.0001


                        Differences of Least Squares Means

                                              Standard
      Effect    irrig     _irrig     Estimate    Error     DF    t Value    Pr > |t|

      irrig     basin     flood       60.7500    28.9930    28      2.10      0.0453
      irrig     basin     spray       66.6250    28.9930    28      2.30      0.0292
      irrig     basin     sprnkler    -1.6250    28.9930    28     -0.06      0.9557
      irrig     basin     trickle     -9.2500    28.9930    28     -0.32      0.7521
      irrig     flood     spray        5.8750    28.9930    28      0.20      0.8409
      irrig     flood     sprnkler   -62.3750    28.9930    28     -2.15      0.0402
      irrig     flood     trickle    -70.0000    28.9930    28     -2.41      0.0225
      irrig     spray     sprnkler   -68.2500    28.9930    28     -2.35      0.0258
      irrig     spray     trickle    -75.8750    28.9930    28     -2.62      0.0141
      irrig     sprnkler  trickle     -7.6250    28.9930    28     -0.26      0.7945
```

For balanced data, LS means for treatments are identical to sample treatment means obtained from the MEANS statement in PROC GLM. Note that the values in the "Estimate" column for the LS means are the same as the IRRIG means in Output 3.19. The output from PROC MIXED gives the standard error of the estimated treatment means by default. Here, the standard error is 42.0652. The standard error of a treatment mean in the random-blocks analysis of a randomized-complete-blocks design is $\sqrt{\dfrac{\hat{\sigma}_b^2 + \hat{\sigma}^2}{r}}$ where r is the number of blocks. In this case, the standard error is $\sqrt{\dfrac{10793 + 3362.38}{8}}$ since $\hat{\sigma}_b^2 = 10793$, $\hat{\sigma}^2 = 3362.38$, and $r=8$. The "Differences of Least Squares Means" table gives the estimated differences, their standard errors, t-statistics and their associated p-values. Note that by definition the t-statistics are **least significant difference** (LSD) tests for pairwise differences. For randomized-complete-blocks designs, the standard error of a treatment difference is $\sqrt{\dfrac{2\hat{\sigma}^2}{r}}$, which for these data equals $\sqrt{\dfrac{2 \times 3362.38}{8}} = 28.993$. You can calculate the LSD statistic by multiplying the standard error of the difference by $t_{(\alpha,\,dfe)}$, the table value of the t-distribution for the error degrees of freedom (DFE) and the desired α-level. For example, for these data, DFE=28, hence $t_{(\alpha=0.05,28)}$=2.048. Thus, the LSD statistic is 28.993 × 2.048=59.378, the same (aside from rounding error) as appears in Output 3.19.

As with the MEANS statement in PROC GLM, you can specify mean comparisons other than the LSD test. The ADJUST= option with the LSMEANS statement in PROC MIXED allows you to adjust the p-value to correspond to different procedures. For example, use these statements to compute p-values for Dunnett's test:

```
proc mixed order=data;
  class bloc irrig;
  model fruitwt=irrig;
  random bloc;
  lsmeans irrig/diff adjust=dunnett;
```

Ordinarily, you do not need to use the ORDER=DATA option in the PROC MIXED statement; it is a special requirement for Dunnett's test with this example. From the description of the problem in Chapter 3, the FLOOD method was intended to be the reference. As with PROC GLM, the default control or reference treatment for the DUNNETT test is the first treatment in alphameric order. In Section 3.5, you used an option with the MEANS statement in PROC GLM to override this default. However, there is no corresponding option in the LSMEANS statements in GLM or MIXED. In order to obtain the desired test, you have to rearrange the data as they are entered in the DATA step so that the FLOOD level of IRRIG appears first, then use the ORDER=DATA option. This changes the order of the treatments for assigning CONTRAST and ESTIMATE coefficients as well as for Dunnett's test. You can look at the "Class Level Information" (not shown) or the order in which the LS means are printed (shown below) to see the order of the treatments MIXED will use. The results appear in Output 4.14.

*Output 4.14
Dunnett-
Adjusted
p-values for
PROC MIXED
Analysis of
Valencia
Orange Data
with FLOOD as
the Control
Irrigation
Method*

```
                              Least Squares Means

                                      Standard
        Effect    irrig      Estimate    Error      DF    t Value    Pr > |t|

        irrig     flood       229.62    42.0652     28      5.46      <.0001
        irrig     basin       290.37    42.0652     28      6.90      <.0001
        irrig     spray       223.75    42.0652     28      5.32      <.0001
        irrig     sprnkler    292.00    42.0652     28      6.94      <.0001
        irrig     trickle     299.62    42.0652     28      7.12      <.0001

                        Differences of Least Squares Means

                                          Standard
     Effect   irrig     _irrig    Estimate    Error     DF    t Value    Pr > |t|

     irrig    basin     flood      60.7500   28.9930    28      2.10     0.0453
     irrig    spray     flood      -5.8750   28.9930    28     -0.20     0.8409
     irrig    sprnkler  flood      62.3750   28.9930    28      2.15     0.0402
     irrig    trickle   flood      70.0000   28.9930    28      2.41     0.0225

             Effect    irrig     _irrig     Adjustment      Adj P

             irrig     basin     flood      Dunnett-Hsu     0.1389
             irrig     spray     flood      Dunnett-Hsu     0.9988
             irrig     sprnkler  flood      Dunnett-Hsu     0.1245
             irrig     trickle   flood      Dunnett-Hsu     0.0728
```

Note the order that the IRRIG levels appear in the "Least Squares Means" table follows from the order the data were entered and the ORDER=DATA option in the PROC MIXED statement. FLOOD was entered first, then the other IRRIG levels in alphabetical order. The "Differences of Least Squares Means" shows only differences allowed by Dunnett's test with FLOOD as the reference treatment. Two sets of *p*-values are shown. The first are the unadjusted *t*-test results, identical to what you would get in an LSD test. The "Dunnett-Hsu" adjustment are the appropriate *p*-values for Dunnett's test.

In Section 3.5, standard errors were not discussed. Least-squares means and their standard errors can be obtained in conjunction with analysis of variance using the LSMEANS statement in PROC GLM. However, GLM does not allow you to compute standard errors of treatment differences, as you can with the DIFF option in MIXED. In addition, GLM and MIXED compute different standard errors for the LS means, revealing the primary distinction between fixed-blocks and random-blocks analysis. The next section discusses these differences.

4.3.2 Differences between GLM and MIXED Randomized-Complete-Blocks Analysis: Fixed versus Random Blocks

For randomized-complete-blocks designs, inference on treatment *differences* is *entirely unaffected* by whether blocks are fixed or random. This is not true with missing data or incomplete-blocks-designs (see Chapter 6). However, for inference on treatment *means*, standard errors, and hence how you interpret the data, can be substantially affected by fixed versus random blocks.

4.3.2.1 Treatment Means

You can obtain LS means and their standard errors in PROC GLM using the statement

```
lsmeans trt / stderr;
```

Unlike PROC MIXED, GLM does not compute the standard error by default. You must use the STDERR option. The results appear in Output 4.15.

Output 4.15
LS Means for
Analysis of
Randomized-
Complete-
Blocks Design
Using PROC
GLM

```
                        The GLM Procedure
                       Least Squares Means

                       fruitwt        Standard
         irrig          LSMEAN          Error      Pr > |t|

         basin        290.375000      20.501170     <.0001
         flood        229.625000      20.501170     <.0001
         spray        223.750000      20.501170     <.0001
         sprnkler     292.000000      20.501170     <.0001
         trickle      299.625000      20.501170     <.0001
```

Note the return to the original order of the data. The LS means are the same as computed by MIXED (Output 4.13) and by the MEANS statement in PROC GLM (Output 3.19). However, whereas MIXED obtained a standard error of 42.0652, the standard error of the mean using GLM is 20.5012. Why the difference? PROC GLM uses the fixed-block formula for the standard error

of the mean, $\sqrt{\dfrac{\hat{\sigma}^2}{r}} = \sqrt{\dfrac{MS(error)}{r}} = \sqrt{\dfrac{3362.38}{8}} = 20.5012$. In fact, PROC GLM's programming

assumes *all* model effects are fixed. Chapter 6 presents the theory underlying PROC GLM in greater detail. While the RANDOM statement allows GLM to determine expected mean squares and to select appropriate mean square ratios accordingly to construct *F*-statistics, the RANDOM statement does *not* affect the way GLM computes standard errors. If you add a RANDOM statement to the GLM program for these data, it will compute expected mean squares, but the standard error of the LS means will remain 20.5012.

With fixed blocks, the definition of a treatment LS mean for the randomized-blocks design is

$\mu + \tau_i + \dfrac{1}{r}\sum_j \beta_j$. With fixed blocks, all uncertainty about the treatment mean is assumed to result

from variation among experimental units within a block. With random blocks, the treatment LS Mean is defined as $\mu + \tau_i$. Therefore, variation among blocks *and* variation among units within a block contribute to uncertainty. You can use the following ESTIMATE statements with PROC MIXED for more insight:

```
estimate 'irrig lsmean' intercept 1 irrig 1 0;
estimate 'irrig narrow lsm' intercept 8 irrig 8 0
         | bloc 1 1 1 1 1 1 1 1/divisor=8;
```

The first ESTIMATE statement uses the coefficients from the definition of the LS mean for BASIN in a random-blocks model. The second uses the fixed-blocks definition of the LS mean for BASIN. Note that for the random-blocks model, the second ESTIMATE statement is actually a BLUP. More precisely, it is a BLUP limiting the estimate of the treatment mean to *only those blocks actually observed*. Hence, it is termed the "narrow" estimate, because it narrows the scope of inference from the entire population of blocks to only the blocks that were observed. Output 4.16 shows the results.

*Output 4.16
BASIN LS
Mean: Usual
Definition and
Narrow
Inference
Definition
from the
ESTIMATE
Statement in
Random-
Blocks
Analysis*

```
                          Estimates

                          Standard
    Label          Estimate    Error    DF    t Value    Pr > |t|

    irrig lsmean     290.37   42.0652   28      6.90      <.0001
    irrig narrow lsm 290.38   20.5012   28     14.16      <.0001
```

The numbers in the "Estimate" column reflect a MIXED round-off idiosyncrasy—they are both 290.375, the same as the LS mean and the MEANS in PROC GLM. The standard error of the first estimate matches the MIXED LS mean; the second matches GLM.

To summarize, the standard error that PROC GLM obtains—that is, the fixed-blocks standard error of the mean—assumes that all uncertainty in the estimated treatment means results exclusively from experimental unit variability. The random-blocks standard error that PROC MIXED obtains assumes that there is additional variation among blocks, over and above experimental unit differences.

One way to view this is to assume you want to use the information from this study to anticipate the mean fruit weight yield you will get at your orchard. The estimated mean for the BASIN treatment is 290.375. You know you will not have a mean yield of exactly 290.375, so you want to put a confidence interval around this estimate. How wide should it be? If you use the fixed-blocks standard error for your confidence interval, you assume your orchard has the same expected yield as the orchards used in the experiment. Only variation among the plots contributes to uncertainty. If you use the random-blocks standard error, you assume your orchard is different from the orchards used in the experiment, and that there is likely to be variation among orchards in addition to variability within the orchard. The block variance is the best measure you have of the variance among orchards.

The validity of the random-blocks confidence interval depends on this last sentence. Does the block variance *really* provide an adequate estimate? Typically, if the blocks are in close proximity and intended mainly to account for local variation, the answer is probably no. On the other hand, if the blocks do a reasonably good job of representing variability in the population, then the block variance can provide the basis for a useful confidence interval for the mean. This latter condition occurs when the blocks are locations, operators, batches, or similar effects, and a plausible, even if not technically random, sampling scheme has been used.

4.3.2.2 Treatment Differences

For fixed blocks, the expected value of the treatment mean is $\mu + \tau_i + \frac{1}{r}\sum_j \beta_j$. For random

blocks, the expected value is $\mu + \tau_i$. The LS mean definitions used in the last section follow from these expectations. Under either definition, the estimate of the treatment mean is $\tau_i - \tau_{i'}$ for any

pair of treatments $i \neq i'$. Therefore, the standard error of a difference is not affected by the question of fixed versus random blocks. It follows that all treatment comparisons, for example, orthogonal contrasts, are similarly unaffected. Unlike PROC MIXED, there is no option in the LSMEANS statement in PROC GLM to compute the standard error of a difference. However, you can use the ESTIMATE statement in PROC GLM. For example, for the Valencia orange data, you can use the statement

```
estimate 'trt diff' irrig 1 -1 0;
```

to compute the estimate and standard error of the difference between the BASIN and FLOOD treatments. In a complete-blocks design, the standard errors for all pairwise differences are equal, so it is sufficient to compute just one. Output 4.17 shows the result.

Output 4.17
Estimate and
Standard
Error of
Treatment
Difference
Using the
GLM
ESTIMATE
Statement

		Estimates			
		Standard			
Label	Estimate	Error	DF	t Value	Pr > |t|
trt diff	60.7500	28.9930	28	2.10	0.0453

You see that the standard error of the difference is identical to that obtained using PROC MIXED for the random-blocks model.

For this reason, just because a design is not suitable for allowing a valid interval estimate of treatment means, this does *not* mean that it is necessarily unsuitable for obtaining valid interval estimates of treatment differences. Because treatment difference estimates do not depend on fixed versus random-blocks issues, well-conceived block designs can always provide valid estimates of treatment effects. However, the requirements for a blocked design to provide believable interval estimates of treatment means are more exacting. Specifically, if you want good interval estimates of the treatment means, ideally you need a random sample of the population of blocks. Failing that, you at least need a sample that plausibly represents variation among blocks in the population.

4.4 The Two-Way Mixed Model

Recall the discussion in Section 3.7, "A Two-Way Factorial Experiment." Assume that you actually have sources of seed from many varieties, perhaps several hundred. Also, suppose the objective of the experiment is to compare the methods across all the varieties in the population of potential varieties. Because it is not feasible to include *all* varieties in the experiment, you randomly choose a manageable number of varieties—say, five, for the purposes of this example—from the population of varieties. Your interest is not specifically in these five varieties, but in the population from which they were drawn. This makes VARIETY a random effect. As a consequence, any effect involving VARIETY is also a random effect. In particular, METHOD*VARIETY is a random effect. Interest remains only in the three methods, so METHOD is still a fixed effect. Since both random and fixed effects are involved, the model is mixed.

The fact that VARIETY is a random effect alters how you should analyze METHOD differences. First of all, VARIETY being a random effect determines how you measure experimental error appropriate for comparing methods. Furthermore, in many applications, you are not interested in simple effects, but only in METHOD main effects. When the response of specific VARIETY levels or simple effects of METHOD given VARIETY *are* of interest, you must take into account the distribution of VARIETY and METHOD*VARIETY effects. As you saw in Section 4.2, the means you use are EBLUPs.

You can test the main effect of METHOD either by using PROC GLM with the RANDOM statement and TEST option to determine the appropriate *F*-statistics, or by using PROC MIXED. You can use both PROC MIXED and, up to a point, PROC GLM, to do mean comparison tests.

Only PROC MIXED can compute correct standard errors for all the means and differences of potential interest. Also, PROC GLM cannot compute EBLUPs; when they are of interest, you must use PROC MIXED.

4.4.1 Analysis of Variance for the Two-Way Mixed Model: Working with Expected Mean Squares to Obtain Valid Tests

A model for the data is

$$y_{ijk} = \mu + \alpha_i + b_j + (ab)_{ij} + e_{ijk}$$

where

$\mu + \alpha_i = \mu_i$ is the mean for method i, averaged across all varieties in the population, $i = 1, 2, 3$.

b_j are random variables representing differences between varieties, assumed i.i.d. $N(0, \sigma_V^2)$; $j = 1, \ldots, 5$.

$(ab)_{ij}$ are random variables representing interaction between methods and varieties, assumed i.i.d. $N(0, \sigma_{MV}^2)$.

e_{ijk} are random variables representing differences in yields among plants of the same variety using the same method, assumed i.i.d. $N(0, \sigma^2)$, with $k = 1, \ldots, 6$.

The random variables b_j, $(ab)_{ij}$, and e_{ijk} are all assumed to be mutually independent.

Note: This formulation of the model is not universally accepted. Other formulations specify other assumptions regarding terms in the model. See Hocking (1973). The main distinction in these formulations pertains to how you define VARIETY variance. This is discussed at greater length in Chapter 6, "Understanding Linear Models Concepts." Here, the concern is comparing METHODs. All formulations of the model lead to the same techniques for comparing METHODs, so the issue of alternative model formulations is not of immediate concern.

The data contain the same sources of variation whether VARIETY is fixed or random, so you can compute the same analysis-of-variance table. But you should use computations from the table differently than when VARIETY was considered fixed. The main effect of differences between METHODs, rather than simple effects, even in the presence of interaction between METHOD and VARIETY, is tested here. Also tested is the comparison A vs B,C between the METHODs. Now, however, the focus is on the main effect of the contrast, even in the presence of interaction.

Run the following statements:

```
proc glm data=fctorial; class method variety;
   model yield = method variety method*variety / ss3;
   contrast 'A vs B,C' method 2 -1 -1;
```

The results appear in Output 4.18.

Output 4.18
Analysis of
Variance for the
Two-Way Mixed
Model

Source	DF	Sum of Squares	Mean Square	F Value	Pr > F
Model	14	1339.024889	95.644635	4.87	<.0001
Error	75	1473.766667	19.650222		
Corrected Total	89	2812.791556			

R-Square	Coeff Var	Root MSE	yield Mean
0.476048	24.04225	4.432857	18.43778

Source	DF	Type III SS	Mean Square	F Value	Pr > F
method	2	953.1562222	476.5781111	24.25	<.0001
variety	4	11.3804444	2.8451111	0.14	0.9648
method*variety	8	374.4882222	46.8110278	2.38	0.0241

Contrast	DF	Contrast SS	Mean Square	F Value	Pr > F
A vs B,C	1	940.7347222	940.7347222	47.87	<.0001

Add the RANDOM statement to specify that VARIETY and METHOD*VARIETY are random effects.

```
random variety method*variety;
```

The RANDOM statement specified here only causes expected mean squares to be computed. It does not affect any of the PROC GLM computations. Note that you must place the RANDOM statement *after* the CONTRAST statement in order to get the expected mean square for the contrast. If you want correct *F*-statistics, you have to specify them in a TEST statement or use the TEST option in the RANDOM statement.

Output 4.19 shows expected mean squares.

Output 4.19
Expected Mean
Squares for
Two-Way
Mixed Models

Source	Type III Expected Mean Square
method	Var(Error) + 6 Var(method*variety) + Q(method)
variety	Var(Error) + 6 Var(method*variety) + 18 Var(variety)
method*variety	Var(Error) + 6 Var(method*variety)

Contrast	Contrast Expected Mean Square
A vs B,C	Var(Error) + 6 Var(method*variety) + Q(method)

In statistical notation, these expected mean squares are as follows:

Source	Expected Mean Squares
METHOD	$\sigma^2 + 6\sigma_{MV}^2 + 30[\sum_i (\alpha_i - \overline{\alpha}.)^2 / 2]$
VARIETY	$\sigma^2 + 6\sigma_{MV}^2 + 18\sigma_V^2$
METHOD*VARIETY	$\sigma^2 + 6\sigma_{MV}^2$
ERROR	σ^2

You can probably see how all of these terms come from Output 4.19 except for the expression $30[\sum_i (\alpha_i - \overline{\alpha}.)^2 / 2]$, which corresponds to Q(METHOD) for the METHOD source of variation.

You can use the Q option if the RANDOM statement of PROC GLM to obtain output you can then use to obtain this expression. Section 4.4.3 presents a more detailed explanation. For now, all you need to know about Q(METHOD) is that it measures differences between the METHOD means. This is true because $\mu_i = \mu + \alpha_i$ and therefore, $\sum_i (\alpha_i - \overline{\alpha}.)^2 = \sum_i (\mu_i - \overline{\mu}.)^2$.

The null hypothesis H_0: $\mu_A = \mu_B = \mu_C$ is true when Q(METHOD)=0. Therefore, when H_0 is true, the expected mean square for METHOD is the same as the expected mean square for METHOD*VARIETY. This tells you to use F=MS(METHOD)/MS(METHOD*VARIETY) to test the null hypothesis. You can obtain this either by adding the TEST option to the RANDOM statement or by using the following TEST statement:

```
test h=method  e=method*variety;
```

At first glance, the expected mean square for the contrast A vs B,C in Output 4.19 appears to be the same as the expected mean square for METHOD, but Q(METHOD) has a different interpretation. For the contrast A vs B,C, Q(METHOD) stands for $30[\mu_A - 0.5(\mu_B + \mu_C)]^2$. The form of the contrast expected mean square tells you to use F=MS (A vs B,C)/MS(METHOD*VARIETY) to test the null hypothesis H_0: $\mu_A - 0.5(\mu_B + \mu_C)$=0. You can do this with the E= option in the CONTRAST statement:

```
contrast 'A vs B,C' method 2 -1 -1 / e=method*variety;
```

Results of the TEST and CONTRAST statements appear in Output 4.20.

Output 4.20
Tests Using
the Proper
Denominator
in the
F-Statistic

```
Dependent Variable: yield

Tests of Hypotheses Using the Type III MS for method*variety as an Error Term

Source                  DF     Type III SS     Mean Square    F Value    Pr > F

method                   2     953.1562222     476.5781111     10.18     0.0063


Tests of Hypotheses Using the Type III MS for method*variety as an Error Term

Contrast                DF     Contrast SS     Mean Square    F Value    Pr > F

A vs B,C                 1     940.7347222     940.7347222     20.10     0.0020
```

The denominator MS(METHOD*VARIETY) of the *F*-values in Output 4.20 is larger than the denominator MS(ERROR) of the *F*-values in Output 4.18. The correct *F*-tests (Output 4.20) are therefore less significant than they appeared to be using Output 4.18 inappropriately. Using MS(METHOD*VARIETY) in the denominator makes inference from the tests valid for all varieties in the population, whereas use of MS(ERROR) in the denominator restricts inference to only the five varieties actually used in the experiment. MS(METHOD*VARIETY) is the experimental error for comparing methods across all varieties in the population, whereas MS(ERROR) is the experimental error for comparing methods across only those varieties used in the experiment.

4.4.2 Standard Errors for the Two-Way Mixed Model: GLM versus MIXED

While you can obtain valid *F*-tests for mixed models with PROC GLM, you *cannot* always obtain valid standard errors. The two-way mixed model is an example of when GLM cannot compute the correct standard errors. The statement

```
estimate 'trt diff' method 1 -1 0;
```

computes the estimate and standard error of $\mu_A - \mu_B$. The result appears in Output 4.21.

Output 4.21
Estimate and
Standard Error
for Method A-B
Difference
Computed by
PROC GLM

Parameter	Estimate	Error	t Value	Pr > |t|
A-B diff	7.31333333	1.14455879	6.39	<.0001

GLM computes the standard error of the difference as $\sqrt{\dfrac{2\hat{\sigma}^2}{5*6}} = \sqrt{\dfrac{2*MSE}{30}} = \sqrt{\dfrac{2*19.65}{30}} = 1.14$.

This would be correct if MS(ERROR) was the appropriate measure of experimental error, but we know that the correct measure is MS(METHOD*VARIETY). Thus, the standard error should be

$\sqrt{\dfrac{MS(METHOD*VARIETY)}{30}} = \sqrt{\dfrac{2(\hat{\sigma}^2 + 6\hat{\sigma}^2_{MV})}{30}}$. PROC GLM does not have an E= option for

the ESTIMATE statement, so the correct standard error is unavailable. You must use PROC MIXED to obtain the correct standard error.

Use the following statements to run the analysis with PROC MIXED:

```
proc mixed data=fctorial;
 class method variety;
  model yield = method;
  random variety method*variety;
  contrast 'A vs B,C' method 2 -1 -1;
```

Output 4.22 shows the results.

Output 4.22
Analysis of the
Two-Way Mixed
Model Using
PROC MIXED

```
                        Covariance Parameter
                             Estimates

                 Cov Parm              Estimate

                 variety                      0
                 method*variety          2.0842
                 Residual               19.6502

                 Type 3 Tests of Fixed Effects

                        Num       Den
          Effect        DF        DF      F Value    Pr > F

          method         2         8       14.82     0.0020

                             Contrasts

                        Num       Den
          Label         DF        DF      F Value    Pr > F

          A vs B,C       1         8       29.26     0.0006
```

Compare these results to Outputs 4.18 through 4.21 using PROC GLM. PROC MIXED provides variance component estimates rather than ANOVA sums of squares and mean squares. From the mean squares in Output 4.18 and the expected mean squares in Output 4.19, you can see that MS(VARIETY)=2.845 is less than MS(METHOD*VARIETY), resulting in a negative estimate of σ_V^2. The default for PROC MIXED is to set negative variance components to 0. Thus, in the output, the "Covariance Parameter Estimates" for VARIETY is 0.

Setting $\sigma_V^2 = 0$ affects the REML estimate of σ_{MV}^2. The ANOVA estimate obtained using the mean squares and their expected values from Outputs 4.18 and 4.19 is 4.527, whereas the REML estimate in Output 4.22 is 2.084. This in turn affects *F*-statistics for METHOD and the A vs B,C contrast. You can see that truncating variance component estimates at zero has a ripple effect on other statistics. Because truncation at zero results in upward bias in variance component estimates, it also causes bias in affected statistics.

Statisticians, quantitative geneticists, and others who use variance component estimates in their work have struggled with the question of how negative variance components should be reported. Some think that negative estimates should be reported as such, whereas others argue that because variance by definition cannot be negative, negative estimates have no meaning and they should be set to zero. Using variance components to construct test statistics adds another dimension to the problem. While you can debate the merits of reporting negative variance components, biased *F*-statistics and standard errors are clearly to be avoided.

PROC MIXED has a number of options that allow you to get the same variance component estimates and hence the same *F*-values that PROC GLM computes. The NOBOUND option uses REML computing algorithms, except that it allows negative variance component estimates to remain negative. The METHOD=TYPE*n* option (*n* =1, 2, or 3) computes ANOVA estimates based on the expected mean squares that result from TYPE I, II, or III sum of squares. For the balanced data sets presented in this chapter, all these options produce the same results. Chapters 5 and 6 discuss the differences between the various types of SS for unbalanced data. As an example, the following SAS statements use the Type III variance component estimates:

```
proc mixed method=type3 data=fctorial;
  class method variety;
   model yield = method;
   random variety method*variety;
   contrast 'A vs B,C' method 2 -1 -1;
```

Output 4.23 shows the results.

```
Type 3 Analysis of Variance

Source              Expected Mean Square                    Error Term

method              Var(Residual) + 6 Var(method*variety)   MS(method*variety)
                    + Q(method)
variety             Var(Residual) + 6 Var(method*variety)   MS(method*variety)
                    + 18 Var(variety)
method*variety      Var(Residual) + 6 Var(method*variety)   MS(Residual)
Residual            Var(Residual)                           .


                         Covariance Parameter
                               Estimates

                      Cov Parm          Estimate

                      variety            -2.4426
                      method*variety      4.5268
                      Residual           19.6502


                    Type 3 Tests of Fixed Effects

                         Num      Den
              Effect     DF       DF      F Value    Pr > F

              method      2        8       10.18     0.0063


                              Estimates

                         Standard
     Label      Estimate     Error      DF    t Value    Pr > |t|

     A-B diff     7.3133     1.7666      8      4.14      0.0033

                              Contrasts

                         Num     Den
          Label          DF      DF     F Value     Pr > F

          A vs B,C        1       8      20.10      0.0020
```

The output gives you the same expected mean squares obtained by the RANDOM statement in PROC GLM. When you use the METHOD=TYPE*n* option, the output also contains the ANOVA table: sum of squares, mean squares, and *F*-statistics. You can see that the "Covariance Parameter Estimates" correspond to the ANOVA estimates. Also, the *F*-tests for the METHOD main effect and the A vs B,C contrast are now the same as those obtained by using PROC GLM with the properly specified E= option. As mentioned above, for balanced data you can obtain the same variance component estimates and *F*-statistics with the NOBOUND option:

```
proc mixed nobound data=fctorial;
```

You can obtain the estimated treatment means and differences with their correct standard errors. In conjunction with the above PROC MIXED METHOD=TYPE*n* or NOBOUND option, use the following statement:

```
lsmeans method/diff;
```

Output 4.24 shows the results.

```
                            Least Squares Means

                                  Standard
         Effect    method    Estimate      Error    DF   t Value   Pr > |t|

         method    A          23.0100     1.0353     8     22.23    <.0001
         method    B          15.6967     1.0353     8     15.16    <.0001
         method    C          16.6067     1.0353     8     16.04    <.0001

                       Differences of Least Squares Means

                                       Standard
       Effect    method   _method   Estimate     Error    DF   t Value   Pr > |t|

       method    A        B          7.3133     1.7666     8      4.14     0.0033
       method    A        C          6.4033     1.7666     8      3.62     0.0067
       method    B        C         -0.9100     1.7666     8     -0.52     0.6204
```

Consistent with the formula for the standard error of a difference, $\sqrt{\dfrac{2(\hat{\sigma}^2 + 6\hat{\sigma}^2_{MV})}{30}}$ given above,

the value in Output 4.23 is $\sqrt{\dfrac{2*(19.650 + 6 \times 4.527)}{30}} = 1.767$. The *t*-values and their associated

p-values are thus valid LSD tests for METHOD main effect differences.

4.4.3 More on Expected Mean Squares: Determining Quadratic Forms and Null Hypotheses for Fixed Effects

In some situations, you may want to obtain detailed information about the hypotheses tested by various *F*-statistics. In balanced cases, this is rarely an issue, because the hypotheses are obvious from the structure of the ANOVA table or the contrast coefficients. In unbalanced data sets, however, the hypotheses may depend on the type of SS you use and the order of the terms in the model. These are considered in Chapters 5 and 6. The purpose of this section is to show how to use optional PROC GLM output to use fixed-effects quadratic forms to determine what hypothesis a given *F*-statistic tests. Using balanced data makes it easier to follow the computations. This section is mainly of interest to graduate students who are learning to use quadratic forms and to readers who are familiar with matrix algebra and want to get a deeper insight into the distinctions among the types of SS.

When you obtain expected mean squares, the basic output gives you a "Q" term for the fixed effects. This "Q" stands for quadratic form. For example, Q(METHOD) is the quadratic form for the fixed-effect METHOD in Outputs 4.19 and 4.23. You can obtain detailed information about the meaning of Q(METHOD) by using the Q option at the end of the PROC GLM RANDOM statement. Output 4.25 shows the results of using the Q option.

```
random variety method*variety / q;
```

Output 4.25
Quadratic
Forms for
Method Effects
from the Q
Option

```
        Quadratic Forms of Fixed Effects in the Expected Mean Squares

             Source: Type III Mean Square for method

                        method A            method B            method C

method A              20.00000000        -10.00000000        -10.00000000
method B             -10.00000000         20.00000000        -10.00000000
method C             -10.00000000        -10.00000000         20.00000000


             Source: Contrast Mean Square for A vs B,C

                        method A            method B            method C

method A              20.00000000        -10.00000000        -10.00000000
method B             -10.00000000          5.00000000          5.00000000
method C             -10.00000000          5.00000000          5.00000000
```

The Q option prints a matrix of the quadratic form for fixed effects, which tells you

$$Q(\text{METHOD}) = \frac{\alpha'A\alpha}{DF}$$

where $\alpha' = (\alpha_A, \alpha_B, \alpha_C)$ is a row vector containing the fixed-effects parameters in the model. A is the matrix of the quadratic form, and DF is the number of degrees of freedom for the effect. You see two matrices printed in Output 4.25, one for the METHOD effect in the analysis-of-variance table and one for the contrast A vs B,C. For the overall METHOD effect, the matrix is

$$A = \begin{bmatrix} 20 & -10 & -10 \\ -10 & 20 & -10 \\ -10 & -10 & 20 \end{bmatrix}$$

This tells you the matrix algebraic expression for the quadratic form is

$$Q(\text{METHOD}) = (\alpha_A, \alpha_B, \alpha_C) \begin{bmatrix} 20 & -10 & -10 \\ -10 & 20 & -10 \\ -10 & -10 & 20 \end{bmatrix} \begin{bmatrix} \alpha_A \\ \alpha_B \\ \alpha_C \end{bmatrix} \left(\frac{1}{2}\right)$$

Some algebraic manipulation yields $Q(\text{METHOD}) = 30\sum_i (\alpha_i - \overline{\alpha}.)^2 / 2$, as indicated earlier. You can go through the same process to get the Q(METHOD) expression for the A vs B,C contrast. Using the Q matrix for the contrast A vs B,C, some straightforward but tedious matrix algebra reveals, as indicated above, that $Q(\text{A vs B,C}) = 30[\mu_A - 0.5(\mu_B + \mu_C)]^2$.

You can use the quadratic form expressions to indicate the null hypothesis tested by an *F*-statistic with the corresponding mean square as its numerator. To do this, determine the values of the fixed parameters that make Q(METHOD)=0. For the overall METHOD effect, you have already seen that Q(METHOD)=0 whenever $\alpha_A = \alpha_B = \alpha_C$, or equivalently, whenever $\mu_A = \mu_B = \mu_C$.

4.5 A Classification with Both Crossed and Nested Effects

Some classifications involve both crossed and nested factors. The example in this section is typical of a study design that is common to many fields of investigation. As you will see by the end of this section, in practical situations, PROC MIXED is better suited to analyze these kinds of data sets. Nonetheless, in order to adequately present the linear model issues associated with these designs, the first part of this section uses PROC GLM.

Output 4.26 presents a data set that illustrates the essential features of designs with crossed and nested factors. An engineer in a semiconductor plant investigated the effect of several modes of a process condition (ET) on the resistance in computer chips. Twelve silicon wafers (WAFER) were drawn from a lot, and three wafers were randomly assigned to each of four modes of ET. Resistance in chips was measured in four positions (POS) on each wafer after processing. The measurement was recorded as the variable RESISTA in a SAS data set named CHIPS.

Output 4.26
Semiconductor
Resistance Data

Obs	resista	et	wafer	pos
1	5.22	1	1	1
2	5.61	1	1	2
3	6.11	1	1	3
4	6.33	1	1	4
5	6.13	1	2	1
6	6.14	1	2	2
7	5.60	1	2	3
8	5.91	1	2	4
9	5.49	1	3	1
10	4.60	1	3	2
11	4.95	1	3	3
12	5.42	1	3	4
13	5.78	2	1	1
14	6.52	2	1	2
15	5.90	2	1	3
16	5.67	2	1	4
17	5.77	2	2	1
18	6.23	2	2	2
19	5.57	2	2	3
20	5.96	2	2	4

Output 4.26
(Continued)
Semiconductor
Resistance
Data

21	6.43	2	3	1
22	5.81	2	3	2
23	5.83	2	3	3
24	6.12	2	3	4
25	5.66	3	1	1
26	6.25	3	1	2
27	5.46	3	1	3
28	5.08	3	1	4
29	6.53	3	2	1
30	6.50	3	2	2
31	6.23	3	2	3
32	6.84	3	2	4
33	6.22	3	3	1
34	6.29	3	3	2
35	5.63	3	3	3
36	6.36	3	3	4
37	6.75	4	1	1
38	6.97	4	1	2
39	6.02	4	1	3
40	6.88	4	1	4
41	6.22	4	2	1
42	6.54	4	2	2
43	6.12	4	2	3
44	6.61	4	2	4
45	6.05	4	3	1
46	6.15	4	3	2
47	5.55	4	3	3
48	6.13	4	3	4

Here are some features of this experiment:

❑ There are two experimental factors, ET and POS, which appear together in all combinations. These factors are crossed because the POS labels 1, 2, 3, and 4 have the same meaning for all levels of ET; POS 1 refers to the same location on a wafer in ET 1 as it does on a wafer in ET 2. The engineer wants to compare mean resistance between levels of ET and between levels of POS. He also wants to determine if differences between levels of ET depend on the value of POS at which they are measured. The data are analyzed in terms of either simple effects or main effects of ET and POS, depending on the presence or absence of interaction between ET and POS. Section 4.5.2 discusses expected mean squares and how to use them to set up appropriate tests for several types of effects.

❑ ET levels are assigned to wafers in a completely randomized design, making WAFER the experimental unit for comparing levels of ET. Wafers are nested within levels of ET.

❑ Levels of POS change between chips on the same wafer, whereas levels of ET change between wafers. So a different measure of experimental error is required to compare positions than is used to compare levels of ET. This is the primary feature that distinguishes this experiment from a standard factorial.

This data set has features of both crossed and nested classification, so it is referred to as crossed-nested. It is similar to a split-plot experiment, with wafer taken as the main-plot unit and chips on a wafer as the sub-plot unit. It also has features in common with repeated-measures experiments, which are discussed in Chapter 9.

A model for the data is

$$y_{ijk} = \mu + \alpha_i + w_{ij} + \beta_k + (\alpha\beta)_{ik} + e_{ijk}$$

where y_{ijk} is the measured resistance at the kth position on the jth wafer in the ith level of ET, $i = 1, 2, 3, 4$; $j = 1, 2, 3$; $k = 1, 2, 3, 4$; $\mu + \alpha_i + \beta_k + (\alpha\beta)_{ik} = \mu_{ik}$ is the mean resistance at position k with ET level i. w_{ij} are random variables representing differences among wafers assigned to the same level of ET. The w_{ij} are assumed i.i.d. $N(0, \sigma_W^2)$. Some writers prefer to use the notation $w(a)_{ij}$ to emphasize the fact that wafers are nested within levels of ET.

e_{ijk} are random variables representing differences among chips on the same wafer. The e_{ijk} are assumed i.i.d. $N(0,\sigma^2)$ and independent of the w_{ij}. The former is a potentially flawed assumption because correlation between chips could depend on their relative proximity on the wafer. If so, then the data should be analyzed using methods appropriate for repeated-measures experiments (see Chapter 9).

4.5.1 Analysis of Variance for Crossed-Nested Classification

An analysis-of-variance table has the form

Source of Variation	DF
ET	3
WAFER(ET)	8
POS	3
ET*POS	9
ERROR = POS*WAFER(ET)	24

4.5.2 Using Expected Mean Squares to Set Up Several Tests of Hypotheses for Crossed-Nested Classification

This section illustrates how to set up several types of tests of hypothesis for an experiment of this type. These include tests of the following null hypotheses:

❑ overall main effect of ET H_0: $\mu_{1.} = \mu_{2.} = \mu_{3.} = \mu_{4.}$

❑ overall main effect of POS H_0: $\mu_{.1} = \mu_{.2} = \mu_{.3} = \mu_{.4}$

❑ main effect contrast of ET (ET1 vs ET2) H_0: $\mu_{1.} = \mu_{2.}$

❑ main effect contrast of POS (POS1 vs POS2) H_0: $\mu_{.1} = \mu_{.2}$

❑ simple effect contrast of POS (POS1 vs POS2 in ET1) H_0: $\mu_{11} = \mu_{12}$

❑ simple effect contrast of ET (ET1 vs ET2 in POS1) H_0: $\mu_{11} = \mu_{21}$

You can use the CONTRAST and RANDOM statements in PROC GLM or PROC MIXED to obtain appropriate tests for these effects. Both allow a high degree of flexibility. GLM is better suited for illustrating key features of linear model theory and methods. Primarily because of standard error considerations discussed in Section 4.5.3, MIXED is better suited to the actual analysis of data. For now, we focus on linear model issues via GLM.

The following SAS statements produce the analysis-of-variance table and contrasts:

```
proc glm data=chips;
  class et wafer pos;
  model resista = et wafer(et) pos et*pos / ss3;
  contrast 'ET1 vs ET2'          et 1 -1 0 0;
  contrast 'POS1 vs POS2'        pos 1 -1 0 0;
  contrast 'POS1 vs POS2 in ET1' pos 1 -1 0 0 et*pos 1 -1;
  contrast 'ET1 vs ET2 in POS1'  et 1 -1 0 0 et*pos 1 0 0 0 -1;
```

Note that the simple effect contrasts POS1 vs POS2 in ET1 and ET1 vs ET2 in POS1 use coefficients of ET*POS in addition to their respective main effects. Table 4.2 and the explanation immediately following detail why these terms are necessary. The analysis-of-variance table and CONTRAST statement results appear in Output 4.27.

Output 4.27
Analysis of Variance for Semiconductor Resistance Data

```
                            The GLM Procedure

Dependent Variable: resista

                              Sum of
Source              DF       Squares     Mean Square   F Value   Pr > F

Model               23     9.32500833    0.40543514      3.65    0.0013

Error               24     2.66758333    0.11114931

Corrected Total     47    11.99259167


           R-Square     Coeff Var     Root MSE     resista Mean

           0.777564     5.553811      0.333391       6.002917


Source              DF    Type III SS   Mean Square   F Value   Pr > F

et                   3     3.11215833    1.03738611      9.33    0.0003
wafer(et)            8     4.27448333    0.53431042      4.81    0.0013
pos                  3     1.12889167    0.37629722      3.39    0.0345
et*pos               9     0.80947500    0.08994167      0.81    0.6125


Contrast            DF    Contrast SS   Mean Square   F Value   Pr > F

ET1 vs ET2           1     0.69360000    0.69360000      6.24    0.0197
POS1 vs POS2         1     0.07706667    0.07706667      0.69    0.4132
POS1 vs POS2 in ET1  1     0.04001667    0.04001667      0.36    0.5541
ET1 vs ET2 in POS1   1     0.21660000    0.21660000      1.95    0.1755
```

The output contains *F*-statistics for all effects in the MODEL statement as well as for all of the effects in the CONTRAST statements. **Note:** These *F*-statistics use the default MS(ERROR) in the denominator. Remember that in PROC GLM, the RANDOM statement does not override the default use of MS(ERROR). You must examine the expected mean squares to determine which of these automatically computed *F*-statistics are valid. For all other tests, you must specify an appropriate error term. As with previous examples in this chapter, you can specify appropriate error terms with a TEST statement or by using the TEST option with the RANDOM statement. For CONTRAST statements, use the E= option.

Now, obtain tests for the fixed effects of ET, POS, and ET*POS in the analysis-of-variance table and tests for the effects specified in the CONTRAST statements. Start by obtaining the expected mean squares for all effects. The following statement gives the results in Output 4.28. Recall that this statement must be placed after the CONTRAST statements in the PROC GLM program given above:

```
random wafer(et);
```

Output 4.28
Expected Mean
Squares for
Semiconductor
Data

```
Source                   Type III Expected Mean Square

et                       Var(Error) + 4 Var(wafer(et)) + Q(et,et*pos)

wafer(et)                Var(Error) + 4 Var(wafer(et))

pos                      Var(Error) + Q(pos,et*pos)

et*pos                   Var(Error) + Q(et*pos)


Contrast                 Contrast Expected Mean Square

ET1 vs ET2               Var(Error) + 4 Var(wafer(et)) + Q(et,et*pos)

POS1 vs POS2             Var(Error) + Q(pos,et*pos)

POS1 vs POS2 in ET1      Var(Error) + Q(pos,et*pos)

ET1 vs ET2 in POS1       Var(Error) + Var(wafer(et)) + Q(et,et*pos)
```

You could use the Q option at the end of the RANDOM statement to get an interpretation of Q(effect) in the expected mean squares. Table 4.2 relates the Q(effect) with the corresponding algebraic expressions in terms of model parameters and means model.

Table 4.2 *Q(effect) in Expected Mean Squares*

Effect Name	Expression in Output	Algebraic Expression for Q(effect)
ET	Q(ET,ET*POS)	$12\sum_{i} [\alpha_i + \overline{(\alpha\beta)}_{i\cdot} - \overline{\alpha}_{\cdot\cdot} - \overline{(\alpha\beta)}_{\cdot\cdot}]^2 = 4\sum_{i} [\mu_{i\cdot} - \overline{\mu}_{\cdot\cdot}]^2$
POS	Q(POS,ET*POS)	$12\sum_{j} [\beta j + \overline{(\alpha\beta)}_{\cdot j} - \overline{\beta}_{\cdot} - \overline{(\alpha\beta)}_{\cdot\cdot}]^2 = 4[\mu_{\cdot j} - \overline{\mu}_{\cdot\cdot}]^2$
ET*POS	Q(ET*POS)	$12\sum_{ij} [(\alpha\beta)_{ij} - \overline{(\alpha\beta)}_{i\cdot} - \overline{(\alpha\beta)}_{\cdot j} + \overline{(\alpha\beta)}_{\cdot\cdot}]^2$ $= 12\sum_{ij} [\mu_{ij} - \mu_{i\cdot} - \mu_{\cdot j} + \mu_{\cdot\cdot}]^2$
ET1 vs ET2	Q(ET,ET*POS)	$12[\alpha_1 + \overline{(\alpha\beta)}_{1\cdot} - \alpha_2 + \overline{(\alpha\beta)}_{2\cdot}]^2 = 4[\overline{\mu}_{1\cdot} - \overline{\mu}_{2\cdot}]^2$
POS1 vs POS2	Q(POS,ET*POS)	$12[\beta_1 + \overline{(\alpha\beta)}_{\cdot 1} - \beta_2 + \overline{(\alpha\beta)}_{\cdot 2}]^2 = 4[\overline{\mu}_{\cdot 1} - \overline{\mu}_{\cdot 2}]^2$
POS1 vs POS2 IN ET1	Q(POS,ET*POS)	$12[\beta_1 + \overline{(\alpha\beta)}_{11} - \beta_2 + \overline{(\alpha\beta)}_{12}]^2 = 4[\overline{\mu}_{11} - \overline{\mu}_{12}]^2$
ET1 vs ET2 IN POS1	Q(ET,ET*POS)	$12[\alpha_1 + \overline{(\alpha\beta)}_{11} - \alpha_2 + \overline{(\alpha\beta)}_{21}]^2 = 4[\overline{\mu}_{11} - \overline{\mu}_{21}]^2$

It may seem strange that the Q(effect) for ET main effects contains the expression ET*POS within the parentheses. The ET*POS expression is present because the quadratic form is a function of the $(\alpha\beta)$ parameters as well as the α parameters. This is because PROC GLM imposes no assumptions on the model parameters, so that $\overline{\mu}_{i.} = \mu + \alpha_i + \overline{\beta}_{.} + \overline{(\alpha\beta)}_{i.}$ and $\overline{\mu}_{..} = \mu + \overline{\alpha}_{.} + \overline{\beta}_{.} + \overline{(\alpha\beta)}_{..}$

Consequently, $\overline{\mu}_{i.} - \overline{\mu}_{..} = \alpha_i - \overline{\alpha}_{.} + \overline{(\alpha\beta)}_{i.} - \overline{(\alpha\beta)}_{..}$; that is, differences between means for two levels of ET are functions of the $(\alpha\beta)_{ik}$ parameters as well as the α_i parameters. The same type of phenomenon holds true for the main effect of POS and the effects in the CONTRAST statements. Note that the quadratic forms of the simple effect contrasts, POS1 vs POS2 in ET1 and ET1 vs ET2 in POS1, correspond to the coefficients you must use in their respective CONTRAST statements.

In each case, you see from the algebraic expression for Q(effect) that the null hypothesis you want to test is H_0: Q(effect)=0. The expected mean squares tell you to use the denominators in the F-statistics as indicated in this table:

Table 4.3 *Required Denominator MS for Tests of Effects*

Effect	Appropriate Denominator for *F*-Statistic
ET	MS(WAFER(ET))
POS	MS(ERROR)
ET*POS	MS(ERROR)
ET1 vs ET2	MS(WAFER(ET))
POS1 vs POS2	MS(ERROR)
POS1 vs POS2 in ET1	MS(ERR0R)
ET1 vs ET2 in POS1	Not directly available

For any tests that use MS(ERROR) the *F*-statistics PROC GLM automatically computes are correct. All others require additional attention. Note that you cannot compute the appropriate test for the ET1 vs ET2 in POS1 contrast using PROC GLM. In fact, you cannot get the correct standard error for the estimate of ET differences for a given POS with PROC GLM either. The method for getting the right test and standard error are explained later in this section and in Section 4.5.3. The correct statistics are easily obtained using PROC MIXED. Section 4.5.4 shows you how to do this. This is the main reason why PROC MIXED is preferred in actual data analysis.

Appropriate *F*-tests for ET main effect and ET1 vs ET2 contrast are obtained with the statements

```
test h=et  e=wafer(et);
contrast 'ET1 vs ET2'  et 1 -1 0 0 / e=wafer(et);
```

Results appear in Output 4.29.

Output 4.29
F-Tests for ET
Effects

```
    Tests of Hypotheses Using the Type III MS for wafer(et) as an Error Term

Source                      DF      Type III SS    Mean Square   F Value   Pr > F

et                           3      3.11215833     1.03738611      1.94    0.2015


    Tests of Hypotheses Using the Type III MS for wafer(et) as an Error Term

Contrast                    DF      Contrast SS    Mean Square   F Value   Pr > F

ET1 vs ET2                   1      0.69360000     0.69360000      1.30    0.2875
```

Compare *F*-statistics in Output 4.29 with those in Output 4.27 for these effects.

Alternatively, you can obtain test statistics for effects in the analysis-of-variance table with the TEST option in the RANDOM statement, as follows:

```
random wafer(et)/test;
```

Output 4.30 shows the results.

Output 4.30
F-Tests from
the TEST
Option

```
          Tests of Hypotheses for Mixed Model Analysis of Variance

Dependent Variable: resista

      Source                   DF    Type III SS   Mean Square   F Value   Pr > F

  *   et                        3     3.112158      1.037386       1.94    0.2015

      Error: MS(wafer(et))      8     4.274483      0.534310
  * This test assumes one or more other fixed effects are zero.


      Source                   DF    Type III SS   Mean Square   F Value   Pr > F

      wafer(et)                 8     4.274483      0.534310       4.81    0.0013
  *   pos                       3     1.128892      0.376297       3.39    0.0345
      et*pos                    9     0.809475      0.089942       0.81    0.6125

      Error: MS(Error)         24     2.667583      0.111149
  * This test assumes one or more other fixed effects are zero.
```

Compare results in Output 4.30 with those in Outputs 4.27 and 4.29. Unfortunately, the TEST option does not compute appropriate tests for effects in CONTRAST statements.

There is no appropriate *F*-test for the contrast ET1 vs ET2 in POS1 directly available. An appropriate denominator for this *F*-statistic would be an estimate of $\sigma^2 + \sigma_W^2$. There is no source of variation in the analysis-of-variance table whose expected mean square is equal to $\sigma^2 + \sigma_W^2$. But you can combine MS(ERROR) and MS(WAFER(ET)) to get an estimate of $\sigma^2 + \sigma_W^2$. You need amounts of both σ^2 and σ_W^2 in an appropriate denominator for the *F*-test; you need one unit of each. Of the expected mean squares for ERROR and WAFER(ET), only the expected mean square for WAFER(ET) contains any σ_W^2; in fact, it contains four units of it. You need to multiply MS(WAFER(ET)) by the appropriate constant to produce one unit of σ_W^2, which, of course, is 1/4. This gives $0.25\sigma^2 + \sigma_W^2$, so you need an additional $.75\sigma^2$, which you can get by adding .75MS(ERROR). So an appropriate denominator of the *F*-statistic for testing for the effect of ET1 vs ET2 in POS1 is

$$\hat{\sigma}^2 + \hat{\sigma}_W^2 \quad =.75\text{MS(ERROR)} + .25\text{MS(WAFER(ET))}$$
$$=.75(0.111).25(0.534)$$
$$= 0.217$$

An appropriate *F*-statistic for testing $H_0: \mu_{11} - \mu_{21} = 0$ is then

$$F \quad =\text{MS(ET1 vs ET2 in POS1)}/(\hat{\sigma}^2 + \hat{\sigma}_W^2)$$
$$=0.2166/0.217$$

Here, *F* is essentially equal to 1.0, indicating that it is nonsignificant. You do not need to refer it to an *F*-distribution to calculate a *p*-value to determine its significance, because *F* must be substantially larger than 1.0 in order to be significant at any meaningful level. Thus, you do not need to know degrees of freedom for the denominator mean square. Normally, however, you would need degrees of freedom in order to assess the level of significance. They can be approximated using Satterthwaite's formula, which is demonstrated in Section 4.5.3, "Satterthwaite's Formula for Approximate Degrees of Freedom."

But first, Table 4.4 gives a summary of all the appropriate *F*-tests from the analysis-of-variance table and CONTRAST statements:

Table 4.4 *Summary of F-Test Results from CONTRAST Statements*

Effect	Appropriate *F*-Statistic	Level of Significance
ET	1.037/0.534 = 1.94	0.202
POS	0.376/0.111 = 3.39	0.034
ET*POS	0.090/0.111 = 0.81	0.612
ET1 vs ET2	0.694/0.534 = 1.30	0.287
POS1 vs POS2	0.077/0.111 = 0.69	0.413
POS1 vs POS2 in ET1	0.040/0.111 = 0.36	0.554
ET1 vs ET2 in POS1	0.217/0.217 = 1.00	NS

4.5.3 Satterthwaite's Formula for Approximate Degrees of Freedom

The denominator of the *F*-statistic for the contrast ET1 vs ET2 in POS1 is a linear combination of mean squares from the analysis-of-variance table. In general, such a linear combination has properties that approximate those of actual mean squares. The number of degrees of freedom for the linear combination of mean squares can be approximated by using a formula attributed to Satterthwaite (1946). This is the subject of the following discussion.

Let MS1, . . . ,MSk be a set of independent mean squares with respective degrees of freedom DF_1, . . . ,DF_k, and let a_1, . . . ,a_k be a set of known constants. Then the linear combination

$$MS = a_1\,MS_1 + a_2\,MS_2 + \ldots + a_k\,MS_k$$

is a synthetic mean square with approximate degrees of freedom equal to

$$DF = \frac{(MS)^2}{\dfrac{(MS_1)^2}{DF_1} + \ldots + \dfrac{(MS_2)^2}{DF_2}}$$

Applying Satterthwaite's formula, you get

$$\hat{\sigma}^2 + \hat{\sigma}_W^2 = 0.75\,MS(ERROR) + .25\,MS[WAFER(ET)]$$

The data used here yield

$$
\begin{array}{lll}
MS_1 = MS(ERROR) = 0.111 & DF_1 = 24 & a_1 = .75 \\
MS_2 = MS(WAFER(ET)) = 0.534 & DF_2 = 8 & a_2 = .25
\end{array}
$$

so

$$DF = \frac{(0.217)^2}{\dfrac{(0.083)^2}{24} + \dfrac{(0.134)^2}{8}}$$

$$= \frac{0.0471}{0.000287 + 0.002245}$$

$$= 18.6$$

SAS has an internal function that can evaluate DF with fractional values. For use with published tables, round down to DF=18. This synthetic mean square with DF=18 is useful for several applications. It is an appropriate denominator for any contrast among levels of ET in a given level of POS, such as the comparison ET1 vs ET2 in POS1 shown in Table 4.3. Other examples include the CONTRAST statement

```
contrast 'ET1,ET2 vs ET3,ET4 at POS2'
    et 1 1 -1 -1  et*pos 0 0 0 0  1 1 -1 -1;
```

In addition to providing appropriate denominators for *F*-statistics, it is equally important to use appropriate mean squares when you compute confidence intervals for means or differences between means. Suppose you wanted a confidence interval on the overall mean for ET1, averaged across WAFER and POS. This mean is

$$\bar{y}_{1..} = \mu + \alpha_1 + \bar{w}_{1.} + \beta_1 + \overline{(\alpha\beta)}_{1.} + \bar{e}_{1..}$$

The random parts are $\bar{w}_{1.}$ and $\bar{e}_{1..}$. Now $\bar{w}_{1.}$ is the mean of three w_{1j}'s (w_{11}, w_{12}, and w_{13}), so $V(\bar{w}_{1.}) = \sigma_W^2/B$.

Additionally, $e_{1..}$ is the mean of twelve e_{ijk}'s, so $V(\bar{e}_{1..}) = \sigma^2/12$. Therefore, the variance of $\bar{y}_{1..}$ is

$$
\begin{aligned}
V(\bar{y}_{1..}) &= V(\bar{w}_{1.}) + V(\bar{e}_{1..}) \\
&= \sigma_W^2/3 + \sigma^2/12 \\
&= (\sigma^2 + 4\sigma_W^2)/12
\end{aligned}
$$

From the table of expected mean squares you have seen that an estimate of $V(\bar{y}_{1..})$ is MS(WAFER(ET))/12. So a 95% confidence interval for the mean is

$$
\bar{y}_{1..} \pm t_{8,0.025} \times \sqrt{\frac{\mathrm{MS[WAFER(ET)]}}{12}}
$$

$$
= 5.632 \pm 2.30(0.210)
$$

$$
= (5.14, 6.12)
$$

Similarly, the mean for POS1 averaged across POS and WAFER is

$$
\bar{y}_{..1} = \mu + \bar{\alpha}_. + \bar{w}_{..} + \beta_1 + \overline{(\alpha\beta)} + \bar{e}_{..1}
$$

The variance of $\bar{y}_{..1}$ is

$$
\begin{aligned}
V(\bar{y}_{..1}) &= V(\bar{w}_{..}) + V(\bar{e}_{..1}) \\
&= \sigma_W^2/12 + \sigma^2/12 \\
&= (\sigma_W^2 + \sigma^2)/12
\end{aligned}
$$

If you look back at the synthetic mean square we computed, you see that the estimate of $V(\bar{y}_{..1})$ is $\{.75\mathrm{MS(ERROR)} + .25\ \mathrm{MS[WAFER(ET)]}\}/12$. So an approximate 95% confidence interval for $\bar{y}_{..1}$ is

$$
\bar{y}_{..1} \pm t_{18,0.025} \times \sqrt{\frac{.75\mathrm{MS(ERROR)} + .25\mathrm{MS[WAFER(ET)]}}{12}}
$$

$$
= 6.02 \pm 2.10(0.134)
$$

$$
= (5.74, 6.30)
$$

4.5.4 PROC MIXED Analysis of Crossed-Nested Classification

The previous sections used PROC GLM to develop the concepts underlying the analysis of models with multiple error terms. As you have seen, several default statistics PROC GLM computes are inappropriate and must be overridden. Other statistics are simply not available. These limitations occur because the underlying programming for PROC GLM is based entirely on fixed-effects-only linear model theory. Its mixed-model features are limited to options that allow you to determine expected mean squares and to certain tests using other than MS(ERROR). On the other hand, PROC MIXED uses more general linear mixed-model theory, meaning that when an effect is defined as random, it is handled as such throughout the entire computational process. Chapter 6 discusses the theory underlying the GLM and MIXED procedures.

Provided you correctly specify the model, MIXED automatically computes the correct test statistics and standard errors. Therefore, when you actually do data analysis, you should use PROC MIXED. This section shows the program statements needed to reproduce the essentials of the analysis developed in Sections 4.5.1 through 4.5.3. Note that MIXED does not, by default, compute an analysis-of-variance table nor the expected mean squares (although you can obtain both using the METHOD=TYPES option). Its focus is on the statistics relevant to analyzing the data.

The needed SAS statements for a PROC MIXED analysis are

```
proc mixed data=chips;
  class et wafer pos;
  model resista = et|pos/ddfm=satterth;
  random wafer(et);
  contrast 'ET1 vs ET2'          et 1 -1 0 0;
  contrast 'POS1 vs POS2'        pos 1 -1 0 0;
  contrast 'POS1 vs POS2 in ET1' pos 1 -1 0 0 et*pos 1 -1;
  contrast 'ET1 vs ET2 in POS1'  et 1 -1 0 0 et*pos 1 0 0 0 -1;
```

The results appear in Output 4.31. Several features of the MIXED program deserve attention. First, recall that you include *ONLY fixed effects* in the MODEL statement; the whole-plot error effect WAFER(ET) *must not* be included. All error terms are random effects and hence appear in the RANDOM statement. The vertical bar between ET and POS in the MODEL statement is SAS shorthand to obtain all main effects and interactions involving the terms connected by the bar. This syntax also works in PROC GLM, PROC GENMOD (see Chapter 10), and several other procedures in SAS that use MODEL statements (but not all of them—check the *SAS/STAT® User's Guide, Version 8*, Volumes 1, 2, and 3, to be sure). The DDFM=SATTERTH option in the MODEL statement causes Satterthwaite's approximation to be used to determine degrees of freedom. A more general degree-of-freedom procedure is the Kenward-Roger option, DDFM=KR. Because Satterthwaite's procedure is a special case of Kenward and Roger's procedure, you will get the same results for the designs discussed in this section. For designs with missing data, the Kenward-Roger option is recommended. The CONTRAST statements are identical for GLM and MIXED, except that the E= option is not needed in PROC MIXED.

Output 4.31
PROC MIXED
Analysis of
Semiconductor
Data

```
                    Covariance Parameter
                        Estimates


            Cov Parm        Estimate

            wafer(et)         0.1058
            Residual          0.1111


        Type 3 Tests of Fixed Effects

                   Num      Den
        Effect     DF       DF     F Value    Pr > F

        et          3        8       1.94     0.2015
        pos         3       24       3.39     0.0345
        et*pos      9       24       0.81     0.6125


                    Contrasts

                       Num     Den
        Label          DF      DF    F Value    Pr > F

        ET1 vs ET2       1       8      1.30    0.2875
        POS1 vs POS2     1      24      0.69    0.4132
        POS1 vs POS2 in ET1   1  24     0.36    0.5541
        ET1 vs ET2 in POS1    1  18.7   1.00    0.3305
```

Note that all the *F*-statistics and *p*-values are identical to the appropriate statistics for the corresponding effects computed in PROC GLM. In addition, PROC MIXED computes the correct *F*-statistic, denominator degrees of freedom, and *p*-value for the ET1 vs ET2 in POS1 contrast, which was unavailable with PROC GLM and would require considerable hand-calculation unless you use MIXED.

You can also obtain standard errors using either ESTIMATE statements or, for means and pairwise differences, the LSMEANS statement with the DIFF option. For example, the following statement computes the estimate and standard error of the ET1 vs ET2 in POS1 simple-effects difference:

```
estimate 'ET1 vs ET2 in POS1' et 1 -1 0 0 et*pos 1 0 0 0 -1;
```

The results appear in Output 4.32. Note that the coefficients for the ESTMATE statement are identical to those you use in the CONTRAST statement. Output 4.32 also shows the result you would get if you used PROC GLM. The GLM result is wrong, and there is no option GLM to correct the problem.

Output 4.32
Estimate of ET
Simple Effect
for POS 1

```
ESTIMATE results using PROC MIXED:

                            Estimates

                          Standard
  Label           Estimate   Error     DF    t Value    Pr > |t|

  ET1 vs ET2 in POS1  -0.3800   0.3803   18.7    -1.00     0.3305

ESTIMATE results using PROC GLM:

                               Standard
  Parameter          Estimate      Error    t Value    Pr > |t|

  ET1 vs ET2 in POS1  -0.38000000  0.27221230   -1.40     0.1755
```

You can see that the two standard errors are different. MIXED uses the correct formula

$$\sqrt{\frac{2(\hat{\sigma}_W^2 + \hat{\sigma}^2)}{3}} = \sqrt{\frac{2(0.1058 + 0.1111)}{3}} = 0.3803.$$ GLM uses MS(ERROR) $= \hat{\sigma}^2$ indiscriminately.

Thus, it computes $\sqrt{\frac{2\hat{\sigma}^2}{3}} = \sqrt{\frac{2(0.1111)}{3}} = 0.2722$. The larger $\hat{\sigma}_W^2$ is, the more seriously this discrepancy misrepresents treatment differences.

Using CONTRAST and ESTIMATE statements for treatment differences can be tedious, especially for simple-effects difference, such as ET1 vs ET2 in POS1 that required you to determine which ET *and* which ET*POS coefficients you need. PROC MIXED allows you to bypass the CONTRAST and ESTIMATE statements. Instead, use the following statement immediately after the RANDOM statement in the MIXED program:

```
lsmeans et|pos/diff;
```

The output contains all means and all possible main effect and simple-effect differences, so it can be quite lengthy. Output 4.33 shows only the results corresponding to the CONTRAST statements discussed above.

Output 4.33
Selected
LSMEAN and
Treatment
Difference
Estimates

```
                              Least Squares Means

                                      Standard
     Effect   et    pos    Estimate    Error     DF    t Value    Pr > |t|

     et       1            5.6258      0.2110      8     26.66      <.0001
     et       2            5.9658      0.2110      8     28.27      <.0001

     pos            1      6.0208      0.1345    18.7    44.78      <.0001
     pos            2      6.1342      0.1345    18.7    45.62      <.0001

     et*pos   1    1       5.6133      0.2689    18.7    20.87      <.0001
     et*pos   1    2       5.4500      0.2689    18.7    20.27      <.0001
     et*pos   2    1       5.9933      0.2689    18.7    22.29      <.0001


                        Differences of Least Squares Means

                                         Standard
     Effect  et  pos  _et  _pos  Estimate   Error    DF   t Value   Pr > |t|

     et      1        2          -0.3400   0.2984     8    -1.14    0.2875
     pos         1         2     -0.1133   0.1361    24    -0.83    0.4132
     et*pos  1   1    1    2       0.1633   0.2722    24     0.60    0.5541
     et*pos  1   1    2    1      -0.3800   0.3803  18.7    -1.00    0.3305
```

You can see that the standard errors of LS means also require Satterthwaite's approximation to get the appropriate degrees of freedom for tests of the null hypothesis that LSMEAN = 0. The output for the differences gives you the same information provided by the ESTIMATE statement. Note that you can get the *F*-statistic for the corresponding CONTRAST by squaring the *t*-statistic. For example, *F* for ET1 vs ET2 contrast is 1.30, which is equal to $(1.14)^2$, the *t*-value squared. For this reason, there is no real need to use the CONTRAST statement unless you want to test linear combinations other than pairwise differences.

Recent versions of both PROC GLM and PROC MIXED provide a SLICE option with the LSMEANS statement for means of factorial combinations. This option provides tests of one factor conditional on a single level of the other factor. For example, a SLICE on POS would test the equality of the ET means for a given level of POS, that is, $H_0: \mu_{1j} = \mu_{2j} = \mu_{3j} = \mu_{4j}$ for the *j*th level of POS. Use the following statement:

```
lsmeans et*pos/slice=pos;
```

You can use the DIFF option and multiple SLICE= statements in the same LSMEANS statement. For example, you could use the following statement

```
lsmeans et*pos/diff slice=(et pos);
```

to do a thorough analysis of the ET*POS effects. Although you can use this statement in both GLM and MIXED, be wary of the GLM results as they may use the wrong error term. Output 4.34 shows the results for the SLICE=POS option.

Output 4.34
Tests of Simple
Effects of ET
Given POS
Using the
SLICE Option
in PROC
MIXED and
PROC GLM

```
SLICE results using PROC MIXED:

                     Tests of Effect Slices

                          Num      Den
         Effect     pos    DF       DF     F Value    Pr > F

         et*pos      1      3      18.7      1.30      0.3038
         et*pos      2      3      18.7      3.19      0.0477
         et*pos      3      3      18.7      0.28      0.8383
         et*pos      4      3      18.7      1.26      0.3181

SLICE results using PROC GLM:

                et*pos Effect Sliced by pos for resista

                          Sum of
         pos      DF       Squares    Mean Square   F Value    Pr > F

          1        3      0.846292     0.282097      2.54      0.0805
          2        3      2.075092     0.691697      6.22      0.0028
          3        3      0.182958     0.060986      0.55      0.6539
          4        3      0.817292     0.272431      2.45      0.0880
```

The GLM *F*-statistics use MS(ERROR), which is incorrect. The appropriate error term, as derived in Section 4.5.2, is 0.75MS(ERROR) + 0.25MS[WAFER(ET)]. You can see that PROC MIXED does this correctly, and uses the Satterthwaite approximation defined by the DDFM=SATTERTH (or DDFM=KR) option in the MODEL statement. In this case, the difference in potential inference is stark: an unwitting GLM user would conclude ET effects, at least at, say, α=0.10, for all positions except POS=3. Using the correct statistics, plausible evidence exists only at POS=2, with a *p*-value of 0.0477, not 0.0028.

The main point of this section is that several crucial statistics are either not computed by PROC GLM or, worse, are computed but not correctly. Therefore, we repeat our recommendation that PROC MIXED be used to analyze all models with multiple error terms.

4.6 Split-Plot Experiments

The split-plot design results from a specialized randomization scheme for a factorial experiment. It is often used when one factor is more readily applied to large experimental units, or main plots, and when another factor can be applied to smaller units, or subplots, within the larger unit. A split-plot design is also useful when more information is needed for comparing the levels of one factor than for comparing the levels of the other factor. In this case, the factor for which more information is needed should be the subplot factor.

A classic example of a split plot is an irrigation experiment where irrigation levels are applied to large areas, and factors such as varieties and fertilizers are assigned to smaller areas within a particular irrigation treatment. Split-plot designs are useful in many other contexts as well. For example, a teaching method may be applied to an entire class, but sections of the class may be assigned to different treatments. It is important to realize that split-plot experiments come in many forms. The whole-plot design may use randomized complete-blocks, as shown in Output 4.35. The semiconductor example in Section 4.5, is a type of split-plot experiment that uses a completely randomized whole-plot design. Incomplete-blocks designs, Latin squares, and other types of row-column designs may also be used for the whole plot. Whenever you analyze a factorial experiment, you should examine the design structure carefully for split-plot features regardless of a researcher's intentions; they are frequently introduced inadvertently.

The proper analysis of a split-plot design must account for the fact that treatments applied to main plots are subject to larger experimental error than those applied to subplots. Hence, different mean squares must be used as denominators for the corresponding *F*-ratios. Also, many mean comparisons of potential interest have error terms that are linear combinations of mean squares. While PROC GLM is useful for determining expected mean squares, PROC MIXED is better suited to analyze split-plot data.

4.6.1 A Standard Split-Plot Experiment

The split-plot example below analyzes the effect on dry weight yields of three bacterial inoculation treatments applied to two cultivars of grasses (A and B). The experiment is a split-plot design with CULT (cultivar) as the main-plot factor and INOC (inoculi) as the subplot factor. INOC has the values CON for control, LIV for live, and DEA for dead. This provides more information for comparing levels of INOC than for comparing levels of CULT. This is desirable because INOC is the factor of primary interest in the experiment. Data for the experiment appear in Output 4.35.

Output 4.35
Data for
Split-Plot
Experiment

Obs	rep	cult	inoc	drywt
1	1	A	CON	27.4
2	1	A	DEA	29.7
3	1	A	LIV	34.5
4	1	B	CON	29.4
5	1	B	DEA	32.5
6	1	B	LIV	34.4
7	2	A	CON	28.9
8	2	A	DEA	28.7
9	2	A	LIV	33.4
10	2	B	CON	28.7
11	2	B	DEA	32.4
12	2	B	LIV	36.4
13	3	A	CON	28.6
14	3	A	DEA	29.7
15	3	A	LIV	32.9
16	3	B	CON	27.2
17	3	B	DEA	29.1
18	3	B	LIV	32.6
19	4	A	CON	26.7
20	4	A	DEA	28.9
21	4	A	LIV	31.8
22	4	B	CON	26.8
23	4	B	DEA	28.6
24	4	B	LIV	30.7

Table 4.5 shows the standard analysis of variance for this experiment.

Table 4.5 *Split-Plot Analysis-of-Variance Table*

Source	DF
replication	3
cultivar	1
replication × cultivar (Error A or whole-plot error)	3
inoculi	2
cultivar × inoculi	

replication × inoculi +

replication × inoculi × cultivar

(Error B or split-plot error)

Note that Table 4.5 contains two error terms. These correspond to the two sizes of experimental units in the design. The experimental unit for cultivar is the replication×cultivar combination. For inoculi, and hence for cultivar×inoculi treatment combinations, the experimental unit is the replication×cultivar×inoculi combination. Usually, determining the experimental units in this fashion is the best way to determine the error terms and hence the random effects that need to be in the mixed model. Also note that no replication×inoculi term appears separately in the model or as an error term in the ANOVA table. This is because, unlike replication×cultivar or replication×cultivar×inoculi, there is no physical unit that corresponds to replication inoculi. No physical unit means no corresponding term in the model or ANOVA table.

Defining the error terms by their corresponding experimental units also tells you what model term is computationally equivalent, and hence how to write the SAS statements. For example, the whole-plot error (called Error A in many texts), the appropriate error term in testing for differences among cultivars, is computationally equivalent to the replication×cultivar interaction. The split-plot error (also known as Error B) is computationally equivalent to the replication×inoculi + replication×inoculi×cultivar interaction. The Error B mean square is the appropriate error term for testing for inoculi and cultivar×inoculi effects. Note that Error B includes the replication×inoculi. Whenever a term does not appear in the model, its sum of squares is pooled with (that is, added to) the simplest term in the model of which it is a subset, in this case replication×inoculi×cultivar.

4.6.1.1 Analysis of Variance Using PROC GLM

The following SAS statements are needed to compute the analysis of variance:

```
proc glm;
   class rep cult inoc;
   model drywt=rep cult rep*cult inoc cult*inoc/ss3;
   test h=cult e=rep*cult;
run;
```

The data are classified according to REP, CULT, and INOC, so these variables are specified in the CLASS statement. The response variable DRYWT appears on the left side of the equation in the MODEL statement, and the terms corresponding to lines in the analysis-of-variance table (Table 4.3) appear on the right side. You can see the similarity to the model used for the semiconductor data. The only difference is that the treatment applied to the larger experimental unit, CULT, was assigned using a randomized-blocks design, whereas ET was assigned to WAFER using a completely randomized design in the semiconductor example. Thus the block effect, REP, appears in this model but not in the semiconductor model. The rationale for the TEST statement follows from the expected mean squares shown in Output 4.36.

Note: You can add the following RANDOM statement to obtain expected mean squares to verify the appropriate tests:

```
random rep rep*cult/test;
```

Results of the analysis of variance and the expected mean squares appear in Output 4.36.

Output 4.36
Expected Mean
Squares for a
Split-Plot
Experiment

Source	Type III Expected Mean Square
rep	Var(Error) + 3 Var(rep*cult) + 6 Var(rep)
cult	Var(Error) + 3 Var(rep*cult) + Q(cult,cult*inoc)
rep*cult	Var(Error) + 3 Var(rep*cult)
inoc	Var(Error) + Q(inoc,cult*inoc)
cult*inoc	Var(Error) + Q(cult*inoc)

You can see that the hypothesis concerning cultivars (H=CULT) should be tested using REP*CULT as the error term (E=REP*CULT). The TEST statement causes the required F=MS(CULT)/MS(REP*CULT) to be computed. As with the semiconductor example, you need either the TEST statement or the TEST option with the RANDOM statement to compute the correct F-statistics, because the F-statistic for CULT in the default is statistically invalid. The default F-values for INOC and CULT*INOC are valid, because the expected mean squares indicate that MS(ERROR) is their proper error term.

Output 4.37
Analysis of
Variance for a
Split-Plot
Experiment

Source	DF	Sum of Squares	Mean Square	F Value	Pr > F
Model	11	157.2083333	14.2916667	20.26	<.0001
Error	12	8.4650000	0.7054167		
Corrected Total	23	165.6733333			

R-Square	Coeff Var	Root MSE	drywt Mean
0.948905	2.761285	0.839891	30.41667

Source	DF	Type III SS	Mean Square	F Value	Pr > F
rep	3	25.3200000	8.4400000	11.96	0.0006
cult	1	2.4066667	2.4066667	3.41	0.0895
rep*cult	3	9.4800000	3.1600000	4.48	0.0249
inoc	2	118.1758333	59.0879167	83.76	<.0001
cult*inoc	2	1.8258333	0.9129167	1.29	0.3098

Tests of Hypotheses Using the Type III MS for rep*cult as an Error Term

Source	DF	Type III SS	Mean Square	F Value	Pr > F
cult	1	2.40666667	2.40666667	0.76	0.4471

These *F*-values indicate no significant CULT*INOC interaction (*F*=1.29, *p*=0.3098). The INOC main effect (*F*=83.76, *p*=0.0001) indicates highly significant differences between INOC means. The appropriate *F*-value, that is, the one printed below the main ANOVA table that uses REP*CULT as the error term, shows no evidence of differences between CULT means (*F*=0.76, *p*=0.4471). *Do not use* the inappropriate *F*-value for CULT of 3.41 from the ANOVA table, which declares differences among CULT means significant at the *p*=0.0895 level.

Note that the same set of valid *F*-statistics would be computed if you use the TEST option of the RANDOM statement in PROC GLM.

4.6.1.2 Analysis with PROC MIXED

Use the following SAS statements for the analysis with PROC MIXED:

```
proc mixed;
   class rep cult inoc;
   model drywt=cult inoc cult*inoc/ddfm=satterth
   random rep rep*cult;
run;
```

To determine what effects to include in the MODEL statement and what effects to include in the RANDOM statement, use the following guidelines:

❑ Treatment main effects and interactions are generally *fixed* and thus belong in the MODEL statement.

❑ Error terms go in the RANDOM statement. Error terms correspond to the experimental unit with respect to a given model effect.

- REP*CULT is the experimental unit for CULT.
- The experimental unit for INOC and for CULT×INOC combinations is a replication×cultivar×inoculi combination.

❑ The smallest experimental unit corresponds to residual error and thus does not appear in either the MODEL or the RANDOM statement.

❑ The specifics of a given design should determine whether to consider the blocking criterion as fixed or random. See Section 4.3 for guidelines. For this example, assume the blocking criterion REP is random.

The results of the basic MIXED analysis appear in Output 4.38.

Output 4.38
PROC MIXED
Analysis of
Split-Plot Data

```
             Covariance Parameter
                  Estimates

             Cov Parm     Estimate

             rep           0.8800
             rep*cult      0.8182
             Residual      0.7054

          Type 3 Tests of Fixed Effects

                     Num    Den
          Effect      DF     DF    F Value   Pr > F

          cult         1      3      0.76    0.4471
          inoc         2     12     83.76    <.0001
          cult*inoc    2     12      1.29    0.3098
```

You can see that the *F*-statistics and their associated *p*-values are identical to those obtained using PROC GLM, assuming the proper TEST options. The ANOVA lines for REP and REP*CULT are replaced by their respective variance component estimates. The variance component estimate for RESIDUAL is equal to the MS(ERROR) from the ANOVA table.

You can add CONTRAST or ESTIMATE statements as needed, as well as the LSMEANS statement, with DIFF and SLICE= options if called for by the objectives. These are subject to the same test statistic and standard error considerations as the semiconductor data in Section 4.5. For example, PROC MIXED computes the appropriate statistics for a CULT difference given a particular INOC, whereas PROC GLM has no option to permit this. To repeat the conclusion of Section 4.5, PROC MIXED is the recommended procedure to analyze split-plot experiments.

Chapter 5 Unbalanced Data Analysis: Basic Methods

5.1 Introduction

Most persons who have analyzed data have experienced "unbalanced data," although the term is hard to precisely define. Usually, it refers to having different numbers of observations in different groups of data. For example, a data set contained final exam scores from a statistics class at a university. The class had 66 students from four colleges, but there were different numbers of students from the various colleges. The set of exam scores would make up an unbalanced one-way classification of data. However, the main problems associated with analyzing unbalanced data do not occur in a one-way classification. In the statistics class, there were both part-time and full-time students within each college, and only coincidentally were there the same number of each from a given college. Thus, the scores of final exams constitute a two-way classification of data with differing numbers of observations in the combinations of college and status. A combination of college and status makes a **cell** of data. With this terminology, we could say that **unbalanced data** refers to data sets with different numbers of observations in the cells. You might consider this situation an "observational study" because the numbers of students in the various groups were not controlled. The instructor would be faced with problems of analyzing unbalanced data if he or she wanted to compare mean exam scores of full-time with part-time students, averaged across colleges. The basic problem is to decide what weight to attach to the mean scores for each status group in each college.

In another example, a pharmaceutical company compared effects of two drugs, A and B, on a clinical measurement called "flush." The study utilized patients in 10 clinics. Multiple clinics were used in order to obtain representation of diverse patient populations. The original plan called for each clinic to assign the two drugs to 15 patients within each clinic. However, there were not enough patients at the clinics, so all patients available were randomly assigned to the two drugs. This was basically done, but a few patients abandoned the trial before completion, leaving unequal numbers of patients on the two drugs within some of the clinics. In addition, the availability of patients varied between the clinics, ranging from three to 28. Thus, even though this was a designed experiment, realities of the situation resulted in unbalanced data. A statistical comparison of mean flush between the two drugs would raise the question of how to assign weights to the individual means.

Analysis of unbalanced data received sporadic attention for several decades, but the attention intensified in the 1970's when computer programs such as PROC GLM became readily accessible. Most of the writing focused on the fixed-effects case, prompted in part by the different types of sums of squares in PROC GLM. Other popular texts that discuss fixed-effects issues include Milliken and Johnson (1991), Hocking (1986), and Searle (1987). Analysis of unbalanced mixed-model data still contains many mysteries. The GLM procedure contains certain capabilities that are adoptions of fixed-effects computations, but there has been relatively little concrete description of how to use them. This prompted a re-evaluation of how to analyze mixed-model data in the late 1980's and PROC MIXED implemented newer methodology based on generalized least squares and likelihood-based methods.

The purpose of this chapter is to illustrate methods that are available in PROC GLM and PROC MIXED for analyzing unbalanced data. In Sections 5.2 and 5.3 you will see the issues of analyzing fixed-effect unbalanced data presented on a conceptual level using the clinical trial example described above. Then, in later sections, you will see ANOVA and generalized least squares and likelihood methods for analysis of unbalanced mixed-model data.

5.2 Applied Concepts of Analyzing Unbalanced Data

The FLUSH measurements from the pharmaceutical study are recorded in a SAS data set named DRUGS. A portion of the data set is printed in Output 5.1. Variables in the data set include STUDY, TRT, PATIENT, FLUSH0, and FLUSH. The values of FLUSH0 were obtained prior to administration of the drugs, but are not used in the discussions in this chapter.

Output 5.1
Data Set
DRUGS

```
                    Unbalanced Two-way Classification

       OBS     STUDY     TRT     PATIENT     FLUSH0     FLUSH

        1       42        A        201        50.5     70.3333
        2       42        A        203        84.5     16.1429
        3       42        B        202        33.5     28.3333
        4       43        A        302        22.0     14.5000
        5       43        A        305        23.0     25.5000
        6       43        A        306        22.0     12.2500
        7       43        A        307        13.0      3.1250
        8       43        A        310        50.5     51.1250
        9       43        A        313        57.0     49.2500
       10       43        A        316        13.5      1.6250
       11       43        A        317        36.5     29.5000
       12       43        A        321        59.0     30.5000
       13       43        A        322        30.5     33.5000
       14       43        A        323        10.5      2.2500
       15       43        A        325        37.0     13.8750
       16       43        A        327        35.5     21.0000
       17       43        A        329        28.0     16.0000
       18       43        B        301        40.5     17.5000
       19       43        B        303        12.5      8.8333
       20       43        B        304        47.5     40.0000
```

Output 5.1
(Continued)
Data Set
DRUGS

21	43	B	308	34.5	23.1429
22	43	B	309	15.5	3.1250
23	43	B	311	43.0	35.8750
24	43	B	314	30.0	31.6250
25	43	B	315	27.5	16.0000
26	43	B	318	62.0	41.1250
27	43	B	319	105.0	44.7500
28	43	B	324	38.5	43.1250
29	43	B	326	7.0	15.0000
30	43	B	328	8.0	4.5000
31	43	B	330	30.5	19.0000
32	44	A	401	46.0	14.8750
33	44	A	405	36.5	2.9231
34	44	A	406	22.5	2.8000
35	44	A	408	21.5	1.3750
36	44	A	409	27.0	22.0000
37	44	A	411	46.5	.
38	44	B	402	14.0	4.7778
39	44	B	403	23.0	3.6667
40	44	B	404	30.0	17.1250
41	44	B	407	19.0	22.3636
42	44	B	410	67.5	18.1667
43	44	B	412	12.0	2.0000
44	45	A	502	60.0	62.0000
45	45	A	503	36.0	13.6250
46	45	A	506	24.0	1.0000
47	45	A	507	29.0	24.1250
48	45	A	510	12.5	11.5000
49	45	A	512	82.5	84.0000
50	45	A	513	31.5	0.6250
51	45	A	515	53.0	45.0000
52	45	A	518	56.0	43.7500
53	45	A	519	23.0	7.3750
54	45	A	520	48.5	43.1429
55	45	A	527	16.0	.
56	45	B	501	34.0	30.0000
57	45	B	504	74.5	38.3750
58	45	B	505	22.0	25.3750
59	45	B	508	7.0	2.8750
60	45	B	509	13.0	8.1250
61	45	B	511	34.5	28.8750
62	45	B	514	20.5	22.6000
63	45	B	516	75.5	37.0000
64	45	B	517	50.0	59.1250
65	45	B	529	27.5	26.8000
66	45	B	530	49.0	33.0000
67	46	A	601	31.0	5.0000
68	46	A	602	53.0	20.8750
69	46	A	605	28.0	16.0000
70	46	A	608	21.5	7.5000
71	46	A	609	11.5	3.3750
72	46	A	611	59.0	35.6250
73	46	B	603	39.0	50.0000
74	46	B	604	65.0	43.0000
75	46	B	606	43.5	41.0000
76	46	B	607	25.0	8.5000
77	46	B	610	26.5	0.5000
78	46	B	629	27.5	15.5000
79	46	B	630	19.5	11.1250

Output 5.2 shows summary statistics for FLUSH for each combination of STUDY and TRT.

Output 5.2
Summary
Statistics for
the FLUSH
Data Set

```
                      Unbalanced Two-way Classification

  Analysis Variable : FLUSH


   STUDY  TRT    N Obs       Mean     N     Std Dev      Minimum      Maximum
  ---------------------------------------------------------------------------
      42  A          2   43.2381000     2   38.3183993   16.1429000   70.3333000
          B          1   28.3333000     1            .   28.3333000   28.3333000

      43  A         14   21.7142857    14   15.8805206    1.6250000   51.1250000
          B         14   24.5429429    14   14.6750757    3.1250000   44.7500000

      44  A          6    8.7946200     5    9.1762473    1.3750000   22.0000000
          B          6   11.3499667     6    8.8404391    2.0000000   22.3636000

      45  A         12   30.5584455    11   27.1736184    0.6250000   84.0000000
          B         11   28.3772727    11   14.9979268    2.8750000   59.1250000

      46  A          6   14.7291667     6   12.2625998    3.3750000   35.6250000
          B          7   24.2321429     7   19.8164006    0.5000000   50.0000000

      47  A          6   20.8777833     6    7.8619415    6.8750000   28.7500000
          B          7   49.0178571     7   30.9056384    4.3750000   92.1250000

      48  A          8   21.7857143     7   13.8723185    7.0000000   42.5000000
          B          8   22.5732500     8   14.7692870    3.8750000   49.2500000

      49  A         10   32.1554000    10   31.0770557    0.1250000   79.7500000
          B         10   27.3953000    10   24.3228740            0   59.6250000

      50  A          4    9.4687500     4    7.4486961    2.1250000   18.7500000
          B          3   71.8750000     3   53.6849898   32.3750000  133.0000000
```

There are different numbers of observations in the TRT-by-STUDY cells, meaning we have unbalanced data. Note that there is at least one observation for each combination of the factors.

5.2.1 ANOVA for Unbalanced Data

The four types of sums of squares available in PROC GLM are designed to deal with unbalanced data. Run the statements

```
proc glm data=drugs;
   class study trt;
   model flush=trt study trt*study / ss1 ss2 ss3;
run;
```

Results appear in Output 5.3. Types I, II, and III are selected with the options ss1 ss2 ss3 on the MODEL statement. Type IV was not selected because Types III and IV are equal for situations such as this that have at least one observation for each factor combination; that is no empty cells. Types I, II, and III have different values for TRT, but Types I and II are the same for STUDY. Types I, II, and III are the same for TRT*STUDY. The technical reasons for the sameness and differences of the Types of sums of squares are explained in Chapter 6.

The primary objective of the study is to compare TRT means. Before making comparisons of the drugs averaged over clinics, note that the TRT*STUDY interaction is significant at the $p=.0178$ level. This means that differences between drug A and drug B vary across clinics. In Output 5.2 you see that drug B has a larger mean than drug A for six of the nine clinics. If there is additional information about the clinics, you might want to try to investigate whether characteristics of the clinics can explain the interaction between STUDY and TRT. Recall in Section 3.7 that

METHODS were compared separately for each VARIETY due to the presence of METHOD*VARIETY interaction.) Depending on the situation, it may or may not be meaningful to compare the drugs averaged across clinics. For the present situation, we have no other information about the clinics. Also, the clinics were used to obtain representation of the drug differences over a set of clinics. Therefore, even in the face of TRT*STUDY interaction, it could be important and meaningful to compare the drugs averaged over clinics. This would be the case if, for example, you must choose one of the drugs to be used at all clinics.

Output 5.3
Three Types of
ANOVA
Tables for the
FLUSH Data
Set

```
                          Unbalanced Two-way Classification

                               The GLM Procedure

Dependent Variable: FLUSH


                                  Sum of
Source                  DF        Squares        Mean Square     F Value     Pr > F

Model                   17      16618.75357       977.57374        2.24      0.0063

Error                  114      49684.09084       435.82536

Corrected Total        131      66302.84440


            R-Square      Coeff Var      Root MSE      FLUSH Mean

            0.250649      80.31125       20.87643       25.99440


Source                  DF       Type I SS       Mean Square     F Value     Pr > F

TRT                      1      1134.560964      1134.560964       2.60      0.1094
STUDY                    8      6971.606045       871.450756       2.00      0.0526
TRT*STUDY                8      8512.586561      1064.073320       2.44      0.0178


Source                  DF       Type II SS      Mean Square     F Value     Pr > F

TRT                      1      1377.550724      1377.550724       3.16      0.0781
STUDY                    8      6971.606045       871.450756       2.00      0.0526
TRT*STUDY                8      8512.586561      1064.073320       2.44      0.0178


Source                  DF       Type III SS     Mean Square     F Value     Pr > F

TRT                      1      1843.572090      1843.572090       4.23      0.0420
STUDY                    8      7081.377266       885.172158       2.03      0.0488
TRT*STUDY                8      8512.586561      1064.073320       2.44      0.0178
```

Next you must decide which of the three *F*-tests from the three types of sums of squares is most appropriate for comparing the drugs. The first consideration is to select a test statistic that tests the hypothesis you want to test. Of course, the hypothesis you want to test should have been prescribed at the planning stage of the study, not in the middle of data analysis.

Let μ_{ij} denote the population for drug i and clinic j. If you are equally interested in each clinic, then a reasonable hypothesis to test is

$$H_0: \mu_{A.} = \mu_{B.}$$

where

$$\mu_{A.} = (1/9)(\mu_{A1} + \mu_{A2} + \mu_{A3} + \mu_{A4} + \mu_{A5} + \mu_{A6} + \mu_{A7} + \mu_{A8} + \mu_{A9})$$

and

$$\mu_{B.} = (1/9)(\mu_{B1} + \mu_{B2} + \mu_{B3} + \mu_{B4} + \mu_{B5} + \mu_{B6} + \mu_{B7} + \mu_{B8} + \mu_{B9})$$

The F-test based on the Type III sum of squares in Output 5.3 gives a test of this hypothesis, which we will refer to as a "Type III hypothesis." This means that if you test this hypothesis at the .05 level using the Type III F-test, the probability you will make a Type I error is exactly .05.

The Type III hypothesis is a statement of equality of the drug means, averaged across the clinics, with equal weights attached to each clinic. Other hypotheses could be formulated attaching different weights to different clinics. Without overriding reasons for attaching different weights, the type III hypothesis is often reasonable. However, there are other considerations that enter into the decision. For example, power of the Type III test can be very low if sample sizes for some cells are small compared to other sample sizes of other cells.

5.2.2 Using the CONTRAST and ESTIMATE Statements with Unbalanced Data

If there are no empty cells you can use CONTRAST and ESTIMATE statements in the same way you used them with balanced data in Chapter 3. For example, you can test the significance of the difference between the drug means with the CONTRAST statement

```
contrast  trtB-trtA' trt -1 1;
```

and you can estimate the difference between the drug means with the statement

```
estimate 'trtB-trtA' trt -1 1;
```

Results of the CONTRAST and ESTIMATE statements appear in Output 5.4.

Output 5.4
Results of the
CONTRAST
and
ESTIMATE
Statements

```
                          Unbalanced Two-way Classification


Dependent Variable: FLUSH

Contrast                  DF      Contrast SS      Mean Square    F Value    Pr > F

trtB-trtA                  1      1843.572090      1843.572090       4.23    0.0420


                                                   Standard
         Parameter                 Estimate          Error      t Value    Pr > |t|

         trtB-trtA                 9.37497409     4.55823028        2.06      0.0420
```

The CONTRAST statement produces the same sum of squares, mean square, F-test and p-value for the difference between drug means that you obtained from the Type III ANOVA F-test in Output 5.4. It is a test of H_0: $\mu_{A.} = \mu_{B.}$.

The difference between the estimates of $\mu_{A.}$ and $\mu_{B.}$ is 9.375, and the standard error of the estimate is 4.558. A t-statistic for testing H_0: $\mu_{A.} = \mu_{B.}$ is $t = 9.375/4.558 = 2.06$. The p-value for the t-statistic is .0420. This also is the same p-value you got from the Type III F-test in Output 5.3. In most situations, the results of t-tests from ESTIMATE statements are equivalent to F-tests from Type III ANOVA.

5.2.3 The LSMEANS Statement

You can calculate means from unbalanced data using the LSMEANS statement. With balanced data, the LSMEANS statement computes ordinary means. LSMEANS for the drugs are obtained from the statement

```
lsmeans trt / pdiff;
```

The PDIFF option is a request for t-tests to compare the LSMEANS. Results appear in Output 5.5.

*Output 5.5
Results of the
LSMEANS
Statement*

```
                    Unbalanced Two-way Classification

                            The GLM Procedure
                          Least Squares Mean

                                                         H0:LSMean1=
                                   Standard    H0:LSMEAN=0    LSMean2
        TRT    FLUSH LSMEAN         Error       Pr > |t|     Pr > |t|

         A      22.5913628        3.0141710      <.0001       0.0420
         B      31.9663369        3.4193912      <.0001
```

The estimate of $\mu_{A.}$ is 22.591, with standard error 3.014, and the estimate of $\mu_{B.}$ is 31.966 with standard error 3.419. The p-value for comparing the means in .0420, which is the same as the p-value you got from the ESTIMATE statement in Output 5.4. In fact, the difference between the means in Output 5.4 is the same as the difference between the LSMEANS in Output 5.5. More information about LSMEANS is given in Chapter 6.

5.2.4 More on Comparing Means: Other Hypotheses and Types of Sums of Squares

In Section 5.2.2 you learned that the F-statistic based on the Type III sum of squares gives a test of the null hypothesis

$$H_0: \mu_{A.} = \mu_{B.}$$

where

$$\mu_{A.} = (1/9)(\mu_{A1} + \mu_{A2} + \mu_{A3} + \mu_{A4} + \mu_{A5} + \mu_{A6} + \mu_{A7} + \mu_{A8} + \mu_{A9})$$

and

$$\mu_{B.} = (1/9)(\mu_{B1} + \mu_{B2} + \mu_{B3} + \mu_{B4} + \mu_{B5} + \mu_{B6} + \mu_{B7} + \mu_{B8} + \mu_{B9})$$

This null hypothesis states that the average of the drug A means is equal to the average of the drug B means, where the averages are computed across the clinics with equal weights for each clinic. In some circumstances you might want to compare averages of the drug means, but with different weights for the clinics. For example, the clinics might serve different patient populations, and you might want to weight the means proportional to the patient population sizes. Let w_j, j = 1, ... ,9, denote the relative population sizes, with $w_1 + \ldots + w_9 = 1$. Then the weighted hypothesis would be

$$H_0: \mu^*_{A.} = \mu^*_{B.} \tag{5.2}$$

where

$$\mu^*_{A.} = (1/9)(w_1\mu_{A1} + \ldots + w_9\mu_{A9})$$

and

$$\mu^*_{B.} = (1/9)(w_1\mu_{B1} + \ldots + w_9\mu_{B9})$$

You could test this hypothesis with the CONTRAST statement. To illustrate, suppose the weights are .03, .20, .09, .17, .1, .1, .1, .14, and .07. The CONTRAST statement would be

```
contrast 'trtB - trtA wtd' trt -1 1
         trt*study -.03 -.20 -.09 -.17 -.1 -.1 -.1 -.14 -.07
                    .03  .20   .09  .17  .1  .1  .1  .14 .07;
```

Results appear in Output 5.6.

Output 5.6
Results of the
CONTRAST
Statement for
Weighted
Hypothesis

```
                     Unbalanced Two-way Classification

                           The GLM Procedure

Dependent Variable: FLUSH

      Contrast         DF    Contrast SS    Mean Square   F Value    Pr > F

  trtB-trtA wtd         1    1829.354286    1829.354286      4.20    0.0428
```

You see that the sum of squares for this CONTRAST statement is different from the sum of squares in Output 5.4 for the equally weighted hypothesis, although not by very much. You also see that the sum of squares for the weighted hypothesis in Output 5.6 is different from the TRT sum of squares for any of the ANOVA tables in Output 5.3. This illustrates that there are different values of sums of squares for TRT depending on the weights assigned to the means. Each type of sum of squares for TRT in Output 5.6 is associated with a certain set of weights. These are explained in detail in Chapter 6.

5.3 Issues Associated with Empty Cells

The data set discussed in Section 5.2 had at least one observation for each cell corresponding to combinations of TRT and STUDY. In the original data set there was another clinic, STUDY=41, that had patients only for drug B. So the cell for that clinic and drug A was empty. Empty cells create another layer of complications in analyzing unbalanced data. In this section we illustrate some of these issues using the original data set, which we call DRUGS1. The first several observations are printed in Output 5.7.

Output 5.7
Partial
Printout of
Data Set
DRUGS1

```
                    Unbalanced Two-way Classification


        Obs    STUDY    TRT    PATIENT    FLUSH0    FLUSH

         1      41       B      102        77.5    72.0000
         2      41       B      104        23.5     5.6250
         3      41       B      105        63.5    81.8750
         4      41       B      106        72.5    83.5000
         5      41       B      107        58.0    75.5000
         6      41       B      108        49.0    13.7500
         7      41       B      109         7.5     9.3750
         8      41       B      110        13.5     7.8750
         9      41       B      111        13.5     6.0000
        10      41       B      112        76.5    61.6000
        11      41       B      113        78.5    98.1250
        12      41       B      114        56.5    46.1250
        13      41       B      115        61.0    24.2500
        14      41       B      116        91.0    64.4000
        15      41       B      117        13.5     7.3333
        16      41       B      118        63.5    79.2500
        17      42       A      201        50.5    70.3333
        18      42       A      203        84.5    16.1429
```

You see that there are 16 observations for STUDY=41 and TRT=B, but no observations for TRT=A. Now we briefly review the effects of the empty cell on the analysis methods shown in Section 5.2.

5.3.1 The Effect of Empty Cells on Types of Sums of Squares

Empty cells create problems with ANOVA computations associated with difficulties in specifying meaningful hypotheses. Run the statements

```
proc glm data=drugs1;
   class study trt;
   model flush=trt study trt*study / ss1 ss2 ss3 ss4;
run;
```

Results appear in Output 5.8.

Output 5.8
Four Types of
ANOVA
Tables for
Data Set with
Empty Cell

```
                    Unbalanced Two-way Classification

                         The GLM Procedure

Dependent Variable: FLUSH

                                 Sum of
        Source            DF     Squares     Mean Square   F Value   Pr > F

        Model             18   22350.89135    1241.71619     2.38    0.0027

        Error            129   67361.38451     522.18128

        Corrected Total  147   89712.27586
```

Output 5.8
(Continued)
Four Types of
ANOVA Tables
for Data Set
with Empty
Cell

		R-Square	Coeff Var	Root MSE	FLUSH Mean		
		0.249140	81.14483	22.85129	28.16111		

Source	DF	Type I SS	Mean Square	F Value	Pr > F
TRT	1	3065.96578	3065.96578	5.87	0.0168
STUDY	9	10772.33900	1196.92656	2.29	0.0202
TRT*STUDY	8	8512.58656	1064.07332	2.04	0.0468

Source	DF	Type II SS	Mean Square	F Value	Pr > F
TRT	1	1377.55072	1377.55072	2.64	0.1068
STUDY	9	10772.33900	1196.92656	2.29	0.0202
TRT*STUDY	8	8512.58656	1064.07332	2.04	0.0468

Source	DF	Type III SS	Mean Square	F Value	Pr > F
TRT	1	1843.57209	1843.57209	3.53	0.0625
STUDY	9	10261.36525	1140.15169	2.18	0.0272
TRT*STUDY	8	8512.58656	1064.07332	2.04	0.0468

Source	DF	Type IV SS	Mean Square	F Value	Pr > F
TRT	1*	1843.572090	1843.572090	3.53	0.0625
STUDY	9*	7462.538828	829.170981	1.59	0.1254
TRT*STUDY	8	8512.586561	1064.073320	2.04	0.0468

* NOTE: Other Type IV Testable Hypotheses exist which may yield different SS.

Compare Output 5.8 with Output 5.3. In general, you see different values in the two tables for the Types I, II and III sums of squares for TRT and STUDY. Also, the Types III and IV sums of squares for STUDY are different from each other in Output 5.8. These differences illustrate the fact that the associated hypotheses are different. Details of the specific hypotheses are discussed in Chapter 6.

5.3.2 The Effect of Empty Cells on CONTRAST, ESTIMATE, and LSMEANS Results

Run the statements

```
estimate 'trtB-trtA' trt -1 1;
contrast 'trtB-trtA' trt -1 1;
lsmeans trt / stderr pdiff;
run;
```

Results appear in Output 5.9.

Output 5.9
Effects of
Empty Cell on
LSMEANS

```
                 Unbalanced Two-way Classification

                        The GLM Procedure
                      Least Squares Means

                                 Standard
       TRT      FLUSH LSMEAN        Error    Pr > |t|

       A          Non-est            .          .
       B        33.3733489       3.4166700   <.0001
```

No output is given from the CONTRAST and ESTIMATE statements because the underlying linear combinations of parameters are *non-estimable*. This message appears in the SAS log. The LSMEAN for TRT=A also is non-estimable because the empty cell was for drug A. Non-estimability is discussed in detail in Chapter 6.

5.4 Some Problems with Unbalanced Mixed-Model Data

In Chapter 4 you read about statistical analysis of data with random effects. The methods discussed there were in the setting of balanced data. The statistical issues concerned construction of *F*-tests and standard errors of estimates that take into account multiple sources of random variation in the data. Applications in Chapter 4 illustrated both analysis-of-variance methods using the GLM procedure, and mixed model methods using the MIXED procedure. In Sections 5.2 and 5.3 you read about statistical analysis of unbalanced data. The critical issues were constructing meaningful linear combinations of model parameters for estimation and hypothesis testing. In the present section we address problems of analyzing unbalanced data with random effects. We must identify statistical procedures that simultaneously define meaningful linear combinations of model parameters and account for multiple sources of random variation. As in the case of balanced data, we illustrate two approaches, *analysis of variance* using the GLM procedure, and *mixed-model methodology* using PROC MIXED.

We return to the clinical trial example of Section 5.1. Now we assume that the clinics were chosen from a population of clinics, and that the objective is to make inference about the drugs that is relevant to the entire population of clinics. Thus, we consider clinics to be a random factor. Ideally, the clinics would be chosen as a random sample from the population of clinics, but this is not realistic. Instead, we assume that the clinics in the data set reasonably represent the population of clinics as would a truly random sample of clinics. The statistical model is

$$y_{ijk} = \mu + \alpha_i + b_j + (\alpha b)_{ij} + e_{ijk}$$

where

y_{ijk} is the FLUSH measurement on the kth patient assigned to drug i in clinic j.

$\mu + \alpha_i$ is the mean FLUSH for drug i.

b_j is the random effect associated with clinic j.

$(\alpha b)_{ij}$ is the random interaction effect associated with drug i and clinic j.

e_{ijk} is the random error associated with the kth patient assigned to drug i in clinic j.

We assume the b_j random variables for clinics are normally and independently distributed with mean 0 and variance σ_{STUDY}^2 and the $(\alpha b)_{ij}$ random variables for DRUG*CLINIC interaction are normally and independently distributed with mean 0 and variance $\sigma_{\text{STUDY*TRT}}^2$. Also, we assume the e_{ijk} random variables for patients are normally and independently distributed with mean 0 and variance σ^2.

This is the same model introduced in Chapter 4 for a two-way classification mixed model. The only distinction is that the number of observations in each clinic-drug cell may change from cell to cell. The objectives are the same as in the balanced case. This is an important point: *The failure to obtain the same number of observations in each cell should not influence the objectives of the research.*

There are basically two approaches to analyzing unbalanced mixed-model data—ANOVA and mixed-model methods. In the context of SAS/STAT procedures, ANOVA means using the GLM procedure, and mixed-model methods means using the MIXED procedure. You saw both approaches applied to balanced data in Chapter 4. Both approaches result in approximate methods for unbalanced mixed-model data. We illustrate ANOVA methods in Section 5.5 and mixed-model methods in Section 5.6.

5.5 Using the GLM Procedure to Analyze Unbalanced Mixed-Model Data

The term "ANOVA" methods refers to adapting analysis-of-variance computations for statistical inference with mixed-model data. The computations have their basis in comparing fixed-effects models, but have been found useful in comparing mixed models. However, there are some troublesome difficulties. First, it is not clear how to choose a mean square to measure the effect we want to test. Second, it also is not clear how to choose a mean square for the denominator of the test. As in the balanced data case, expected mean squares are used to determine appropriate denominators for F-tests, but coefficients for variance components do not match with coefficients in the numerator. Third, the two mean squares are usually not independent, so the ratio does not have a true F-distribution. The same difficulties carry over to contrasts and standard errors for linear combinations.

5.5.1 Approximate *F*-Statistics from ANOVA Mean Squares with Unbalanced Mixed-Model Data

Run the statements

```
proc glm data=drugs;
   class study trt;
   model flush=trt study trt*study / ss1 ss2 ss3;
run;
```

Results appear in Output 5.10.

Output 5.10
Three Types of
ANOVA Tables
for the FLUSH
Data Set

```
                    Unbalanced Two-way Classification

                          The GLM Procedure

Dependent Variable: FLUSH

                                Sum of
       Source           DF      Squares      Mean Square   F Value   Pr > F

       Model            17    16618.75357      977.57374     2.24    0.0063

       Error           114    49684.09084      435.82536

       Corrected Total 131    66302.84440


       Source           DF    Type I SS      Mean Square   F Value   Pr > F

       TRT               1    1134.560964    1134.560964     2.60    0.1094
       STUDY             8    6971.606045     871.450756     2.00    0.0526
       TRT*STUDY         8    8512.586561    1064.073320     2.44    0.0178


       Source           DF    Type II SS     Mean Square   F Value   Pr > F

       TRT               1    1377.550724    1377.550724     3.16    0.0781
       STUDY             8    6971.606045     871.450756     2.00    0.0526
       TRT*STUDY         8    8512.586561    1064.073320     2.44    0.0178


       Source           DF    Type III SS    Mean Square   F Value   Pr > F

       TRT               1    1843.572090    1843.572090     4.23    0.0420
       STUDY             8    7081.377266     885.172158     2.03    0.0488
       TRT*STUDY         8    8512.586561    1064.073320     2.44    0.0178
```

The Types I, II, and III sums of squares are the same as in Output 5.3. The first task is to choose a mean square for the numerator of an *F*-statistic to test for the difference between drug means. The technical considerations in doing so are very different from those faced with fixed-effects models. Essentially, we want to select a mean square that measures the effect we want to test with the least amount of random variation. In general, it is not clear which of Types I, II, or III mean squares to use for this purpose. See Littell (1996) for details. Without further justification, we will use the Type III mean square for TRT as the numerator of an *F*-statistic. However, we return to this problem in Section 6.5.1.

The next task is to select a denominator for the test. While this choice is not totally clear, at least we have some useful criteria for the choice in the expected mean squares. You learned in Chapter 4 how to use the expected mean squares to select a mean square for the denominator of an *F*-statistic whose expectation matches the expectation of the numerator mean square under the null hypothesis. Run the RANDOM statement to obtain the expected mean squares:

```
random study trt*study / test;
```

Results appear in Output 5.11.

Output 5.11
Results of the
RANDOM
Statement

```
                    Unbalanced Two-way Classification

                           The GLM Procedure

       Source                Type III Expected Mean Square

       TRT                   Var(Error) + 4.6613 Var(TRT*STUDY) + Q(TRT)

       STUDY                 Var(Error) + 7.0585 Var(TRT*STUDY) + 14.117 Var(STUDY)

       TRT*STUDY             Var(Error) + 7.0585 Var(TRT*STUDY)


               Tests of Hypotheses for Mixed Model Analysis of Variance

   Dependent Variable: FLUSH

       Source              DF     Type III SS    Mean Square   F Value   Pr > F

       TRT                  1    1843.572090    1843.572090      2.17    0.1674

       Error            11.689   9943.710652     850.710718
     Error: 0.6604*MS(TRT*STUDY) + 0.3396*MS(Error)
```

You see that the expected mean square for TRT is $\sigma^2 + 4.66\,\sigma^2_{\text{STUDY*TRT}} + \phi^2(\text{TRT})$. The Q option (see Chapter 4) could be used to discover that $\phi^2(\text{TRT})=20.97\,(\alpha_1 - \alpha_2)^2$. Thus, under the null hypothesis H_0: $\alpha_1 - \alpha_2 = 0$, the expected mean square for TRT is $\sigma^2 + 4.66\,\sigma^2_{\text{STUDY*TRT}}$. We want to obtain another mean square with this expectation to use as the denominator for the F-statistic. Unfortunately, none is directly available, so a combination of mean squares must be used. The TEST option in the RANDOM statement instructs GLM to compute such a combination, shown in Output 5.11 as 0.66*MS(STUDY*TRT) + 0.34*MS(Error), with Satterthwaite's approximate DF=11.69. The approximate F-statistic has value $F=2.17$ and significance probability $p=0.1674$. Thus, the difference between drugs is less significant when making inference to the population of clinics instead of to the set of clinics in the data set.

You should remember that the significance probability for an ANOVA F-test is only approximate due to the complications of unbalanced data. The F-statistic does not have a true F-distribution for two reasons: One, the denominator is a linear combination of mean squares, but is not distributed as a constant-times-a-chi-squared random variable. Two, the numerator and denominator of the F-ratio are not independent. Nonetheless, statistics obtained in this manner are very useful and sometimes provide the only available means of statistical inference.

Expected mean squares also can be used to estimate variance components, as you learned in Chapter 4 with balanced data. To do this, equate the mean squares to their expectations and solve for the values of the variances estimates. These are called **ANOVA** variance component estimates. Here are the equations to solve to obtain the ANOVA estimates:

$$\hat{\sigma}^2 + 7.06\,\hat{\sigma}^2_{\text{STUDY*TRT}} + 14.12\,\hat{\sigma}^2_{\text{STUDY}} \quad = \quad 885.17$$
$$\hat{\sigma}^2 + 7.06\,\hat{\sigma}^2_{\text{STUDY*TRT}} \quad = \quad 1064.07$$
$$\hat{\sigma}^2 \quad = \quad 435.83$$

The last equation gives $\hat{\sigma}^2 = 435.83$. Next, substitute $\hat{\sigma}^2 = 435.83$ into the second equation to obtain $435.83 + 7.06\,\hat{\sigma}^2_{\text{STUDY*TRT}} = 1064.07$ and solve for $\hat{\sigma}^2_{\text{STUDY*TRT}} = 88.99$. Finally, substitute $\hat{\sigma}^2 = 435.83$ and $\hat{\sigma}^2_{\text{STUDY*TRT}} = 88.99$ into the first equation to obtain $435.83 + 7.06\,(88.99) + 14.12\,\hat{\sigma}^2_{\text{STUDY}} = 885.17$. Solving this equation gives $\hat{\sigma}^2_{\text{STUDY}} = -12.67$. Since the variances are positive numbers by definition, a negative estimate is not satisfactory. Zero is often substituted instead of the negative estimate. However, this has ripple effects on other issues, such as the standard errors of estimates and test statistics that utilize the variance component estimates in their computations. From this perspective, there are legitimate reasons for *not* routinely setting negative estimates to zero. This problem is not limited to unbalanced data. Refer to Section 4.4.2.

5.5.2 Using the CONTRAST, ESTIMATE, and LSMEANS Statements in GLM with Unbalanced Mixed-Model Data

You must be very careful in using the CONTRAST, ESTIMATE, and LSMEANS statements with mixed-model data because their output is not automatically modified to accommodate random effects when you specify a RANDOM statement. In some cases it is not possible to appropriately modify the results. These comments pertain to both the balanced and unbalanced situations.

Run the statements

```
contrast 'trtA-trtB' trt 1 -1;
estimate 'trtA-trtB' trt 1 -1;
lsmeans trt / pdiff;
random study trt*study;
run;
```

Results appear in Output 5.12.

Output 5.12
Results of
CONTRAST,
ESTIMATE,
and LSMEANS
Statements
with the
RANDOM
Statement

```
                        Unbalanced Two-way Classification

 Contrast              DF      Contrast SS     Mean Square    F Value    Pr > F

 trtB-trtA              1      1843.572090     1843.572090      4.23     0.0420


                                       Standard
 Parameter             Estimate          Error     t Value   Pr > |t|

 trtB-trtA           9.37497409      4.55823028       2.06     0.0420



                             Least Squares Means

                                                                  H0:LSMean1=
                                      Standard    H0:LSMEAN=0      LSMean2
         TRT    FLUSH LSMEAN             Error     Pr > |t|        Pr > |t|

          A       22.5913628         3.0141710      <.0001          0.0420
          B       31.9663369         3.4193912      <.0001


         Contrast                   Contrast Expected Mean Square

         trtB-trtA             Var(Error) + 4.6613 Var(TRT*STUDY) + Q(TRT)
```

You see that the *F*-test for trtB-trtA is the same as in the fixed case in Output 5.4. The expected mean square for the CONTRAST statement is printed when the RANDOM statement follows the CONTRAST statement. The expected mean square for trtB-trtA in Output 5.12 indicates an appropriate denominator for the *F*-statistic to be $\sigma^2 + 4.66\,\sigma^2_{\text{STUDY*TRT}}$. (Since there is only one degree of freedom for TRT, the sum of squares for the contrast trtB-trtA is the same as the Type III sum of squares for TRT in the ANOVA table. This would not be true with more degrees of freedom for TRT.) There is no mean square with this expectation. Thus, you cannot directly obtain an *F*-statistic for the contrast that has an appropriate denominator. You can use the expected mean squares to determine an appropriate combination of mean squares and then compute the *F*-statistic by hand. In this case, we know from the ANOVA results in Section 5.5.1 that the appropriate combination is 0.66*MS(STUDY*TRT) + 0.34*MS(Error), and has Satterthwaite's approximate DF = 11.69. The appropriate *F*-statistic is then F=1843.57 / (0.66(1064.07) + 0.34(435.82)) = 2.17 and it has significance probability *p*=0.1674.

The ESTIMATE statement cannot be modified to accommodate random effects. Therefore, the standard error and *t*-statistic for ESTIMATE statement are usually invalid with mixed-model data.

You can specify an option E= *effect* in the LSMEANS statement, where *effect* is an effect in the MODEL statement. This sometimes is useful for declaring an appropriate mean square to compute standard errors of differences for LSMEANS, but almost certainly does not specify appropriate computations for standard errors of individual LSMEANS. With unbalanced data there usually is no effect whose expected mean square provides the correct linear combination of variance components.

In summary, the expected mean squares for CONTRAST statements can be useful for determining appropriate combinations of mean squares for the denominator of *F*-statistics. Unfortunately, these are not computed automatically with the TEST option on the RANDOM statement. Standard errors for ESTIMATE and LSMEANS statements cannot be computed correctly in most cases.

Many of the shortcomings of the GLM procedure for analyzing unbalanced mixed-model data can be overcome by using the MIXED procedure, as you will see in the next section.

5.6 Using the MIXED Procedure to Analyze Unbalanced Mixed-Model Data

The MIXED procedure is used in the same way with unbalanced data as it is with balanced data. Run the statements

```
proc mixed data=drugs;
   class study trt;
   model flush=trt / ddfm=satterth;
   random study study*trt;
   contrast 'trtB-trtA';
   estimate 'trtB-trtA';
   lsmeans trt;
run;
```

Results appear in Output 5.13.

Output 5.13
Results of the
MIXED
Procedure
with
Unbalanced
Data

```
                    Unbalanced Two-way Classification

                          The Mixed Procedure

                          Model Information

           Data Set                    WORK.DRUGS
           Dependent Variable          FLUSH
           Covariance Structure        Variance Components
           Estimation Method           REML
           Residual Variance Method    Profile
           Fixed Effects SE Method     Model-Based
           Degrees of Freedom Method   Satterthwaite


                      Covariance Parameter
                            Estimates

                      Cov Parm        Estimate

                      STUDY                  0
                      TRT*STUDY        75.3629
                      Residual          447.57

                  Type 3 Tests of Fixed Effects

                          Num     Den
            Effect        DF      DF     F Value    Pr > F

            TRT            1      9.3     1.88      0.2028

                            Estimates

                            Standard
     Label       Estimate    Error     DF    t Value    Pr > |t|

     trtB-trtA    7.8198     5.7076    9.3    1.37      0.2028

                            Contrasts

                          Num     Den
            Label         DF      DF     F Value    Pr > F

            trtB-trtA      1      9.3     1.88      0.2028

                      Least Squares Means

                              Standard
     Effect    TRT    Estimate   Error     DF    t Value   Pr > |t|

     TRT        A     22.3593    4.0316    9.6    5.55     0.0003
     TRT        B     30.1791    4.0401    9      7.47     <.0001

                                Standard
   Effect   TRT   _TRT    Estimate    Error     DF   t Value   Pr > |t|

   TRT       A     B      -7.8198    5.7076    9.3   -1.37     0.2028
```

First of all, you see that the REML estimate of σ^2_{STUDY} is 0. (Recall that the ANOVA estimate you would obtain using the expected mean squares in Output 5.11, was negative.) The REML estimates of $\sigma^2_{STUDY*TRT}$ and σ^2 are 75.36 and 447.57, respectively.

You see that the *F*-statistic is equal to 1.88, with *p*-value equal to 0.2028 in the "Type 3 Tests of Fixed Effects." This is the test for the TRT null hypothesis $H_0: \alpha_1 - \alpha_2 = 0$. The results are similar to the test using expected mean squares from the GLM procedure in Output 5.11.

The ESTIMATE statement produces an estimated difference equal to 7.82, with standard error 5.71. The resulting *t*-statistic for testing the null hypothesis H_0: $\alpha_1 - \alpha_2 = 0$ has value $t=1.37$ and significance probability $p=0.2079$.

The CONTRAST statement produces results equivalent to the *F*-test in "Type 3 Tests of Fixed Effects."

The LSMEANS statement produces LS means of 22.36 for drug A and 30.18 for treatment B. These are slightly different from the LS means produced by GLM in Output 5.12. More importantly, the standard errors of LS means in Output 5.13 are larger than standard errors in Output 5.12 because the MIXED procedure correctly computes the standard errors. Likewise, the difference between LS means in Output 5.13 is less significant than in Output 5.12 because the MIXED procedure computes a *t*-statistic that is more nearly valid that does the GLM procedure. Details on these computations are described in Chapter 6.

5.7 Using the GLM and MIXED Procedures to Analyze Mixed-Model Data with Empty Cells

Refer once more to the DRUGS1 data set, which has no data for TRT=A in STUDY=41 (See Output 5.7). Run the statements

```
proc glm data=drugs1;
   class study trt;
   model flush=trt study trt*study / ss1 ss2 ss3 ss4;
   random study trt*study / test;
run;
```

You would get exactly the same ANOVA results from the MODEL statement that you saw in Output 5.7.

The RANDOM statement produces Types III and IV expected mean squares, as shown in Output 5.14

Output 5.14
Results of the
RANDOM
Statement with
Empty Cells

```
                         Unbalanced Two-way Classification

                              The GLM Procedure

       Source                Type III Expected Mean Square

       TRT                   Var(Error) + 4.6613 Var(TRT*STUDY) + Q(TRT)

       STUDY                 Var(Error) + 7.8109 Var(TRT*STUDY) + 14.111 Var(STUDY)

       TRT*STUDY             Var(Error) + 7.0585 Var(TRT*STUDY)
                         Unbalanced Two-way Classification
```

*Output 5.14
(Continued)
Results of the
RANDOM
Statement with
Empty Cells*

```
                            The GLM Procedure
              Tests of Hypotheses for Mixed Model Analysis of Variance


   Source                      DF     Type III SS    Mean Square    F Value    Pr > F

   TRT                          1     1843.572090    1843.572090      2.09     0.1724

   Error                   12.498          10999     880.038506
   Error: 0.6604*MS(TRT*STUDY) + 0.3396*MS(Error)


       Source                   Type IV Expected Mean Square

       TRT                      Var(Error) + 4.6613 Var(TRT*STUDY) + Q(TRT)

       STUDY                    Var(Error) + 7.0961 Var(TRT*STUDY) + 13.16 Var(STUDY)

       TRT*STUDY                Var(Error) + 7.0585 Var(TRT*STUDY)
                            Unbalanced Two-way Classification

              Tests of Hypotheses for Mixed Model Analysis of Variance

   Source                      DF     Type IV SS     Mean Square    F Value    Pr > F

   TRT                          1     1843.572090    1843.572090      2.09     0.1724

   Error                   12.498          10999     880.038506
   Error: 0.6604*MS(TRT*STUDY) + 0.3396*MS(Error)
```

Types III and IV expected mean squares are the same for TRT because there are only two levels of the factor. Types III and IV expected mean squares for STUDY differ only slightly. A greater prevalence of empty cells would tend to cause a greater difference between all aspects of Type III and Type IV, including the expected mean squares. Additional detail is presented in Chapter 6 on the Type III and Type IV distinction.

The F-tests based on the Types III and IV mean squares for TRT are the same, with $F=2.09$ and $p=0.1724$. The results could differ if TRT had more levels. As discussed following Output 5.11, there are no definitive reasons for using one of these tests instead of the other. This point is discussed further in Chapter 6.

CONTRAST, ESTIMATE, and LSMEANS statements with PROC GLM would produce the same results as in the fixed-effects case because the LSMEAN for TRT A is non-estimable. (See Output 5.9.)

The MIXED procedure is used in the same way with unbalanced data as it is with balanced data, even with empty cells. Run the statements

```
proc mixed data=drugs;
   class study trt;
   model flush=trt / ddfm=satterth;
   random study study*trt;
   contrast 'trtB-trtA';
   estimate 'trtB-trtA';
   lsmeans trt;
run;
```

Edited results appear in Output 5.15.

Output 5.15
Results of the
MIXED
Procedure
with
Unbalanced
Data

```
                        The Mixed Procedure

                        Model Information

                     Covariance Parameter
                           Estimates

                Cov Parm          Estimate

                STUDY                    0
                TRT*STUDY          77.0369
                Residual           530.50


                 Type 3 Tests of Fixed Effects

                        Num      Den
            Effect       DF       DF     F Value    Pr > F

            TRT           1     12.4       2.96     0.1103


                           Estimates

                         Standard
      Label       Estimate      Error      DF    t Value   Pr > |t|

      trtB-trtA     9.9035     5.7585    12.4      1.72     0.1103


                           Contrasts

                        Num      Den
            Label        DF       DF     F Value    Pr > F

            trtB-trtA     1     12.4       2.96     0.1103


                     Least Squares Means

                              Standard
      Effect   TRT   Estimate    Error     DF    t Value   Pr > |t|

      TRT       A     22.3908    4.2200   13.4      5.31     0.0001
      TRT       B     32.2943    3.9182   11.4      8.24    <.0001
```

Results of tests, estimates and standard errors are similar, but not identical, to those in the case of no empty cell in Output 5.13. Do not expect this to always happen. Generally speaking, with more prevalent empty cells, you can expect more different results.

There is a very important point to be observed in the comparison of results for unbalanced data analysis with and without empty cells. Empty cells cause non-estimability of certain LSMEANS and linear functions of model parameters. Thus, ESTIMATE and CONTRAST statements will not produce output if they specify non-estimable linear functions. This occurred when using PROC GLM with the data set DRUGS1, both for the fixed and mixed-model analyses, because GLM makes the same essential computations with or without a RANDOM statement. In other words, estimability is judged by GLM considering all effects fixed, even though a RANDOM statement is used. PROC MIXED, on the contrary, judges estimability only in terms of fixed effects. That is why complete results were presented for the LSMEANS, ESTIMATE, and CONTRAST statements in Output 5.15.

5.8 Summary and Conclusions about Using the GLM and MIXED Procedures to Analyze Unbalanced Mixed-Model Data

The GLM and MIXED procedures both have certain capabilities for analysis of mixed-model data, as described in Chapter 4. The GLM capabilities are oriented around analysis of variance, based on an ordinary least squares fit of the model, in which random effects are treated as fixed effects. The RANDOM statement in PROC GLM produces expected mean squares, which can be used to construct F-statistics for tests of hypotheses. In balanced data situations, these F-statistics are often "exact," meaning that the distribution of the statistic, under the null hypothesis, has a true F-distribution. In unbalanced data applications, the distributions are only approximate, but still useful, and must be used with caution. Moreover, there are no definitive guidelines for selecting a "type" of sum of squares for the numerator of the F-statistic. Standard errors computed by PROC GLM for LS means, differences between LS means, and ESTIMATE statements are generally unreliable. There are methods for determining appropriate standard errors of estimates from ESTIMATE and LSMEANS statements using the CONTRAST and RANDOM statements (Littell and Linda 1990; Milliken and Johnson 1994, Chapter 28), but these are tedious and are not feasible for many users.

The MIXED procedure, on the other hand, uses true mixed-model methodology. It builds the parameters for the random effects into the statistical model through the covariance structure, using either the RANDOM or REPEATED statement. Test statistics, and estimates, and standard errors of estimates for fixed effects are computed from principles of generalized least squares, with random effects parameters replaced by their estimates (see Chapter 6). Estimates computed in this manner are called *estimated* generalized least squares estimates. They are unbiased, and their standard errors are computed on the basis of a valid formula, except that the standard errors do not account for the variation in the random effects parameter estimates. In most cases, this is not a serious problem. Test statistics for fixed effects are also computed using basically sound methodology, with the same exception that variation due to estimation of the random effects parameters is ignored. Determination of degrees of freedom for variation estimates is complicated, especially in unbalanced data. PROC MIXED allows several options for assessing degrees of freedom.

Chapter 6 Understanding Linear Models Concepts

6.1 Introduction

The purpose of this chapter is to provide detailed information about how PROC GLM and PROC MIXED work for certain applications. Certainly, this is not a complete documentation. Rather, it provides enough information to you for a basic understanding and a basis for further reading.

Both GLM and MIXED utilize "dummy" variables, which are also called "indicator" variables in mathematics. They are created whenever a CLASS statement is specified. The primary distinction between GLM and MIXED in this regard is that MIXED separates the sets of dummy variables into a group for fixed effects and a group for random effects, whereas the primary computations of GLM use the dummy variables as representing fixed effects. The general linear model approach uses dummy variables in a regression model. Although this technique is useful in all situations, it is primarily applied to analysis of variance with **unbalanced** data, where the direct computation of sums of squares fails, and to analysis of covariance and associated techniques.

While the dummy variable approach is capable of handling a vast array of applications, it also presents some complications that must be overcome. Two of the principal complications regarding fixed effects are

❏ specifying model parameters and their estimates

❏ setting up meaningful combinations of parameters for testing and estimation.

Both of these are concerned with *estimable functions*. These complications must be dealt with in computer programs using general linear models. The purpose of this chapter is to explain, with the use of fairly simple examples, how the GLM procedure deals with the complications. A more technical description of GLM features is given in the *SAS/STAT User's Guide, Version 8,* Volume 2.

This chapter describes the essence of general linear model and mixed-model computations. It is more or less self-contained, and you will notice some overlap with previous and subsequent chapters. In particular, the CONTRAST and ESTIMATE statements are discussed in Chapter 3, "Analysis of Variance for Balanced Data," and the RANDOM statement is discussed in Chapter 4, "Analyzing Data with Random Effects." This present chapter delves more deeply into some of the same topics. Section 6.2 provides essential concepts of using dummy variables in the context of a one-way classification. Section 6.3 does the same for a two-way classification with both factors fixed. Then Section 6.4 discusses technical issues for mixed models.

6.2 The Dummy-Variable Model

This section presents the analysis-of-variance model using dummy variables, methods for specifying model parameters, and the methods used by PROC GLM. For simplicity, an analysis-of-variance model with one-way classification that results from a completely randomized design illustrates the discussion. In application, however, such a structure might be adequately (and more efficiently) analyzed by using the ANOVA procedure (see Section 3.4, "Analysis of One-Way Classification of Data").

6.2.1 The Simplest Case: A One-Way Classification

Data for the one-way classification consist of measurements classified according to a one-dimensional criterion. An example of this kind of structure is a set of student exam scores, where each student is taught by one of three teachers. The exam scores are thus grouped or classified according to TEACHER. The most straightforward model for data of this type is

$$y_{ij} = \mu_i + \varepsilon_{ij}$$

where

y_{ij}	represents the jth measurement in the ith group.
μ_i	represents the population mean for the ith group.
ε_{ij}	represents the random error with mean=0 and variance=σ^2.
$i = 1, \ldots, t$	where t equals the number of groups.
$j = 1, \ldots, n_i$	where n_i equals the number of observations in the ith group.

This is called the means or μ-model because it uses the means $\mu_1, \ldots, \mu_t$ as the basic parameters in the mathematical expression for the model (Hocking and Speed 1975). The corresponding estimates of these parameters are

$$\hat{\mu}_1 = \bar{y}_{1.}$$

$$\centerdot$$

$$\centerdot$$

$$\centerdot$$

$$\hat{\mu}_t = \bar{y}_{t.}$$

where $\bar{y}_{i.} = \left(\Sigma_j y_{ij} \right) / n_i$ is the mean of n_i observations in group i.

In these situations, the statistical inference of interest is often about differences between the means of the form $(\mu_i - \mu_{i'})$ or between the means and some reference or baseline value μ. Therefore, many statistical textbooks present a model for the one-way structure that employs these differences as basic parameters. This is the familiar analysis-of-variance model illustrated in Section 2.3.4:

$$y_{ij} = \mu + \tau_i + \varepsilon_{ij}$$

where μ equals the reference value and

$$\tau_i = \mu_i - \mu$$

Thus, the means can be expressed as

$$\mu_i = \mu + \tau_i$$

This relates the set of t means $\mu_i, \ldots, \mu_t$, to a set of $t+1$ parameters, $\mu, \tau_1, \ldots, \tau_t$. Therefore, this model is said to be *overspecified*. Consequently, the parameters, $\mu, \tau_1, \ldots, \tau_t$ are not well defined. For any set of values of $\mu_1, \ldots, \mu_t$, there are infinitely many choices for $\mu, \tau_1, \ldots, \tau_1$, which satisfy the basic equations $\mu_1 = \mu + \tau_1, i = 1, \ldots, t$. The choice may depend on the situation at hand, or it may not be necessary to fully define the parameters.

For the implementation of the dummy-variable model, the analysis-of-variance model

$$y_{ij} = \mu + \tau_i + \varepsilon_{ij}$$

is rewritten as a regression model

$$y_{ij} = \mu + \tau_1 x_1 + \ldots + \tau_t x_t + \varepsilon_{ij}$$

where the dummy variables $x_1, \ldots, x_t$ are defined as follows:

x_1 equals 1 for an observation in group 1 and 0 otherwise.

x_2 equals 1 for an observation in group 2 and 0 otherwise.

.

.

.

x_t equals 1 for an observation in group t and 0 otherwise.

In matrix notation, the model equations for the data become

$$
\mathbf{Y} =
\begin{bmatrix}
y_{11} \\
\cdot \\
\cdot \\
\cdot \\
y_{1n_1} \\
y_{21} \\
\cdot \\
\cdot \\
\cdot \\
y_{2n_2} \\
\cdot \\
\cdot \\
\cdot \\
y_{t1} \\
\cdot \\
\cdot \\
\cdot \\
y_{tn_t}
\end{bmatrix}
=
\begin{bmatrix}
1\,1\,0\,\ldots\,0 \\
\cdot\;\;\cdot\;\;\cdot\;\;\;\;\cdot \\
\cdot\;\;\cdot\;\;\cdot\;\;\;\;\cdot \\
\cdot\;\;\cdot\;\;\cdot\;\;\;\;\cdot \\
1\,1\,0\,\ldots\,0 \\
1\,0\,1\,\ldots\,0 \\
\cdot\;\;\cdot\;\;\cdot\;\;\;\;\cdot \\
\cdot\;\;\cdot\;\;\cdot\;\;\;\;\cdot \\
\cdot\;\;\cdot\;\;\cdot\;\;\;\;\cdot \\
1\,0\,1\,\ldots\,0 \\
\cdot\;\;\cdot\;\;\cdot\;\;\;\;\cdot \\
\cdot\;\;\cdot\;\;\cdot\;\;\;\;\cdot \\
\cdot\;\;\cdot\;\;\cdot\;\;\;\;\cdot \\
1\,0\,0\,\ldots\,1 \\
\cdot\;\;\cdot\;\;\cdot\;\;\;\;\cdot \\
\cdot\;\;\cdot\;\;\cdot\;\;\;\;\cdot \\
\cdot\;\;\cdot\;\;\cdot\;\;\;\;\cdot \\
1\,0\,0\,\ldots\,1
\end{bmatrix}
\begin{bmatrix}
\beta_0 \\
\beta_1 \\
\cdot \\
\cdot \\
\cdot \\
\beta_t
\end{bmatrix}
+
\begin{bmatrix}
\varepsilon_{11} \\
\cdot \\
\cdot \\
\cdot \\
\varepsilon_{1n_1} \\
\varepsilon_{21} \\
\cdot \\
\cdot \\
\cdot \\
\varepsilon_{2n_2} \\
\cdot \\
\cdot \\
\cdot \\
\varepsilon_{t1} \\
\cdot \\
\cdot \\
\cdot \\
\varepsilon_{tn_t}
\end{bmatrix}
= \mathbf{X}\beta + \varepsilon
$$

Thus, the matrices of the normal equations are

$$
\mathbf{X'X} =
\begin{bmatrix}
n & n_1 & n_2 & \ldots & n_t \\
n_1 & n_1 & 0 & \ldots & 0 \\
n_2 & 0 & n_2 & \ldots & 0 \\
\cdot & \cdot & \cdot & & \cdot \\
\cdot & \cdot & \cdot & & \cdot \\
\cdot & \cdot & \cdot & & \cdot \\
n_t & 0 & 0 & \ldots & n_t
\end{bmatrix}
, \quad
\mathbf{X'Y} =
\begin{bmatrix}
Y_{..} \\
Y_{1.} \\
Y_{2.} \\
\cdot \\
\cdot \\
\cdot \\
Y_{t.}
\end{bmatrix}
$$

where $Y_{i.}$ and $Y_{..}$ are totals corresponding to $\bar{y}_{i.}$ and $\bar{y}_{..}$. The normal equations $(\mathbf{X'X})\hat{\beta} = \mathbf{X'Y}$ are equivalent to the set

$$\hat{\mu} + \hat{\tau}_1 = \bar{y}_{1.}$$
$$\hat{\mu} + \hat{\tau}_2 = \bar{y}_{2.}$$
$$\vdots$$
$$\hat{\mu} + \hat{\tau}_t = \bar{y}_{t.}$$

Because there are only t-equations, there is no unique solution for the $(t+1)$ estimates $\hat{\mu}, \hat{\tau}_1, \ldots, \hat{\tau}_t$. Corresponding to this, the $\mathbf{X'X}$ matrix describing the set of normal equations is of dimension $(t+1)\times(t+1)$ and of rank t. In this model the first row of $\mathbf{X'X}$ is equal to the sum of the other t-rows. The same relationship exists among the columns of $\mathbf{X'X}$. Therefore, $\mathbf{X'X}$ is said to be of less than full rank.

6.2.2 Parameter Estimates for a One-Way Classification

There are two popular methods for obtaining estimates with a less-than-full-rank model. Restrictions can be imposed on the parameters to obtain a full-rank model, or a generalized inverse of $\mathbf{X'X}$ can be obtained. PROC GLM uses the latter method. This section reviews both methods in order to put the approach used by PROC GLM into perspective.

The restrictions method is based on the fact that any definition of one of the parameters in the model (say the reference parameter) causes the other parameters to be uniquely defined. The definition can be restated in the form of a restriction. Another view of the term restriction is to define the parameters to have a unique interpretation. The corresponding estimates are then required to coincide with the definition of the parameters.

One type of restriction is to define one of the τ_i equal to 0, say $\tau_t = 0$. In this case, μ becomes the mean of the tth group $\mu_\tau = \mu + \tau_t = \mu$, and τ_i becomes the difference between the mean for the ith group and the mean for the tth group, $\tau_i = \mu_i - \mu = \mu_i - \mu_t$.

The corresponding restriction on the solution to the normal equations is to require $\hat{\tau}_t = 0$. Requiring $\hat{\tau}_t = 0$. leads automatically to a unique set of values for the remaining set of estimates $\hat{\mu}, \hat{\tau}_1, \ldots, \hat{\tau}_{t-1}$. This occurs because τ_t is dropped from the linear model. Consequently, the column corresponding to τ_t is dropped from the $\mathbf{X}$ matrix, producing the following model equation:

$$
\begin{bmatrix}
y_{11} \\
\cdot \\
\cdot \\
\cdot \\
y_{1n_1} \\
\cdot \\
\cdot \\
y_{t-1,1} \\
\cdot \\
\cdot \\
\cdot \\
y_{t-1,n_{t-1}} \\
y_{t1} \\
\cdot \\
\cdot \\
\cdot \\
y_{tn_t}
\end{bmatrix}
=
\begin{bmatrix}
1\,1 \ldots 0 \\
\cdot\ \ \cdot\ \ \ \cdot \\
\cdot\ \ \cdot\ \ \ \cdot \\
\cdot\ \ \cdot\ \ \ \cdot \\
1\,1 \ldots 0 \\
\cdot\ \ \cdot\ \ \ \cdot \\
\cdot\ \ \cdot\ \ \ \cdot \\
1\,0 \ldots 1 \\
\cdot\ \ \cdot\ \ \ \cdot \\
\cdot\ \ \cdot\ \ \ \cdot \\
\cdot\ \ \cdot\ \ \ \cdot \\
1\,0 \ldots 1 \\
1\,0 \ldots 0 \\
\cdot\ \ \cdot\ \ \ \cdot \\
\cdot\ \ \cdot\ \ \ \cdot \\
\cdot\ \ \cdot\ \ \ \cdot \\
1\,0 \ldots 0
\end{bmatrix}
\begin{bmatrix}
\mu \\
\tau_1 \\
\cdot \\
\cdot \\
\cdot \\
\tau_{t-1}
\end{bmatrix}
+
\begin{bmatrix}
\varepsilon_{11} \\
\cdot \\
\cdot \\
\cdot \\
\varepsilon_{1n_1} \\
\cdot \\
\cdot \\
\varepsilon_{t-1,1} \\
\cdot \\
\cdot \\
\cdot \\
\varepsilon_{t-1,n_{t-1}} \\
\varepsilon_{t-1} \\
\cdot \\
\cdot \\
\cdot \\
\varepsilon_{tn_t}
\end{bmatrix}
$$

The solution to the corresponding normal equation $(\mathbf{X'X})\hat{\beta} = \mathbf{X'Y}$, where $\mathbf{X'X}$ is now nonsingular, results in

$$\hat{\mu} = \bar{y}_{t.}$$
$$\hat{\tau}_1 = \bar{y}_{1.} - \bar{y}_{t.}$$
$$\hat{\tau}_2 = \bar{y}_{2.} - \bar{y}_{t.}$$
$$\cdot$$
$$\cdot$$
$$\cdot$$
$$\hat{\tau}_{(t-1)} = \bar{y}_{(t-1).} - \bar{y}_{t.}$$

Another approach defines μ to be equal to the mean of $\mu_1, \mu_2, \ldots, \mu_t$; —that is, $\mu = (\mu_1 + \mu_2 + \ldots + \mu_t)/t$. Then μ is called the grand mean and the τ_i are called the group effects. From this definition of μ, it follows that $\Sigma_i \tau_i = 0$. Consequently,

$$\tau_1 = -\tau_1 - \tau_2 - \ldots - \tau_{t-1}$$

Therefore, observations $y_{tj} = \mu + \tau_t + \varepsilon_{ij}$ in the tth group can be written

$$y_{tj} = \mu - \tau_1 - \tau_2 - \ldots - \tau_{t-1} + \varepsilon_{ij}$$

The parameter τ_t is dropped from the model, which now becomes

$$
\begin{bmatrix}
y_{11} \\
\cdot \\
\cdot \\
\cdot \\
y_{1n_1} \\
\cdot \\
\cdot \\
\cdot \\
y_{t-1,1} \\
\cdot \\
\cdot \\
\cdot \\
y_{t-1,n_{t-1}} \\
y_{t1} \\
\cdot \\
\cdot \\
\cdot \\
y_{tn_t}
\end{bmatrix}
=
\begin{bmatrix}
1\ 1\ 0\ \ldots 0 \\
\cdot\ \cdot\ \cdot\ \ \ \cdot \\
\cdot\ \cdot\ \cdot\ \ \ \cdot \\
1\ 1\ 0\ \ldots 0 \\
\cdot\ \cdot\ \cdot\ \ \ \cdot \\
\cdot\ \cdot\ \cdot\ \ \ \cdot \\
\cdot\ \cdot\ \cdot\ \ \ \cdot \\
1\ 0\ 0\ \ldots 1 \\
\cdot\ \cdot\ \cdot\ \ \ \cdot \\
\cdot\ \cdot\ \cdot\ \ \ \cdot \\
\cdot\ \cdot\ \cdot\ \ \ \cdot \\
1\ 0\ \ 0 \ldots 1 \\
1 -1 -1 \ldots -1 \\
\cdot\ \cdot\ \cdot\ \ \ \cdot \\
\cdot\ \cdot\ \cdot\ \ \ \cdot \\
\cdot\ \cdot\ \cdot\ \ \ \cdot \\
1 -1 -1 \ldots -1
\end{bmatrix}
\begin{bmatrix}
\mu \\
\tau_1 \\
\tau_2 \\
\cdot \\
\cdot \\
\cdot \\
\tau_{t-1}
\end{bmatrix}
+
\begin{bmatrix}
\varepsilon_{11} \\
\cdot \\
\cdot \\
\cdot \\
\varepsilon_{1n_1} \\
\cdot \\
\cdot \\
\cdot \\
\varepsilon_{t-1,1} \\
\cdot \\
\cdot \\
\cdot \\
\varepsilon_{t-1,n_{t-1}} \\
\varepsilon_{t1} \\
\cdot \\
\cdot \\
\cdot \\
\varepsilon_{tn_t}
\end{bmatrix}
$$

The solution to the corresponding normal equation yields

$$\hat{\mu} = (\bar{y}_{1.} + \ldots + \bar{y}_{t.}) / t$$
$$\hat{\tau}_1 = \bar{y}_{1.} - \bar{y}_{..}$$
$$\hat{\tau}_2 = \bar{y}_{2.} - \bar{y}_{..}$$

$$\cdot$$

$$\cdot$$

$$\cdot$$

$$\hat{\tau}_{t-1} = \bar{y}_{(t-1).} - \bar{y}_{..}$$

and the implementation of the condition $\tau_1 = -\tau_1 - \tau_2 - \ldots - \tau_{t-1}$ yields

$$\hat{\tau}_t = \bar{y}_{t.} - \bar{y}_{..}$$

The use of generalized inverses and estimable functions may be preferable for a variety of reasons. In the restrictions method, it might not be clear which particular restriction is desired. In cases of empty cells in multiway classifications, it can be difficult to define the parameters. In fact, it is often hard to identify the empty cells in large, multiway classifications, let alone to define a set of parameters that adequately describe all pertinent effects and interactions. The generalized-inverse approach partially removes the burden of defining parameters from the data analyst.

Section 2.4.4, "Using the Generalized Inverse," showed that there is no unique solution to a system of equations with a less-than-full-rank coefficient matrix and introduced the generalized inverse to obtain a nonunique solution. Although the set of parameter estimates produced using the generalized inverse is not unique, there is a class of linear functions of parameters called **estimable functions** for which unique estimates do exist. For example, the function $(\tau_i - \tau_j)$ is estimable: its least-squares estimate is the same regardless of the particular solution obtained for the normal equations. For a discussion of the definition of estimable functions as it relates to the theory of linear models, see Graybill (1976) or Searle (1971).

PROC GLM uses a generalized inverse to obtain a solution that produces one set of estimates. The technique, in some respects, is parallel to using a set of restrictions that set some of the parameter estimates to 0. Quantities to be estimated or comparisons to be made are specified, and PROC GLM determines whether or not the estimates or comparisons represent estimable functions. PROC GLM then provides estimates, standard errors, and test statistics.

For certain applications, there is more than one set of hypotheses that can be tested. To cover these situations, PROC GLM provides four types of sums of squares and associated F-statistics and also gives additional information to assist in interpreting the hypotheses tested.

6.2.3 Using PROC GLM for Analysis of Variance

Using PROC GLM for analysis of variance is similar to using PROC ANOVA; the statements listed for PROC ANOVA in Section 3.3.2, "Using the ANOVA and GLM Procedures," are also used for PROC GLM. In addition to the statements listed for PROC ANOVA, the following SAS statements can be used with PROC GLM:

CONTRAST *'label' effect values< . . . effect values> < / options>*;
ESTIMATE *'label' effect values< . . . effect values> < / options>*;
ID *variables*;
LSMEANS *effects< / options>*;
OUTPUT *<OUT=SAS-data-set> keyword= names < . . . keyword=names>*;
RANDOM *effects< / options>*;
WEIGHT *variable*;

The CONTRAST statement provides a way of obtaining custom hypotheses tests. The ESTIMATE statement can be used to estimate linear functions of the parameters. The LSMEANS (least-squares means) statement specifies effects for which least-squares estimates of means are computed. The uses of these statements are illustrated in Section 6.2.4, "Estimable Functions in the One-Way Classification," and Section 6.3.6, "MEANS, LSMEANS, CONTRAST, and ESTIMATE Statements in the Two-Way Layout." The RANDOM statement specifies which effects in the model are random (see Section 6.4.1, "Proper Error Terms"). When predicted values are requested as a MODEL statement option, values of the variable specified in the ID statement are printed for identification beside each observed, predicted, and residual value. The OUTPUT statement produces an output data set that contains the original data set values along with predicted and residual values. The WEIGHT statement is used when a weighted residual sum of squares is needed. For more information, refer to Chapter 24 in the *SAS/STAT User's Guide, Version 8,* Volume 2.

Implementing PROC GLM for an analysis-of-variance model is illustrated by an example of test scores made by students in three classes taught by three different teachers. The data appear in Output 6.1.

```
                        The SAS System

            Obs      teach    score1     score2

             1       JAY        69         75
             2       JAY        69         70
             3       JAY        71         73
             4       JAY        78         82
             5       JAY        79         81
             6       JAY        73         75
             7       PAT        69         70
             8       PAT        68         74
             9       PAT        75         80
            10       PAT        78         85
            11       PAT        68         68
            12       PAT        63         68
            13       PAT        72         74
            14       PAT        63         66
            15       PAT        71         76
            16       PAT        72         78
            17       PAT        71         73
            18       PAT        70         73
            19       PAT        56         59
            20       PAT        77         83
            21       ROBIN      72         79
            22       ROBIN      64         65
            23       ROBIN      74         74
            24       ROBIN      72         75
            25       ROBIN      82         84
            26       ROBIN      69         68
            27       ROBIN      76         76
            28       ROBIN      68         65
            29       ROBIN      78         79
            30       ROBIN      70         71
            31       ROBIN      60         61
```

In terms of the analysis-of-variance model described above, the τ_i are the parameters associated with the different teachers (TEACH)—τ_1 is associated with JAY, τ_2 with PAT, and τ_3 with ROBIN. The following SAS statements are used to analyze SCORE2:

```
proc glm;
   class teach;
   model score2=teach / solution xpx i;
```

In this example, the CLASS variable TEACH identifies the three classes. In effect, PROC GLM establishes a dummy variable (1 for presence, 0 for absence) for each level of each CLASS variable. In this example, the CLASS statement causes PROC GLM to create dummy variables corresponding to JAY, PAT, and ROBIN, resulting in the following **X** matrix:

INTERCEPT JAY PAT ROBIN

$$
\mathbf{X} = \begin{array}{cccc}
\mu & \tau_1 & \tau_2 & \tau_3 \\
1 & 1 & 0 & 0 \\
\cdot & \cdot & \cdot & \cdot \\
\cdot & \cdot & \cdot & \cdot \\
\cdot & \cdot & \cdot & \cdot \\
1 & 1 & 0 & 0 \\
1 & 0 & 1 & 0 \\
\cdot & \cdot & \cdot & \cdot \\
\cdot & \cdot & \cdot & \cdot \\
\cdot & \cdot & & \cdot & \cdot \\
1 & 0 & 1 & 0 \\
1 & 0 & 0 & 1 \\
\cdot & \cdot & \cdot & \cdot \\
\cdot & \cdot & \cdot & \cdot \\
\cdot & \cdot & \cdot & \cdot \\
1 & 0 & 0 & 1
\end{array}
$$

6 rows for Jay's group

14 rows for Pat's group

11 rows for Robin's group

Note that the columns for the dummy variables are in alphabetical order; the column positioning depends only on the values of the CLASS variable. For example, the column for JAY would appear after the columns for PAT and ROBIN if the value JAY were replaced by ZJAY.[1]

The MODEL statement has the same purpose in PROC GLM as it does in PROC REG and PROC ANOVA. Note that the MODEL statement contains the SOLUTION option. This option is used because PROC GLM does not automatically print the estimated parameter vector when a model contains a CLASS statement. The results of the SAS statements shown above appear in Output 6.2.

---

[1] You can use the ORDER= option in the PROC GLM statement to alter the column position.

Output 6.2
One-Way
Analysis of
Variance from
PROC GLM

```
                              The GLM Procedure

Dependent Variable: score2

                                  Sum of
Source                   DF      Squares     Mean Square   F Value   Pr > F

Model                     2     49.735861     24.867930      0.56    0.5776

Error                    28   1243.941558     44.426484

Corrected Total          30   1293.677419


          R-Square     Coeff Var     Root MSE     score2 Mean

          0.038445     9.062496      6.665320      73.54839


Source                   DF     Type I SS     Mean Square   F Value   Pr > F

teach                     2   49.73586091     24.86793046     0.56    0.5776


Source                   DF    Type III SS    Mean Square   F Value   Pr > F

teach                     2   49.73586091     24.86793046     0.56    0.5776


                                              Standard
       Parameter            Estimate            Error      t Value    Pr > |t|

       Intercept         72.45454545 B       2.00966945      36.05     <.0001
       teach   JAY         3.54545455 B      3.38277775       1.05     0.3036
       teach   PAT         0.90259740 B      2.68553376       0.34     0.7393
       teach   ROBIN       0.00000000 B          .             .         .

NOTE: The X'X matrix has been found to be singular, and a generalized inverse was used to
solve the normal equations. Terms whose estimates are followed by the letter 'B' are not
uniquely estimable.
```

The first portion of the output, as in previous examples, shows the statistics for the overall model.

The second portion partitions the model sum of squares (MODEL SS) into portions corresponding to factors defined by the list of variables in the MODEL statement. In this model there is only one factor, TEACH, so the Type I and Type III SS are the same as the MODEL SS. Type II and Type IV have no special meaning here and would be the same as Type I and Type III.

The final portion of the output contains the parameter estimates obtained with the generalized inverses. Specifying XPX and I in the list of options in the MODEL statement causes the **X'X** and **(X'X)⁻** matrices to be printed. Results appear in Output 6.3.

Output 6.3
XX and (XX)
Matrices for a
One-Way
Classification

```
                              The GLM Procedure

                              The X'X Matrix

                  Intercept      teach JAY      teach PAT     teach ROBIN        score2

Intercept            31              6             14             11            2280
teach JAY             6              6              0              0             456
teach PAT            14              0             14              0            1027
teach ROBIN          11              0              0             11             797
score2             2280            456           1027            797          168984
                              The SAS System   14:53 Wednesday, November 14, 2001    3


                        X'X Generalized Inverse (g2)

                  Intercept      teach JAY      teach PAT     teach ROBIN        score2

Intercept        0.0909090909   -0.090909091   -0.090909091          0     72.454545455
teach JAY       -0.090909091     0.2575757576    0.0909090909         0      3.5454545455
teach PAT       -0.090909091     0.0909090909    0.1623376623         0      0.9025974026
teach ROBIN              0                0               0            0              0
score2          72.454545455     3.5454545455    0.9025974026         0   1243.9415584
```

For this example, the matrix $\mathbf{X'Y}$ is

$$\mathbf{X'Y} = \begin{matrix} \text{SCORE2 total overall} \\ \text{SCORE2 total for JAY} \\ \text{SCORE2 total for PAT} \\ \text{SCORE2 total for ROBIN} \end{matrix} = \begin{matrix} 2280 \\ 456 \\ 1027 \\ 797 \end{matrix}$$

Taking $(\mathbf{X'X})^-$ from the PROC GLM output and using $\mathbf{X'Y}$ above, the solution $\hat{\beta} = (\mathbf{X'X})^- \mathbf{X'Y}$ is

$$\begin{bmatrix} \hat{\beta}_0 \\ \hat{\beta}_1 \\ \hat{\beta}_2 \\ \hat{\beta}_3 \end{bmatrix} = \begin{bmatrix} .0909 & -.0909 & -.0909 & .0000 \\ -.0909 & .2575 & .0909 & .0000 \\ -.0909 & .0909 & .1623 & .0000 \\ .0000 & .0000 & .0000 & .0000 \end{bmatrix} \begin{bmatrix} 2280 \\ 456 \\ 1027 \\ 797 \end{bmatrix} = \begin{bmatrix} 72.45 \\ 3.54 \\ 0.90 \\ 0.00 \end{bmatrix}$$

As pointed out in Section 2.4.4., the particular generalized inverse used by PROC GLM causes the last row and column of $(\mathbf{X'X})^-$ to be set to 0. This yields a set of parameter estimates equivalent in this example to the set given by the restriction that $\tau_3 = 0$. Using the principles discussed in Section 6.2.2, "Parameter Estimates for a One-Way Classification," it follows that the INTERCEPT μ is actually the mean for the reference group ROBIN. The estimate $\hat{\tau}_1$ labeled JAY is the difference between the mean for Jay's group and the mean for Robin's group, and similarly, the estimate $\hat{\tau}_2$ labeled PAT is the mean for Pat's group minus the mean for Robin's group. Finally, the estimate $\hat{\tau}_3$ labeled ROBIN, which is set to 0, can be viewed as the mean for Robin's group minus the mean for Robin's group.

Remember that these estimates are not unique—that is, they depend on the alphabetical order of the values of the CLASS variable. This fact is recognized in the output by denoting the estimates as biased, which is explained in the note after the listing of estimates.

The other MODEL statement options (P, CLM, CLI, and TOLERANCE), as well as the BY, ID, WEIGHT, FREQ, and OUTPUT statements, are not affected by the use of CLASS variables and may be used as described in Section 2.2.4, "The SS1 and SS2 Options: Two Types of Sums of Squares" and Section 2.2.5, "Tests of Subsets and Linear Combinations of Coefficients."

6.2.4 Estimable Functions in a One-Way Classification

It is often the case that the particular parameter estimates obtained by the SOLUTION option in PROC GLM are not the estimates of interest, or there may be additional functions of the parameters that you want to estimate. You can specify such other estimates with PROC GLM.

An estimable function is a member of a special class of linear functions of parameters (see Section 2.2.4). An estimable function of the parameters has a definite interpretation regardless of how the parameters themselves are specified. Denote with $\mathbf{L}$ a vector of coefficients $\left(L_1, L_2, \ldots, L_t, L_{t+1}\right)$. Then $\mathbf{L}\boldsymbol{\beta} = L_1\mu + L_2\tau_1 + \ldots + L_{t+1}\,\tau_t$ is a linear function of the model parameters and is estimable (for this example) if it can be expressed as a linear function of the means $\mu_1, \ldots, \mu_t$. Let $\hat{\boldsymbol{\beta}}$ be a solution to the normal equation. The function $\mathbf{L}\boldsymbol{\beta}$ is estimated by $\mathbf{L}\hat{\boldsymbol{\beta}}$, the corresponding linear function of the parameters. If $\mathbf{L}\boldsymbol{\beta}$ is estimable, then $\mathbf{L}\hat{\boldsymbol{\beta}}$ will have the same value regardless of the solution obtained from the normal equations. In the example,

$$\hat{\boldsymbol{\beta}} = \begin{matrix} \bar{\mu} \\ \hat{\tau}_1 \\ \hat{\tau}_2 \\ \hat{\tau}_3 \end{matrix} = \begin{matrix} 72.454 & \text{INTERCEPT} \\ 3.545 & \text{JAY} \\ 0.902 & \text{PAT} \\ 0.000 & \text{ROBIN} \end{matrix}$$

To illustrate, define

$$\mathbf{L} = \begin{bmatrix} 1 & 1 & 0 & 0 \end{bmatrix}$$

Then $\mathbf{L}\hat{\boldsymbol{\beta}} = \hat{\mu} + \hat{\tau}_1 = \hat{\mu}_1 = 76.0$, which is the estimate of the mean score of Jay's group. Alternately, let

$$\mathbf{L} = \begin{bmatrix} 0 & +1 & -1 & 0 \end{bmatrix}$$

Then $\mathbf{L}\hat{\boldsymbol{\beta}} = \left(\hat{\tau}_1 - \hat{\tau}_2\right) = \hat{\mu}_1 - \hat{\mu}_2 = 2.643$, the estimated mean difference between `Jay`'s and `Pat`'s groups. Because both of these are estimable functions, identical estimates would be obtained using a different generalized inverse—for example, if different names for the teachers changed the order of the dummy variables.

Variances of these estimates can be readily obtained with standard formulas that involve elements of the generalized inverse (see Section 2.2.4).

You can obtain the general form of the estimable functions with the E option in the MODEL statement

```
model score2 = teach / e ;
```

Output 6.4 shows you that L4, the coefficient or τ_3, must be equal to L1 – L2 – L3. Equivalently, L1 = L2 + L3 + L4. That is, the coefficient on μ must be the sum of the coefficients on τ_1, τ_2 and τ_3.

Output 6.4
Obtaining the
General Form
of Estimable
Functions
Using the E
Option

```
                    The GLM Procedure

            General Form of Estimable Functions

            Effect                Coefficients

            Intercept             L1

            teach      JAY        L2
            teach      PAT        L3
            teach      ROBIN      L1-L2-L3
```

PROC GLM calculates estimates and variances for several special types of estimable functions with LSMEANS, CONTRAST, or ESTIMATE statements as well as estimates of user-supplied functions.

The LSMEANS statement produces the least-squares estimates of CLASS variable means—these are sometimes referred to as adjusted means. For the one-way structure, these are simply the ordinary means. In terms of model parameter estimates, they are $\hat{\mu} + \hat{\tau}_i$. The following SAS statement lists the least-squares means for the three teachers for all dependent variables in the MODEL statement:

```
lsmeans teach / options;
```

The available options in the LSMEANS statement are

STDERR   prints the standard errors of each estimated least-squares mean and the *t*-statistic for a test of the hypothesis that the mean is 0.

PDIFF   prints the *p*-values for the tests of equality of all pairs of CLASS means.

E   prints a description of the linear function used to obtain each least-squares mean; this has importance in more complex situations.

E=   specifies an effect in the model to use as an error term.

ETYPE=   specifies the type (1, 2, 3, or 4) of the effect specified in the E= option.

SINGULAR=   tunes the estimability checking.

For more information, refer to Chapter 24 in the *SAS/STAT User's Guide, Version 8*, Volume 2. Output 6.5 shows results from the following SAS statement:

```
lsmeans teach / stderr pdiff;
```

Output 6.5
Results of the
LSMEANS
Statement

```
                              The GLM Procedure
                            Least Squares Means

                       score2        Standard                      LSMEAN
          teach        LSMEAN          Error      Pr > |t|         Number

          JAY        76.0000000      2.7211053     <.0001             1
          PAT        73.3571429      1.7813816     <.0001             2
          ROBIN      72.4545455      2.0096694     <.0001             3


                   Least Squares Means for effect teach
                   Pr > |t| for H0: LSMean(i)=LSMean(j)

                        Dependent Variable: score2

              i/j            1            2            3

               1                       0.4233       0.3036
               2         0.4233                     0.7393
               3         0.3036       0.7393


  NOTE: To ensure overall protection level, only probabilities associated with pre-planned
        comparisons should be used.
```

The least-squares mean for JAY is computed as $\hat{\mu} + \hat{\tau}_1 = 72.45 + 3.54$. Note that this linear function has coefficients L1=1, L2=1, L3=0 and L4=0, so it meets the estimability condition L1 = L2 + L3 + L4.

Least-squares means should not, in general, be confused with ordinary means, which are available with a MEANS statement. The MEANS statement produces simple, unadjusted means of all observations in each class or treatment. Except for one-way designs and some nested and balanced factorial structures that are normally analyzed with PROC ANOVA, these unadjusted means are generally not equal to the least-squares means. Note that for this example, the least-squares means are the same as the means obtained with the MEANS statement. (The MEANS statement is discussed in Section 3.4.2.)

A contrast is a linear function such that the elements of the coefficient vector sum to 0 for each effect. PROC GLM can be instructed to calculate a sum of squares and associated *F*-test due to one or more contrasts.

As an example, assume that teacher JAY used a special teaching method. You might then be interested in testing whether Jay's students had mean scores different from the students of the other teachers, and whether PAT and ROBIN, using the same method, produced different mean scores. The corresponding contrasts are shown below:

Multipliers for TEACH

Contrast	JAY	PAT	ROBIN
JAY vs others	−2	+1	+1
PAT vs ROBIN	0	−1	+1

Taking $\beta = \left(\mu,\ \tau_1,\ \tau_2,\ \tau_3 \right)$, the contrasts are

$$\begin{aligned} \mathbf{L}\beta &= -2\mu_1 + \mu_2 + \mu_3 \quad (\text{JAY vs others}) \\ &= -2\tau_1 + \tau_2 + \tau_3 \end{aligned}$$

and

$$\begin{aligned} \mathbf{L}\beta &= -\mu_2 + \mu_3 \quad (\text{PAT vs ROBIN}) \\ &= -\tau_2 + \tau_3 \end{aligned}$$

The corresponding CONTRAST statements are as follows:

```
contrast 'JAY vs others' teach -2 1 1;
contrast 'PAT vs ROBIN' teach 0 -1 1;
```

The results appear in Output 6.6.

Output 6.6
Results of the
CONTRAST
and
ESTIMATE
Statements

```
                          The GLM Procedure

Contrast                 DF    Contrast SS    Mean Square    F Value    Pr > F

JAY vs others             1    46.19421179    46.19421179       1.04    0.3166
PAT vs ROBIN              1     5.01844156     5.01844156       0.11    0.7393


                                             Standard
         Parameter              Estimate        Error    t Value    Pr > |t|

         LSM JAY              76.0000000     2.72110530      27.93     <.0001
         LSM PAT              73.3571429     1.78138157      41.18     <.0001
         LSM ROBIN           72.4545455     2.00966945      36.05     <.0001
```

Keep the following points in mind when using the CONTRAST statement:

❏ You must know how many classes (categories) are present in the effect and in what order they are sorted by PROC GLM. If there are more effects (classes) in the data than the number of coefficients specified in the CONTRAST statement, PROC GLM adds trailing zeros. In other words, there is no check to see if the proper number of classes has been specified.

❏ The name or label of the contrast must be 20 characters or less.

❏ Available CONTRAST statement options are

E   prints the entire **L** vector.

E=*effect*   specifies an alternate error term.

ETYPE=*n*   specifies the type (1, 2, 3, or 4) of the E=*effect*.

❏ Multiple degrees-of-freedom contrasts can be specified by repeating the effect name and coefficients as needed. Thus, the statement

```
contrast 'ALL' teach -2 1 1, teach 0 -1 1;
```

produces a two DF sum of squares due to both contrasts. This feature can be used to obtain partial sums of squares for effects through the reduction principle, using sums of squares from multiple degrees-of-freedom contrasts that include and exclude the desired contrasts.

❑ If a non-estimable contrast has been specified, a message to that effect appears in the SAS log.

❑ Although only $(t-1)$ linearly independent contrasts exist for t classes, any number of contrasts can be specified.

❑ The contrast sums of squares are not partial of (adjusted for) other contrasts that may be specified for the same effect (see the fourth point above).

❑ The CONTRAST statement is not available with PROC ANOVA; thus, the computational inefficiency of PROC GLM for analyzing balanced data may be justified if contrasts are required. However, contrast variables can be defined in a DATA step and estimates and statistics can be obtained by a full-rank regression analysis.

The ESTIMATE statement is used to obtain statistics for estimable functions other than least-squares means and contrasts, although it can also be used for these. For the current example, the ESTIMATE statement is used to re-estimate the least-squares means.

The respective least-squares means for JAY, PAT, and ROBIN estimate $\mu_1 = \mu + \tau_1$, $\mu_2 = \mu + \tau_2$, and $\mu_3 = \mu + \tau_3$. The following statements duplicate the least-squares means:

```
estimate 'LSM JAY' intercept 1 teach 1;
estimate 'LSM PAT' intercept 1 teach 0 1;
estimate 'LSM ROBIN' intercept 1 teach 0 0 1;
```

Note the use of the term INTERCEPT (referring to μ) and the fact that the procedure supplies trailing zero-valued coefficients. The results of these statements appear after the listing of parameter estimates at the bottom of Output 6.6 for convenient comparison with the results of the LSMEANS statement.

6.3 Two-Way Classification: Unbalanced Data

The major applications of the two-way structure are the two-factor factorial experiment and the randomized blocks. These applications usually have balanced data. In this section, the two-way classification with unbalanced data is explored. This introduces new questions, such as how means and sums of squares should be computed.

6.3.1 General Considerations

The two-way classification model is

$$y_{ijk} = \mu + \alpha_i + \beta_j + (\alpha\beta)_{ij} + \varepsilon_{ijk}$$

where

y_{ijk} equals the kth observed score for the (i, j)th cell.

α_i equals the effect of the ith level of factor A.

β_j equals an effect of the jth level of factor B.

$(\alpha\beta)_{ij}$ equals the interaction effect for the ith level of factor A and the jth level of factor B.

ε_{ijk} equals the random error associated with individual observations.

The model can be defined without the interaction term when appropriate. Let n_{ij} denote the number of observations in the cell for level i of A and level j of B. If μ_{ij} denotes the population cell mean for level i of A and level j of B, then

$$\mu_{ij} = \mu + \alpha_i + \beta_j + (\alpha\beta)_{ij}$$

At this point, no further restrictions on the parameters are assumed.

The computational formulas for PROC ANOVA that use the various treatment means provide correct statistics for balanced data—that is, data with an equal number of observations ($n_{ij}=n$ for all i, j) for each treatment combination. When data are not balanced, sums of squares computed by PROC ANOVA can contain functions of the other parameters of the model, and thereby produce biased results.

To illustrate the effects of unbalanced data on the estimation of differences between means and computation of sums of squares, consider the data in this two-way table:

		B	
		1	2
A	1	7, 9	5
	2	8	4, 6

Within level 1 of B, the cell mean for each level of A is 8—that is, $\bar{y}_{11.} = (7+9)/2 = 8$ and $\bar{y}_{21.} = 8$. Hence, there is no evidence of a difference between the levels of A within level 1 of B. Similarly, there is no evidence of a difference between levels of A within level 2 of B, because $\bar{y}_{12.} = 5$ and $\bar{y}_{22.} = (4+6)/2 = 5$. Therefore, you may conclude that there is no evidence in the table of a difference between the levels of A. However, the marginal means for A are

$$\bar{y}_{1..} = (7+9+5)/3 = 7$$

and

$$\bar{y}_{2..} = (8+4+6)/3 = 6$$

The difference of 7–6=1 between these marginal means may be erroneously interpreted as measuring an overall effect of the factor A. Actually, the observed difference between the marginal means for the two levels of A measures the effect of factor B in addition to the effect of factor A. This can be verified by expressing the observations in terms of the analysis-of-variance model $y_{ijk} = \mu + \alpha_i + \beta_j$.(For simplicity, the interaction and error terms have been left out of the model.)

B

	1	2
	$7 = \mu + \alpha_1 + \beta_1$	
1		$5 = \mu + \alpha_1 + \beta_1$
	$9 = \mu + \alpha_1 + \beta_2$	
A		$4 = \mu + \alpha_2 + \beta_2$
2	$8 = \mu + \alpha_2 + \beta_1$	
		$6 = \mu + \alpha_2 + \beta_2$

The difference between marginal means for A_1 and A_2 is shown to be

$$
\begin{aligned}
\bar{y}_{1..} - \bar{y}_{2..} &= (1/3)[(\alpha_1 + \beta_1) + (\alpha_1 + \beta_1) + (\alpha_1 + \beta_2)] \\
&\quad -(1/3)[(\alpha_2 + \beta_1) + (\alpha_2 + \beta_2) + (\alpha_2 + \beta_2)] \\
&= (\alpha_1 - \alpha_2) + (1/3)(\beta_1 - \beta_2)
\end{aligned}
$$

Thus, instead of estimating $(\alpha_1 - \alpha_2)$, the difference between the marginal means of A estimates $(\alpha_1 - \alpha_2)$ plus a function of the factor B parameters $(\beta_1 - \beta_2)/3$. In other words, the difference between the A marginal means is biased by factor B effects.

The null hypothesis about A that would normally be tested is

$$H_0 : \alpha_1 - \alpha_2 = 0$$

However, for this example, the sum of squares for A computed by PROC ANOVA can be shown to equal $3(\bar{y}_{1..} - \bar{y}_{2..})^2 / 2$. Hence, the PROC ANOVA F-test for A actually tests the hypothesis

$$H_0 : (\alpha_1 - \alpha_2) + (\beta_1 - \beta_2)/3 = 0$$

which involves the factor B difference $(\beta_1 - \beta_2)$ in addition to the factor A difference $(\alpha_1 - \alpha_2)$.

In terms of the μ model $y_{ijk} = \mu_{ij} + \varepsilon_{ijk}$, you usually want to estimate $(\mu_{11} + \mu_{12})/2$ and $(\mu_{21} + \mu_{22})/2$ or the difference between these quantities. However, the A marginal means for the example are

$$\bar{y}_{1..} = (2\mu_{11} + \mu_{12})/3 + \bar{\varepsilon}_{1..}$$

and

$$\bar{y}_{2..} = (\mu_{21} + 2\mu_{22})/3 + \bar{\varepsilon}_{2..}$$

These means estimate $2(\mu_{11} + \mu_{22})/3$ and $(\mu_{21} + \mu_{22})/3$, which are functions of the cell frequencies and might not be meaningful.

In summary, a major problem in the analysis of unbalanced data is the contamination of differences between factor means by effects of other factors. The solution to this problem is to adjust the means to remove the contaminating effects.

6.3.2 Sums of Squares Computed by PROC GLM

PROC GLM recognizes different theoretical approaches to analysis of variance by providing four types of sums of squares and associated statistics. The four types of sums of squares in PROC GLM are called Type I, Type II, Type III, and Type IV (SAS Institute). The four types of sums of squares are explained in general, conceptual terms, followed by more technical descriptions.

Type I sums-of-squares retain the properties discussed in Chapter 2, "Regression." They correspond to adding each source (factor) sequentially to the model in the order listed. For example, the Type I sum of squares for the first factor listed is the same as PROC ANOVA would compute for that effect. It reflects differences between unadjusted means of that factor as if the data consist of a one-way structure. The Type I SS may not be particularly useful for analysis of unbalanced multiway structures but may be useful for nested models, polynomial models, and certain tests involving the homogeneity of regression coefficients (see Chapter 7, "Analysis of Covariance"). Also, comparing Type I and other types of sums of squares provides some information on the effect of the lack of balance.

Type II sums of squares are more difficult to understand. Generally, the Type II SS for an effect U, which may be a main effect or interaction, is adjusted for an effect V if and only if V does not contain U. Specifically, for a two-factor structure with interaction, the main effects, A and B, are not adjusted for the A*B interactions because the symbol A*B contains both A and B. Factor A is adjusted for B because the symbol B does not contain A. Similarly, B is adjusted for A, and the A*B interaction is adjusted for the two main effects.

Type II sums of squares for the main effects A and B are mainly appropriate for situations in which no interaction is present. These are the sums of squares presented in many major statistical textbooks. Their method of computation is often referred to as the method of fitting constants.

The Type II analysis relates to the following general guidelines often given in applied statistical texts. First, test for the significance of the A*B interaction. If A*B is insignificant, delete it from the model and analyze main effects, each adjusted for the other. If A*B is significant, then abandon main-effects analysis and focus your attention on simple effects.

Note that for full-rank regression models, the Type II sums of squares are adjusted for cross-product terms. This occurs because, for example,

$$y = \beta_0 + \beta_1 x_1 + \beta_2 x_2 + \beta_3 x_1 x_2 + \varepsilon$$

where the product $x_1 x_2$ is dealt with simply as another independent variable with no concept of order of the term.

The Type III sums of squares correspond to Yates's weighted squares of means analysis. Their principal use is in situations that require a comparison of main effects even in the presence of interaction. Type III sums of squares are partial sums of squares. In this sense, each effect is adjusted for all other effects. In particular, main effects A and B are adjusted for the interaction A*B if all these terms are in the model. If the model contains only main effects, then Type II and Type III analyses are the same. See Steel and Torrie (1980), Searle (1971), and Speed et al. (1978) for further discussion of the method of fitting constants and the method of weighted squares of means.

The Type IV functions were designed primarily for situations where there are empty cells. The principles underlying the Type IV sums of squares are quite involved and can be discussed only in a framework using the general construction of estimable functions. It should be noted that the Type IV functions are not necessarily unique when there are empty cells, but the functions are identical to those provided by Type III when there are no empty cells.

You can request four sums of squares in PROC GLM as options in the MODEL statement. For example, the following SAS statement specifies the printing of Type I and Type IV sums of squares:

```
model . . . / ss1 ss4;
```

Any or all types may be requested. If no sums of squares are specified, PROC GLM computes the Type I and Type III sums of squares by default.

The next two sections interpret the different sums of squares in terms of reduction notation and the μ-model.

6.3.3 Interpreting Sums of Squares in Reduction Notation

The types of sums of squares can be explained in terms of the reduction notation that is developed for regression models in Chapter 2. This requires writing the model as a regression model using dummy variables, with certain restrictions imposed on the parameters to give them unique interpretation.

As an example, consider a 2×3 factorial structure with n_{ij} observations in the cell in row i, column j. The equation for the model is

$$y_{ijk} = \mu + \alpha_i + \beta_j + \alpha\beta_{ij} + \varepsilon_{ijk}$$

where $i=1, 2, j=1, 2, 3$, and $k=1, \ldots, n_{ij}$. Assume $n_{ij}>0$ for all i, j. An expression of the form $R(\alpha | \mu, \beta)$ means the same as $R(\alpha_1, \alpha_2 | \mu, \beta_1, \beta_2, \beta_3)$. The sums of squares printed by PROC GLM can be interpreted in reduction notation most easily under the restrictions

$$\Sigma_i \alpha_i = \Sigma_j \beta_j = \Sigma_i \alpha\beta_{ij} = \Sigma_j \alpha\beta_{ij} = 0 \qquad (6.1)$$

that is, by taking an **X** matrix with full-column rank given by

$$
\begin{array}{cccccc}
\mu & \alpha_1 & \beta_1 & \beta_2 & \alpha\beta_{11} & \alpha\beta_{12}
\end{array}
$$

$$
X = \begin{bmatrix}
1 & 1 & 1 & 0 & 1 & 0 \\
\cdot & \cdot & \cdot & \cdot & \cdot & \\
\cdot & \cdot & \cdot & \cdot & \cdot & \cdot \\
\cdot & \cdot & \cdot & \cdot & \cdot & \\
1 & 1 & 1 & 0 & 1 & 0 \\
1 & 1 & 0 & 1 & 0 & 1 \\
\cdot & \cdot & \cdot & \cdot & \cdot & \cdot \\
\cdot & \cdot & \cdot & \cdot & \cdot & \\
\cdot & \cdot & \cdot & \cdot & \cdot & \cdot \\
1 & 1 & 0 & 1 & 0 & 1 \\
1 & 1 & -1 & -1 & -1 & -1 \\
\cdot & \cdot & \cdot & \cdot & \cdot & \\
\cdot & \cdot & \cdot & \cdot & \cdot & \\
\cdot & \cdot & \cdot & \cdot & \cdot & \cdot \\
1 & 1 & -1 & -1 & -1 & -1 \\
1 & -1 & 1 & 0 & -1 & 0 \\
\cdot & \cdot & \cdot & \cdot & \cdot & \cdot \\
\cdot & \cdot & \cdot & \cdot & \cdot & \\
\cdot & \cdot & \cdot & \cdot & \cdot & \cdot \\
1 & -1 & 1 & 0 & -1 & 0 \\
1 & -1 & 0 & 1 & 0 & -1 \\
\cdot & \cdot & \cdot & \cdot & \cdot & \\
\cdot & \cdot & \cdot & \cdot & \cdot & \cdot \\
\cdot & \cdot & \cdot & \cdot & \cdot & \\
1 & -1 & 0 & 1 & 0 & -1 \\
1 & -1 & -1 & -1 & 1 & 1 \\
\cdot & \cdot & \cdot & \cdot & \cdot & \cdot \\
\cdot & \cdot & \cdot & \cdot & \cdot & \\
\cdot & \cdot & \cdot & \cdot & \cdot & \\
1 & -1 & -1 & -1 & 1 & 1
\end{bmatrix}
\begin{array}{l}
\\ n_{11} \text{ rows for observations in cell 11} \\ \\ \\ \\ n_{12} \text{ rows for observations in cell 12} \\ \\ \\ \\ n_{13} \text{ rows for observations in cell 13} \\ \\ \\ \\ n_{21} \text{ rows for observations in cell 21} \\ \\ \\ \\ n_{22} \text{ rows for observations in cell 22} \\ \\ \\ \\ n_{23} \text{ rows for observations in cell 23}
\end{array}
$$

With this set of restrictions or definitions of the parameters, the sums of squares that result from the following MODEL statement are summarized below:

```
model y=a b a*b / ss1 ss2 ss3 ss4;
```

Effect	Type I	Type II	Type III = Type IV
A	$R(\alpha\mid\mu)$	$R(\alpha\mid\mu,\beta)$	$R(\alpha\mid\mu,\beta,\alpha\beta)$
B	$R(\beta\mid\mu,\alpha)$	$R(\beta\mid\mu,\alpha)$	$R(\beta\mid\mu,\alpha,\alpha\beta)$
A*B	$R(\alpha\beta\mid\mu,\alpha,\beta)$	$R(\alpha\beta\mid\mu,\alpha,\beta)$	$R(\alpha\beta\mid\mu,\alpha,\beta)$

You should be careful when using reduction notation with less-than-full-rank models. If no restrictions had been specified on the model for the two-way structure above, then $R(\alpha|\mu, \beta, \alpha\beta) = 0$ because the columns of the $\mathbf{X}$ matrix corresponding to the α_i would be linearly dependent on the columns corresponding to μ and the $\alpha\beta_{ij}$.

In addition, the dependence of reduction notation on the restrictions imposed cannot be overemphasized. For example, imposing the restriction

$$\alpha_2 = \beta_3 = \alpha\beta_{21} = \alpha\beta_{22} = \alpha\beta_{13} = \alpha\beta_{23} = 0 \tag{6.2}$$

results in a different value for $R(\alpha|\mu, \beta, \alpha\beta)$. Although the restrictions of equation 6.1 are those that correspond to the sums of squares computed by PROC GLM, the restrictions of equation 6.2 are those that correspond to the (biased) parameter estimates computed by PROC GLM.

There is a relationship between the four types of sums of squares and the four types of data structures in a two-way classification. The relationship derives from the principles of adjustment that the sums-of-squares types obey. Letting n_{ij} denote the number of observations in level i of factor A and level j of factor B, the four types of data structures are

❑   equal cell frequencies: n_{ij} =common value for all i, j

❑   proportionate cell frequencies: $n_{ij}/n_{il} = n_{kj}/n_{kl}$ for all i, j, k, l

❑   disproportionate, nonzero cell frequencies: $n_{ij}/n_{il} \neq n_{kj}/n_{kl}$ for some i, j, k, l, but $n_{ij} > 0$ for all i, j

❑   empty cell(s): n_{ij} =0 for some i, j.

The display below shows the relationship between sums-of-squares types and data structure types pertaining to the following MODEL statement:

```
model y=a b a*b / ss1 ss2 ss3 ss4;
```

For example, writing III=IV indicates that Type III is equal to Type IV.

Data Structure Type

	1 (Equal n_{ij})	2 (Proportionate n_{ij})	3 (Disproportionate, nonzero n_{ij})	4 (Empty Cell)
A	I=II=III=IV	I=II, III=IV	III=IV	
B	I=II=III=IV	I=II, III=IV	I=II, III=IV	I=II
A*B	I=II=III=IV	I=II=III=IV	I=II=III=IV	I=II=III=IV

6.3.4 Interpreting Sums of Squares in the μ-Model Notation

The μ model for the two-way structure takes the form

$$y_{ijk} = \mu_{ij} + \varepsilon_{ijk} \tag{6.3}$$

The parameters of the μ model relate to the parameters of the standard analysis-of-variance model according to the equation

$$\mu_{ij} = \mu + \alpha_i + \beta_j + \alpha\beta_{ij}$$

This relation holds regardless of any restriction that may be imposed upon the α, β, and $\alpha\beta$ parameters. The advantage of using the μ-model notation over standard analysis-of-variance notation is that all of the μ_{ij} parameters are clearly defined without specifying restrictions; thus, a hypothesis stated in terms of the μ_{ij} can be easily understood.

Speed et al. (1978) give interpretations of the different types of sums of squares (I, II, III, and IV) computed by PROC GLM using the μ-model notation. It is assumed that all $n_{ij} > 0$, making Type III equal to Type IV.

Using their results, the sums of squares obtained from the following MODEL statement are expressed in terms of the μ_{ij}, as given in Table 6.1.

```
model response = a b a*b / ss1 ss2 ss3 ss4;
```

Table 6.1 *Interpretation of Sums of Squares in the μ-Model Notation*

Type	Effect
	Effect A
I	$(\Sigma_j \, n_{1j}\mu_{1j}) / n_{1.} = \ldots = (\Sigma_j \, n_{aj}\mu_{aj}) / n_{a.}$
II	$\Sigma_j n_{1j}\mu_{1j} = \Sigma_i\Sigma_j \left(n_{1j}n_{ij}\mu_{ij} \, / \, n_{.j}\right), \ldots, \Sigma_j n_{aj}\mu_{aj}$ $= \Sigma_i\Sigma_j \left(n_{aj}n_{ij}\mu_{ij} \, / \, n_{.j}\right)$
III & IV	$\mu_{11} + \ldots + \mu_{1b} = \ldots = \mu_{a1} + \ldots + \mu_{ab}$
	that is, $\overline{\mu}_{1.} = \ldots = \overline{\mu}_{a.}$
	where $\overline{\mu}_{i.} = \Sigma_j \mu_{ij} \, / \, b$
	Effect B
I & II	$\Sigma_i \, n_{i1}\mu_{i1} = \Sigma_i\Sigma_j \, (n_{i1}n_{ij}\mu_{ij} \, / \, n_{i.}, \ldots, \Sigma_i n_{ib}\mu_{ib} = \Sigma_i\Sigma_j \, (n_{ib}n_{ij}\mu_{ij}) \, / \, n_{i.})$
III & IV	$\mu_{11} + \ldots + \mu_{a1} = \ldots \, \mu_{1b} + \ldots + \mu_{ab}$
	that is, $\overline{\mu}_{.1} = \ldots = \overline{\mu}_{.b}$
	where $\overline{\mu}_{ij} = \Sigma_i \mu_{ij} \, / \, a$
	Effect A*B
I, II, III, IV	$\mu_{ij} - \mu_{im} - \mu_{lj} + \mu_{lm} = 0$ for all i, j, l, m

Table 6.1 shows that the tests can be expressed in terms of equalities of weighted cell means, only some of which are easily interpretable. Considering the Type I A effect, the weights $n_{ij}/n_{i.}$ attached to μ_{ij} are simply the fraction of the $n_{i.}$ observations in level i of A that were in level j of B. If these weights reflect the distribution across the levels of B in the population of units in level i of A, then the Type I test may have meaningful interpretation. That is, suppose the population of units in level i of A is made up of a fraction ρ_{i1} of units in level 1 of B, of a fraction ρ_{i2} of units in level 2 of B, and so on, where $\rho_{i1} + \ldots + \rho_{ib} = 1$. Then it may be reasonable to test

$$H_0: \Sigma_j \rho_{1j}\mu_{1j} = \ldots = \Sigma_j \rho_{aj}\mu_{aj}$$

which would be the Type I test in case $n_{ij}/n_{i.} = \rho_{ij.}$.

Practical interpretation of the Type II weights is more difficult—refer to Section 6.3.7.2, "Interpreting Sums of Squares Using Estimable Functions." Recall that the Type II tests are primarily for main effects with no interaction. You can see from Table 6.1 that the Type II hypothesis clearly depends on the n_{ij}, the numbers of observations in the cells.

The interpretation of Type III and Type IV tests is clear because all weights are unity. When the hypotheses are stated in terms of the μ-model, the benefit of the Type III test is more apparent because the Type III hypothesis does not depend on the n_{ij}, the numbers of observations in the cells. Type I and Type II hypotheses do depend on the n_{ij}, and this may not be desirable.

For example, suppose a scientist sets up an experiment with ten plants in each combination of four levels of nitrogen (N) and three levels of lime (P). Suppose also that some plants die in some of the cells for reasons unrelated to the effects of N and P, leaving some cells with $n_{ij} < 10$. A hypothesis test concerning the effects of N and P, which depends on the values of n_{ij}, would be contaminated by the accidental variation in the n_{ij}. The scientific method declares that the hypotheses to be tested should be stated before data are collected. It would be impossible to state Type I and Type II hypotheses prior to data collection because the hypotheses depend on the n_{ij}, which are known only after data are collected.

Note that the Type I and Type II hypotheses are different for effect A but the same for effect B. This occurs because the Type I sums of squares are model-order dependent. Being sequential, the Type I sums of squares are A (unadjusted), B (adjusted for A), and A*B (adjusted for A and B). Thus, the Type I sums of squares for the effects A and B listed in the MODEL statement in the order A, B, A*B would not be the same as in the order B, A, A*B. The Type II hypotheses are not model-order dependent because, for the two-way structure, both Type II main-effect sums of squares are adjusted for each other—that is, A (adjusted for B), B (adjusted for A), and A*B (adjusted for A and B). These Type II sums of squares are the partial sums of squares if no A*B interaction is specified in the model, in which case Type II, Type III, and Type IV would be the same.

Another interpretation of these tests is given by Hocking and Speed (1980), who point out that Type I, Type II, and Type III=Type IV tests for effect A each represent a test of

$$H_0: \mu_{11} + \ldots + \mu_{1b} = \ldots = \mu_{a1} + \ldots + \mu_{ab}$$

subject to certain conditions on the cell means. The conditions are

Type I	no B effect, $\mu_{.1} = \ldots = \mu_{.k}$, and
	no A*B effect, $\mu_{ij} - \mu_{im} - \mu_{lj} + \mu_{lm} = 0$ for all i, j, l, m
Type II	no A*B effect, $\mu_{ij} - \mu_{im} - \mu_{lj} + \mu_{lm} = 0$ for all i, j, l, m
Type III=Type IV	none (provided $n_{ij} > 0$ for all ij).

6.3.5 An Example of Unbalanced Two-Way Classification

This example is a two-factor layout with data presented by Harvey (1975). Two types of feed rations (factor A) are given to calves from three different sires (factor B). The observed dependent variable y_{ijk} (variable Y) is the coded amount of weight gained by each calf. Because unequal numbers of calves of each sire are fed each ration, this is an unbalanced experiment. However, there are observations for each ration-sire combination; that is, there are no empty cells. The data appear in Output 6.7.

Output 6.7
Data for an
Unbalanced
Two-Way
Classification

Obs	a	b	y
1	1	1	5
2	1	1	6
3	1	2	2
4	1	2	3
5	1	2	5
6	1	2	6
7	1	2	7
8	1	3	1
9	2	1	2
10	2	1	3
11	2	2	8
12	2	2	8
13	2	2	9
14	2	3	4
15	2	3	4
16	2	3	6
17	2	3	6
18	2	3	7

The analysis-of-variance model for these data is

$$y_{ijk} = \mu + \alpha_i + \beta_j + \alpha\beta_{ij} + \varepsilon_{ijk}$$

where

i   equals 1, 2.

j   equals 1, 2, 3.

i and *j* have elements as defined in Section 6.3.1, "General Considerations." The model contains twelve parameters, which are more than can be estimated uniquely by the six cell means that are the basis for estimating parameters. The analysis is implemented with the following SAS statements:

```
proc glm;
    class a b;
    model y=a b a*b / ss1 ss2 ss3 ss4 solution;
```

The statements above cause PROC GLM to create the following twelve dummy variables:

❑   1 dummy variable for the mean (or intercept)

❑   2 dummy variables for factor A (ration)

❑   3 dummy variables for factor B (sire)

❑   6 dummy variables for the interaction A*B (all six possible pairwise products of the variables from factor A with those from factor B).

The options requested are SOLUTION, and for purposes of illustration, all four types of sums of squares. The results appear in Output 6.8.

Output 6.8
Sums of
Squares for an
Unbalanced
Two-Way
Classification

```
                              The GLM Procedure

Dependent Variable: y

                                    Sum of
   Source                  DF       Squares     Mean Square    F Value    Pr > F

   Model                    5    63.71111111    12.74222222       5.87    0.0057

   Error                   12    26.06666667     2.17222222

   Corrected Total         17    89.77777778


               R-Square      Coeff Var      Root MSE      y Mean

               0.709653      28.83612       1.473846      5.111111


   Source                  DF      Type I SS     Mean Square    F Value    Pr > F

   a                        1     7.80277778      7.80277778       3.59    0.0824
   b                        2    20.49185393     10.24592697       4.72    0.0308
   a*b                      2    35.41647940     17.70823970       8.15    0.0058


   Source                  DF     Type II SS     Mean Square    F Value    Pr > F

   a                        1    15.85018727     15.85018727       7.30    0.0193
   b                        2    20.49185393     10.24592697       4.72    0.0308
   a*b                      2    35.41647940     17.70823970       8.15    0.0058


   Source                  DF    Type III SS     Mean Square    F Value    Pr > F

   a                        1     9.64065041      9.64065041       4.44    0.0569
   b                        2    30.86591760     15.43295880       7.10    0.0092
   a*b                      2    35.41647940     17.70823970       8.15    0.0058


   Source                  DF     Type IV SS     Mean Square    F Value    Pr > F

   a                        1     9.64065041      9.64065041       4.44    0.0569
   b                        2    30.86591760     15.43295880       7.10    0.0092
   a*b                      2    35.41647940     17.70823970       8.15    0.0058

                                               Standard
         Parameter            Estimate            Error    t Value    Pr > |t|

         Intercept         5.400000000 B       0.65912400      8.19      <.0001
         a       1         -4.400000000 B      1.61451747     -2.73      0.0184
         a       2         0.000000000 B            .            .          .
         b       1         -2.900000000 B      1.23310809     -2.35      0.0366
         b       2         2.933333333 B       1.07634498      2.73      0.0184
         b       3         0.000000000 B            .            .          .
         a*b     1 1       7.400000000 B       2.18606699      3.39      0.0054
         a*b     1 2       0.666666667 B       1.94040851      0.34      0.7371
         a*b     1 3       0.000000000 B            .            .          .
         a*b     2 1       0.000000000 B            .            .          .
         a*b     2 2       0.000000000 B            .            .          .
         a*b     2 3       0.000000000 B            .            .          .

NOTE: The X'X matrix has been found to be singular, and a generalized inverse was used to
solve the normal equations. Terms whose estimates are followed by the letter 'B' are not
uniquely estimable.
```

The first portion shows the statistics for the overall model. The overspecification of the model is obvious: The twelve dummy variables generate only six degrees of freedom (five for the terms listed in the MODEL statement plus one for the intercept).

The next portion of the output shows the four types of sums of squares. Note that Types III and IV give identical results. This is because $n_{ij} > 0$ for all i, j.

As noted in Section 6.3.3, "Interpreting Sums of Squares in Reduction Notation," the parameter estimates printed by PROC GLM are a solution to the normal equations corresponding to the restriction in equation (6.2.) The same condition applies to the parameter estimates

$$\hat{\alpha}_2 = \hat{\beta}_3 = \widehat{\alpha\beta}_{13} = \widehat{\alpha\beta}_{21} = \widehat{\alpha\beta}_{22} = \widehat{\alpha\beta}_{23} = 0 \tag{6.4}$$

These values, which now equal 0, appear in Output 6.8.

Section 6.3.4, "Interpreting Sums of Squares in the μ-Model Notation," also shows that the parameters of the μ model relate to the parameters of the standard analysis-of-variance model (see equation 6.3). A corresponding relation holds between the respective parameter estimates, namely

$$\bar{y}_{ij} = \hat{\mu}_{ij} = \hat{\mu} + \hat{\alpha}_i + \hat{\beta}_j + \widehat{\alpha\beta}_{ij} \tag{6.5}$$

Putting equations (6.4) and (6.5) together as shown in the table below gives the interpretation of the parameter estimates printed by PROC GLM. The table below also shows the relationship between means and parameter estimates.

	Sire 1	**Sire 2**	**Sire 3**
Ration 1	$\hat{\mu}_{11} = \bar{y}_{11.}$ $= \hat{\mu} + \hat{\alpha}_1 + \hat{\beta}_1 + \widehat{\alpha\beta}_{11}$	$\hat{\mu}_{12} = \bar{y}_{12.}$ $= \hat{\mu} + \hat{\alpha}_1 + \hat{\beta}_2 + \widehat{\alpha\beta}_{12}$	$\hat{\mu}_{13} = \bar{y}_{13.}$ $= \hat{\mu} + \hat{\alpha}_1$
Ration 2	$\hat{\mu}_{21.} = \bar{y}_{21.}$ $= \hat{\mu} + \hat{\beta}_1$	$\hat{\mu}_{22} = \bar{y}_{22.}$ $= \hat{\mu} + \hat{\beta}_2$	$\hat{\mu}_{23} = \bar{y}_{23.}$ $= \hat{\mu}$

Note especially the following items in Output 6.8:

❏ The intercept $\hat{\mu}$ printed by PROC GLM is the cell mean $\bar{y}_{23.}$ for the lower right-hand cell (ration 2, sire 3).

❏ The estimate $\bar{\alpha}_1 = -2.4$ is the difference between the cell means for the two rations fed to sire 3, $\hat{\alpha}_1 = \bar{y}_{13.} - \bar{y}_{23.}$.

❏ The interaction parameter estimate $\widehat{\alpha\beta}_{11} = 5.4$ is the interaction of ration 1 and ration 2 by sire 1 and sire 3, $\widehat{\alpha\beta}_{11} = \bar{y}_{11.} - \bar{y}_{13.} - \bar{y}_{21.} + \bar{y}_{23.}$. Generally, in a two-way layout with a rows and b columns, the interaction parameter estimates $\alpha\beta_{ij}$ and measures the interaction of rows i and a by columns j and b.

6.3.6 The MEANS, LSMEANS, CONTRAST, and ESTIMATE Statements in a Two-Way Layout

The parameter estimates printed by PROC GLM are the result of a computational algorithm and may or may not be the estimates with the greatest practical value. However, there is no single choice of estimates (corresponding to a particular generalized inverse or set of restrictions) that satisfies the requirements of all applications. In most instances, specific estimable functions of these parameter estimates, like the estimates obtained with the LSMEANS, CONTRAST, and ESTIMATE statements, can be used to provide more useful estimates. The CONTRAST and ESTIMATE statements for balanced data applications are discussed in Chapter 2.

The CONTRAST, LSMEANS, and ESTIMATE statements are similar for one- and two-way models, but principles and interpretations become more complex. Consider the results from the following SAS statements:

```
proc glm;
    class a b;
    model y=a b a*b;
    means a b;
    lsmeans a b / stderr;
    contrast 'A EFFECT' a -1 1;
    contrast 'B 1 vs 2 & 3' b -2 1 1;
    contrast 'B 2 vs 3' b 0 -1 1;
    contrast 'ALL B' b -2 1 1, b 0 -1 1;
    contrast 'A*B 2 vs 3' a*b 0 1 -1 0 -1 1;
    estimate 'B2, B3 MEAN' intercept 1 a .5 .5 b 0 .5 .5
                                a*b 0 .25 .25 0 .25 .25;
    estimate 'A in B1' a -1 1 a*b -1 0 0 1;
```

The MEANS statement provides the raw or unadjusted main-effect and interaction means. The LSMEANS statement produces least-squares (adjusted) means for main effects together with their standard errors. The results of these two statements are combined in Output 6.9. (PROC GLM prints these results on separate pages.)

Output 6.9
Results of the MEANS and LSMEANS Statements for a Two-Way Classification

```
                          The GLM Procedure

        Level of            --------------y--------------
        a          N              Mean           Std Dev

        1          8         4.37500000        2.13390989
        2         10         5.70000000        2.35937845


        Level of            --------------y--------------
        b          N              Mean           Std Dev

        1          4         4.00000000        1.82574186
        2          8         6.00000000        2.50713268
        3          6         5.00000000        1.54919334
```

*Output 6.9
(Continued)
Results of the
MEANS and
LSMEANS
Statements for
a Two-Way
Classification*

```
                        The GLM Procedure
                       Least Squares Means

                                      Standard
           a         y LSMEAN          Error      Pr > |t|

           1        3.70000000       0.64055339    <.0001
           2        5.41111111       0.49940294    <.0001


                                      Standard
           b         y LSMEAN          Error      Pr > |t|

           1        4.00000000       0.73692303    0.0002
           2        6.46666667       0.53817249    <.0001
           3        3.20000000       0.80725874    0.0019
```

The raw and least-squares means are different for all levels except B1, which is balanced with respect to factor A.

Quantities estimated by the raw means and least-squares means can be expressed in terms of the μ model. For level 1 of factor A, the raw mean (4.625) is an estimate of $(2\mu_{11} + \mu_{12} + \mu_{13})/(2 + 5 + 1)$, whereas the least-squares mean (4.367) is an estimate of $(\mu_{11} + \mu_{12} + \mu_{13})/3$. The raw means estimate weighted averages of the μ_{ij} whose weights are a function of sample sizes. The least-squares means estimate unweighted averages of the μ_{ij}.

The results of the five CONTRAST statements appear in Output 6.10.

*Output 6.10
Contrast in a
Two-Way
Classification*

```
Dependent Variable: y

Contrast              DF    Contrast SS    Mean Square    F Value    Pr > F

A effect               1     9.64065041     9.64065041       4.44    0.0569
B 1 vs 2 & 3           1     1.93798450     1.93798450       0.89    0.3635
B 2 vs 3               1    24.62564103    24.62564103      11.34    0.0056
ALL B                  2    30.86591760    15.43295880       7.10    0.0092
A*B 2 vs 3             1     0.25641026     0.25641026       0.12    0.7371
```

The first four CONTRAST statements are similar to those presented for the one-way structure. Note that when a contrast uses all available degrees of freedom for the factor (such as the ALL B contrast), the sums of squares are the same as the Type III sums of squares for the factor.

The fifth CONTRAST statement requests the interaction between the factor A contrast and the B2 vs 3 contrast. It is constructed by computing the product of corresponding main-effect contrasts for each AB treatment combination. The procedure is illustrated in the table below:

Construction of Interaction Contrast

		Level of Factor B		
		1	2	3
Level of Factor A	Factor A Contrast		Factor B Contrast	
		0	1	−1
1	1	0	1	−1
2	−1	0	−1	1

Main-effect contrasts are given on the top and left, and interaction contrasts (products of marginal entries) are given in the body of the table. These are inserted into the CONTRAST statement in the same order of interaction cells as indicated by the CLASS statement (levels of B within levels of A).

In terms of the μ model, the hypothesis tested by the *F*-statistic for this interaction contrast is

$$H_0 : \mu_{12} - \mu_{13} - \mu_{22} + \mu_{23} = 0$$

The two ESTIMATE statements request estimates of linear functions of the model parameters. The first function to be estimated is the average of the cell means for levels 2 and 3 of factor B. The other statement requests an estimate of the effect of factor A within level 1 of factor B, or an estimate of $\mu_{21} - \mu_{11}$. Output 6.11 summarizes results from the ESTIMATE statements.

Output 6.11
Parameter
Estimates in a
Two-Way
Classification

```
                                      Standard
    Parameter              Estimate      Error    t Value   Pr > |t|

    B2, B3 MEAN          4.83333333   0.48510213     9.96    <.0001
    A in B1             -3.00000000   1.47384606    -2.04    0.0645
```

Expressing each comparison in terms of model parameters α_i, β_j, $(\alpha\beta)_{ij}$ is the key to filling in the coefficients of the CONTRAST and ESTIMATE statements. Consider the ESTIMATE A in B1 statement, which is used to estimate $\mu_{21} - \mu_{11}$. Writing this expression as a function of the model parameters by substituting

$$\mu_{ij} = \mu + \alpha_i + \beta_j + \alpha\beta_{ij}$$

yields

$$\mu_{21} - \mu_{11} = -\alpha_1 + \alpha_2 - \alpha\beta_{11} + \alpha\beta_{21}$$

The $-\alpha_1 + \alpha_2$ term tells you to insert A – 1 1 into the ESTIMATE statement. There are no β's in the function, so no B expression appears in the ESTIMATE statement. The $-\alpha\beta_{11} + \alpha\beta_{21}$ term tells you to insert A*B – 1 0 0 1 0 0, or, equivalently, A*B–1 0 0 1 into the ESTIMATE statement. The ordering in the statement

```
class a b;
```

specifies that the ordering of the coefficients following A*B corresponds to $\alpha\beta_{11}$ $\alpha\beta_{12}$ $\alpha\beta_{13}$ $\alpha\beta_{21}$ $\alpha\beta_{22}$ $\alpha\beta_{23}$. The SAS statement

```
class b a;
```

would indicate an ordering that corresponds to $\alpha\beta_{11}$ $\alpha\beta_{21}$ $\alpha\beta_{12}$ $\alpha\beta_{22}$ $\alpha\beta_{13}$ $\alpha\beta_{23}$.

Now consider the CONTRAST A statement. The hypothesis to be tested is

$$H_0 : -(\mu_{11} + \mu_{12} + \mu_{13})/3 + (\mu_{21} + \mu_{22} + \mu_{23})/3 = 0$$

Substituting $\mu_{ij} + \alpha_i + \beta_j + \alpha\beta_{ij}$ gives the equivalent hypothesis:

$$H_0 : -\alpha_1 + \alpha_2 - (\alpha\beta_{11} + \alpha\beta_{12} + \alpha\beta_{13}) / 3$$
$$+ (\alpha\beta_{21} + \alpha\beta_{22} + \alpha\beta_{23}) / 3 = 0$$

Again the $-\alpha_1 + \alpha_2$ term in the function tells you to insert A − 1 1. This brings up an important usage note: Specifying A − 1 1 causes the coefficients of the A*B interaction term to be automatically included by PROC GLM. That is, the SAS statement

```
contrast 'A' a -1 1;
```

is equivalent to the statement

```
contrast 'A' a -1 1 a*b -.333333 -.333333 -.333333
.333333 .333333 .333333;
```

Similarly, the SAS statement

```
estimate 'A EFFECT' a -1 1;
```

provides an estimate of

$$-(\mu_{11} + \mu_{12} + \mu_{13}) / 3 + (\mu_{21} + \mu_{22} + \mu_{23}) / 3$$

without explicitly specifying the coefficients of the $\alpha\beta_{ij}$ terms. However, you should note that specifying the $\alpha\beta_{ij}$ coefficients does not cause PROC GLM to automatically include coefficients for α_i or β_j. For example, the term A − 1 1 must appear in the ESTIMATE A in B1 statement. Similarly, a contrast to test $H_0: -\mu_{11} + \mu_{21} = 0$ requires the following statement:

```
contrast 'A in B1' a -1 1 a*b -1 0 0 1;
```

The A − 1 1 term must be included.

6.3.7 Estimable Functions for a Two-Way Classification

The previous section discussed the application of the CONTRAST statement, which employs the concept of estimable functions. PROC GLM can display the construction of estimable functions as an optional request in the MODEL, LSMEANS, CONTRAST, and ESTIMATE statements. This section discusses the construction of estimable functions and their relation to the sums of squares and associated hypotheses available in the GLM procedure, and to CONTRAST, ESTIMATE, and LSMEANS statements. The presentation of estimable functions consists of results obtained using the unbalanced factorial data given in Output 6.7. For more thorough discussions of these principles, see Graybill (1976) and Searle (1971).

6.3.7.1 The General Form of Estimable Functions

The general form of estimable functions is a vector of elements that are the building blocks for generating specific estimable functions. The number of unique symbols in the vector represents the maximum number of linearly independent coefficients estimated by the model, which is equal to the rank of the $\mathbf{X'X}$ matrix. In the GLM procedure this is obtained by the E option in the MODEL statement

```
proc glm;
   class a b;
   model y=a b a*b / e solution;
```

Table 6.2 gives the vector of coefficients of the general form of estimable functions for our example. There are only six elements (L1, L2, L4, L5, L7, L8), which correspond to the number of degrees of freedom in the model (including the intercept). The number of elements for an effect corresponds to the degrees of freedom for that effect; for example, L4 and L5 are introduced opposite the effect B, indicating B has 2 degrees of freedom.

Table 6.2 *General Form of Estimable Functions*

Effect			Parameters*	Coefficients
Intercept			μ	L1
A	1		α_1	L2
	2		α_2	L1-L2
B	1		β_1	L4
	2		β_2	L5
	3		β_3	L1 – L4 – L5
A*B	1	1	$\alpha\beta_{11}$	L7
	1	2	$\alpha\beta_{12}$	L8
	1	3	$\alpha\beta_{13}$	L2 – L7 – L8
	2	1	$\alpha\beta_{21}$	L4 – L7
	2	2	$\alpha\beta_{22}$	L5 – L8
	2	3	$\alpha\beta_{23}$	L1 – L2 – L4 – L5 + L7 + L8

* These are implied by the output but not printed in this manner.

According to Table 6.2, any estimable function $\mathbf{L}\beta$ must be of the form

$$\mathbf{L}\beta = L1\mu + L2\alpha_1 + (L1-L2)\alpha_2 + L4\alpha\beta_1 + L5\beta_2 + (L1 - L_4 - L5)\beta_3$$
$$+ L7\alpha\beta_{11} + L8\alpha\beta_{12} + (L2-L7-L8)\alpha\beta_{13} + (L4-L7)\alpha\beta_{21}$$
$$+ (L5-L8)\alpha\beta_{22} + (L1-L2-L4-L5 + L7 + L8)\alpha\beta_{23} \tag{6.6}$$

for some specific values of L1 through L8. The various tests in PROC GLM test hypotheses of the form H_0: $\mathbf{L}\beta = 0$.

Coefficients for any specific estimable function are constructed by assigning values to the individual L's. For example, setting L2=1 and all others equal to 0 provides the estimable function $\alpha_1 - \alpha_2 + \alpha\beta_{13} - \alpha\beta_{23}$. It is clear, however, that no estimable function can be constructed in this manner to equal 1 or 2 individually. That is, no matter what values you choose for L1 through L8, you cannot make $\mathbf{L}\beta = \alpha_1$ or $\mathbf{L}\beta = \alpha_2$. This is because α_1 and α_2 are non-estimable functions; without additional restrictions there is no linear function of the data whose expected value is α_1 or α_2.

6.3.7.2 Interpreting Sums of Squares Using Estimable Functions

The coefficients required to construct estimable functions for each effect in the MODEL statement are available for any type of sum of squares requested as an option in the MODEL statement. For example,

```
model y = a b a*b / e e1 e2 e3;
```

will provide the general form, the coefficients of estimable functions for Types I, II, and III, and the corresponding sums of squares for each effect listed in the MODEL statement.

Table 6.3 gives the coefficients of the Type I, Type II, and Type III estimable functions associated with factor A. Types III and IV are identical for this example because all $n_{ij} > 0$.

Table 6.3 *Estimable Functions for Factor A*

Effect			Type I Parameters	Type II Coefficients	Type III Coefficients
Intercept			0	0	0
A	1		L2	L2	L2
	2		–L2	–L2	–L2
B	1		0.05*L2	0	0
	2		0.325*L2	0	0
	3		–0.375*L2	0	0
A*B	1	1	0.25*L2	0.2697*L2	0.3333*L2
	1	2	0.625*L2	0.5056*L2	0.3333*L2
	1	3	0.125*L2	0.2247*L2	0.3333*L2
	2	1	–0.2*L2	–0.2697*L2	–0.3333*L2
	2	2	–0.3*L2	–0.5056*L2	–0.3333*L2
	2	3	–0.5*L2	–0.2247*L2	–0.3333*L2

All coefficients involve only one element (L2), since the A effect has only 1 degree of freedom. Estimable functions are constructed by assigning specific values to the elements. For factor A, with only one variable, the best choice is L2=1. Application to the Type I coefficients generates the estimable function

$$L\beta = \alpha_1 - \alpha_2$$
$$+ \ 0.05\beta_1 + 0.325\beta_2 - 0.375\beta_2$$
$$+ \ 0.25\alpha\beta_{11} + 0.625 \ \alpha\beta_{12} + 0.125\alpha\beta_{13}$$
$$- \ 0.2\alpha\beta_{21} - 0.3\alpha\beta_{22} - 0.5\alpha\beta_{23}$$

Thus, using the Type I sum of squares in the numerator of an *F*-statistic tests the hypothesis $L\beta = 0$ for this particular $L\beta$. In addition to $\alpha_1 - \alpha_2$, this $L\beta$ also involves the function of coefficients of factor B

$$0.5\beta_1 + 0.325\beta_2 - 0.375\beta_3$$

as well as a function of the interaction parameters. Actually, this is to be expected, since the Type I function for A is unadjusted—it is based on the difference between the two A factor means $(\bar{y}_{1..} - \bar{y}_{2..})$.

As explained in Section 6.3.1, the mean for A is

$$\bar{y}_{1..} = 1/8[(2\bar{y}_{11.} + 5 \ \bar{y}_{12.} + \bar{y}_{13.})]$$

Each cell mean $\bar{y}_{ij.}$ is an estimate of the function $\mu + \alpha_{ii} + \beta_j + \alpha\beta_{ij}$. Omitting for this discussion the interaction parameters, $\bar{y}_{1..}$ is an estimate of

$$1/8 \ [2(\mu + \alpha_1 + \beta_1) + 5(\mu + \alpha_1 + \beta_2) + (\mu + \alpha_1 + \beta_3)]$$
$$= \mu + \alpha_1 + (0.25\beta_1 + 0.625\beta_2 + 0.125\beta_3)$$

Likewise $\bar{y}_{2..}$ is an estimate of

$$\mu + \alpha_2 + (0.2\beta_1 + 0.3\beta_2 + 0.5\beta_3)$$

Hence, $(\bar{y}_{1..} - \bar{y}_{2..})$ is an estimate of

$$a_1 - \alpha_2 + 0.05\beta_1 + 0.325\beta_2 - 0.375\beta_3$$

which is the function provided by Type I.

The coefficients associated with A*B provide the coefficients of the interaction terms in the Type I estimable functions. The coefficients associated with A*B are useful for expressing the estimable functions and interpreting the tests in terms of the μ model. Recall that

$$\mu_{ij} = \mu + \alpha_i + \beta_j + \alpha\beta_{ij}$$

A little algebra shows that any estimable $L\beta$ with coefficients as given in Table 6.3 can also be written as

$$L\beta = L7\mu_{11} + L8\mu_{12} + (L2 - L7 - L8)\mu_{13} + (L4 - L7)\mu_{21}$$
$$+ (L5 - L8)\mu_{22} + (L1 - L2 - L4 - L5 + L8)\mu_{23}$$

$$(6.7)$$

This is easily verified by starting with $\mathbf{L\beta}$ in equation (6.7), replacing each μ_{ij} with $\mu + \alpha_i + \beta_j + \alpha\beta_{ij}$, and combining terms to end up with the original expression for $\mathbf{L\beta}$ in equation 6.6. For example, after factoring out L2, we see that the Type I estimable function for A is

$$\mathbf{L\beta} = L2(0.25\mu_{11} + 0.625\mu_{12} + 0.125\mu_{13} - 0.2\mu_{21} - 0.3\mu_2 - 0.5\mu_{23})$$

Thus the Type I *F*-test for A tests the hypothesis $H_0: \mathbf{L\beta} = 0$, or equivalently,

$$H_0 : 0.25 \ \mu_{11} + 0.625\mu_{12} + 0.125\mu_{12} = 0.2\mu_{21} + 0.3\mu_{22} + 0.5\mu_{23}$$

This is the hypothesis that is tested in Table 6.1. Since the coefficients are functions of the frequencies of the cells, the hypothesis might not be particularly useful.

Applying the same method to the Type II coefficients for A, we have, after setting L2=1,

$$\mathbf{L\beta} = .2697\mu_{11} + .5056\mu_{12} + .2247\mu_{13} - .2697\mu_{21} - .5056\mu_{22} - .2247\mu_{23}$$
$$= .2697(\mu_{11} - \mu_{21}) + .5056(\mu_{12} - \mu_{22}) + .2247(\mu_{13} - \mu_{23})$$

This expression sheds some light on the meaning of the Type II coefficients. Recall that the Type II SS are based on a main-effects model. With no interaction we have, for example, $\mu_{11} - \mu_{21} = \mu_{12} - \mu_{22} = \mu_{13} - \mu_{23}$. Let Δ denote the common value of these differences. The Type II coefficients are the coefficients of the best linear unbiased estimate $\hat{\Delta}$ of Δ given by

$$\hat{\Delta} = \Sigma w_j \ (\bar{y}_{1j.} - \bar{y}_{2j.})$$

where

$$w_j = \frac{n_{1j}n_{2j} \ / \ (n_{1j} + n_{2j})}{\Sigma_k (n_{1k}n_{2k} \ / \ (n_{1k} + n_{2k}))}$$

For example,

$$.2697 = \frac{(2)(2) \ / \ (2 + 2)}{(2)(2) \ / \ (2 + 2) + (5)(3) \ / \ (5 + 3) + (1)(5) \ / \ (1 + 5)}$$

Note that these are functions of cell frequencies and thus do not necessarily generate meaningful hypotheses.

Type III (and Type IV) estimable functions for A likewise (Table 6.2) do not involve the parameters of the B factor. Further, in terms of the parameters of the cell means (μ model)

$$\mathbf{L\beta} = 1/3(\mu_{11} + \mu_{12} + \mu_{13}) - 1/3(\mu_{21} + \mu_{22} + \mu_{23})$$

Thus the Type III *F*-statistic tests $H_0: \bar{\mu}_{1.} = \bar{\mu}_{2.}$ as stated in Table 6.1. Note again that this hypothesis does not involve the cell frequencies, n_{ij}.

Table 6.3 gives the coefficients of estimable functions for factor B. There are 2 degrees of freedom for factor B, thus two elements, L4 and L5. Consider first the Type III coefficients because they are more straightforward. The Type III F-test for factor B is testing simultaneously that any two linearly independent functions are equal to 0; functions are obtained by selecting two choices of values for L4 and L5.

The simplest choices are to take L4=1, L5=0 and L4=0, L5=1. This gives the estimable functions

$$\mathbf{L_1\beta} = \beta_1 - \beta_3 + (\alpha\beta_{11} - \alpha\beta_{13} + \alpha\beta_{21} - \alpha\beta_{23}) / 2$$

and

$$\mathbf{L_2\beta} = \beta_2 - \beta_3 + (\alpha\beta_{12} - \alpha\beta_{13} + \alpha\beta_{22} - \alpha\beta_{23}) / 2$$

In terms of the μ model, this gives

$$\mathbf{L_1\beta} = (\mu_{11} - \mu_{13} + \mu_{21} - \mu_{23}) / 2$$

and

$$\mathbf{L_2\beta} = (\mu_{12} - \mu_{13} + \mu_{22} - \mu_{23}) / 2$$

Thus, the Type III F-statistic tests

$$H_0 : \bar{\mu}_{.1} = \bar{\mu}_{.3} \quad \text{and} \quad \bar{\mu}_{.2} = \bar{\mu}_{.3}$$

or, in equivalent form

$$H_0 : \bar{\mu}_{.1} = \bar{\mu}_{.2} = \bar{\mu}_{.3} \tag{6.8}$$

Another set of choices is L4=1, L5=1 and L4=L5=−1. These lead to

$$H_0 : \bar{\mu}_{.1} = \bar{\mu}_{.2} \quad \text{and} \quad (\bar{\mu}_{.1} + \bar{\mu}_{.2}) / 2 = \bar{\mu}_{.3}$$

which is also equivalent to equation (6.8). Therefore, both sets of choices lead to the same H_0. It is significant that the H_0 in equation (6.8) is independent of cell frequencies and, thus, is desirable for the usual case where cell frequencies are unrelated to the effects of the factors. Table 6.4 gives the coefficients of the Type I & Type II and Type III & Type IV estimable functions associated with factor B.

Table 6.4 *Estimable Functions for Factor B*

Effect		Type I & Type II Coefficients	Type III & Type IV Coefficients
Intercept		0	0
A	1	0	0
	2	0	0
B	1	L4	L4
	2	L5	L5
	3	−L4–L5	−L4–L5

Table 6.4 *(Continued)* *Estimable Functions for Factor B*

Effect			Type I & Type II Coefficients	Type III & Type IV Coefficients
A*B	1	1	0.401*L4-0.1236*L5	0.5*L4
	1	2	−0.1658*L4+0.3933*L5	0.5*L5
	1	3	−0.2416*L4-0.2697*L5	−0.5*L4-0.5*L5
	2	1	0.5899*L4+0.1236*L5	0.5*L4
	2	2	−0.1685*L4+0.6067*L5	0.5*L5
	2	3	−0.7584*L4-0.7303*L5	−0.5*L4–0.5*L5

Recall that since B followed A in the MODEL statement, the Type I SS for B is the same as the Type II SS for B. The coefficients are again a function of the cell frequencies. The nature of the function is not easy to determine but is similar to the Type II coefficients for factor A (see Table 6.1).

As a matter of computational interest, the Type II estimable functions for B are equal to the Type III estimable functions if there is no interaction. For then $\alpha\beta_{ij} = 0$, and Table 6.4 shows that the coefficients for α_i and β_j are the same for Types II and III. This is not to say that the Type II SS and Type III SS will be equal, but rather that they give tests of the same hypothesis when there is no interaction. If, indeed, there is no interaction, then the Type II F-test is more powerful than the Type III F-test. The assumption of no interaction is, however, probably rarely satisfied in nature.

Table 6.5 *Estimable Functions for A*B*

Effect		Coefficients for All Types
Intercept		0
A	1	0
	2	0
B	1	0
	2	0
	3	0
A*B	1	L7
	1	L8
	1	−L7−L8
	2	−L7
	2	−L8
	2	L7+L8

Table 6.5 gives the coefficients of the estimable function for the A*B interaction and, again, two elements are available. In this case all types of effects give the same results, since for each type the interaction effects are adjusted for all other effects. The estimable functions can be readily interpreted if the coefficients are recorded in the 2×3 cell format implied by the factorial array. For example, let L7 = −1 and L8 = 0; the resulting function can be illustrated as follows:

		B		
		1	2	3
A	1	−1	0	+1
	2	+1	0	−1

This is the interaction in the 2×2 subtable consisting of the columns for B1 and B3, or the interaction of the contrast $(\alpha_1 - \alpha_2)$ with the contrast $(\beta_1 - \beta_2)$.

6.3.7.3 Estimating Estimable Functions

Estimates of estimable functions can be obtained by multiplying the vector of coefficients by the vector of parameter estimates using the SOLUTION option. For example, letting L2=1 in Table 6.3 for Type I results in the vector

$$\mathbf{L} = (0 \ \ 1 \ \ -1 \ \ .05 \ \ .325 \ \ -.375 \ \ -.375 \ \ .25 \ \ .625 \ \ .125 \ \ -.2 \ \ -3 \ \ -.5)$$

The vector of parameter estimates (see Output 6.7) is

$$\hat{\boldsymbol{\beta}} = (5.4 \ \ -2.4 \ \ .0 \ \ -2.9 \ \ 2.933 \ \ .0 \ \ 5.4 \ \ -1.33 \ \ .0 \ \ .0 \ \ .0 \ \ .0)$$

The estimate $\mathbf{L}\hat{\boldsymbol{\beta}} = 1.075$ is equal to $\bar{y}_{1..} - \bar{y}_{2..}$, the unadjusted treatment difference. Likewise, using the Type III coefficients gives an estimate of 1.044, which is the difference between the two least-squares means of the A factor (see Table 6.8).

Variances of these estimates can be obtained by the standard formula for the variance of a linear function. The estimated variance of the estimates is $s^2 \left(\mathbf{L} \left(\mathbf{X}'\mathbf{X} \right)^{-} \mathbf{L}' \right)$, where s^2, the estimated error variance, is the residual mean square from the overall analysis of variance, and $(\mathbf{X}'\mathbf{X})^{-}$ is the generalized inverse of the $\mathbf{X}'\mathbf{X}$ matrix generated by the dummy variables. The square root of this variance provides the standard error of the estimated function; hence, a *t*-test is readily constructed.

Since the LSMEANS, CONTRAST, and ESTIMATE statements offer methods of estimating and testing the most desired functions, the preceding technique is seldom employed. However, if these statements produce functions that are nonestimable, the generation of estimates from scratch may provide otherwise unobtainable estimates.

6.3.7.4 Interpreting LSMEANS, CONTRAST, and ESTIMATE Results Using Estimable Functions

Sometimes it may be useful to examine the construction of estimable functions associated with the LSMEANS, CONTRAST, and ESTIMATE statements. Information on the construction of these functions is available by specifying E as one of the options in the statement. (Don't confuse this with the E=*effect* option, which specifies an alternate error term.) Output 6.12 shows the results from the E option:

```
proc glm;
    class a b;
    model y=a b a*b / solution;
    contrast 'B 2 vs 3' b 0 -1 1 / e;
    estimate 'A in B1' a -1 1 a*b -1 0 0 1 / e;
    lsmeans a / stderr e;
```

Output 6.12
Estimable
Functions for
the LSMEANS,
CONTRAST,
and
ESTIMATE
Statements

```
                   Coefficients for Contrast A*B 2 vs 3

                                        Row 1

           Intercept                      0

           a          1                   0
           a          2                   0

           b          1                   0
           b          2                   0
           b          3                   0

           a*b        1 1                  0
           a*b        1 2                  1
           a*b        1 3                 -1
           a*b        2 1                  0
           a*b        2 2                 -1
           a*b        2 3                  1


                   Coefficients for Estimate A in B1

                                        Row 1

           Intercept                      0

           a          1                  -1
           a          2                   1

           b          1                   0
           b          2                   0
           b          3                   0

           a*b        1 1                 -1
           a*b        1 2                  0
           a*b        1 3                  0
           a*b        2 1                  1
           a*b        2 2                  0
           a*b        2 3                  0


                 Coefficients for a Least Square Means

                                           a Level
        Effect                          1              2

        Intercept                       1              1
        a          1                    1              0
        a          2                    0              1
        b          1             0.33333333     0.33333333
        b          2             0.33333333     0.33333333
        b          3             0.33333333     0.33333333
        a*b        1 1           0.33333333         0
        a*b        1 2           0.33333333         0
        a*b        1 3           0.33333333         0
        a*b        2 1               0          0.33333333
        a*b        2 2               0          0.33333333
        a*b        2 3               0          0.33333333
```

The hypothesis tested by the CONTRAST statement is

$$H_0: -\beta_2 + \beta_3 - .5\alpha\beta_{12} + .5\alpha\beta_{13} - .5\alpha\beta_{22} + .5\alpha\beta_{23} = 0$$

or in the μ-model notation

$$-.5\mu_{12} + .5\mu_{13} - .5\mu_{22} + .5\mu_{23}$$

Note that the coefficients of the interaction effects are supplied by the procedure.

The function estimated by the ESTIMATE statement is

$$\mathbf{L}\beta = -\alpha_1 + \alpha_2 - \alpha\beta_{11} + \alpha\beta_{21}$$

or in μ-model notation

$$\mathbf{L}\beta = -\mu_{11} + \mu_{21}$$

The least-squares means for A1 estimates is

$$\mu + \alpha_1 + (\beta_1 + \beta_2 + \beta_3 + \alpha\beta_{11} + \alpha\beta_{12} + \alpha\beta_{13})/3$$

or, in μ-model notation

$$(\mu_{11} + \mu_{12} + \mu_{13})/3$$

6.3.8 Empty Cells

Analyzing multifactor data with empty (or missing) cells often gives results of questionable value, since the data contain insufficient information to estimate the parameters of the model. (See Freund (1980) for a discussion of this problem.) The absence of data in one or more cells makes it very difficult to establish guidelines for imposing restrictions or generating appropriate estimable functions. PROC GLM helps investigate alternate estimable functions for such analyses, but no packaged program, including PROC GLM, provides a single best solution for all situations.

The problem of empty cells is illustrated by deleting the A=1, B=3 cell from the 2*3 factorial data in Output 6.7.

Table 6.6 gives the general form of estimable functions, showing that the data with the empty cells generate only five parameters. This is because there are only five cells with data, so that only five (linearly independent) parameters are estimable.

For this example, only the Type III and Type IV estimable functions will be discussed. The following statements are used:

```
proc glm;
   class a b;
   model y=a b a*b / e3 e4 solution;
   lsmeans a b / e stderr;
```

Table 6.6 *General Form of Estimable Functions*

Effect			Coefficients
Intercept			L1
A	1		L2
	2		L1 – L2
B	1		L4
	2		L5
	3		L1 – L4 – L5
A*B	1	1	L7
	1	2	L2 – L7
	2	1	L4 – L7
	2	2	–L2 + L5 + L7
	2	3	L1 – L4 – L5

The coefficients of the functions for the A effect appear in Table 6.7. Type III and Type IV are identical for A, since A has only two levels.

Table 6.7 *Estimable Functions for Factor A*

Effect			Type III & Type IV Coefficients
Intercept	1		0
A	1		L2
	2		–L2
B	1		0
	2		0
	3		0
A*B	1	1	0.5*L2
	1	2	0.5*L2
	2	1	–0.5*L2
	2	2	–0.5*L2
	2	3	0

As in a classification with complete data, Type III and Type IV functions for A do not involve the parameters of factor B. Setting L2=1 shows that the A effect is equal to

$$\mathbf{L}\beta = .5(\mu_{11} + \mu_{12}) - .5 \ (\mu_{21} + \mu_{22})$$

This is shown in this two-way, μ-model diagram:

		B		
		1	2	3
A	1	.5	.5	.
	2	−.5	−.5	0

No information from the A_2B_3 cell is used, since there is no matching data from the A_1B_3 cell.

Table 6.8 gives the Type III and Type IV coefficients associated with the B factor. (Notice that there are no coefficients for μ, α_1, and α_2.)

Table 6.8 *Estimable Functions for Factor B*

Effect			Type III Coefficients	Type IV Coefficients
Intercept			0	0
A	1		0	0
	2		0	0
B	1		L4	L4
	2		L5	L5
	3		−L4−L5	−L4−L5
A*B	1	1	0.25*L4−0.25*L5	0
	1	2	−0.25*L4+0.25*L5	0
	2	1	0.75*L4+0.25*L5	L4
	2	2	0.25*L4+0.75*L5	L5
	2	3	−L4−L5	−L4−L5

First, consider the Type III coefficients. A set of two estimable functions tested with the Type III F statistic for B can be obtained by setting L4=1, L5=0 and L4=0, L5=1, which test $H_0: (\beta_1 - \beta_3 = 0)$ and $(\beta_2 - \beta_3 = 0)$, respectively. The coefficients of the cell means in terms of the μ model are shown in the diagram:

	(L4=1, L5=0)				(L4=0, L5=1)		
Type III A	.25	−.25		A	−.25	.25	
	.75	.25	−1.0		.25	.75	−1.0

The choices L4=1, L5=0 and L4=0, L5=1 did not result in very appealing comparisons, since they involve cell means from levels of B that are not part of the desired hypotheses. Therefore, make another selection. Taking L4=1, L5=1 and L4=1, L5=1, more interesting coefficients result.

	(L4=1, L5=1)		
Type III A	0	0	
	1	1	−2

	(L4=1, L5=−1)		
A	.5	−.5	
	.5	−.5	0

The hypotheses being tested are

$$H_0: \mu_{21} + \mu_{22} - 2\mu_{23} = 0 \quad \text{and} \quad .5(\mu_{11} + \mu_{21} - \mu_{12} - \mu_{22}) = 0$$

or equivalently

$$H_0: .5(\mu_{21} + \mu_{22}) = \mu_{23} \quad \text{and} \quad \bar{\mu}_{.1} = \bar{\mu}_{.2}$$

Thus, the Type III *F*-test for B simultaneously compares B3 with the average of B1 and B2 within the level of A that has complete data and compares B1 with B2 averaged across the levels of A. Remember that the first selection of coefficients (L4=1, L5=0 and L4=0, L5=1) provides an equivalent H_0; it is just more difficult to understand.

For the same Type III test with no missing cells, the Type III hypothesis for B is

$$H_0: \mu_{.1} = \mu_{.2} = \mu_{.3}$$

or equivalently

$$H_0: \bar{\mu}_{.1} = \bar{\mu}_{..}, \ \bar{\mu}_{.2} = \bar{\mu}_{..}, \ \bar{\mu}_{.3} = \bar{\mu}_{..}$$

For the example with the empty A_1B_3 cell, the parameter μ_{13} is not estimable unless further conditions are imposed. (If there are no data from a population, the mean of that population cannot be estimated unless some relationship between this mean and other means is established.) The Type III hypothesis is equivalent to

$$H_0: \mu_{.1} = \mu_{.2} = \mu_{.3}$$

subject to $\mu_{13} = .5(\mu_{11} + \mu_{12})$.

Now consider the Type IV coefficients. Taking L4=1, L5=0 and L4=0, L5=1 results in the diagram

	(L4=1, L5=0)		
Type III A	0	0	
	1	0	−1

	(L4=0, L5=1)		
A	0	0	
	0	1	−1

Thus, the Type IV H_0 is clearly different from the Type III H_0, because the Type IV H_0 does not involve the means in level 1 of A, namely μ_{11} and μ_{12}, even though there are data in these cells.

Output 6.13 shows that Type III and Type IV sums of squares are indeed different (Type III SS=16.073, Type IV SS=41.733).

*Output 6.13
Comparison of
the Type III
and Type IV
Sums of
Squares with
Empty Cells*

```
                              The GLM Procedure

                          Class Level Information

                    Class          Levels    Values

                    a                 2       1 2

                    b                 3       1 2 3


                        Number of observations    18

NOTE: Due to missing values, only 17 observations can be used in this analysis.
                      The SAS System   14:53 Wednesday, November 14, 2001   30
                            The GLM Procedure

Dependent Variable: y


                                      Sum of
        Source              DF        Squares       Mean Square    F Value    Pr > F

        Model                4      45.81568627     11.45392157      5.27     0.0110

        Error               12      26.06666667      2.17222222

        Corrected Total     16      71.88235294


                  R-Square      Coeff Var      Root MSE       y Mean

                  0.637370      27.53339       1.473846      5.352941


        Source              DF     Type III SS     Mean Square    F Value    Pr > F

        a                    1      0.35072464      0.35072464      0.16     0.6949
        b                    2     16.07330642      8.03665321      3.70     0.0560
        a*b                  1     29.56811594     29.56811594     13.61     0.0031


        Source              DF      Type IV SS     Mean Square    F Value    Pr > F

        a                   1*      0.35072464      0.35072464      0.16     0.6949
        b                   2*     41.73333333     20.86666667      9.61     0.0032
        a*b                  1     29.56811594     29.56811594     13.61     0.0031


* NOTE: Other Type IV Testable Hypotheses exist which may yield different SS.


                                        Standard
        Parameter              Estimate     Error      t Value   Pr > |t|

        Intercept            5.400000000 B  0.65912400     8.19    <.0001
        a        1          -3.733333333 B  1.07634498    -3.47    0.0046
        a        2           0.000000000 B     .            .        .
        b        1          -2.900000000 B  1.23310809    -2.35    0.0366
        b        2           2.933333333 B  1.07634498     2.73    0.0184
        b        3           0.000000000 B     .            .        .
        a*b      1 1         6.733333333 B  1.82503171     3.69    0.0031
        a*b      1 2         0.000000000 B     .            .        .
        a*b      2 1         0.000000000 B     .            .        .
        a*b      2 2         0.000000000 B     .            .        .
        a*b      2 3         0.000000000 B     .            .        .


NOTE: The X'X matrix has been found to be singular, and a generalized inverse was used to
solve the normal equations. Terms whose estimates are followed by the letter 'B' are not
uniquely estimable.
```

The message in Output 6.13,

*NOTE: Other Type IV Testable Hypotheses exist which may yield different SS.

is a warning that the Type IV estimable functions (and consequently the associated hypotheses and sums of squares) are not unique. This refers to the phenomenon that Type IV estimable functions depend on the location of the empty cell.

Suppose, for example, that the values of the levels of B are changed, say, from B=1 to B=9, B=2 to B=8, and B=3 to B=7. Since PROC GLM sorts on the levels, the first and third columns are interchanged, placing the empty cell in the upper-left-hand corner. An examination of the estimable functions (which involve all cells that contain data) reveals these coefficients:

	B				B		
	7	8	9		7	8	9
		0	0		0	.5	−.5
Type III A	1	0	−1	A	0	.5	−.5
	old	old	old		old	old	old
	B=3	B=2	B=1		B=3	B=2	B=1

Now we repeat the analysis with the missing cells placed in all possible locations. Note that the data are not changed. The same data are missing, only subscripts are changed to locate the missing cell in a different position.

The partitioning of the sums of squares from these analyses is given in the first two columns of Table 6.8. Two items are of interest are that

❑ the Type III and Type IV sums of squares always give identical results for the sum of squares due to the A factor.

❑ the Type IV sums of squares due to B depend on the location of the missing cell with respect to the B factor and are never the same as the Type III sums of squares.

The right-hand portion of Table 6.8 gives estimates of functions whose analogs with no missing data would be B1 vs B3 and B2 vs B3, obtained by taking L4=1, L5=0 and L4=0, L5=1 from the table. Estimates of the A effects are not given since they are not affected by changing the location of the missing cell. Note that the Type III and Type IV estimates disagree even in some cases when the function does not involve the missing cell. When the missing cell is in the last level of B, they always differ, in which case the Type IV function uses only information in rows not containing the empty cell.

Table 6.9 gives the estimable function for the interaction effect. Since there is only one degree of freedom, there is one function, involving only the complete portion of the factorial structure.

Table 6.9 *Sums of Squares and Estimates for B Effect*

Location of Missing Cell	Sums of Squares Type IV		B1 vs B3 (L4=1, L5=0) Type IV				B2 vs B3 (L4=0, L5=1) Type IV			
	A	B	Basis			Estimate	Basis			Estimate
1,3	.351	41.733	0	0		−2.900	0	0		2.933
			1	0	−1		1	0	−1	
2,3	.351	41.733	0	0		−2.900	0	0		2.933
			1	0	−1		0	1	−1	
1,1	.351	23.386	.5	−.5		.467	0	0		2.933
			.5	+.5	−1		0	1	−1	
2,1	.351	23.386	.5	−.5		.467	0	0		2.933
			.5	.5	−1		0	1	−1	
1,2	.351	18.978	0	0		−2.900	−.5	.5		−4.33
			1	0	−1		.5	.5	−1	
2,2	.351	23.286	.5	−.5		.467	0	0		2.933
			.5	.5	−1		0	1	−1	
All Type III	.351	16.073	.25	−.25	x	−1.217	−.25	.25	x	1.250
			.75	.25	−1		.25	.75	−1	

Table 6.10 *Estimable Functions for A*B*

Effect			All Types Coefficients
Intercept			0
A	1		0
	2		0
B	1		0
	2		0
	3		0
A*B	1	1	L7
	1	2	−L7
	2	1	−L7
	2	2	L7
	2	3	0

Now consider these CONTRAST, ESTIMATE, and LSMEANS statements used on the data with an empty $A_1 B_3$ cell:

```
proc glm;
    class a b;
    model y=a b a*b;
    contrast 'B 2 vs 3' b 0 -1 1 / e;
    estimate 'A in B1' a -1 1 a*b -1 0 1 / e;
    lsmeans a / stderr e;
```

The estimable functions resulting from the E option appear in Output 6.14.

Output 6.14
Estimable Functions for the CONTRAST, ESTIMATE, and LSMEANS Statements

```
                        Coefficients for Contrast B 2 vs 3

                                        Row 1


                  Intercept                    0


                  a         1                  0
                  a         2                  0


                  b         1                  0
                  b         2                 -1
                  b         3                  1


                  a*b      1 1                 0
                  a*b      1 2                -0.5
                  a*b      2 1                 0
                  a*b      2 2                -0.5
                  a*b      2 3                 1



                    Coefficients for Estimate A in B1

                                        Row 1


                  Intercept                    0


                  a         1                 -1
                  a         2                  1


                  b         1                  0
                  b         2                  0
                  b         3                  0


                  a*b      1 1                -1
                  a*b      1 2                 0
                  a*b      2 1                 1
                  a*b      2 2                 0
                  a*b      2 3                 0



                   Coefficients for a Least Square Means

                                              a Level
          Effect                             1            2

          Intercept                          1            1
          a         1                        1            0
          a         2                        0            1
          b         1                  0.33333333   0.33333333
          b         2                  0.33333333   0.33333333
          b         3                  0.33333333   0.33333333
          a*b      1 1                        0.5          0
          a*b      1 2                        0.5          0
          a*b      2 1                        0     0.33333333
          a*b      2 2                        0     0.33333333
          a*b      2 3                        0     0.33333333
```

a	y LSMEAN	Standard Error	Pr > |t|
1	Non-est	.	.
2	5.41111111	0.49940294	<.0001

Parameter	Estimate	Standard Error	t Value	Pr > |t|
A in B1	-3.00000000	1.47384606	-2.04	0.0645

The function specified by the CONTRAST statement B2 vs 3 is designated nonestimable in the SAS log (not shown) because it involves level 3 of B, which contained the empty cell. Therefore, no statistical computation is printed. A more technical reason for the nonestimable CONTRAST function is ascertainable from the coefficients printed in Output 6.14. They show that the function is

$$\mathbf{L\beta} = -\beta_2 + \beta_3 - .5\alpha\beta_{12} - .5\alpha\beta_{22} + \alpha\beta_{23}$$

In order for a function $\mathbf{L\beta}$ to be estimable, it must be equal to a linear function of the μ_{ij}. This is equivalent to the condition that the coefficients of the μ_{ij} are equal to the corresponding coefficients of $\alpha\beta_{ij}$ in $\mathbf{L\beta}$. Thus, the condition for the $\mathbf{L\beta}$ in the CONTRAST statement to be estimable is that $\mathbf{L\beta}$ is equal to

$$-.5\mu_{12} - .5\mu_{22} + \mu_{23}$$

but

$$-.5\mu_{12} - .5\mu_{22} + \mu_{23} = -.5\alpha_1 + .5\alpha_2 +$$
$$[-\beta_2 + \beta_3 - .5\alpha\beta_{12} - .5\alpha\beta_{22} + \alpha\beta_{23}]$$

which contains α_2 and is thus not the same function as $\mathbf{L\beta}$.

Compare the estimable functions in Output 6.14 with those in Output 6.12 to see further effects of the empty cell. Later in this section, you see that this CONTRAST statement does produce an estimable function if no interaction is specified in the model.

The function specified in the ESTIMATE statement is estimable because it involves only cells that contain data; namely A_1B_1 and A_2B_1 (see Output 6.14). From the printed coefficients in Output 6.14, the function is evidently

$$-\alpha_1 + \alpha_2 - \alpha\beta_{11} + \alpha\beta_{21}$$

In terms of the μ model the function is

$$-\mu_{11} + \mu_{21}$$

and is therefore estimable because it is expressible as a linear function of the μ_{ij} for cells containing data.

Caution is required in using the ESTIMATE or CONTRAST statements with empty cells. Compare the ESTIMATE statements that produced Output 6.12 and Output 6.14. If we used the ESTIMATE statement that produced Output 6.12 with a data set that had the A_1B_3 cell empty, the following nonestimable function would result:

Table 6.11 *Coefficients Produced by the ESTIMATE Statement*

Effect			Coefficients
Intercept			0
A	1		−1
	2		1
B	1		0
	2		0
	3		0
A*B	1	1	−1
	1	2	0
	2	1	0
	2	2	1
	2	3	0

That is, specifying A*B −1 0 0 1 places the −1 as a coefficient on $\alpha\beta_{22}$ instead of $\alpha\beta_{21}$ as desired. When interaction parameters are involved, it is necessary to know the location of the empty cells if you are using the ESTIMATE statement.

The estimable functions from the LSMEANS statement show the least-squares mean for A1 to be nonestimable and the least-squares mean for A2 to be estimable. The principles involved are the same as those already discussed with respect to the CONTRAST and ESTIMATE statements.

To illustrate the computations, consider the estimability of functions produced by CONTRAST, ESTIMATE, and LSMEANS statements in the context of a MODEL statement that contains no interaction:

```
proc glm;
   class a b;
   model y=a b / solution ss1 ss2 ss3 ss4;
   contrast 'B 2 vs 3' b 0 -1 1 / e;
   estimate 'A EFFECT' a -1 1 / e;
   lsmeans a / e;
```

Partial results appear in Output 6.16. None of the functions are nonestimable even though the A_1B_3 cell has no data. This is because the functions involve only the parameters in the model

$$Y_{ijk} = \mu + \alpha_i + \beta_j + \varepsilon_{ijk}$$

under which the means $\left(\mu_{ij} = \mu + \alpha_i + \beta_j\right)$ are all estimable, even μ_{13}. Any linear function of the μ_{ij} is estimable, and all the functions in Output 6.16 are linear functions of the μ_{ij}. For example, the A1 least-squares mean is

$$\mathbf{L\beta} = \mu + \alpha_1 + .3333(\beta_1 + \beta_2 + \beta_3)$$
$$= .3333(3\mu + 3\alpha_1 + \beta_1 + \beta_2 + \beta_3)$$
$$= .3333(\mu_{11} + \mu_{12} + \mu_{13})$$

Output 6.16
Coefficients
with an
Empty Cell
and No
Interaction

```
                          The GLM Procedure

                  Coefficients for Contrast B 2 vs 3

                                    Row 1

              Intercept                   0

              a         1                 0
              a         2                 0

              b         1                 0
              b         2                -1
              b         3                 1


                          The GLM Procedure

                 Coefficients for Estimate A EFFECT

                                    Row 1

              Intercept                   0

              a         1                -1
              a         2                 1

              b         1                 0
              b         2                 0
              b         3                 0


                 Coefficients for a Least Square Means

                                             a Level
        Effect                            1              2

        Intercept                         1              1
        a         1                       1              0
        a         2                       0              1
        b         1                 0.33333333     0.33333333
        b         2                 0.33333333     0.33333333
        b         3                 0.33333333     0.33333333
```

Table 6.12 summarizes results of the CONTRAST, ESTIMATE, and LSMEANS statements.

Table 6.12 *Summary of Results from CONTRAST, ESTIMATE, and LSMEANS Output for a Two-Way Layout with an Empty Cell and No Interaction*

Contrast	DF	SS	F	Pr>F
B2 vs 3	1	4.68597211	1.09	0.3144

Parameter	Estimate	T for H_0 Parameter=0	Pr> $\mid$ T $\mid$	STD Error of Estimate
A	−1.39130435	−1.14	0.2747	1.22006396

		Least-Squares Means		
A	Y LSMEAN	Std Err LSMEAN	Pr > $\mid$ T $\mid$ H_0: LSMEAN=0	
1	4.26376812	0.92460046	0.0005	
2	5.65507246	0.69479997	0.0001	

Understanding the results from an analysis with empty cells is admittedly difficult, another reminder that the existence of empty cells precludes a universally correct analysis. The problem is that empty cells leave a gap in the data and make it difficult to estimate interactions and to adjust for interaction effects. If the interaction is not requested, the empty cell causes fewer problems. Of course, the analysis is incorrect if the interaction is present in the data.

For situations that require analyses including interaction effects, the GLM procedure does not claim to always have the *correct* answer, but it does provide information about the nature of the estimates and allows the experimenter to decide whether these results have any real meaning.

6.4 Mixed-Model Issues

Previous sections in this chapter were concerned with fixed-effects models, in which all parameters are measures of the effects of given levels of the factors. For such models, F-tests of hypotheses on those parameters use the residual mean square in the denominator. In some models, however, one or more terms represent a random variable that measures the effect of a random sample of levels of the corresponding factor. Such terms are called random effects; models containing random effects only are called **random models**, whereas models containing both random and fixed effects are called **mixed models**. See Steel and Torrie (1980), especially Sections 7.5 and 16.6.

6.4.1 Proper Error Terms

For situations in which the proper error term is known by the construction of the design, PROC GLM provides the TEST statement (see Section 4.6.1, "A Standard Split-Plot Experiment"). PROC GLM also allows specification of appropriate error terms in MEANS, LSMEANS, and CONTRAST statements. For situations that are not obvious, PROC GLM gives a set of expected mean squares that can be used to indicate proper denominators of F-statistics. In some cases, these may have to be computed by hand. To illustrate the use and interpretation of these tools, consider a variation of the split-plot design involving the effect on yield of different irrigation treatments and cultivars.

Irrigation treatments are more easily applied to larger areas (main plots), whereas different cultivars may be planted in smaller areas (subplots). In this example, consider three irrigation treatments (IRRIG) assigned in a completely random manner to nine main-plot units (REPS). REPS are each split into two subplots, and two cultivars (CULT), **A** and **B**, are randomly assigned to the subplots.

The appropriate partitioning of sums of squares is

Source	DF
IRRIG	2
REPS in IRRIG	6
CULT	1
IRRIG*CULT	2
ERROR	6
TOTAL	17

The proper error term for irrigation is REPS within IRRIG because the REPS are the experimental units for the IRRIG factor. The other effects are tested by ERROR (which is actually CULT*REPS in IRRIG). Data for the irrigation experiment appear in Output 6.17.

Output 6.17
Data for a
Split-Plot
Experiment

```
                          The SAS System

            Obs    irrig    reps    cult    yield

             1       1       1       A      27.4
             2       1       1       B      29.7
             3       1       2       A      34.5
             4       1       2       B      29.4
             5       1       3       A      32.5
             6       1       3       B      34.4
             7       2       1       A      28.9
             8       2       1       B      28.7
             9       2       2       A      33.4
            10       2       2       B      28.7
            11       2       3       A      32.4
            12       2       3       B      36.4
            13       3       1       A      28.6
            14       3       1       B      29.7
            15       3       2       A      32.9
            16       3       2       B      27.2
            17       3       3       A      29.1
            18       3       3       B      32.6
```

The analysis is implemented as follows:

```
proc glm;
    class irrig reps cult;
    model yield=irrig reps(irrig)
                cult irrig*cult;
    test h=irrig e=reps(irrig);
    contrast 'IRRIG 1 vs IRRIG 2' irrig 1 -1 / e=reps(irrig);
```

Note the use of the nested effect in the MODEL statement (See Section 4.2, "Nested Classifications").

The TEST statement requests that the IRRIG effect be tested against the REP within the IRRIG mean square.

The CONTRAST statement requests a comparison of the means for irrigation methods 1 and 2. The appropriate error mean square for testing the contrast is REPS(IRRIG) for the same reason that this was the appropriate error term for testing the IRRIG factor. The results appear in Output 6.18.

Output 6.18
F-Tests with
Correct
Denominators

```
                                    The GLM Procedure

                                      Sum of
     Source                  DF       Squares     Mean Square   F Value   Pr > F

     Model                   11    31.91611111    2.90146465      1.34    0.3766

     Error                    6    13.02666667    2.17111111

     Corrected Total         17    44.94277778


     Source                  DF     Type III SS   Mean Square   F Value   Pr > F

     irrig                    2    13.06777778    6.53388889      3.01    0.1244
     reps(irrig)              6    13.32000000    2.22000000      1.02    0.4896
     cult                     1     3.12500000    3.12500000      1.44    0.2755
     irrig*cult               2     2.40333333    1.20166667      0.55    0.6017


        Tests of Hypotheses Using the Type III MS for reps(irrig) as an Error Term

     Source                  DF     Type III SS   Mean Square   F Value   Pr > F

     irrig                    2    13.06777778    6.53388889      2.94    0.1286


        Tests of Hypotheses Using the Type III MS for reps(irrig) as an Error Term

     Contrast                DF     Contrast SS   Mean Square   F Value   Pr > F

     IRRIG 1 vs IRRIG 2       1     0.70083333    0.70083333      0.32    0.5946
```

6.4.2 More on Expected Mean Squares

Expected mean squares are algebraic expressions specifying the functions of the model parameters that are estimated by the mean squares resulting from partitioning the sums of squares. Generally, these expected mean squares are linear functions of elements that represent the

❏   error variance

❏   functions of variances of random effects

❏   functions of sums of squares and products (quadratic forms) of fixed effects.

The underlying principle of an *F*-test on a set of fixed-effects parameters is that the expected mean square for the denominator contains a linear function of variances of random effects, whereas the expected mean square for the numerator contains the same function of these variances plus a quadratic form of the parameters being tested. If no such matching pair of variance functions is available, no proper test exists; however, approximate tests are available.

For fixed models, all mean squares estimate the residual error variance plus a quadratic form (variance) of the parameters in question. Hence, the proper denominator for all tests is the error term. Expected mean squares are usually not required for fixed models.

For a mixed model, the expected mean squares are requested by a RANDOM statement in PROC GLM, specifying the model effects that are random.

For the irrigation example, REPS within IRRIG is random. Regarding IRRIG and CULT as fixed effects in the example, a model is

$$y_{ijk} = \mu + \alpha_i + w_{ij} + \beta_k + \alpha\beta_{ik} + e_{ijk} \tag{6.9}$$

where α_i, β_k, and $\alpha\beta_{ik}$ are the main effect of IRRIG, the main effect of CULT, and the interaction effect of IRRIG*CULT. The w_{ij} term is the random effect of REPS within IRRIG.

Use the following SAS statements to obtain an analysis for this model:

```
proc glm;
    class irrig reps cult;
    model yield=irrig reps(irrig) cult cult*irrig / ss3;
    contrast 'IRRIG 1 vs IRRIG 2' irrig 1-1 / e=reps(irrig);
    random reps(irrig) / q;
```

The RANDOM statement specifies that REPS(IRRIG) is a random effect. The Q option requests that the coefficients of the quadratic form in the expected mean squares (EMS) are printed. The analysis of variance is the same as in Output 6.18.

The expected mean squares appear in Output 6.19.

Output 6.19
Expected Mean Squares, and Quadratic Form for IRRIG

```
                        The SAS System

                       The GLM Procedure

        Source                 Type III Expected Mean Square

        irrig                  Var(Error) + 2 Var(reps(irrig)) + Q(irrig,irrig*cult)

        reps(irrig)            Var(Error) + 2 Var(reps(irrig))

        cult                   Var(Error) + Q(cult,irrig*cult)

        irrig*cult             Var(Error) + Q(irrig*cult)


        Contrast               Contrast Expected Mean Square

        IRRIG 1 vs IRRIG 2     Var(Error) + 2 Var(reps(irrig)) + Q(irrig,irrig*cult)
```

Output 6.19
(Continued)
Expected Mean
Squares, and
Quadratic
Form for
IRRIG

Quadratic Forms of Fixed Effects in the Expected Mean Squares				
Source: Type III Mean Square for irrig				
	irrig 1	irrig 2	irrig 3	Dummy010	Dummy011
irrig 1	4.00000000	-2.00000000	-2.00000000	2.00000000	2.00000000
irrig 2	-2.00000000	4.00000000	-2.00000000	-1.00000000	-1.00000000
irrig 3	-2.00000000	-2.00000000	4.00000000	-1.00000000	-1.00000000
Dummy010	2.00000000	-1.00000000	-1.00000000	1.00000000	1.00000000
Dummy011	2.00000000	-1.00000000	-1.00000000	1.00000000	1.00000000
Dummy012	-1.00000000	2.00000000	-1.00000000	-0.50000000	-0.50000000
Dummy013	-1.00000000	2.00000000	-1.00000000	-0.50000000	-0.50000000
Dummy014	-1.00000000	-1.00000000	2.00000000	-0.50000000	-0.50000000
Dummy015	-1.00000000	-1.00000000	2.00000000	-0.50000000	-0.50000000

	Dummy012	Dummy013	Dummy014	Dummy015
irrig 1	-1.00000000	-1.00000000	-1.00000000	-1.00000000
irrig 2	2.00000000	2.00000000	-1.00000000	-1.00000000
irrig 3	-1.00000000	-1.00000000	2.00000000	2.00000000
Dummy010	-0.50000000	-0.50000000	-0.50000000	-0.50000000
Dummy011	-0.50000000	-0.50000000	-0.50000000	-0.50000000
Dummy012	1.00000000	1.00000000	-0.50000000	-0.50000000
Dummy013	1.00000000	1.00000000	-0.50000000	-0.50000000
Dummy014	-0.50000000	-0.50000000	1.00000000	1.00000000
Dummy015	-0.50000000	-0.50000000	1.00000000	1.00000000

Because this experiment is completely balanced, all four types of expected mean squares are identical. Consider the EMSs for IRRIG. The results in Output 6.19 translate into

$$\sigma_e^2 + 2\sigma_w^2 + (\text{quadratic form in } \alpha\text{'s and } \alpha\beta\text{s}) / (a - 1)$$

where $\sigma_e^2 = V(e)$, $\sigma_w^2 = V(w)$, and a=number of levels of IRRIG. (The a–1 term is not explicitly indicated by PROC GLM.) The quadratic form is a measure of differences among the irrigation means

$$\mu_{i..} = E(\bar{y}_{i..}) = \mu + \alpha_i + \beta_k + \alpha\beta_{ik}$$

The coefficients appear in Output 6.19. (Although not shown here, the labels DUMMY010-DUMMY015 would be previously indicated in the output to correspond to $\alpha\beta_{11}, \alpha\beta_{12}, \alpha\beta_{21}, \alpha\beta_{22}, \alpha\beta_{31}$, and $\alpha\beta_{32}$.) In matrix notation, the quadratic form is $\boldsymbol{\alpha}'\mathbf{A}\boldsymbol{\alpha}$, where

$$\mathbf{A} = \begin{bmatrix} 4 & -2 & -2 & 2 & 2 & -1 & -1 & -1 & -1 \\ -2 & 4 & -2 & -1 & -1 & 2 & 2 & -1 & -1 \\ -2 & -2 & 4 & -1 & -1 & -1 & -1 & 2 & 2 \\ 2 & -1 & -1 & 1 & 1 & -.5 & -.5 & -.5 & -.5 \\ 2 & -1 & -1 & 1 & 1 & -.5 & -.5 & -.5 & -.5 \\ -1 & 2 & -1 & -.5 & -.5 & 1 & 1 & -.5 & -.5 \\ -1 & 2 & -1 & -.5 & -.5 & 1 & 1 & -.5 & -.5 \\ -1 & -1 & 2 & -.5 & -.5 & -.5 & -.5 & 1 & 1 \\ -1 & -1 & 2 & -.5 & -.5 & -.5 & -.5 & 1 & 1 \end{bmatrix}$$

and

$$\alpha' = \alpha_1 \quad \alpha_2 \quad \alpha_3 \quad \alpha\beta_{11} \quad \alpha\beta_{12} \quad \alpha\beta_{21} \quad \alpha\beta_{22} \quad \alpha\beta_{31} \quad \alpha\beta_{32}$$

This is the general expression for the quadratic form in which no constraints are assumed on the parameters. In this representation, irrigation mean differences are functions not only of the α_i but also of the $\alpha\beta_{ij}$, that is,

$$\mu_{1..} - \mu_{2..} = \alpha_1 - \alpha_2 + \alpha\beta_{1.} - \alpha\beta_{2.}$$

Thus, the quadratic form measuring these combined differences involves the $\alpha\beta_{ij}$s. In most texts, these expected mean squares are presented with the constraints $\alpha_1 + \alpha_2 + \alpha_3 = 0$ and $\alpha\beta_{1.} + \alpha\beta_{2.} + \alpha\beta_{3.} = 0$. When these constraints are imposed, the quadratic form reduces to

$$\alpha'A\alpha = 6\alpha_1^2 + 6\alpha_2^2 + 6\alpha_3^2$$

To see this, note that the contribution from the first row of **A** is

$$\alpha_1(4\alpha_1 - 2\alpha_2 - 2\alpha_3 + 2\alpha\beta_{11} + 2\alpha\beta_{12} - \alpha\beta_{21} - \alpha\beta_{22} - \alpha\beta_{31} - \alpha\beta_{32})$$
$$= \alpha_1(4\alpha_1 + 2(-\alpha_2 - \alpha_3 + 2\alpha\beta_{1.} - \alpha\beta_{2.} - \alpha\beta_{3.})$$
$$= \quad \alpha_1(4\alpha_1 + 2\alpha_1 - 0 - 0 - 0)$$
$$= 6\alpha_1^2$$

Similarly, row 2 and row 3 yield $6\alpha_2^2$ and $6\alpha_3^2$, respectively. Row 4 gives

$$\alpha\beta_{11}(2\alpha_1 - \alpha_2 - \alpha_3 + \alpha\beta_{11} + \alpha\beta_{12} - .5\alpha\beta_{21} - .5\alpha\beta_{22} - .5\alpha\beta_{31} - .5\alpha\beta_{32})$$
$$= \alpha\beta_{11}(3\alpha_1)$$

and row 5 gives $\alpha\beta_{12}(3\alpha_1)$. Thus, the sum of rows 4 and 5 is 0. Similarly, the net contribution from rows 6 through 9 is 0.

Under the null hypothesis of no difference between irrigation methods, the expected mean squares for IRRIG becomes $\sigma_e^2 + 4\sigma_w^2$. But this is the same as the expected mean squares for REPS(IRRIG). Therefore, REPS(IRRIG) is, in fact, the correct denominator in the F-test for IRRIG.

Now, view this example as if the irrigation treatments are in a randomized-blocks design instead of a completely random design—that is, assume REPS is crossed with IRRIG rather than nested in IRRIG. If both IRRIG and CULT are fixed effects, then a model is

$$y_{ijk} = \mu + \varphi_j + \alpha_i + w_{ij} + \beta_k + \alpha\beta_{ik} + \varepsilon_{ijk} \tag{6.10}$$

where $\alpha_i, \beta_k, \alpha\beta_{ik}$, and ε_{ijk} have the same meaning as in the model shown in equation (6.9). The φ_j term is the random block (REPS) effect, with $V(\varphi_j) = \sigma_\rho^2$, and w_{ij} is the random main-plot error, that is, the random block by irrigation (REP*IRRIG) effect. The model in equation (6.10) is equivalent to model 12.3 of Steel and Torrie (1980, p. 245) with their A=IRRIG and B=CULT. Now it is commonly presumed that any interaction effect must be a random effect if it involves a random main effect. In terms of equation (6.10), this presumption implies that if φ_j is random, then w_{ij} must be random. However, PROC GLM does not operate under this presumption, and, therefore, both REPS and REPS*IRRIG must be explicitly designated as random in the RANDOM statement in order to obtain expected mean squares corresponding to φ_j and w_{ij} both as random. In other words, if only REPS appears in the RANDOM statement, then the expected mean squares printed by PROC GLM would correspond to w_{ij} as a fixed effect.

Output 6.20 contains the expected mean squares resulting from the following statements:

```
proc glm;
   class reps irrig cult;
   model yield=reps irrig reps*irrig cult cult*irrig / ss1;
   random reps reps*irrig;
```

*Output 6.20
Expected Mean
Squares for
Split Plot:
Fixed Main-
Plot and
Subplot
Factors*

```
                          The SAS System

                         The GLM Procedure

   Source                  Type III Expected Mean Square

   irrig                   Var(Error) + 2 Var(irrig*reps) + Q(irrig,irrig*cult)

   reps                    Var(Error) + 2 Var(irrig*reps) + 6 Var(reps)

   irrig*reps              Var(Error) + 2 Var(irrig*reps)

   cult                    Var(Error) + Q(cult,irrig*cult)

   irrig*cult              Var(Error) + Q(irrig*cult)
```

The line for IRRIG in the output translates to

$$\delta_e^2 + 2\delta_w^2 + (\text{quadratic form in } \alpha\text{'s and } \alpha\beta\text{'s}) \ / \ 2$$

and, as discussed above, the quadratic form reduces to $6\Sigma\alpha_i^2$ if $\Sigma_i\alpha_i = \Sigma_i \ \alpha\beta_{ij} = 0$. This matches the expected mean squares given in Steel and Torrie (1980, p. 394) for this design.

6.4.3 An Issue of Model Formulation Related to Expected Mean Squares

Before leaving this example, you should understand one more point concerning the expected mean squares in mixed models. Suppose that CULT is a random effect, but IRRIG remains fixed. Following Steel and Torrie (1980), the table below shows the expected mean squares:

Source	Expected Mean Squares
REPS	$\sigma_\varepsilon^2 + 2\sigma_\gamma^2 + 6\sigma_\rho^2$
IRRIG	$\sigma_\varepsilon^2 + 2\sigma_\gamma^2 + 3(3/2)\sigma_{\alpha\beta}^2 + 6\Sigma_i \alpha_i^2/2$
Error(A)=REPS*IRRIG	$\sigma_\varepsilon^2 + 2\sigma_\gamma^2$
CULT	$\sigma_\varepsilon^2 + 9\sigma_\beta^2$
CULT*IRRIG	$\sigma_\varepsilon^2 + 3(3/2)\sigma_{\alpha\beta}^2$
Error(B)	σ_ε^2

Output 6.21 contains the expected mean squares output from the following statements:

```
proc glm;
    class reps irrig cult;
    model yield=reps irrig reps*irrig cult
                irrig*cult / ss3;
    random reps reps*irrig cult cult*irrig;
```

Output 6.21 Expected Mean Squares for a Split-Plot Design: Fixed Main-Plot and Random Subplot Factors

```
                            The SAS System                        69

                          The GLM Procedure

Source                 Type III Expected Mean Square

irrig                  Var(Error) + 3 Var(irrig*cult) + 2 Var(irrig*reps) + Q(irrig)

reps                   Var(Error) + 2 Var(irrig*reps) + 6 Var(reps)

irrig*reps             Var(Error) + 2 Var(irrig*reps)

cult                   Var(Error) + 3 Var(irrig*cult) + 9 Var(cult)

irrig*cult             Var(Error) + 3 Var(irrig*cult)
```

The only quadratic form for fixed effects is Q(IRRIG) in the IRRIG line, which corresponds to $6\Sigma\alpha_i^2$, assuming $\Sigma\alpha_i = 0$. The IRRIG line also contains Var(IRRIG*CULT) and Var(IRRIG*REPS), in agreement with the expected mean squares of Steel and Torrie given above. The lines in the output for REPS, IRRIG*REPS, and IRRIG*CULT also agree with these expected mean squares. However, although the line for CULT in Output 6.21 contains Var(IRRIG*CULT), the expected mean squares given by Steel and Torrie for CULT B does not contain $\sigma^2_{\alpha\beta}$. The exclusion by Steel and Torrie is in agreement with the general principle that if U is a fixed effect and V is a random effect, then the expected mean squares for U contains σ^2_{U*V}, but the expected mean squares for V does not contain σ^2_{U*V}.

This principle is an item of controversy among statisticians and relates to formulation and parameterization of the model. Basically, exclusion or inclusion of $\sigma^2_{\alpha\beta}$ in the line for CULT depends on variance and covariance definitions in the model. See Hocking (1973, 1985) for detailed accounts of modeling ramifications for the two-way mixed model. The PROC GLM output for a two-way mixed model corresponds to the results shown by Hocking (1973) for Model III in his Table 2. Refer to Hartley and Searle (1969), who point out that exclusion of $\sigma^2_{\alpha\beta}$ is inconsistent with results commonly reported for the unbalanced case. More recently, other authors, including Samuels, Casella, and McCabe (1991), Lentner, Arnold, and Hinkleman (1989), and McLean, Sanders, and Stroup (1991), have discussed this modeling problem, but it is not resolved.

6.5 ANOVA Issues for Unbalanced Mixed Models

We stated in Chapter 5 that there are no definitive guidelines for using ANOVA methods for analyzing unbalanced mixed-model data. This is true even in the simplest case of a two-way classification. In this section we discuss in greater detail the issues involved, although we are not able to resolve all of them. One of the main problems is to choose numerator and denominator means squares to construct approximate F-ratios for hypotheses about fixed effects.

6.5.1 Using Expected Mean Squares to Construct Approximate *F*-Tests for Fixed Effects

Refer again to the unbalanced clinical trial data set. We consider the studies to be random and drugs fixed. Run the following statements to obtain Types I, II, and III sums of squares and their expected mean squares:

```
proc glm data=drugs;
   class trt study;
   model flush = trt study trt*study / e1 e2 e3;
   random study trt*study / q test;
run;
```

ANOVA results appear in Output 6.22.

Output 6.22
Three Types of
Sums of
Squares for an
Unbalanced
Two-Way
Classification

```
                        Unbalanced Two-way Classification

                              The GLM Procedure

Dependent Variable: FLUSH

                                    Sum of
        Source              DF      Squares     Mean Square   F Value   Pr > F

        Model               17    16618.75357     977.57374     2.24    0.0063

        Error              114    49684.09084     435.82536

        Corrected Total    131    66302.84440


        Source              DF     Type I SS     Mean Square   F Value   Pr > F

        TRT                  1    1134.560964   1134.560964     2.60    0.1094
        STUDY                8    6971.606045    871.450756     2.00    0.0526
        TRT*STUDY            8    8512.586561   1064.073320     2.44    0.0178


        Source              DF     Type II SS    Mean Square   F Value   Pr > F

        TRT                  1    1377.550724   1377.550724     3.16    0.0781
        STUDY                8    6971.606045    871.450756     2.00    0.0526
        TRT*STUDY            8    8512.586561   1064.073320     2.44    0.0178


        Source              DF     Type III SS   Mean Square   F Value   Pr > F

        TRT                  1    1843.572090   1843.572090     4.23    0.0420
        STUDY                8    7081.377266    885.172158     2.03    0.0488
        TRT*STUDY            8    8512.586561   1064.073320     2.44    0.0178
```

First of all, we consider the choice of a mean square for TRT to use in the numerator of an approximate *F*-statistic. You see that the values of MS(TRT) range from 1134.6 to Type I to 1843.6 for Type III. But it is not legitimate to choose the mean square based on its observed value. Instead, the choice should be made based on the **expected** mean squares, which describe what the mean squares measure. Expected mean squares are computed from the RANDOM statement. But in order to get the Types I and II expected mean squares the e1 and e2 options must be specified in the MODEL statement. The Types I-III expected mean squares are shown in Output 6.23.

Source	Type I Expected Mean Square
TRT	Var(Error) + 9.1461 Var(TRT*STUDY) + 0.04 Var(STUDY) + Q(TRT)
STUDY	Var(Error) + 7.1543 Var(TRT*STUDY) + 14.213 Var(STUDY)
TRT*STUDY	Var(Error) + 7.0585 Var(TRT*STUDY)

Source	Type II Expected Mean Square
TRT	Var(Error) + 9.1385 Var(TRT*STUDY) + Q(TRT)
STUDY	Var(Error) + 7.1543 Var(TRT*STUDY) + 14.213 Var(STUDY)
TRT*STUDY	Var(Error) + 7.0585 Var(TRT*STUDY)

Source	Type III Expected Mean Square
TRT	Var(Error) + 4.6613 Var(TRT*STUDY) + Q(TRT)
STUDY	Var(Error) + 7.0585 Var(TRT*STUDY) + 14.117 Var(STUDY)
TRT*STUDY	Var(Error) + 7.0585 Var(TRT*STUDY)

Basically, we want to choose a mean square for TRT that results in the most powerful test for differences between drug means. The mean squares are quadratic forms of normally distributed data, and therefore they are approximately distributed as a constant times non-central chi-square random variables. We want to select the mean square with the largest non-centrality parameter. The non-centrality parameters are equal to $Q(TRT)/[(Var(Error) + k_1 Var(TRT*STUDY) + k_2 Var(STUDY))]$. Except for Type I, k_2 is equal to 0, and it is very small (0.04). Thus, the choice boils down to selecting the mean square with largest value of $Q(TRT)/k_1$. The quantities $Q(TRT)$ k_1 are available from the Q option on the MODEL statement, shown in Output 6.24.

Output 6.24
Three Types of
Expected
Mean Squares
for an
Unbalanced
Two-Way
Classification

```
                        Source: Type I Mean Square for TRT

                               TRT A                 TRT B

            TRT A              32.99242424          -32.99242424
            TRT B             -32.99242424           32.99242424


                        Source: Type II Mean Square for TRT

                               TRT A                 TRT B

            TRT A              32.80309690          -32.80309690
            TRT B             -32.80309690           32.80309690


                        Source: Type III Mean Square for TRT

                               TRT A                 TRT B

            TRT A              20.97586951          -20.97586951
            TRT B             -20.97586951           20.97586951
```

In terms of model parameters, for each type of mean square, Q(TRT) is equal to $\alpha' A \alpha$, where $\alpha' = (\alpha_1, \alpha_1)$ and A is the matrix shown in Output 6.23. Since there are only two treatments, all elements in A have the same value (except for sign), and Q(TRT) is determined by this number. Thus the mean square with largest value of that number divided by k_1 will have the largest non-centrality parameter. For this example the values are

Type I: $32.99/9.15 = 3.61$

Type II: $32.81/9.14 = 3.59$

Type III: $20.98/4.66 = 4.50$

There is not much to choose between Types I and II, but the value for Type III is approximately 20% larger than the others. Therefore, Type III has the largest non-centrality parameter.

6.6 GLS and Likelihood Methodology Mixed Model

The MIXED procedure uses generalized least squares (GLS) and likelihood methodology to construct test statistics, estimates, and standard errors of estimates. In this section we present a brief description of the methods. Although the methods are based on sound criteria, they usually do not result in "exact" results, in the sense that *p*-values and confidence coefficients are only approximate.

6.6.1 An Overview of Generalized Least Squares Methodology

The statistical model on which PROC MIXED is based is

$$Y = X\beta + ZU + \varepsilon,$$

where Y is the vector of data, $X\beta$ is the set of linear combinations for fixed effects, ZU is a set of linear combinations of random effects, and ε is a vector of residual errors. The matrices X and Z

contain known constants, often values of independent variables or values of indicator variables for classification variables. The vector β contains fixed but unknown constants, such as regression parameters or differences between treatment means. The random vector $\mathbf{U}$ has expectation $E(\mathbf{U}) = \mathbf{0}$ and covariance $V(\mathbf{U}) = \mathbf{G}$. The random vector ε has expectation $E(\varepsilon) = \mathbf{0}$ and covariance $V(\varepsilon) = \mathbf{R}$. With these specifications, the data vector $\mathbf{Y}$ has expectation $E(\mathbf{Y}) = \mathbf{X}\beta$ and variance $V(\mathbf{Y}) = \mathbf{ZGZ}' + \mathbf{R}$. In most applications, we assume $\mathbf{U}$ and ε are normally distributed, which makes $\mathbf{Y}$ normally distributed, also.

Let $\mathbf{V} = \mathbf{ZGZ}' + \mathbf{R}$ be the covariance matrix of $\mathbf{Y}$. Then GLS methodology gives us the following results: The best estimator of β is

$$\mathbf{b} = (\mathbf{X}'\mathbf{V}^{-1}\mathbf{X})^{-1}\mathbf{X}'\mathbf{V}^{-1}\mathbf{Y} \qquad (6.11)$$

and the variance of $\mathbf{b}$ is

$$V(\mathbf{b}) = (\mathbf{X}'\mathbf{V}^{-1}\mathbf{X})^{-1} \qquad (6.12)$$

It follows that the best estimator of a linear combination, or set of linear combinations, $\mathbf{L}\beta$ is $\mathbf{Lb}$ and its variance is $\mathbf{L}(\mathbf{X}'\mathbf{V}^{-1}\mathbf{X})^{-1}\mathbf{L}'$. These results provide the basis for statistical inference about model parameters that is used in PROC MIXED. Here are some of the most common inference procedures:

Standard error of an estimate $\mathbf{Lb}$: $\qquad (\mathbf{L}(\mathbf{X}'\mathbf{V}^{-1}\mathbf{X})^{-1}\mathbf{L}')^{1/2} \qquad (6.13)$

Test about linear combinations H_0: $\mathbf{L}\beta = \mathbf{0}$: $\quad \chi^2 = \mathbf{Lb}(\mathbf{L}(\mathbf{X}'\mathbf{V}^{-1}\mathbf{X})^{-1}\mathbf{L}')\mathbf{b}'\mathbf{L}' \qquad (6.14)$

These statistical inference methods are exact. However, they cannot be used in most practical applications because they require knowledge of the covariance matrix $\mathbf{V}$. Therefore, $\mathbf{V}$ must be estimated from the data. In most cases, the elements of $\mathbf{V}$ are functions of a small number of parameters. These are called *covariance* parameters, although individual parameters might not represent covariances, strictly speaking. For example, in a repeated measures analysis, $\mathbf{V}$ might involve variation between subjects and variation between repeated measures within subjects. One of the parameters might be the variance between subjects, and other parameters might represent variances and covariances within subjects. (See Chapter 8.)

Equations (6.11) through (6.14) assume $(\mathbf{X}'\mathbf{V}^{-1}\mathbf{X})^{-1}$. If $(\mathbf{X}'\mathbf{V}^{-1}\mathbf{X})$ is singular, then a g-inverse is computed instead.

The structure of $\mathbf{V}$ is specified with REPEATED and RANDOM statements. The REPEATED statement determines $\mathbf{R}$ and the RANDOM statement determines $\mathbf{Z}$ and $\mathbf{G}$. Parameters in $\mathbf{R}$ and $\mathbf{G}$ and estimated, and the values of the estimates are inserted into $\mathbf{R}$ and $\mathbf{G}$. The method of maximum likelihood, or a variation of it, is usually used to estimate the covariance parameters. Then the estimates of $\mathbf{R}$ and $\mathbf{G}$ are used to get the estimate of $\mathbf{V}$, and it is inserted into expressions in (6.11)–(6.14). As a result, the "exactness" no longer holds, and most results become approximate. In addition, t-distributions are used instead of z-distributions for confidence intervals, and F-distributions are used instead of chi-square distributions for tests of fixed effects.

There are more consequences due to estimating the parameters in $\mathbf{V}$. The degrees of freedom for the t- and F-statistics must be estimated using Satterthwaite-type methods. Also, standard errors of estimates are biased downward because the estimates contain variations induced by estimation of $\mathbf{V}$.

6.6.2 Some Practical Issues about Generalized Least Squares Methodology

The methods employed by PROC MIXED are based on sound principles as prescribed by the specified model. Of course, the model must be valid in order for results to be valid. Assuming the model is valid, the inferential methods are not exact only because covariance parameters must be estimated. Nonetheless, standard errors of estimates from the MODEL statement, and from the ESTIMATE and LSMEANS statements, use the correct basic mathematical expression. Therefore, these standard errors are credible, whereas standard errors computed by PROC GLM in mixed-model applications are usually suspect because they are not based on the correct basic mathematical expression. Instead, they are based on expressions for fixed-effects models.

PROC MIXED treats random effects as random and fixed effects as fixed. These concepts are built into the model. This takes considerable guesswork out of using PROC MIXED that occurs when using PROC GLM. Here are two examples. One, in choosing a numerator mean square for a test of fixed effects, you do not have to be concerned about the involvement of random effects. You simply choose a type of hypothesis you want to test based on considerations of only the fixed effects. Two, estimability is judged on the basis of fixed effects alone, without regard to the random effects.

Analysis of Covariance

7.1 Introduction

Analysis of covariance can be described as a combination of the methods of regression and analysis of variance. Regression models use **direct** independent variables—that is, variables whose values appear directly in the model—for example, the linear regression $y = \beta_0 + \beta_1 x$. Analysis of variance models use **class** variables—that is, the independent variable's classifications appear in the model—for example, the one-way ANOVA model $y = \mu + \tau_i$. In more theoretical terms, ANOVA *class* variables set up dummy 0-1 columns in the X matrix (see Chapter 6). Analysis-of-covariance models use both direct and class variables. A simple example combines linear regression and one-way ANOVA, yielding $y = \mu + \tau_i + \beta x$.

Analysis of covariance uses at least two measurements on each unit: the response variable y, and another variable x, called a **covariable**. You may have more than one covariable. The basic objective is to use information about y that is contained in x in order to refine inference about the response. This is done primarily in three ways:

- [ ] In all applications, variation in y that is associated with x is removed from the error variance, resulting in more precise estimates and more powerful tests.

- [ ] In some applications, group means of the y variable are adjusted to correspond to a common value of x, thereby producing an equitable comparison of the groups.

- [ ] In other applications, the regression of y on x for each group is of intrinsic interest, either to predict the effect of x on y for each group, or to compare the effect of x on y among groups.

Textbook discussions of covariance analysis focus on the first two points, with the main goal of establishing differences among adjusted treatment means. By including a related variable that accounts for substantial variation in the dependent variable of interest, you can reduce error. This increases the precision of the model parameter estimates. Textbooks discuss **separate slopes** versus **common slope** models—that is, covariance models with a separate regression slope coefficient for each treatment group, versus a single slope coefficient for all treatment groups. In these discussions, the main role of the separate slope model is to test for differences in slopes among the treatments. Typically, this test should be conducted as a preliminary step before an analysis of covariance, because, aside from carefully defined exceptions, the validity of comparing adjusted means using the analysis of covariance requires that the slopes be homogeneous. The beginning sections of this chapter present the textbook approach to analysis of covariance

There are broader uses of covariance models, such as the study of partial regression coefficients adjusted for treatment effects. Applied to factorial experiments with qualitative and quantitative factors, covariance models provide a convenient alternative to orthogonal polynomial and related contrasts that are often tedious and awkward. Later sections of this chapter present these methods.

To give a practical definition, analysis of covariance refers to models containing both continuous variables and group indicators (CLASS variables in the GLM procedure). Because CLASS variables create less-than-full-rank models, covariance models are typically more complex and hence involve more difficulties in interpretation than regression-only or ANOVA-only models. These issues are addressed throughout this chapter.

7.2 A One-Way Structure

Analysis of covariance can be applied in any data classification whenever covariables are measured. This section deals with the simplest type of classification, the one-way structure.

7.2.1 Covariance Model

The simplest covariance model is written

$$y_{ij} = \mu + \tau_i + \beta\left(x_{ij} - \overline{x}_{..}\right) + \varepsilon_{ij}$$

and combines a one-way treatment structure with parameters τ_i, one independent covariate x_{ij}, and associated regression parameter β.

Two equivalent models are

$$y_{ij} = \beta_0 + \tau_i + \beta x_{ij} + \varepsilon_{ij}$$

where $\beta_0 = \left(\mu - \beta\overline{x}_{..}\right)$, and

$$y_{ij} = \beta_{0i} + \beta x_{ij} + \varepsilon_{ij}$$

where β_{0_i} is the intercept for the ith treatment. This expression reveals a model that represents a set of parallel lines; the common slope of the lines is β, and the intercepts are $\beta_{0_i} = (\beta_0 + \tau_i)$. The model contains all the elements of an analysis-of-variance model of less-than-full rank, requiring restrictions on the τ_i or the use of generalized inverses and estimable functions. Note, however, that the regression coefficient β is not affected by the singularity of the $\mathbf{X'X}$ matrix; hence, the estimate of β is unique.

The following example, using data on the growth of oysters, illustrates the basic features of analysis of covariance. The goal is to determine

❑   if exposure to water heated artificially affects growth

❑   if the position in the water column (surface or bottom) affects growth.

Four bags with ten oysters in each bag are randomly placed at each of five stations in the cooling water canal of a power-generating plant. Each location, or station, is considered a treatment and is represented by the variable TRT. Each bag is considered to be one experimental unit. Two stations are located in the intake canal, and two stations are located in the discharge canal, one at the surface (TOP), the other at the bottom (BOTTOM) of each location. A single mid-depth station is located in a shallow portion of the bay near the power plant. The treatments are described below:

Treatment (TRT)	Station
1	INTAKE-BOTTOM
2	INTAKE-SURFACE
3	DISCHARGE-BOTTOM
4	DISCHARGE-SURFACE
5	BAY

Stations in the intake canal act as controls for those in the discharge canal, which has a higher temperature. The station in the bay is an overall control in case some factor other than the heat difference due to water depth or location is responsible for an observed change in growth rate.

The oysters are cleaned and measured at the beginning of the experiment and then again about one month later. The initial weight and the final weight are recorded for each bag. The data appear in Output 7.1.

Output 7.1
Data for
Analysis of
Covariance

Obs	trt	rep	initial	final
1	1	1	27.2	32.6
2	1	2	32.0	36.6
3	1	3	33.0	37.7
4	1	4	26.8	31.0
5	2	1	28.6	33.8
6	2	2	26.8	31.7
7	2	3	26.5	30.7
8	2	4	26.8	30.4
9	3	1	28.6	35.2
10	3	2	22.4	29.1
11	3	3	23.2	28.9
12	3	4	24.4	30.2
13	4	1	29.3	35.0
14	4	2	21.8	27.0
15	4	3	30.3	36.4
16	4	4	24.3	30.5
17	5	1	20.4	24.6
18	5	2	19.6	23.4
19	5	3	25.1	30.3
20	5	4	18.1	21.8

You can address the objectives given above by analysis of covariance. The response variable is final weight, but the analysis must also account for initial weight. You can do this using initial weight as the covariate. The following SAS statements are required to compute the basic analysis:

```
proc glm;
   class trt;
   model final=trt initial / solution;
```

The CLASS statement specifies that TRT is a classification variable. The variable INITIAL is the covariate. The MODEL statement defines the model $y_{ij} = \beta_0 + \tau_i + \beta x_{ij} + \varepsilon_{ij}$. Specifying the SOLUTION option requests printing of the coefficient vector.

Results of these statements appear in Output 7.2.

Output 7.2
Results of
Analysis of
Covariance

```
                          The GLM Procedure

Dependent Variable: final

                          Sum of
Source            DF      Squares    Mean Square   F Value   Pr > F

Model              5   354.4471767    70.8894353    235.05   <.0001

Error             14     4.2223233     0.3015945

Corrected Total   19   358.6695000


            R-Square      Coeff Var      Root MSE      FINAL Mean

            0.988228       1.780438      0.549176      30.84500


Source            DF    Type I SS     Mean Square   F Value   Pr > F

trt                4   198.4070000    49.6017500    164.47    <.0001
initial            1   156.0401767   156.0401767    517.38    <.0001


Source            DF    Type III SS   Mean Square   F Value   Pr > F

trt                4    12.0893593     3.0223398     10.02    0.0005
initial            1   156.0401767   156.0401767    517.38    <.0001


                                       Standard
Parameter            Estimate            Error    t Value   Pr > |t|

Intercept        2.494859769 B        1.02786287     2.43    0.0293
trt        1    -0.244459378 B        0.57658196    -0.42    0.6780
trt        2    -0.280271345 B        0.49290825    -0.57    0.5786
trt        3     1.654757698 B        0.42943036     3.85    0.0018
trt        4     1.107113519 B        0.47175112     2.35    0.0342
trt        5     0.000000000 B             .           .         .
initial          1.083179819          0.04762051    22.75    <.0001

NOTE: The X'X matrix has been found to be singular, and a generalized inverse was
used to solve the normal equations. Terms whose estimates are followed by the
letter 'B' are not uniquely estimable.
```

Consider the Type I and Type III SS (Type II and Type IV would be the same as Type III here). The Type I SS for TRT is the **unadjusted** treatment sum of squares. The ERROR SS for a simple analysis of variance can be reconstructed by subtracting the Type I SS from the TOTAL SS, for example,

Source	DF	SS	MS	F
TRT	4	198.407	49.602	4.642
ERROR	15	160.263	10.684	
TOTAL	19	358.670		

The resulting *F*-value indicates that *p* is less than .01. Thus, a simple analysis of variance leads to concluding that statistically significant treatment differences in final weight exist even when initial weights are not considered.

Now compare these results with the analysis of covariance. The Type III TRT SS is 12.089 whereas the Type I TRT SS equals the one-way ANOVA TRT SS of 198.407. The Type III TRT SS reflects differences among treatment means that have been adjusted to a common value of the covariate, INITIAL. In analysis of covariance, the TYPE III TRT SS is the *adjusted* treatment sum of squares; the Type I TRT SS is the **unadjusted** treatment sum of squares because it reflects the difference among treatment means prior to adjustment for the covariate. In this example, the unadjusted TRT SS is much larger than the adjusted one. However, the reduction in error mean squares from 10.684 to 0.302 allows an increase in the F-statistic from 4.642 in the simple analysis of variance to 10.02 in Output 7.2. The power of the test for treatment differences increases when the covariate is included because most of the error in the simple analysis of variance is due to variation in INITIAL values.

The last part of Output 7.2 contains the SOLUTION vector. In this one-factor case, the TRT estimates are obtained by setting the estimates for the last treatment (TRT 5) to 0. Therefore, the INTERCEPT estimate is the intercept for TRT 5, and the other four treatment effects are differences between each TRT and TRT 5. Because TRT 5 is the control, the output estimates, standard errors, and t-tests are for treatment versus control. Note that the means of TRT 3 and TRT 4 in the discharge canal differ from TRT 5.

The coefficient associated with INITIAL is the pooled within-groups regression coefficient relating FINAL to INITIAL. The coefficient estimate is a weighted average of the regression coefficients of FINAL on INITIAL, estimated separately for each of the five treatment groups. This coefficient estimates that a difference of 1.083 units in FINAL is associated with a one-unit difference in INITIAL.

7.2.2 Means and Least-Squares Means

A MEANS statement requests the unadjusted treatment means of all continuous (non-CLASS) variables in the model, that is, the response variable and the covariate. You can suppress printing the covariate means by using the DEPONLY option. These means are not strictly relevant to an analysis of covariance unless they are used to determine the effect of the covariance adjustment. The DUNCAN and WALLER options, among others, for multiple-comparisons tests are also available, but they are not useful here.

The LSMEANS (least-squares means) statement produces the estimates that are usually called adjusted treatment means. They are defined as

$$\bar{y}_i - \hat{\beta}(\bar{x}_{i.} - \bar{x}_{..}), \text{ or, equivalently, } \hat{\beta}_0 + \hat{\tau}_i + \hat{\beta}\bar{x}_{..}$$

Consistent with the MODEL statement in the SAS statements above, this example uses the latter form. Recall that you also obtain adjusted means for the unbalanced two-way classification using the same LSMEANS statement, which is

```
lsmeans trt / stderr tdiff;
```

The TDIFF option requests LSD tests among the adjusted means. You can use the ADJUST= option to obtain alternative multiple comparison tests, depending on the relative seriousness of Type I and Type II error. These issues were discussed in more detail in Section 3.3.3. The LSMEANS and TDIFF results appear in Output 7.3.

Output 7.3
Results of
Analysis of
Covariance:
Adjusted
Treatment
Means (Least-
Squares
Means)

```
                         Least Squares Means

                                  Standard                   LSMEAN
           trt     final LSMEAN     Error      Pr > |t|      Number

            1       30.1531125    0.3339174     <.0001          1
            2       30.1173006    0.2827350     <.0001          2
            3       32.0523296    0.2796295     <.0001          3
            4       31.5046854    0.2764082     <.0001          4
            5       30.3975719    0.3621988     <.0001          5

                 t for H0: LSMean(i)=LSMean(j) / Pr > |t|

      i/j          1           2          3          4          5

       1                    0.087941   -4.1466    -3.22289   -0.42398
                            0.9312      0.0010     0.0061     0.6780
       2       -0.08794               -4.76003    -3.55771   -0.56861
                0.9312                  0.0003     0.0032     0.5786
       3       4.146599    4.76003                 1.378002   3.853378
                0.0010      0.0003                 0.1898     0.0018
       4       3.222892    3.557715   -1.378                  2.346817
                0.0061      0.0032     0.1898                 0.0342
       5       0.42398     0.568608   -3.8533    -2.34682
                0.6780      0.5786      0.0018     0.0342

NOTE: To ensure overall protection level, only probabilities associated with pre-
planned comparisons should be used.
```

The estimated least-squares means are followed by their standard errors, which are printed because of the STDERR option. The *t*-values and associated significance probabilities for all pairwise tests of treatment differences are printed because of the TDIFF option.

The table below shows the unadjusted and adjusted means for the response variable FINAL and the mean of the covariate, INITIAL for each treatment group:

TRT	FINAL Unadjusted Means	Adjusted Least-Squares Means	INITIAL Covariate Mean
1	34.475	30.153	29.750
2	31.650	30.117	27.175
3	30.850	32.052	24.650
4	32.225	31.504	26.425
5	25.025	30.398	20.800

Figure 7.1 illustrates the distinction between adjusted and unadjusted means. The five linear regressions for each treatment are parallel, each with a common slope. Four points appear on each regression line: the end points are the predicted values of the response variable FINAL for the minimum and maximum values of INITIAL in the data set. The other two are 1) the predicted value of FINAL at the INITIAL mean for that treatment, and 2) the predicted value of FINAL at the mean value of the covariate INITIAL for the entire data set (shown by the solid vertical line at the mean value of INITIAL = 25.76). The former are the unadjusted sample means of FINAL; the latter are the adjusted, or LS means. The light-shaded vertical lines represent the mean of the covariate INITIAL at each treatment; the light-shaded horizontal lines correspond to the unadjusted sample means.

Figure 7.1 *Regressions for Five Oyster Data Treatments Showing Means and LS Means*

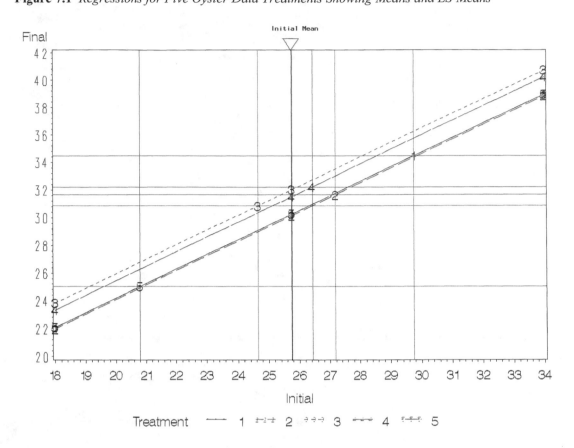

You can see that there are large changes between the unadjusted to adjusted treatment means for the variable FINAL. These changes result from the large treatment differences in the covariable INITIAL. Apparently, the random assignment of oysters to treatments did not result in equal mean initial weights. Some treatments, particularly TRT 5, received smaller oysters than other treatments. This biases the unadjusted treatment means. Computation of the adjusted treatment means is intended to remove the bias.

Note: Although the purpose of adjusted means is to remove bias resulting from unequal covariate means among the treatments, adjusted means are not always appropriate. The basic rule is that if the covariate means themselves depend on the treatments, adjustment is likely to be misleading, whereas if there is no reason to believe covariate means depend on treatment, *failing to adjust* is likely to be misleading. In the oyster growth example, there is no reason initial weights of oysters should depend on TRT. Therefore, you should use adjusted means. Consider, however, a typical example in plant breeding. Plant yield is affected by the population density of the plants. Accounting for plant density is essential to reduce error variance and hence allow manageable experiments to provide adequate power and precision. However, different plant varieties have inherently different plant densities, for a number of well-known reasons. Adjusting mean yield to a common plant density would distort differences among varieties you would actually see under realistic conditions. In this case, you should use *unadjusted* means. However, analysis of covariance is still useful because it improves precision.

You can use ESTIMATE statements to provide further insight into the mean and LS mean. The following SAS statements illustrate treatments 1 and 2 and their difference:

```
estimate 'trt 1 adj mean'
      intercept 1 trt 1 0 0 0 0 initial 25.76;
estimate 'trt 2 adj mean'
      intercept 1 trt 0 1 0 0 0 initial 25.76;
estimate 'adj trt diff' trt 1 -1 0 0 0;
```

The overall mean of the covariable INITIAL is $\bar{x}_{..}=25.76$, hence ESTIMATE computes the adjusted mean $\hat{\beta}_0 + \hat{\tau}_i + \hat{\beta}(25.76)$. Because $\hat{\beta}_0$ and $\hat{\beta}\bar{x}_{..}$ are the same for all adjusted means, the adjusted treatment difference is $\hat{\tau}_1 - \hat{\tau}_2$. The unadjusted means, $\bar{y}_{i.}$ estimate $\beta_0 + \tau_i + \beta\bar{x}_{i.}$ where $\bar{x}_{i.}$ is the sample mean of the covariate for the *i*th treatment. Use the following ESTIMATE statements to compute the unadjusted means:

```
estimate 'trt 1 unadj mean'
      intercept 1 trt 1 0 0 0 0 initial 29.75;
estimate 'trt 2 unadj mean'
      intercept 1 trt 0 1 0 0 0 initial 27.175;
estimate 'unadj diff' trt 1 -1 0 0 0 initial 2.575;
```

For the unadjusted means, $\hat{\beta}\bar{x}_{i.}$ is different for each treatment, so the unadjusted treatment difference estimates $\hat{\tau}_1 - \hat{\tau}_2 + \hat{\beta}(\bar{x}_{1.} - \bar{x}_{2.})$. This shows how the unadjusted means are confounded with the $\hat{\beta}\bar{x}_{i.}$. The results of both sets of ESTIMATE statements appear in Output 7.4.

Output 7.4
ESTIMATE
Statements for
Adjusted and
Unadjusted
Means

Parameter	Estimate	Standard Error	t Value	Pr > |t|
trt 1 adj mean	30.1531125	0.33391743	90.30	<.0001
trt 2 adj mean	30.1173006	0.28273504	106.52	<.0001
adj diff	0.0358120	0.40722674	0.09	0.9312
trt 1 unadj mean	34.4750000	0.27458811	125.55	<.0001
trt 2 unadj mean	31.6500000	0.27458811	115.26	<.0001
unadj diff	2.8250000	0.38832623	7.27	<.0001

You can see that the apparent difference between treatments 1 and 2 in unadjusted means results from their different mean initial weights. When these are adjusted, the difference disappears. Notice that, with an equal number of observations per treatment, the standard errors of the unadjusted treatment means are equal, whereas for the adjusted means, they depend on the difference between $\bar{x}_{i.}$ and $\bar{x}_{..}$. As a final note, you can modify the LSMEANS statement to obtain results similar to Output 7.3, but for the unadjusted means. You use the statement

```
lsmeans trt / bylevel stderr tdiff;
```

The BYLEVEL option causes $\bar{x}_{i.}$ to be used in place of $\bar{x}_{..}$ in computing the LS mean. These results are not shown, but would simply complete the results in Output 7.4 for the unadjusted means and differences.

7.2.3 Contrasts

This section illustrates comparing means with contrasts, using the oyster growth example discussed in Section 7.2.1, "Covariance Model." The five treatments can also be looked upon as a

2×2 factorial (BOTTOM/TOPXINTAKE/DISCHARGE) plus a CONTROL. The adjusted treatment means from the analysis of covariance can be analyzed further with four orthogonal contrasts implemented by the following CONTRAST statements:

```
contrast 'CONTROL VS. TREATMENT' TRT  -1 -1 -1 -1  4;
contrast 'BOTTOM VS. TOP'        TRT  -1  1 -1  1  0;
contrast 'INTAKE VS. DISCHARGE'  TRT  -1 -1  1  1  0;
contrast 'BOT/TOP*INT/DIS'       TRT   1 -1 -1  1  0;
```

The output that results from these statements follows the partitioning of sums of squares in Output 7.5. Note that the only significant contrast is INTAKE VS. DISCHARGE. Also, note that these are comparisons among *adjusted* means. If the objectives of your study compel you to define contrasts among unadjusted means, these must include the contribution from the $\hat{\beta}\bar{x}_{i\cdot}$, which do not cancel out as do the $\hat{\beta}\bar{x}_{\cdot\cdot}$ for the adjusted means. For example, for the first contrast above, CONTROL VS. TREATMENT, you need to include

$$\hat{\beta}(-\bar{x}_{1\cdot} - \bar{x}_{2\cdot} - \bar{x}_{3\cdot} - \bar{x}_{4\cdot} + 4\bar{x}_{5\cdot}) = \hat{\beta}(-24.85)$$ and thus the required SAS statement

```
contrast 'CTL V TRT UNADJUSTED'
    TRT  -1 -1 -1 -1 4  INITIAL -24.85;
```

Notice that the significance level of the unadjusted contrast far exceeds that of the adjusted CONTROL VS. TREATMENT contrast given the large discrepancy between adjusted and unadjusted means, especially for the CONTROL, noted earlier.

Equivalent results can be obtained with the ESTIMATE statement, which also gives the estimated coefficients for the contrasts. All options for CONTRAST and ESTIMATE statements discussed in Chapter 3, "Analysis of Variance for Balanced Data," and in Chapter 6, "Understanding Linear Models Concepts," apply here. Although constructed to be orthogonal, these contrasts are not orthogonal to the covariable; hence, their sums of squares do not add to the adjusted treatment sums of squares.

Output 7.5
Results of
Analysis of
Covariance:
Orthogonal
Contrasts Plus
an Unadjusted
Example

Contrast	DF	Contrast SS	Mean Square	F Value	Pr>F
CONTROL VS. TREATMENT	1	0.52000411	0.52000411	1.72	0.2103
BOTTOM VS. TOP	1	0.33879074	0.33879074	1.12	0.3071
INTAKE VS. DISCHARGE	1	8.59108077	8.59108077	28.49	0.0001
BOT/TOP*INT/DIS	1	0.22934155	0.22934155	0.76	0.3979
CTL V TRT UNADJ	1	169.9923582	169.9923582	563.65	<.0001

7.2.4 Multiple Covariates

Multiple covariates are specified as continuous (non-CLASS) variables in the MODEL statement. If the CLASS variable is designated as the first independent variable, the Type I sums of squares for individual covariates can be added to get the adjusted sums of squares due to all covariates. The Type III sums of squares are the fully adjusted sums of squares for the individual regression coefficients as well as those for the adjusted treatment means.

7.3 Unequal Slopes

Section 7.2 presented analysis of covariance assuming the *equal slopes* model. The **unequal slopes** model is a natural extension of covariance analysis. Both models are usually applied to data characterized by treatment groups and one or more covariates. The unequal slopes model allows you to test for heterogeneity of slopes—that is, it tests whether or not the regression coefficients are constant over groups. The analysis in Section 7.2 assumes constant regression coefficients and is invalid if this assumption fails. You can draw valid inference from the unequal slopes model, but it requires considerable care. This section presents the test for heterogeneity of slopes and inference strategies appropriate for the unequal slopes model.

Extending the one independent variable (covariate) and one-way treatment structure used in Section 7.2, the *unequal slopes analysis-of-covariance model* can be written

$$y_{ij} = \beta_{0_i} + \beta_{1i} x_{ij} + \varepsilon$$

where i denotes the treatment group. The hypothesis of equal slopes is

$$H_0: \beta_{1i} = \beta_{1i'} \text{ for all } i \neq i'$$

An alternate formulation of the model is

$$y_{ij} = \beta_0 + \alpha_i + \beta_1 x_{ij} + \delta_i x_{ij} + \varepsilon$$

where β_0 and β_1 are overall intercept and slope coefficients, and α_i and δ_i are coefficients for the treatment effect on intercept and slope, respectively, thus comparing the formulations of the model, $\beta_{0_i} = \beta_0 + \alpha_i$ and $\beta_{1i} = \beta_1 + \delta_i$. Under the alternate formulation, the hypothesis of equal slopes becomes

$$H_0: \delta_i = 0 \text{ for all } i = 1, 2, \ldots, t$$

Note that any possible intercept differences are irrelevant to both hypotheses.

Regression relationships that differ among treatment groups actually reflect an interaction between the treatment groups and the covariates. In fact, the GLM procedure specifies and analyzes this phenomenon as an interaction. Thus, if you use the following statements, the expression X*A produces the appropriate statistics for estimating different regressions of Y on X for the different values, or classes, specified by A:

```
proc glm;
   class a;
   model y=a x x*a / solution;
```

This MODEL fits the formulation $y_{ij} = \beta_0 + \alpha_i + \beta_1 x_{ij} + \delta_i x_{ij} + \varepsilon$. The α_i correspond to A, $\beta_1 x_{ij}$ corresponds to X, and $\delta_i x_{ij}$ corresponds to X*A. In this application, the Type I sums of squares for this model provide the most useful information:

X is the sum of squares due to a single regression of Y on X, ignoring the group.

A is the sum of squares due to different intercepts (adjusted treatment differences), assuming equal slopes.

X*A is an additional sum of squares due to different regression coefficients for the groups specified by the factor A.

The associated sequence of tests provides a logical stepwise analysis to determine the most appropriate model. Equivalent results can also be obtained by fitting the nested effects formulation $y_{ij} = \beta_{0_i} + \beta_{1_i}x_{ij} + \varepsilon$. Use the following statements:

```
proc glm;
   class a;
   model y=a x(a) / noint solution;
```

Here, the β_{0_i} correspond to A, and the β_{1_i} correspond to X(A). This formulation is more convenient than the alternative for obtaining estimates of the slopes for each treatment group. You can write a CONTRAST statement that generates a test equivalent to X*A in the previous model. However, X(A) does not test for the heterogeneity of slopes. Instead, it tests the hypothesis that *all* regression coefficients are 0. Also, A tests the hypothesis that all intercepts are equal to 0. For models like this one, the Type III (or Type IV) sums of squares have little meaning; it is not instructive to consider the effect of the CLASS variable over and above the effect of different regressions.

7.3.1 Testing the Heterogeneity of Slopes

This section uses the oyster growth data from Output 7.1 to demonstrate the test for the homogeneity of slopes. In a practical situation, you would do this testing before proceeding to the inference based on equal slopes shown in Section 7.2. Use the following SAS statements to obtain the relevant test statistics:

```
proc glm;
  class trt;
  model final=trt initial trt*initial;
```

Output 7.6 shows the results.

Output 7.6
Unequal
Slopes
Analysis of
Covariance for
Oyster Growth
Data

Source	DF	Sum of Squares	Mean Square	F Value	Pr > F
Model	9	355.8354908	39.5372768	139.51	<.0001
Error	10	2.8340092	0.2834009		
Corrected Total	19	358.6695000			

Source	DF	Type I SS	Mean Square	F Value	Pr > F
initial	1	342.3578175	342.3578175	1208.03	<.0001
trt	4	12.0893593	3.0223398	10.66	0.0012
initial*trt	4	1.3883141	0.3470785	1.22	0.3602

The Type I sums of squares show that

❑ the INITIAL weight has an effect on FINAL weight (F=1208.03, p<0.0001)

❑ TRT has an effect on FINAL weight, at any given INITIAL weight (F=10.66, p=0.0012)

❑ there is no significant difference in the INITIAL/FINAL relationship among the different levels of TRT (F=1.22, p=0.3602). That is, there is no evidence to contradict the null hypothesis of homogeneous slopes.

The last result is especially important, as it validates the analysis using the equal slopes model. The first two results essentially reiterate the results in Section 7.2.

7.3.2 Estimating Different Slopes

In many cases, the individual regression slopes for each treatment contain useful information. Output 7.7 contains data to illustrate this. The data are from a study of the relationship between the price of oranges and sales per customer. The hypothesis is that sales vary as a function of price differences for different stores (STORE) and days of the week (DAY). The price is varied daily for two varieties of oranges. The variables P1 and P2 denote the prices for the two varieties, respectively. Q1 and Q2 are the sales per customer of the corresponding varieties.

Output 7.7
Orange Sales
Data

Obs	STORE	DAY	P1	P2	Q1	Q2
1	1	1	37	61	11.3208	0.0047
2	1	2	37	37	12.9151	0.0037
3	1	3	45	53	18.8947	7.5429
4	1	4	41	41	14.6739	7.0652
5	1	5	57	41	8.6493	21.2085
6	1	6	49	33	9.5238	16.6667
7	2	1	49	49	7.6923	7.1154
8	2	2	53	53	0.0017	1.0000
9	2	3	53	45	8.0477	24.2176
10	2	4	53	53	6.7358	2.9361
11	2	5	61	37	6.1441	40.5720
12	2	6	49	65	21.7939	2.8324
13	3	1	53	45	4.2553	6.0284
14	3	2	57	57	0.0017	2.0906
15	3	3	49	49	11.0196	13.9329
16	3	4	53	53	6.2762	6.5551
17	3	5	53	45	13.2316	10.6870
18	3	6	53	53	5.0676	5.1351
19	4	1	57	57	5.6235	3.9120
20	4	2	49	49	14.9893	7.2805
21	4	3	53	53	13.7233	16.3105
22	4	4	53	45	6.0669	23.8494
23	4	5	53	53	8.1602	4.1543
24	4	6	61	37	1.4423	21.1538
25	5	1	45	45	6.9971	6.9971
26	5	2	53	45	5.2308	3.6923
27	5	3	57	57	8.2560	10.6679
28	5	4	49	49	14.5000	16.7500
29	5	5	53	53	20.7627	15.2542
30	5	6	53	45	3.6115	21.5442
31	6	1	53	53	11.3475	4.9645
32	6	2	53	45	9.4650	11.7284
33	6	3	53	53	22.6103	14.8897
34	6	4	61	37	0.0020	19.2000
35	6	5	49	65	20.5997	2.3468
36	6	6	37	37	28.1828	17.9543

In this example, consider the variety 1 only—that is, the response variable, sales, for variety 1 is Q1 and the covariable, price, is P1. Examples in Section 7.4 also involve variety 2. Here, you compute the unequal slopes covariance model using the SAS statements

```
proc glm;
  class day;
  model q1=p1 day p1*day/solution;
```

Output 7.7 shows the results of this analysis.

Output 7.8
Unequal
Slopes
Analysis of
Covariance
for Orange
Sales Data

Source	DF	Sum of Squares	Mean Square	F Value	Pr > F
Model	11	1111.522562	101.047506	4.64	0.0008
Error	24	522.153228	21.756384		
Corrected Total	35	1633.675790			

Source	DF	Type I SS	Mean Square	F Value	Pr > F
P1	1	516.5921408	516.5921408	23.74	<.0001
DAY	5	430.5384175	86.1076835	3.96	0.0093
P1*DAY	5	164.3920040	32.8784008	1.51	0.2236

Parameter		Estimate	Standard Error	t Value	Pr > |t|
Intercept		73.27263578 B	13.48373708	5.43	<.0001
P1		-1.22521164 B	0.26520396	-4.62	0.0001
DAY	1	-54.59714671 B	19.73545845	-2.77	0.0107
DAY	2	-34.78570099 B	20.25105926	-1.72	0.0987
DAY	3	-27.94295765 B	29.42842946	-0.95	0.3518
DAY	4	-24.12342640 B	21.39334761	-1.13	0.2706
DAY	5	4.62631110 B	30.62842608	0.15	0.8812
DAY	6	0.00000000 B	.	.	.
P1*DAY	1	1.00474758 B	0.39410534	2.55	0.0176
P1*DAY	2	0.60164207 B	0.39876566	1.51	0.1444
P1*DAY	3	0.61415851 B	0.57034268	1.08	0.2923
P1*DAY	4	0.42959726 B	0.41510986	1.03	0.3110
P1*DAY	5	0.02936476 B	0.57034268	0.05	0.9594
P1*DAY	6	0.00000000 B	.	.	.

From Output 7.8, you can see that there is no evidence to reject the null hypothesis of equal slopes (*F* for P1*DAY is 1.51 and $p=0.2236$). Ordinarily, you would then proceed as with the oyster growth data, using an equal slopes model. The Type I results here indicate a significant effect of price (P1) on sales and a significant effects of DAY on sales at any given price. For these data, however, a closer look at the estimates of the daily regression coefficients reveals additional information. Although the differences in the daily regressions are not statistically significant, it is instructive to look at their estimates.

The estimated daily regression slope coefficients are $\hat{\beta}_{1_i} = \hat{\beta}_1 + \hat{\delta}_i$. The estimate corresponding to P1 is $\hat{\beta}_1$ and P1*DAY is $\hat{\delta}_i$. For example, for DAY 1, the estimated slope is $\hat{\beta}_{1_i} = \hat{\beta}_1 + \hat{\delta}_1 = -1.2252 + 1.0047 = -0.2205$. You can use ESTIMATE statements to obtain the daily regression coefficients:

```
estimate 'P1:DAY 1' p1 1 p1*day 1 0 0 0 0 0;
estimate 'P1:DAY 2' p1 1 p1*day 0 1 0 0 0 0;
estimate 'P1:DAY 3' p1 1 p1*day 0 0 1 0 0 0;
estimate 'P1:DAY 4' p1 1 p1*day 0 0 0 1 0 0;
estimate 'P1:DAY 5' p1 1 p1*day 0 0 0 0 1 0;
estimate 'P1:DAY 6' p1 1 p1*day 0 0 0 0 0 1;
```

The results appear in Output 7.9.

Output 7.9
Estimated
Regression
Coefficients
for Each
TRT

Parameter	Estimate	Standard Error	t Value	Pr > |t|
P1:DAY 1	-0.22046406	0.29152337	-0.76	0.4569
P1:DAY 2	-0.62356957	0.29779341	-2.09	0.0470
P1:DAY 3	-0.61105313	0.50493329	-1.21	0.2380
P1:DAY 4	-0.79561438	0.31934785	-2.49	0.0200
P1:DAY 5	-1.19584688	0.50493329	-2.37	0.0263
P1:DAY 6	-1.22521164	0.26520396	-4.62	0.0001

Note that these estimated coefficients are larger in absolute value toward the end of the week. This is quite reasonable given the higher level of overall sales activity near the end of the week, which may result in a proportionately larger response in sales to changes in price. Thus, it is likely that a coefficient specifically testing for a linear trend in price response during the week would be significant.

You can obtain the results in Output 7.7 more conveniently using the nested-effects formation, whose SAS statements are

```
proc glm;
  class day;
  model q1=day p1(day)/noint solution;
  contrast 'equal slopes' p1(day) 1 0 0 0 0 -1,
                          p1(day) 0 1 0 0 0 -1,
                          p1(day) 0 0 1 0 0 -1,
                          p1(day) 0 0 0 1 0 -1,
                          p1(day) 0 0 0 0 1 -1;
```

The CONTRAST statement contains one independent comparison of the daily regressions per degree of freedom among daily regressions. In this case, there are 6 days and hence 5 DF. If all five comparisons are true, then H_0: all β_{1i} equal must be true. Hence the contrast is equivalent to the test generated by P1*DAY in Output 7.8. Output 7.10 contains the results.

Output 7.10
Analysis of
Orange Sales
Data Using a
Nested
Covariance
Model

Source	DF	Type I SS	Mean Square	F Value	Pr > F
DAY	6	4008.414213	668.069035	30.71	<.0001
P1(DAY)	6	861.125290	143.520882	6.60	0.0003

Source	DF	Type III SS	Mean Square	F Value	Pr > F
DAY	6	1250.581757	208.430293	9.58	<.0001
P1(DAY)	6	861.125290	143.520882	6.60	0.0003

Contrast	DF	Contrast SS	Mean Square	F Value	Pr > F
equal slopes	5	164.3920040	32.8784008	1.51	0.2236

Parameter		Estimate	Standard Error	t Value	Pr > |t|
DAY	1	18.67548906	14.41100810	1.30	0.2073
DAY	2	38.48693478	15.10940884	2.55	0.0177
DAY	3	45.32967813	26.15762403	1.73	0.0959
DAY	4	49.14920937	16.60915881	2.96	0.0068
DAY	5	77.89894687	27.50071487	2.83	0.0092
DAY	6	73.27263578	13.48373708	5.43	<.0001
P1(DAY)	1	-0.22046406	0.29152337	-0.76	0.4569
P1(DAY)	2	-0.62356957	0.29779341	-2.09	0.0470
P1(DAY)	3	-0.61105313	0.50493329	-1.21	0.2380
P1(DAY)	4	-0.79561437	0.31934785	-2.49	0.0200
P1(DAY)	5	-1.19584687	0.50493329	-2.37	0.0263
P1(DAY)	6	-1.22521164	0.26520396	-4.62	0.0001

As noted earlier, the Type I and Type III sums of squares provide no useful information. P1(DAY) tests the null hypothesis that all daily regression slopes are equal to zero. You can see that the contrast for equal slopes is identical to the equal slopes test given by P1*DAY in Output 7.8. The parameter estimate is the most useful output. You can see that the daily regression coefficients, P1(DAY), are identical to the estimates obtained via the ESTIMATE statements in Output 7.9, although Output 7.10 is easier to compute because no ESTIMATE statements are required. Output 7.9 also gives the intercept of the regressions for each day. For example, using DAY 1 and P1(DAY) 1, you can see that the regression for DAY 1 is Q1 = 18.675 − 0.2205×P1.

Using more than one independent variable is straightforward and can determine which variables have a different coefficient for each treatment group. More complex designs are not difficult to implement but may be difficult to interpret.

7.3.3 Testing Treatment Differences with Unequal Slopes

Tests among adjusted means with the equal slopes model apply to all values of the covariable. For example, Section 7.2.2 showed that the difference between the adjusted mean of TRT 1 and 2 is $(\mu + \tau_1 + \beta\,\bar{x}_{..}) - (\mu + \tau_2 + \beta\,\bar{x}_{..}) = \tau_1 - \tau_2$. If you use any other value of the covariable in the expression, the difference is still $\tau_1 - \tau_2$. However, for the unequal slope model, the adjusted mean is $\mu + \alpha_i + \beta_{1_i} x$, and hence the difference depends on x. Typically $\bar{x}_{..}$ is used for x. If you compare TRT 1 and 2 at $\bar{x}_{..}$, the difference is $\alpha_1 - \alpha_2 + (\beta_{1_1} - \beta_{1_2})\bar{x}_{..}$. If you evaluate the treatment difference at a different value of x, the difference changes. Thus, in the unequal slopes model, you can compare treatments, but only conditional upon a specified value of the covariable.

PROC GLM and PROC MIXED offer a great deal of flexibility for unequal slopes models, but you must be careful, because some defaults do not necessarily result in sensible tests.

Output 7.11 contains the Type III sums of squares and the LS means for the unequal slopes models for the orange sales data. The SAS statements are

```
proc glm;
  class day;
  model q1=p1 day p1*day;
  lsmeans day;
```

Output 7.11
Type III SS and
Default LS
Means for
Unequal
Slopes
Covariance
Analysis of
Orange Sales
Data

Source	DF	Type III SS	Mean Square	F Value	Pr > F
P1	1	554.7860985	554.7860985	25.50	<.0001
DAY	5	201.1717701	40.2343540	1.85	0.1412
P1*DAY	5	164.3920040	32.8784008	1.51	0.2236

Least Squares Means

DAY	Q1 LSMEAN
1	7.3828299
2	6.5463159
3	14.0301792
4	8.3960731
5	16.6450125
6	10.5145730

The least-squares means are computed using the overall covariable mean, $\bar{x}_{..}$, which for these data is equal to 51.222. You can use the AT MEANS option with the LSMEANS statement to print the value of the covariable means being used. The SAS statement is

```
lsmeans day/at means;
```

Output 7.12 shows the results.

Output 7.12
Least-Squares Means for Orange Sales Data Using the AT MEANS Option

```
          Least Squares Means at P1=51.22222

              DAY       Q1 LSMEAN

               1         7.3828299
               2         6.5463159
               3        14.0301792
               4         8.3960731
               5        16.6450125
               6        10.5145730
```

Suppose you want to test the null hypothesis of equal adjusted means—that is, among the means adjusted to a common value $\bar{x}_{..}$. Clearly, the Type I sum of squares for DAY from Output 7.8 does not do this: it tests means for a common slope, not adjusted for P1*DAY, the differences among slopes. What about the Type III sum of squares for DAY in Output 7.11? The Type III sum of squares tests the means adjusted not to $\bar{x}_{..}$ but to $x=0$. You can see this from the following CONTRAST statement:

```
contrast 'trt' day 1 -1 0 0 0 0,
               day 1 0 -1 0 0 0,
               day 1 0 0 -1 0 0,
               day 1 0 0 0 -1 0,
               day 1 0 0 0 0 -1;
```

Output 7.13 shows the results.

Output 7.13
Contrast Testing the Equality of Sales Means Adjusted to Covariable Price=0

Contrast	DF	Contrast SS	Mean Square	F Value	Pr > F
trt	5	201.1717701	40.2343540	1.85	0.1412

You can see that the $F=1.85$ and p-value of 0.1412 are identical to the Type III SS results in Output 7.11.

You can use the following LSMEANS statement to compute adjusted means at $x=0$ to see what you are testing:

```
lsmeans day/at p1=0;
```

Output 7.14 shows the results.

Output 7.14
Orange Sales
Adjusted
Means at Price
Covariable=0

```
                   Least Squares Means at P1=0


              DAY      Q1 LSMEAN

               1       18.6754891
               2       38.4869348
               3       45.3296781
               4       49.1492094
               5       77.8989469
               6       73.2726358
```

Recalling Output 7.10, these adjusted means are in fact the intercepts of the separate daily regression equations. Testing makes no sense in this context, because the oranges are not going to be sold for a price P1=0—that is, they are not going to be given away.

You can test the adjusted means at $\bar{x}_{..}$ using the following CONTRAST statement:

```
contrast 'trt' day 1 -1 0 0 0 0
        p1*day 51.2222 -51.2222 0 0 0 0,
          day 1 0 -1 0 0 0 p1*day 51.2222 0 -51.2222 0 0 0,
          day 1 0 0 -1 0 0 p1*day 51.2222 0 0 -51.2222 0 0,
          day 1 0 0 0 -1 0 p1*day 51.2222 0 0 0 -51.2222 0,
          day 1 0 0 0 0 -1 p1*day 51.2222 0 0 0 0 -51.2222;
```

This statement uses a set of independent comparisons for the unequal slopes model difference, whose form is $\alpha_1 - \alpha_2 + (\beta_{1_1} - \beta_{1_2})\bar{x}_{..}$. The number of comparisons in the contrast equals the DF for DAY. Output 7.15 gives the results.

Output 7.15
Contrast to
Test the
Equality of
Means Adjusted to
Mean Covariable
Price=51.22

Contrast	DF	Contrast SS	Mean Square	F Value	Pr > F
trt	5	376.3758925	75.2751785	3.46	0.0170

You can change the value of the AT P1= option in the LSMEANS statement and the coefficients for P1*DAY in the above contrast to test the equality of the adjusted means at any value of the covariable deemed reasonable. Alternatively, you can center the covariable in the DATA step. For example, you can define a new covariable $X=P1-51.22$ and use X in place of P1 in the analysis. The default Type III sum of squares tests the equality of adjusted means at $X=0$, which corresponds to P1=51.22, the overall mean. The crucial thing to keep in mind with unequal slopes models is that the treatment difference changes with the covariable, so the test is only valid if it is done at a value of the covariable agreed to be reasonable.

7.4 A Two-Way Structure without Interaction

Analysis of covariance can be applied to other experimental and treatment structures. This section illustrates covariance analysis of a two-factor factorial experiment with two covariates.

This example uses the data from the study of the relationship between the price of oranges and sales per customer described in Section 7.3 and presented in Output 7.7. Recall that the data set had two varieties. The section uses only response variable Q_1, sales per customer for the first variety to illustrate the main ideas. You can easily adapt these methods to Q_2, the response variable for the second variety.

A model for the sales of oranges for variety 1 is

$$Q_1 = \mu + \tau_i + \delta_j + \beta_1 P_1 + \beta_2 P_2 + e$$

where

Q_1 is the sales per customer for the first variety.

τ_i is the effect of the *i*th STORE, $i = 1, 2, \ldots, 6$.

δ_j is the effect of the *j*th DAY, $j = 1, 2, \ldots, 6$.

β_1 is the coefficient of the relationship between sales Q_1 and P_1 (the price of one variety of oranges).

β_2 is the coefficient of the relationship between Q_1 and P_2 (the price of the other variety of oranges).

e is the random error term.

Note that because there is no replication, the interaction between STORE and DAY must be used as the error term. In this example, the primary focus is on the influence of price, P1 and P2, and on sales, Q_1. The DAY and STORE differences are of secondary importance.

To implement the model, use following SAS statements:

```
proc glm;
   class store day;
   model q1 q2=store day p1 p2 / solution;
   lsmeans day / stderr;
```

The results appear in Output 7.16.

Output 7.16
Results of
Analysis of
Covariance:
Two-Way
Structure
without
Interaction

```
                                The GLM Procedure

Dependent Variable: q1

                                    Sum of
Source                   DF         Squares    Mean Square    F Value    Pr > F

Model                    12     1225.367548     102.113962       5.75    0.0002

Error                    23      408.308242      17.752532

Corrected Total          35     1633.675790


             R-Square    Coeff Var     Root MSE       q1 Mean

             0.750068    41.23842      4.213375      10.21711

Source                   DF       Type I SS    Mean Square    F Value    Pr > F

store                     5     313.4198071     62.6839614       3.53    0.0163
day                       5     250.3972723     50.0794545       2.82    0.0396
p1                        1     622.0082168    622.0082168      35.04    <.0001
p2                        1      39.5422519     39.5422519       2.23    0.1492


Source                   DF     Type III SS    Mean Square    F Value    Pr > F

store                     5     223.8326734     44.7665347       2.52    0.0583
day                       5     433.0968700     86.6193740       4.88    0.0035
p1                        1     538.1688512    538.1688512      30.32    <.0001
p2                        1      39.5422519     39.5422519       2.23    0.1492

                                           Standard
          Parameter        Estimate          Error    t Value    Pr > |t|

          Intercept     51.69987930 B     9.79103443       5.28    <.0001
          store   1      -7.64532641 B     2.69194414      -2.84    0.0093
          store   2      -5.60226472 B     2.46416942      -2.27    0.0327
          store   3      -7.36284806 B     2.46416942      -2.99    0.0066
          store   4      -4.36498239 B     2.48754952      -1.75    0.0926
          store   5      -5.02052157 B     2.43612208      -2.06    0.0508
          store   6       0.00000000 B          .            .        .
          day     1      -5.83036664 B     2.51932754      -2.31    0.0299
          day     2      -4.89997548 B     2.44708866      -2.00    0.0572
          day     3       2.26978922 B     2.54028189       0.89    0.3808
          day     4      -2.65249315 B     2.44667751      -1.08    0.2895
          day     5       4.04702055 B     2.55655852       1.58    0.1271
          day     6       0.00000000 B          .            .        .
          p1             -0.83036470       0.15081334      -5.51    <.0001
          p2              0.14884706       0.09973319       1.49    0.1492

NOTE: The X'X matrix has been found to be singular, and a generalized inverse
      was used to solve the normal equations. Terms whose estimates are
      followed by the letter 'B' are not uniquely estimable
```

Output 7.16
(Continued)
Results of
Analysis of
Covariance:
Two-Way
Structure
without
Interaction

```
                        Least Squares Means

                                  Standard
         DAY      Q1 LSMEAN          Error      Pr > |t|

          1       5.5644154       1.7680833      0.0045
          2       6.4948065       1.7289585      0.0010
          3      13.6645712       1.7515046      <.0001
          4       8.7422889       1.7339197      <.0001
          5      15.4418026       1.7858085      <.0001
          6      11.3947820       1.7667260      <.0001
```

In addition to the details previously discussed, these also are of interest:

❑ The Type I SS for P1 and P2 can be summed to obtain a test for the partial contribution of both prices:

$$F_{2,23} = [(622.01 + 39.54)/2]/(17.75) = 74.54$$

❑ The Type III SS show that all effects are highly significant except P2, the price of the competing orange.

❑ Each coefficient estimate is the mean difference between each CLASS variable value (STORE, DAY) and the last CLASS variable value, because there is no interaction.

❑ The P1 coefficient is negative, indicating the expected negatively sloping price response (demand function). The P2 coefficient, although not significant, has the expected positive sign for the price response of a competing product.

❑ Least-squares means are requested only for DAY, which shows the expected higher sales toward the end of the week.

Contrasts and estimates of linear functions could, of course, be requested with this analysis.

7.5 A Two-Way Structure with Interaction

The most complex covariance model discussed in this chapter is a two-factor factorial with two stages of subsampling. Output 7.17 shows data from a study whose objective is to estimate y, the weight of usable lint from x, the total weight of cotton bolls. In addition, the researcher wants to see if lint estimation is affected by varieties of cotton (VARIETY) and the distance between planting rows (SPACING), using x, the boll weight (BOLLWT), as a covariate in the analysis of y, the lint weight. The study is a factorial experiment with two levels of VARIETY (37 and 213) and two levels of SPACING (30 and 40). There are two plants for each VARIETYXSPACING treatment combination, and there are from five to nine bolls per plant (PLANT).

Output 7.17
Data for Analysis
of Covariance:
Two-Way
Structure with
Interaction

Obs	variety	spacing	plant	bollwt	lint
1	37	30	3	8.4	2.9
2	37	30	3	8.0	2.5
3	37	30	3	7.4	2.7
4	37	30	3	8.9	3.1
5	37	30	5	5.6	2.1
6	37	30	5	8.0	2.7
7	37	30	5	7.6	2.5
8	37	30	5	5.4	1.5
9	37	30	5	6.9	2.5
10	37	40	3	4.5	1.3
11	37	40	3	9.1	3.1
12	37	40	3	9.0	3.1
13	37	40	3	8.0	2.3
14	37	40	3	7.2	2.2
15	37	40	3	7.6	2.5
16	37	40	3	9.0	3.0
17	37	40	3	2.3	0.6
18	37	40	3	8.7	3.0
19	37	40	5	8.0	2.6
20	37	40	5	7.2	2.5
21	37	40	5	7.6	2.4
22	37	40	5	6.9	2.2
23	37	40	5	6.9	2.5
24	37	40	5	7.6	2.4
25	37	40	5	4.7	1.4
26	213	30	3	4.6	1.7
27	213	30	3	6.8	1.7
28	213	30	3	3.5	1.3
29	213	30	3	2.4	1.0
30	213	30	3	3.0	1.0
31	213	30	5	2.8	0.5
32	213	30	5	3.6	0.9
33	213	30	5	6.7	1.9
34	213	40	0	7.4	2.1
35	213	40	0	4.9	1.0
36	213	40	0	5.7	1.0
37	213	40	0	3..0	0.7
38	213	40	0	4.7	1.5
39	213	40	0	5.0	1.3
40	213	40	0	2.8	0.4
41	213	40	0	5.2	1.2
42	213	40	0	5.6	1.0
43	213	40	3	4.5	1.0
44	213	40	3	5.6	1.2
45	213	40	3	2.0	0.7
46	213	40	3	1.2	0.2
47	213	40	3	4.2	1.2
48	213	40	3	5.3	1.2
49	213	40	3	7.0	1.7

The model for the analysis is

$$y_{ijkl} = \mu + v_i + \tau_i + (v\tau)_{ij} + \gamma(v\tau)_{ijk} + \beta x_{ijkl} + \varepsilon_{ijkl}$$

where

y_{ijkl} equals the weight of the lint for the ith VARIETY, jth SPACING, kth PLANT of the (i, j)th cell, and lth boll of each plant.

μ is the intercept.

v_i is the effect of the ith VARIETY.

τ_j is the effect of the jth SPACING.

$(v\tau)_{ij}$ is the VARIETY×SPACING interaction.

$\gamma(v\tau)_{ijk}$ is the effect of the kth plant in the (i, j)th VARIETY and SPACING combination.

x_{ijkl} is the total weight of each boll, the covariate.

β is the regression effect of the covariate.

ε_{ijkl} equals the error variation among bolls within plants.

The primary focus of this study is on estimating lint weight from boll weight (that is, the regression) and only later in determining if this relationship is affected by VARIETY and SPACING factors. In the SAS program to analyze the data, the order of variables in the MODEL statement is changed so that the Type I sums of squares provide the appropriate information:

```
proc glm;
  class variety spacing plant;
  model lint=bollwt variety spacing variety*spacing
     plant(variety*spacing) / solution;
  random plant(variety*spacing)/test;
```

Note that the RANDOM statement with the TEST option has been added because the plant-to-plant variation provides the appropriate error term. Results of the analysis appear in Output 7.18.

Because PLANT(VARIETY*SPACING) is a random effect, an alternative is to use PROC MIXED. You use the following SAS statements:

```
proc mixed;
 class variety spacing plant;
 model lint=bollwt variety spacing
      variety*spacing/solution;
 random plant(variety*spacing);
```

The results for this analysis appear in Output 7.19. Littell et al. (1996) discuss analysis of covariance for mixed models in much greater detail.

Output 7.18
Results of Analysis of Covariance: Two-Way Structure with Interaction

```
Dependent Variable: lint
                                       Sum of
Source                     DF          Squares    Mean Square   F Value   Pr > F

Model                       8       31.16009287    3.89501161     80.70   <.0001
Error                      40        1.93051938    0.04826298
Corrected Total            48       33.09061224

Source                     DF       Type I SS     Mean Square   F Value   Pr > F

 bollwt                     1       29.06931406   29.06931406    602.31   <.0001
 variety                    1        1.26353553    1.26353553     26.18   <.0001
 spacing                    1        0.46664798    0.46664798      9.67   0.0034
 variety*spacing            1        0.09326994    0.09326994      1.93   0.1722
 plant(variet*spacin)       4        0.26732535    0.06683134      1.38   0.2565

Source                     DF       Type III SS   Mean Square   F Value   Pr > F

bollwt                      1       11.11855999   11.11855999    230.37   <.0001
variety                     1        0.94242614    0.94242614     19.53   <.0001
spacing                     1        0.37483940    0.37483940      7.77   0.0081
variety*spacing             1        0.04785515    0.04785515      0.99   0.3253
plant(variet*spacin)        4        0.26732535    0.06683134      1.38   0.2565

          Tests of Hypotheses for Mixed Model Analysis of Variance

Source                     DF       Type III SS   Mean Square   F Value   Pr > F

bollwt                      1       11.118560     11.118560      230.37   <.0001
Error: MS(Error)           40        1.930519      0.048263

    Source                 DF       Type III SS   Mean Square   F Value   Pr > F

 *  variety                 1        0.942426      0.942426       16.27   0.0021

    Error                  10.657    0.617126      0.057907
Error: 0.5194*MS(plant(variet*spacin)) + 0.4806*MS(Error)
* This test assumes one or more other fixed effects are zero.

    Source                 DF       Type III SS   Mean Square   F Value   Pr > F

 *  spacing                 1        0.374839      0.374839        5.76   0.0660

    Error                  4.6073    0.300008      0.065116
Error: 0.9076*MS(plant(variet*spacin)) + 0.0924*MS(Error)
* This test assumes one or more other fixed effects are zero.

Source                     DF       Type III SS   Mean Square   F Value   Pr > F

variety*spacing             1        0.047855      0.047855        0.74   0.4324

Error                      4.6791    0.303859      0.064939
Error: 0.8981*MS(plant(variet*spacin)) + 0.1019*MS(Error)
                                                  Standard
Parameter                            Estimate      Error      t Value   Pr > |t|

Intercept                          -.2724440749 B  0.11934010   -2.28    0.0278
bollwt                              0.3056076686    0.02013479   15.18   <.0001
variety                37           0.4232705043 B  0.12964467    3.26    0.0022
variety               213           0.0000000000 B    .            .       .
spacing                30           0.0379572553 B  0.15161542    0.25    0.8036
spacing                40           0.0000000000 B    .            .       .
variety*spacing        37 30        0.0236449357 B  0.19897993    0.12    0.9060
variety*spacing        37 40        0.0000000000 B    .            .       .
variety*spacing       213 30        0.0000000000 B    .            .       .
variety*spacing       213 40        0.0000000000 B    .            .       .
plant(variet*spacin)  3  37 30      0.0892286888 B  0.15033417    0.59    0.5562
plant(variet*spacin)  5  37 30      0.0000000000 B    .            .       .
plant(variet*spacin)  3  37 40     -.0271310434 B  0.11085696    -0.24    0.8079
plant(variet*spacin)  5  37 40      0.0000000000 B    .            .       .
plant(variet*spacin)  3 213 30      0.3337196850 B  0.16055649    2.08    0.0441
plant(variet*spacin)  5 213 30      0.0000000000 B    .            .       .
plant(variet*spacin)  0 213 40     -.0984914494 B  0.11151946    -0.88    0.3824
plant(variet*spacin)  3 213 40      0.0000000000 B    .            .       .

NOTE: The X'X matrix has been found to be singular, and a generalized inverse
      was used to solve the normal equations. Terms whose estimates are
      followed by the letter 'B' are not uniquely estimable.
```

The Type I SS for BOLLWT is what would be obtained by a simple linear regression of LINT on BOLLWT. If you ran this simple linear regression, you would get an R^2 value of (29.069/33.091)=0.878, a residual mean square of (33.091−29.069)/47)=0.08557, and an *F*-statistic of 339.69, thus indicating a strong relationship of lint weight to boll weight.

The Type III SS from the RANDOM statement with the TEST option shows a non-significant contribution from the VARIETY*SPACING interaction (*F*=0.74, *p*-value 0.4324). The VARIETY effect (*F*=16.27, *p*-value 0.0021) is statistically significant, whereas the SPACING main effect (*F*=5.76, *p*-value 0.0660) is only marginally significant. Note that the error terms for the VARIETY and SPACING main effects and interaction use linear combinations of PLANT(VARIETY*SPACING) and ERROR mean squares. This follows from the complex set of expected means squares that result in analysis of covariance.

It might seem that it would be simpler to assume from inspection of the analysis of covariance sources of variance that PLANT(VARIETY*SPACING) is the proper error term for VARIETY, SPACING, and VARIETY*SPACING and use the statement

```
test h=variety spacing variety*spacing
     e=plant(variety*spacing);
```

in place of the RANDOM statement. If you do this, the resulting test statistics will be affected and there is the possibility of drawing erroneous conclusions. Now consider the results obtained using PROC MIXED.

Output 7.19
Analysis of
Covariance
Results Using
PROC MIXED

```
                        Covariance Parameter Estimates

             Cov Parm                      Estimate

             plant(variet*spacin)                 0
             Residual                       0.04995


                       Solution for Fixed Effects

                                         Standard
   Effect          variety  spacing  Estimate   Error   DF   t Value   Pr > |t|

   Intercept                          -0.3210   0.1078    4    -2.98     0.0408
   bollwt                              0.3041   0.01990   40    15.28    <.0001
   variety           37                0.4671   0.09351    4     5.00     0.0075
   variety          213                     0       .      .       .        .
   spacing                    30       0.3013   0.09720    4     3.10     0.0362
   spacing                    40            0       .      .       .        .
   variety*spacing   37       30      -0.1844   0.1350     4    -1.37     0.2436
   variety*spacing   37       40            0       .      .       .        .
   variety*spacing  213       30            0       .      .       .        .
   variety*spacing  213       40            0       .      .       .        .


                     Type 3 Tests of Fixed Effects

                            Num     Den
               Effect        DF      DF    F Value    Pr > F

               bollwt         1      40    233.51     <.0001
               variety        1       4     18.21      0.0130
               spacing        1       4      9.68      0.0358
               variety*spacing 1      4      1.87      0.2436
```

Compared to Output 7.18 for PROC GLM, the exact parameter estimates and *F*-values differ somewhat. This is partly because PROC MIXED recovers information from the random-model effects, similar to the recovery of interblock information for incomplete-blocks designs discussed in Chapter 4, and partly because the variance component estimate for PLANT(VARIETY*SPACING) is zero. MIXED's default option for computing *F*-values in the presence of zero or negative variance component estimates is somewhat different than the *F*-ratios derived from the expected mean square by PROC GLM.

Section 4.4.2 discussed the possible bias to *F*-statistics that the MIXED default of setting negative variance component estimates to zero may introduce. This is especially evident in this example with the main effect test for SPACING. The *p*-value here is 0.0358, whereas it was 0.0660 with the GLM analysis. The MIXED result reflects potential bias. Here, as in Section 4.4.2, you can avoid this problem by using the METHOD=TYPE3 option. The results are not shown here, but they are very similar to the results in Output 7.18. Slight discrepancies result from the recovery of random-effects information present in PROC MIXED but not in PROC GLM.

Both the GLM and MIXED results suggest dropping the terms VARIETY*SPACING and PLANT(VARIETY*SPACING). The justification for VARIETY*SPACING is the same, that is, the *F*-test. However, with MIXED, there is no test for PLANT(VARIETY*SPACING). The zero variance component estimate provides equivalent justification. Given these results, the model has more factors than necessary; hence, their coefficient estimates are not necessary and can make further inference needlessly awkward. Drop VARIETY*SPACING and PLANT(VARIETY*SPACING), and use the following GLM statements to re-estimate:

```
model lint=bollwt variety spacing / solution;
lsmeans variety spacing / stderr;
```

You can use equivalent MIXED statements and obtain the same results. The results appear in Output 7.20. They differ only slightly from those in Outputs 7.18 and 7.19.

Output 7.20
Results of a
Simplified
Covariance
Analysis: Two-
Way Structure
with
Interaction

Source	DF	Sum of Squares	Mean Square	F Value	Pr > F
Model	3	30.79949757	10.26649919	201.65	<.0001
Error	45	2.29111467	0.05091366		
Corrected Total	48	33.09061224			

Source	DF	Type I SS	Mean Square	F Value	Pr > F
bollwt	1	29.06931406	29.06931406	570.95	<.0001
variety	1	1.26353553	1.26353553	24.82	<.0001
spacing	1	0.46664798	0.46664798	9.17	0.0041

Source	DF	Type III SS	Mean Square	F Value	Pr > F
bollwt	1	11.57173388	11.57173388	227.28	<.0001
variety	1	1.19732512	1.19732512	23.52	<.0001
spacing	1	0.46664798	0.46664798	9.17	0.0041

Parameter		Estimate	Standard Error	t Value	Pr > |t|
Intercept		-.2769483300 B	0.10384452	-2.67	0.0106
bollwt		0.3014429094	0.01999507	15.08	<.0001
variety	37	0.4106564020 B	0.08468173	4.85	<.0001
variety	213	0.0000000000 B	.	.	.
spacing	30	0.2052058951 B	0.06778167	3.03	0.0041
spacing	40	0.0000000000 B	.	.	.

NOTE: The X'X matrix has been found to be singular, and a generalized inverse was used to solve the normal equations. Terms whose estimates are followed by the letter 'B' are not uniquely estimable.

Output 7.20
(Continued)
Results of a
Simplified
Covariance
Analysis: Two-
Way Structure
with
Interaction

```
                     Least Squares Means

                                   Standard
        var       lint LSMEAN        Error      Pr > |t|

        37        2.00805710       0.05320406    <.0001
        213       1.59740070       0.05523778    <.0001

                                   Standard
        spac      lint LSMEAN        Error      Pr > |t|

        30        1.90533185       0.05479483    <.0001
        40        1.70012595       0.03988849    <.0001
```

This model specifies a single regression coefficient that relates LINT to BOLLWT (0.3014), but with different intercepts for the four treatment combinations. These intercepts can be constructed from the SOLUTION vector by summing appropriate component values.

For example, for VARIETY=37, SPACING=30, the model estimate is

$$y = \mu + v_1 + \tau_1 + \beta_x = -.2769 + .4107 + .2052 + .3014x$$

For the other treatment combinations, the results are

VARIETY	SPACING	Values for Model Estimate, Where x = BOLLWT, the Covariate
37	40	$0.1338 + 0.3014x$
213	30	$-0.0715 + 0.3014x$
213	40	$-0.2769 + 0.3014x$

Note that these results can be obtained with ESTIMATE statements. Least-squares means appear in Output 7.20; other statistics can be obtained but are not necessary in this situation.

7.6 Orthogonal Polynomials and Covariance Methods

A standard method for analyzing treatments with quantitative levels is to decompose the treatment sum of squares using orthogonal polynomial contrasts. That is, contrasts whose coefficients measure the linear, quadratic, and higher-order regression effects associated with treatment level. Most statistical methods textbooks have tables of orthogonal polynomial coefficients for balanced data with equally spaced treatment levels. You can use the ORPOL function in PROC IML to determine coefficients when you have a design that standard tables do not cover, such as unequally spaced designs. Section 7.6.2 shows you how to use the ORPOL function.

In many practical applications, orthogonal polynomials are awkward to use. Often, you want to estimate the regression equation, not merely decide what is "significant." Even with treatment designs covered by standard tables, extracting the regression equation from orthogonal polynomials is laborious. For factorial experiments, interest usually centers on interaction. That is, are the regressions over the quantitative factor the same for all levels of the other factor? Except for very simple factorial treatment designs, trying to use orthogonal polynomials to measure interaction can become a daunting task.

This section presents analysis-of-covariance methods that are equivalent to orthogonal polynomial contrasts. The main advantage of the covariance, or direct regression, approach is that in most cases, it is easier to implement using SAS.

7.6.1 A 2x3 Example

Output 7.21 contains data from an experiment designed to compare response to increasing dosage for two types of drug. There were three levels of the actual dosage, DOSE in the SAS data set—1, 10, and 100 units. The data were analyzed using LOGDOSE, the base 10 logs of the dosages. Note that the levels of LOGDOSE are equally spaced. The experiment was conducted as a randomized-complete-blocks design. BLOC denotes the blocks and Y denotes the response variable.

Output 7.21
Data for a
Type-Dose
Factorial
Orthogonal
Polynomial
Example

Obs	bloc	type	dose	logdose	y
1	1	1	1	0	63
2	1	2	1	0	59
3	1	1	10	1	62
4	1	2	10	1	62
5	1	1	100	2	62
6	1	2	100	2	68
7	2	1	1	0	50
8	2	2	1	0	49
9	2	1	10	1	49
10	2	2	10	1	55
11	2	1	100	2	48
12	2	2	100	2	58
13	3	1	1	0	53
14	3	2	1	0	47
15	3	1	10	1	52
16	3	2	10	1	51
17	3	1	100	2	51
18	3	2	100	2	50
19	4	1	1	0	52
20	4	2	1	0	48
21	4	1	10	1	54
22	4	2	10	1	49
23	4	1	100	2	55
24	4	2	100	2	72

The analysis of variance of these data is as follows:

SOURCE OF VARIATION	DF
Block	3
Type	1
Log Dose	2
Type × Log Dose	2
Error	15

The dose main effect can be partitioned into linear and quadratic components by orthogonal polynomials whose contrast coefficients are

	COEFFICIENT FOR LOGDOSE		
CONTRAST	0	1	2
Linear	−1	0	1
Quadratic	−1	2	−1

Similarly, the Type×Log Dose interaction can be partitioned into a Linear×Type and a Quadratic×Type component. Because the log dose levels are equally spaced, you can look up the contrast coefficients shown above. Most statistical methods texts have such a table. If you don't have a table readily available, or if you want to use contrasts for a situation not in the table, for instance partitioning DOSE rather than Log Dose into linear and quadratic components, you can use the ORPOL function in PROC IML demonstrated in Section 7.6.2.

For analysis with LOGDOSE, use the following SAS statements:

```
proc glm;
 class bloc type logdose;
 model y=bloc type|logdose;
 contrast 'linear logdose' logdose -1 0 1;
 contrast 'quadratic logdose' logdose -1 2 -1;
 contrast 'linear logdose x type' type*logdose 1 0 -1 -1 0 1;
 contrast 'quad logdose x type' type*logdose 1 -2 1 -1 2 -1;
```

Output 7.22 contains the results.

Output 7.22
Analysis of
Variance for
Type-Dose
Data

```
                               Sum of
Source               DF        Squares    Mean Square  F Value   Pr > F

Model                 8     816.500000     102.062500     6.06    0.0014
Error                15     252.458333      16.830556
Corrected Total      23    1068.958333


Source               DF     Type I SS     Mean Square  F Value   Pr > F

bloc                  3     538.7916667   179.5972222    10.67    0.0005
type                  1      12.0416667    12.0416667     0.72    0.4109
logdose               2     121.5833333    60.7916667     3.61    0.0524
type*logdose          2     144.0833333    72.0416667     4.28    0.0338


Source               DF     Type III SS   Mean Square  F Value   Pr > F

bloc                  3     538.7916667   179.5972222    10.67    0.0005
type                  1      12.0416667    12.0416667     0.72    0.4109
logdose               2     121.5833333    60.7916667     3.61    0.0524
type*logdose          2     144.0833333    72.0416667     4.28    0.0338


Contrast             DF     Contrast SS   Mean Square  F Value   Pr > F

linear logdose        1     115.5625000   115.5625000     6.87    0.0193
quadratic logdose     1       6.0208333     6.0208333     0.36    0.5587
linear logdose x type 1     138.0625000   138.0625000     8.20    0.0118
quad logdose x type   1       6.0208333     6.0208333     0.36    0.5587
```

From this output you can see that there is a significant Type×Log Dose interaction. The Linear Logdose×Type interaction explains most of the interaction. A look at the Type×Log Dose least-squares means (Output 7.23) reveals why. For Type 1, LOGDOSE does not affect mean response, whereas for Type 2, mean response increases approximately linearly with increasing LOGDOSE.

Output 7.23
Least-Squares
Means for
Type-Dose
Data

```
                  Least Squares Means

        type      logdose        y LSMEAN

         1           0          54.5000000
         1           1          54.2500000
         1           2          54.0000000
         2           0          50.7500000
         2           1          54.2500000
         2           2          62.0000000
```

By inspection, a linear regression for each type appears sufficient to explain the LOGDOSE effect. You can use the following SAS code to formally confirm this:

```
proc glm;
   class bloc type logdose;
   model y=bloc type logdose(type);
   contrast 'lin in type 1' logdose(type)  1  0 -1   0  0  0;
   contrast 'lin in type 2' logdose(type)  0  0  0   1  0 -1;
   contrast 'quad in type 1' logdose(type) 1 -2  1   0  0  0;
   contrast 'quad in type 2' logdose(type) 0  0  0   1 -2  1;
```

Output 7.24 shows the results.

Output 7.24
Orthogonal
Polynomial
Contrast
Results within
Each Type

Contrast	DF	Contrast SS	Mean Square	F Value	Pr > F
lin in type 1	1	0.5000000	0.5000000	0.03	0.8655
lin in type 2	1	253.1250000	253.1250000	15.04	0.0015
quad in type 1	1	0.0000000	0.0000000	0.00	1.0000
quad in type 2	1	12.0416667	12.0416667	0.72	0.4109

You can see that neither type has a significant quadratic regression. The *F*-values are 0.0 and 0.72, respectively for Types 1 and 2. Type 2 does have a highly significant linear regression, as shown by the Lin In Type 2 contrast. The *F*-value is 15.04, and the associated *p*-value is 0.0015. You would then fit the linear regression equation over LOGDOSE for Type 1.

7.6.2 Use of the IML ORPOL Function to Obtain Orthogonal Polynomial Contrast Coefficients

The contrast coefficients used in the previous section are the standard orthogonal polynomial coefficients for equally spaced treatments. You can find these coefficients in tables contained in many statistical methods textbooks. Suppose, however, that you want to partition treatment effects into linear effects, quadratic effects, and so forth, but your treatment levels are *not* equally spaced. Or suppose that you simply don't have convenient access to a table of standard orthogonal polynomial contrasts. The interactive matrix algebra procedure PROC IML has a function, ORPOL, that computes orthogonal polynomial contrasts for any set of quantitative treatment levels. The function is simple to use, and does not require knowledge of matrix algebra.

To illustrate the ORPOL function, suppose you want to use DOSE rather than LOGDOSE in the example in Section 7.6.1. The levels are 1, 10, and 100, which are not equally spaced so you won't find the correct coefficients in any table. Use the following PROC IML statements:

```
proc iml;
  levels={1,10,100};
  coef=orpol(levels);
  print coef;
```

Output 7.25 shows the results. The LEVELS={level 1, level 2,...} statement defines a variable named LEVELS that contains a list of the treatment levels. Note that the treatment levels are separated by commas. The name of the variable is your choice; here it is called LEVELS, but you can give it any name you like within the conventions of allowable SAS variable names. The variable named COEF will contain the contrast coefficients; its name is your choice as well. You set it equal to the ORPOL function and put the variable with the treatment levels in parentheses. The PRINT statement causes the variable COEF to be printed.

Output 7.25
Contrast
Coefficients for
Unequally
Spaced Levels
of DOSE Using
ORPOL

```
                       COEF

     0.5773503 -0.464991  0.6711561
     0.5773503 -0.348743 -0.738272
     0.5773503  0.8137335 0.0671156
```

The first column of numbers, all 0.577, is the contrast for the mean, which is rarely, if ever, used. The second column gives you the coefficients for the linear contrasts. The third column gives you the quadratic contrast coefficients. Note that for each contrast, there is one coefficient per treatment (in this case dose) level. Also, there is one contrast per treatment degree of freedom. If you had four treatment levels, there would be four coefficients per contrast and a cubic as well as a linear and quadratic contrast.

The information from Output 7.25 allows you to write the appropriate CONTRAST statements. For example, for linear and quadratic dose effects, use the statements:

```
contrast 'linear dose' dose -0.465 -0.349 0.814;
contrast 'quadratic dose'  dose 0.671 -0.738 0.067;
```

Pay attention to the sum of the contrasts. Occasionally, a rounding error will cause the sum of the coefficients to not equal zero; you might get 0.001, for example. This will cause both the GLM and MIXED procedures to declare the contrast non-estimable and you will get no output. Simply adjust one number so the coefficients sum to exactly zero. The impact of this adjustment on the resulting computations is negligible.

You can use ORPOL for equally spaced treatments. For example, add the following statements to the IML program given above to compute the coefficients for the equally spaced levels of LOGDOSE:

```
log_lev=log10(levels);
coef=orpol(log_lev);
print log_lev;
print coef;
fuzzed_coef=fuzz(coef);
print fuzzed_coef;
```

Output 7.26 shows the results. The LOG10 function takes the base 10 log of each element of the vector variable LEVELS, defined above. The name of the new variable is your choice; here it is called LOG_LEV. Occasionally, machine-rounding error causes numbers that are supposed to be zero to be computed as very small nonzero numbers, as with the coefficient for the second treatment level in the linear contrast (second column). The FUZZ function cleans up these rounding errors and sets them to zero, as shown in the variable FUZZED_COEF, the COEF variable with FUZZ applied.

Output 7.26
ORPOL
Results for
Equally
Spaced
Treatment
Level
(LOGDOSE)

```
                              LOG_LEV

                                 0
                                 1
                                 2

                               COEF

          0.5773503 -0.707107 0.4082483
          0.5773503 8.194E-17 -0.816497
          0.5773503 0.7071068 0.4082483

                            FUZZED_COEF

          0.5773503 -0.707107 0.4082483
          0.5773503         0 -0.816497
          0.5773503 0.7071068 0.4082483
```

Note that the coefficients are given the **orthonormal** form, that is, the squared coefficients for each contrast sum to one, for example, $(-0.707)^2 + 0^2 + (0.707)^2 = 1$ for the linear contrast (second column). You can rescale these coefficients to integer values, for example, -1, 0, and 1 for the linear contrast and 1, -2, and 1 for the quadratic, without affecting the sum of squares or F-values for the contrasts. This gives you the same coefficients you used in the previous section for LOGDOSE.

7.6.3 Use of Analysis of Covariance to Compute ANOVA and Fit Regression

In the previous sections you saw how to assess regression over treatment levels using orthogonal polynomial contrasts and how to use ORPOL to obtain the needed contrast coefficients. However, orthogonal polynomials can be awkward to use, especially when you want to estimate the regression equation in addition to merely partitioning variation for testing purposes. In practical situations, orthogonal polynomials are something of a holdover from pre-computer-era statistical analysis. You can use analysis-of-covariance methods to compute the same statistics you got using orthogonal polynomials, as well as additional statistics that are often useful. This section uses the example from Section 7.6.1 to illustrate.

You can reproduce the essential elements of the analysis in Output 7.22 using analysis-of-covariance methods. First, you need to define a new variable in the DATA step for the square of LOGDOSE. In the SAS program between, the new variable is called LOGD2, defined in the DATA step as LOGDOSE*LOGDOSE. Then use the following SAS statements:

```
proc glm;
  class bloc type;
  model y=bloc type logdose logd2
     type*logdose type*logd2;
```

The results appear in Output 7.27. Notice that LOGDOSE does *not* appear in the CLASS statement. You use it, as well as LOGD2, as a *direct* regression variable, exactly as you would if LOGDOSE was a covariable and you suspected heterogeneous quadratic regressions of y on the covariable.

Output 7.27 ANOVA for Type-Dose Data Using Analysis-of-Covariance Methods

Source	DF	Type I SS	Mean Square	F Value	Pr > F
bloc	3	538.7916667	179.5972222	10.67	0.0005
type	1	12.0416667	12.0416667	0.72	0.4109
logdose	1	115.5625000	115.5625000	6.87	0.0193
logd2	1	6.0208333	6.0208333	0.36	0.5587
logdose*type	1	138.0625000	138.0625000	8.20	0.0118
logd2*type	1	6.0208333	6.0208333	0.36	0.5587

Source	DF	Type III SS	Mean Square	F Value	Pr > F
bloc	3	538.7916667	179.5972222	10.67	0.0005
type	1	28.1250000	28.1250000	1.67	0.2157
logdose	1	0.3894231	0.3894231	0.02	0.8811
logd2	1	6.0208333	6.0208333	0.36	0.5587
logdose*type	1	0.8125000	0.8125000	0.05	0.8291
logd2*type	1	6.0208333	6.0208333	0.36	0.5587

Comparing Output 7.25 to Output 7.22, you can see that the Type I SS for LOGDOSE and LOGD2 are identical to the contrast results for Linear Logdose and Quadratic Logdose. The sums of squares are both 115.5625 and 6.0208, respectively. Also, the TYPE*LOGDOSE and TYPE*LOGD2 Type I sum of squares are identical to the Linear Logdose×Type and Quadratic Logdose×Type contrasts from Output 7.21. In both cases, the significant difference between linear regressions of *y* on LOGDOSE for the two types is the main result. Note that the Type III sums of squares produce nonsense results for the LINEAR LOGDOSE main effect and interaction terms because they are both adjusted for quadratic effects. In general, the Type III SS should be ignored when using analysis of covariance in lieu of orthogonal polynomials.

Outputs 7.22 and 7.25 both lead to the conclusion that you should fit a linear regression over LOGDOSE for each type. You can use the following SAS statements to do so:

```
proc glm;
  class bloc type;
  model y= bloc type logdose(type)/solution;
  estimate 'beta-0, type 1'
    intercept 4 bloc 1 1 1 1 type 4 0/divisor=4;
  estimate 'beta-0, type 2'
    intercept 4 bloc 1 1 1 1 type 0 4/divisor=4;
```

These statements are similar to those used in Section 7.3 to fit the unequal slopes. The additional ESTIMATE statements allow you to compute the β_{0_i} terms for the *i*th type, which the MODEL statement implicitly defines to be the sum of the intercept, the average block effect, and the type effect. The results appear in Output 7.28.

Output 7.28 ANOVA for Type-Dose Data Using Analysis-of-Covariance Methods

Parameter		Estimate	Error	t Value	Pr > |t|
beta-0, type 1		54.5000000	1.80039484	30.27	<.0001
beta-0, type 2		50.0416667	1.80039484	27.79	<.0001

Parameter		Estimate	Standard Error	t Value	Pr > |t|
Intercept		50.08333333 B	2.27733935	21.99	<.0001
bloc	1	7.66666667 B	2.27733935	3.37	0.0037
bloc	2	-3.50000000 B	2.27733935	-1.54	0.1427
bloc	3	-4.33333333 B	2.27733935	-1.90	0.0741
bloc	4	0.00000000 B	.	.	.
type	1	4.45833333 B	2.54614280	1.75	0.0980
type	2	0.00000000 B	.	.	.
logdose(type)	1	-0.25000000	1.39457984	-0.18	0.8598
logdose(type)	2	5.62500000	1.39457984	4.03	0.0009

NOTE: The X'X matrix has been found to be singular, and a generalized inverse was used to solve the normal equations. Terms whose estimates are followed by the letter 'B' are not uniquely estimable.

From Output 7.25, the main results are the regression equations. For Type 1, the equation is *y*=54.5 − 0.25*LOGDOSE; for Type 2 it is *y*=50.042+5.625*LOGDOSE.

PROC MIXED makes it somewhat more convenient to obtain the regression equations. You can use the SAS statements

```
proc mixed;
  class bloc type;
  model y=type logdose(type)/noint solution;
  random bloc;
```

The results appear in Output 7.29.

Output 7.29
Estimate of
Linear
Regression
Equations for
Each Type

Parameter		Estimate	Standard Error	t Value	Pr > |t|
type	1	54.50000000	2.89268414	18.84	<.0001
type	2	50.04166667	2.89268414	17.30	<.0001
logdose(type)	1	-0.25000000	2.24066350	-0.11	0.9123
logdose(type)	2	5.62500000	2.24066350	2.51	0.0208

The RANDOM BLOC statement and the NOINT option cause the β_{0_i} terms to be estimated directly, rather than requiring ESTIMATE statements. You can assume that RANDOM BLOC does not affect the estimates, but it does change the standard errors of the regression coefficients. For PROC MIXED, the standard errors are 2.89 and 2.24 for the intercept and slope, respectively. These are valid if assuming blocks to be random is reasonable. The PROC GLM results, 1.80 and 1.39, assume fixed blocks.

Chapter **8** Repeated-Measures Analysis

8.1 Introduction

Repeated measures refers to multiple measures on the same experimental unit. Usually, repeated measures are made over time, but they can be over space. A very common situation is for treatments to be applied to experimental units in a completely randomized design. Then measurements are made at several different times. For example, a pharmaceutical company examined effects of three drugs on respiratory ability of asthma patients. The drugs were randomly assigned to 24 patients each. The assigned drug was administered to each patient. Then a standard measure of respiratory ability called FEV1 was measured hourly for eight hours following treatment. FEV1 was also measured immediately prior to administration of the drugs. A statistical model for the data is

$$y_{ijk} = \mu + \alpha_i + \gamma_k + (\alpha\gamma)_{ik} + e_{ijk} \tag{8.1}$$

where

y_{ijk} is the FEV1 measurement at hour k on the jth patient assigned to drug i.

$\mu + \alpha_i + \gamma_k + (\alpha\gamma)_{ik}$ is the mean FEV1 for drug i at hour k, containing effects for drug, hour, and interaction DRUG*HOUR interaction.

e_{ijk} is the random error associated with the jth patient assigned to drug i at hour k.

Equation (8.1) is the same as the model equation for a standard factorial experiment with main effects of drug and hour, and DRUG*HOUR interaction. In fact, the basic objectives of repeated-measures data analysis are essentially those of any factorial experiment. We want to examine interaction between the factors, and assess either simple or main effects of the factors. But the distinguishing feature of a repeated-measures model is in the variance and covariance structure of the errors, e_{ijk}. Although drugs were randomly assigned to patients, the levels of the repeated-measures factor, in this case hour, is not randomly assigned to units. Thus we cannot reasonably assume that the random errors, e_{ijk}, are independent. There are often two aspects of covariance structure in the errors. First, two FEV1 measures on the same subject are likely to be more nearly the same than two measures on different subjects. Thus, measures on the same subject are usually positively correlated simply because they possess common effects from that subject. This is basically the phenomenon of measures on the same whole-plot unit in a split-plot experiment. Second, two FEV1 measures made close in time on the same subject are likely to be more highly correlated than two measures made far apart in time. This distinguishes repeated-measures covariance from split-plot covariance structure. In a split-plot experiment, levels of the sub-plot factor are randomized to sub-plot units within whole-plot units, resulting in equal correlation between all pairs of measures in the same whole-plot unit.

The practical consequence of these issues is that the correlation structure of repeated measures must be considered in statistical analysis techniques. There are numerous methods for analyzing repeated-measures data, and they deal with the covariance structure in different ways. In this chapter you will see three general methods. First is **univariate ANOVA**. This approach treats repeated-measures data as split-plot data. It considers the experimental units to which treatments are assigned to be the whole-plot units. It considers the experimental units *at a particular time* to be the sub-plot units. This approach accommodates the first aspect of repeated-measures covariance (between-subject variation) but not the second (measures close in time are more highly correlated than measures far apart in time). Univariate ANOVA was a standard method for analyzing repeated-measures data because it was the only method for which computer software was generally available. Because of the split-plot connection, the method was sometimes called a *split-plot in time* ANOVA. If correlations between measures on the same subject are the same regardless of time proximity, then the univariate ANOVA is a perfectly good method of analysis.

The second general method of repeated-measures analysis is **analysis of contrasts**. This method entails computing one or more linear combinations of data on each of the subjects, and then analyzing the linear combinations as data. One of the most common applications of analysis of contrasts is to compute the slopes of regression lines for each subject, where the response variable is regressed on time. The slope is a measure of the effect of time on each subject, and analyzing these slopes as data is a means of assessing the effects of treatments on these time effects. Analysis of contrasts is a device for avoiding the covariance problem in repeated-measures analysis. The method does not directly accommodate the covariance structure and thus is not an optimal method, but it is often quite adequate. The REPEATED statement in PROC GLM implements analysis of contrasts.

The third general method applies **mixed-model methodology** in a two-stage approach. The first stage attempts to estimate the covariance structure. Then the estimate of the covariance is substituted into the mixed model, and generalized least-squares methodology is used to assess treatment and time effects. From a conceptual point of view, this approach is very appealing because it directly addresses the covariance structure and attempts to accommodate it in the statistical model. However, assessing and estimating the covariance structure is more easily said than done. Only recently has computer software been available to enable use of mixed-model methodology. As you might expect, this is the approach implemented in PROC MIXED.

We now describe and display the FEV1 data more explicitly. The three drugs are labeled A, C, and P. Drug A was a standard drug used to treat asthma. Drug C was a potential competitor that was developed by the pharmaceutical company. Drug P was a placebo consisting of the carrier that was used in the other two drugs. The data are presented in two forms, one in Output 8.1, the other in Output 8.2. The description and purpose of each form of the data set is given below. Profile plots for the three drugs are shown in Figure 8.1.

Figure 8.1 *Profile Plot of the Drug Means over Time for FEV1 Repeated-Measures Data*

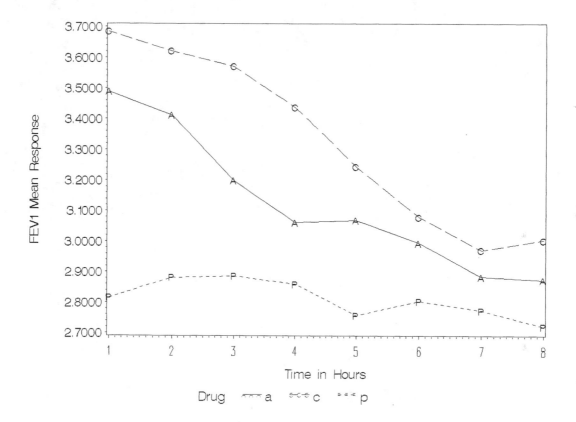

A SAS data set named FEV1MULT is displayed is Output 8.1. The data set is in a multivariate arrangement in the sense that each observation in the data set FEV1MULT contains all the FEV1 readings for a particular patient. The FEV1 values are contained in variables named FEV11H through FEV18H corresponding to measures at hours 1 through 8, respectively. The variable BASEFEV1 contains FEV1 values measures immediately prior to treatment. The data set FEV1MULT will be used to illustrate analysis of contrasts with the REPEATED statement in PROC GLM.

Output 8.1
Data Set
FEV1MULT

```
                                            FEV1 Data
                                      Multivariate Arrangement

OBS   PATIENT  BASEFEV1  FEV11H  FEV12H  FEV13H  FEV14H  FEV15H  FEV16H  FEV17H  FEV18H  DRUG

  1     201     2.46     2.68    2.76    2.50    2.30    2.14    2.40    2.33    2.20    a
  2     202     3.50     3.95    3.65    2.93    2.53    3.04    3.37    3.14    2.62    a
  3     203     1.96     2.28    2.34    2.29    2.43    2.06    2.18    2.28    2.29    a
  4     204     3.44     4.08    3.87    3.79    3.30    3.80    3.24    2.98    2.91    a
  5     205     2.80     4.09    3.90    3.54    3.35    3.15    3.23    3.46    3.27    a
  6     206     2.36     3.79    3.97    3.78    3.69    3.31    2.83    2.72    3.00    a
  7     207     1.77     3.82    3.44    3.46    3.02    2.98    3.10    2.79    2.88    a
  8     208     2.64     3.67    3.47    3.19    2.19    2.85    2.68    2.60    2.73    a
  9     209     2.30     4.12    3.71    3.57    3.49    3.64    3.38    2.28    3.72    a
 10     210     2.27     2.77    2.77    2.75    2.75    2.71    2.75    2.52    2.60    a
 11     211     2.44     3.77    3.73    3.67    3.56    3.59    3.35    3.32    3.18    a
 12     212     2.04     2.00    1.91    1.88    2.09    2.08    1.98    1.70    1.40    a
 13     214     2.77     3.36    3.42    3.28    3.30    3.31    2.99    3.01    3.08    a
 14     215     2.96     4.31    4.02    3.38    3.31    3.46    3.49    3.38    3.35    a
 15     216     3.11     3.88    3.92    3.71    3.59    3.57    3.48    3.42    3.63    a
 16     217     1.47     1.97    1.90    1.45    1.45    1.24    1.24    1.17    1.27    a
 17     218     2.73     2.91    2.99    2.87    2.88    2.84    2.67    2.69    2.77    a
 18     219     3.25     3.59    3.54    3.17    2.92    3.48    3.05    3.27    2.96    a
 19     220     2.73     2.88    3.06    2.75    2.71    2.83    2.58    2.68    2.42    a
 20     221     3.30     4.04    3.94    3.84    3.99    3.90    3.89    3.89    2.98    a
 21     222     2.85     3.38    3.42    3.28    2.94    2.96    3.12    2.98    2.99    a
 22     223     2.72     4.49    4.35    4.38    4.36    3.77    4.23    3.83    3.89    a
 23     224     3.68     4.17    4.30    4.16    4.07    3.87    3.87    3.85    3.82    a
 24     232     2.49     3.73    3.51    3.16    3.26    3.07    2.77    2.92    3.00    a
 25     201     2.30     3.41    3.48    3.41    3.49    3.33    3.20    3.07    3.15    c
 26     202     2.91     3.92    4.02    4.04    3.64    3.29    3.10    2.70    2.69    c
 27     203     2.08     2.52    2.44    2.27    2.23    2.01    2.26    2.34    2.44    c
 28     204     3.02     4.43    4.30    4.08    4.01    3.62    3.23    2.46    2.97    c
 29     205     3.26     4.55    4.58    4.44    4.04    4.33    3.87    3.75    3.81    c
 30     206     2.29     4.25    4.37    4.10    4.20    3.84    3.43    3.79    3.74    c
 31     207     1.96     3.00    2.80    2.59    2.42    1.61    1.83    1.21    1.50    c
 32     208     2.70     4.06    3.98    4.06    3.93    3.61    2.91    2.07    2.67    c
 33     209     2.50     4.37    4.06    3.68    3.64    3.17    3.37    3.20    3.25    c
 34     210     2.35     2.83    2.79    2.82    2.79    2.80    2.76    2.64    2.69    c
 35     211     2.34     4.06    3.68    3.59    3.27    2.60    2.72    2.22    2.68    c
 36     212     2.20     2.82    1.90    2.57    2.30    1.67    1.90    2.07    1.76    c
 37     214     2.78     3.18    3.13    3.11    2.97    3.06    3.27    3.24    3.33    c
 38     215     3.43     4.39    4.63    4.19    4.00    4.01    3.66    3.47    3.22    c
 39     216     3.07     3.90    3.98    4.09    4.03    4.07    3.56    3.83    3.75    c
 40     217     1.21     2.31    2.19    2.21    2.09    1.75    1.72    1.80    1.36    c
 41     218     2.60     3.19    3.18    3.15    3.14    3.08    2.96    2.97    2.85    c
 42     219     2.61     3.54    3.45    3.25    3.01    3.07    2.65    2.47    2.55    c
 43     220     2.48     2.99    3.02    3.02    2.94    2.69    2.66    2.68    2.70    c
 44     221     3.73     4.37    4.20    4.17    4.19    4.07    3.86    3.89    3.89    c
 45     222     2.54     3.26    3.39    3.27    3.20    3.32    3.09    3.25    3.15    c
 46     223     2.83     4.72    4.97    4.99    4.96    4.95    4.82    4.56    4.49    c
 47     224     3.47     4.27    4.50    4.34    4.00    4.11    3.93    3.68    3.77    c
 48     232     2.79     4.10    3.85    4.27    4.01    3.78    3.14    3.94    3.69    c
 49     201     2.14     2.36    2.36    2.28    2.35    2.31    2.62    2.12    2.42    p
 50     202     3.37     3.03    3.02    3.19    2.98    3.01    2.75    2.70    2.84    p
 51     203     1.88     1.99    1.62    1.65    1.68    1.65    1.85    1.96    1.30    p
 52     204     3.10     3.24    3.37    3.54    3.31    2.81    3.58    3.76    3.05    p
 53     205     2.91     3.35    3.92    3.69    3.97    3.94    3.63    2.92    3.31    p
 54     206     2.29     3.04    3.28    3.17    2.99    3.31    3.21    2.98    2.82    p
 55     207     2.20     2.46    3.22    2.65    3.02    2.25    1.50    2.37    1.94    p
 56     208     2.70     2.85    2.81    2.96    2.69    2.18    1.91    2.21    1.71    p
 57     209     2.25     3.45    3.48    3.80    3.60    2.83    3.17    3.22    3.13    p
 58     210     2.48     2.56    2.52    2.67    2.60    2.68    2.64    2.65    2.61    p
 59     211     2.12     2.19    2.44    2.41    2.55    2.93    3.08    3.11    3.06    p
 60     212     2.37     2.14    1.92    1.75    1.58    1.51    1.94    1.84    1.76    p
 61     214     2.73     2.57    3.08    2.62    2.91    2.71    2.39    2.42    2.73    p
 62     215     3.15     2.90    2.80    3.17    2.39    3.01    3.22    2.75    3.14    p
 63     216     2.52     3.02    3.21    3.17    3.13    3.38    3.25    3.29    3.35    p
 64     217     1.48     1.35    1.15    1.24    1.32    0.95    1.24    1.04    1.16    p
 65     218     2.52     2.61    2.59    2.77    2.73    2.70    2.72    2.71    2.75    p
 66     219     2.90     2.91    2.89    3.01    2.74    2.71    2.86    2.95    2.66    p
 67     220     2.83     2.78    2.89    2.77    2.77    2.69    2.65    2.84    2.80    p
 68     221     3.50     3.81    3.77    3.78    3.90    3.80    3.78    3.70    3.61    p
 69     222     2.86     3.06    2.95    3.07    3.10    2.67    2.68    2.94    2.89    p
 70     223     2.42     2.87    3.08    3.02    3.14    3.67    3.84    3.55    3.75    p
 71     224     3.66     3.98    3.77    3.65    3.81    3.77    3.89    3.63    3.74    p
 72     232     2.88     3.04    3.00    3.24    3.37    2.69    2.89    2.89    2.76    p
```

A SAS data set named FEV1UNI is displayed in Output 8.2, which contains the FEV1 data in a univariate arrangement, meaning that all FEV1 values at hours 1 through 8 are contained in a single variable named FEV1. In addition, FEV1UNI contains a variable named HOUR with values 1 through 8, indicating the measurement times. The data set FEV1UNI will be used to illustrate the univariate ANOVA and mixed-model approaches to analysis of repeated-measures data.

Output 8.2
Partial Listing
of Data Set
FEV1UNI

```
                            FEV1 Data
                       Univariate Arrangement


      OBS     PATIENT    BASEFEV1    DRUG    HOUR    FEV1

       1        201        2.46       a       1      2.68
       2        201        2.46       a       2      2.76
       3        201        2.46       a       3      2.50
       4        201        2.46       a       4      2.30
       5        201        2.46       a       5      2.14
       6        201        2.46       a       6      2.40
       7        201        2.46       a       7      2.33
       8        201        2.46       a       8      2.20
       9        202        3.50       a       1      3.95
      10        202        3.50       a       2      3.65
      11        202        3.50       a       3      2.93
      12        202        3.50       a       4      2.53
      13        202        3.50       a       5      3.04
      14        202        3.50       a       6      3.37
      15        202        3.50       a       7      3.14
      16        202        3.50       a       8      2.62

     561        224        3.66       p       1      3.98
     562        224        3.66       p       2      3.77
     563        224        3.66       p       3      3.65
     564        224        3.66       p       4      3.81
     565        224        3.66       p       5      3.77
     566        224        3.66       p       6      3.89
     567        224        3.66       p       7      3.63
     568        224        3.66       p       8      3.74
     569        232        2.88       p       1      3.04
     570        232        2.88       p       2      3.00
     571        232        2.88       p       3      3.24
     572        232        2.88       p       4      3.37
     573        232        2.88       p       5      2.69
     574        232        2.88       p       6      2.89
     575        232        2.88       p       7      2.89
     576        232        2.88       p       8      2.76
```

8.2   The Univariate ANOVA Method for Analyzing Repeated Measures

The univariate ANOVA method for analyzing repeated measures data refers to using methods for the types of random-effects models you saw in Chapter 4. For the FEV1 data, there would be a random effect for patients within drugs. Thus a model is

$$y_{ijk} = \mu + \alpha_i + b_{ij} + \gamma_k + (\alpha\gamma)_{ik} + e_{ijk} \tag{8.2}$$

where

y_{ijk} is the FEV1 measurement at hour k on the jth patient assigned to drug i,

$\mu + \alpha_i + \gamma_k + (\alpha\gamma)_{ik}$ is the mean FEV1 for drug i at hour k,

b_{ij} is the random effect associated with patient j in drug i, assumed to be i.i.d. $N(0,\sigma_B^2)$,

e_{ijk} is the random error associated with the jth patient assigned to drug i at hour k, assumed to be i.i.d. $N(0,\sigma^2)$.

Also, the b_{ij} and e_{ijk} are assumed to be independent of one another. In repeated-measures analysis, the b_{ij} are often called the **between-subjects effects** and the e_{ijk} are called the **within-subjects effects**. It follows from this model formulation that

$$E(y_{ijk}) = \mu + \alpha_i + \gamma_k + (\alpha\gamma)_{ik}$$

$$V(y_{ijk}) = V(b_{ijk} + e_{ijk}) = V(b_{ijk}) + V(e_{ijk}) = \sigma_B^2 + \sigma^2$$

and

$$\mathrm{cov}(y_{ijk}, y_{ijl}) = \sigma_B^2$$

In other words, the variance of an observation is the sum of the between-subject and the within-subject variance components, and the covariance between two observations taken on the same subject at different times is equal to the between-subjects variance component. Notice that this presumes the covariance is the same for *all* pairs of observations, regardless of their proximity in time. This assumption often is not realistic.

If these assumptions hold, then analysis-of-variance methods are valid for repeated-measures data.

8.2.1 Using GLM to Perform Univariate ANOVA of Repeated-Measures Data

Run the statements

```
proc glm data=fev1uni;
   class drug patient hour;
   model fev1=drug patient(drug) hour drug*hour/ss3;
   random patient(drug)/test;
run;
```

Results appear in Output 8.3. You see the MODEL SS = 296.64 partitioned into sources of variation for each term in the MODEL statement. *F*-tests are computed for each of these sources of variation using MS(ERROR) in the denominator. As in a split-plot situation, some of these are valid and some are not. In particular, the *F*-test for DRUG, which has $F=204.21$ and $p<.0001$, is *not* correct.

The RANDOM statement produces expected mean squares for each source of variation, and the TEST option produces *F*-tests that are valid *provided the model assumption* $e_{ijk} \sim \mathrm{N.I.D.}(0,\sigma^2)$ *is correct*. You see that the appropriate test for DRUG is F = MS(DRUG)/MS(PATIENT(DRUG)) = 3.5952, with $p=0.0327$. The appropriate tests for HOUR and DRUG*HOUR use MS(ERROR) in the denominator.

```
                                      FEV1 Data
                                   Univariate ANOVA

                             General Linear Models Procedure
                                Class Level Information

Class      Levels    Values

DRUG          3       a c p

PATIENT      24       201 202 203 204 205 206 207 208 209 210 211 212 214 215 216 217 218 219
                      220 221 222 223 224 232

HOUR          8       1 2 3 4 5 6 7 8

                          Number of observations in data set = 576

Dependent Variable: FEV1

Source              DF          Sum of Squares          Mean Square    F Value    Pr > F

Model               92            296.64552326           3.22440786     51.08     0.0001

Error              483             30.49025937           0.06312683

Corrected Total    575            327.13578264

            R-Square                  C.V.             Root MSE          FEV1 Mean

            0.906796               8.138859            0.25125053        3.08704861

Source              DF          Type III SS             Mean Square    F Value    Pr > F

DRUG                 2            25.78256701           12.89128351     204.21     0.0001
PATIENT(DRUG)       69           247.41249063            3.58568827      56.80     0.0001
HOUR                 7            17.17039931            2.45291419      38.86     0.0001
DRUG*HOUR           14             6.28006632            0.44857617       7.11     0.0001

Source              Type III Expected Mean Square

DRUG                Var(Error) + 8 Var(PATIENT(DRUG)) + Q(DRUG,DRUG*HOUR)

PATIENT(DRUG)       Var(Error) + 8 Var(PATIENT(DRUG))

HOUR                Var(Error) + Q(HOUR,DRUG*HOUR)

DRUG*HOUR           Var(Error) + Q(DRUG*HOUR)

                     Tests of Hypotheses for Mixed Model Analysis of Variance

Source: DRUG *
Error: MS(PATIENT(DRUG))
                                      Denominator    Denominator
        DF      Type III MS               DF             MS          F Value    Pr > F
         2      12.891283507              69        3.5856882699       3.5952    0.0327
* - This test assumes one or more other fixed effects are zero.

Source: HOUR *
Error: MS(Error)
                                      Denominator    Denominator
        DF      Type III MS               DF             MS          F Value    Pr > F
         7      2.4529141865             483        0.063126831       38.8569    0.0001
* - This test assumes one or more other fixed effects are zero.

Source: DRUG*HOUR
Error: MS(Error)
                                      Denominator    Denominator
        DF      Type III MS               DF             MS          F Value    Pr > F
        14      0.4485761657             483        0.063126831        7.1060    0.0001
```

8.2.2 The CONTRAST, ESTIMATE, and LSMEANS Statements in Univariate ANOVA of Repeated-Measures Data

Other data analysis tools in GLM, such as the CONTRAST, ESTIMATE, and LSMEANS statements may be used as described in Chapter 4. However, as was the case for split-plot experiments, users should recognize their shortcomings and not assume that standard errors or tests of hypotheses are valid simply because they were printed by GLM. For example, the statement

```
lsmeans drug / pdiff stderr e=patient(drug);
```

computes means and standard errors for drugs, and tests of hypotheses comparing drugs in each pair of drugs, averaged over hours. Results appear in Output 8.4.

Output 8.4
LSMEANS in
Univariate
ANOVA of
FEV1 Data

```
                                    FEV1 Data
                                 Univariate ANOVA
Least Squares Means

   Standard Errors and Probabilities calculated using the Type III MS for PATIENT(DRUG) as
an
                                    Error term

     DRUG          FEV1        Std Err      Pr > |T|    Pr > |T|  H0: LSMEAN(i)=LSMEAN(j)
                   LSMEAN      LSMEAN       H0:LSMEAN=0  i/j    1       2       3

       a        3.12276042   0.13665819      0.0001      1     .     0.2956  0.1123
       c        3.32645833   0.13665819      0.0001      2   0.2956    .     0.0096
       p        2.81192708   0.13665819      0.0001      3   0.1123  0.0096    .
```

Verification that MS(PATIENT(DRUG)) is the appropriate error term for the standard errors and comparisons of means must be done by hand.

CONTRAST statements can be used to compare means averaged over hours or at individual hours. The RANDOM statement will provide expected means squares to indicate appropriate tests. Run the statements

```
contrast 'a-c overall'   drug 1 -1 0;
contrast 'a-c at hour 1' drug 1 -1 0
   drug* hour 1 0 0 0 0 0 0   -1 0 0 0 0 0 0 0;
contrast 'a-c*hour' drug*hour  1 0 0 0 0 0 0 -1
                              -1 0 0 0 0 0 0 1,
                    drug*hour  0 1 0 0 0 0 0 -1
                               0 -1 0 0 0 0 0 1,
                    drug*hour  0 0 1 0 0 0 0 -1
                               0 0 -1 0 0 0 0 1,
                    drug*hour  0 0 0 1 0 0 0 -1
                               0 0 0 -1 0 0 0 1,
                    drug*hour  0 0 0 0 1 0 0 -1
                               0 0 0 0 -1 0 0 1,
                    drug*hour  0 0 0 0 0 1 0 -1
                               0 0 0 0 0 -1 0 1,
                    drug*hour  0 0 0 0 0 0 1 -1
                               0 0 0 0 0 0 -1 1;
              random patient(drug);
run;
```

Expected mean squares are shown in Output 8.5.

Output 8.5
Expected Mean Squares for Contrasts in Univariate ANOVA of FEV1 Data

```
                                      FEV1 Data
                                   Univariate ANOVA
Contrast                 Contrast Expected Mean Square

a-c overall              Var(Error) + 8 Var(PATIENT(DRUG)) + Q(DRUG,DRUG*HOUR)

a-c at hour 1            Var(Error) + Var(PATIENT(DRUG)) + Q(DRUG,DRUG*HOUR)

a-b*hour                 Var(Error) + Q(DRUG*HOUR)
```

Results in Output 8.5 show that MS(PATIENT(DRUG)) is the appropriate error term for comparing drugs A and C *averaged over hours*, but there is no available mean square appropriate for comparing drugs A and C *at a particular hour*. Also, expected mean squares indicate that Var(Error) is the appropriate error term for testing interaction between the A-C contrast and hour, but the validity of this test, as with the test for DRUG*HOUR interaction, depends on covariance structure of the repeated measures. Thus, results for only the first CONTRAST statement should be used without suspicion. Nonetheless, we show results for all contrasts. To obtain the appropriate test for A-C OVERALL, run the statement

```
contrast 'a-c overall'   drug 1 -1 0 / e=patient(drug);
```

Results appear in Output 8.6.

Output 8.6
Results of the CONTRAST Statement in Univariate ANOVA of FEV1 Data

```
                               FEV1 Data
                            Univariate ANOVA

Contrast          DF        Contrast SS        Mean Square   F Value   Pr > F

a-b at hour 1     1         0.46216875         0.46216875    7.32      0.0071
a-b*hour          7         1.10406432         0.15772347    2.50      0.0157


Tests of Hypotheses using the Type III MS for PATIENT(DRUG) as an error term

Contrast          DF        Contrast SS        Mean Square   F Value   Pr > F

a-c overall       1         3.98331276         3.98331276    1.11      0.2956
```

The *p*-value for comparing A with C is 0.2956. This is the same as the *p*-value in Output 8.4 from the comparison of LSMEANS. The *p*-value for comparing A with C at hour 1 is 0.0071, which is probably not valid because the denominator of the *F* did not contain the between-subjects variance, as it should. The *p*-value for testing interaction between the A-C contrast and hour is 0.0157; however, this *p*-value is suspect because of the strong possibility of a within-subjects covariance structure.

8.3 Multivariate and Univariate Methods Based on Contrasts of the Repeated Measures

The univariate ANOVA methods you saw in Section 8.2 are valid if the covariances of the repeated measures on the same subject obey certain conditions. For example, if measures at each time have equal variances and the correlations between any two measures are equal, then the conditions are met. Repeated measures with equal variances and correlations are said to have **compound symmetric** (CS) covariance structure. In some situations the compound symmetric assumption is reasonably met. But in general it is unwise to assume compound symmetry holds without justification.

Actually, compound symmetry is a sufficient condition, but not necessary. The necessary condition is called the **Huynh-Feldt** (H-F) condition. It permits unequal variances, but the covariances have a rather rigid form. The H-F condition is equivalent to differences between each pair of measures having equal variances. While the H-F condition is mathematically more general than CS, it is not often more general from a practical point of view.

Multivariate methods can be used when the H-F condition is not met. They can be implemented using the REPEATED statement in PROC GLM. In fact, the REPEATED statement produces many useful statistics that are not truly multivariate. The REPEATED statement actually implements several methods that are essentially univariate. The common feature of the REPEATED statement techniques, multivariate and univariate, is that they are applied to *differences* between measures on the same subject.

8.3.1 Univariate ANOVA of Repeated Measures at Each Time

It is often useful to perform analyses at each time point simply to become more familiar with the data and to make diagnostic assessment. Run the statements

```
proc glm data=fev1mult;
  class drug;
  model fev11h fev12h fev11h fev12h fev11h fev12h fev11h fev12h
      = drug;
  contrast 'a-c' drug 1 -1 0;
```

The results are summarized in Output 8.7. The actual GLM output does not have this format.

Output 8.7
A Summary of
ANOVA at
Each Hour

		FEV11H	FEV12H	FEV13H	FEV14H	FEV15H	FEV16H	FEV17H	FEV18H
					FEV1 Data				
Source	DF	MS	MS	MS	MS	MS	MS	MS	MS
Drug	2	4.997	3.490	2.822	2.064	1.458	0.477	0.236	0.484
Error	69	0.454	0.516	0.492	0.493	0.577	0.490	0.499	0.503
a-c	1	0.462	0.520	1.661	1.695	1.459	0.086	0.091	0.205
F=MS(Drug)/MSE		11.01	6.76	5.74	4.19	2.53	0.97	0.47	0.96
		(.0001)	(.0021)	(.0050)	(.0193)	(.0815)	(.3828)	(.6252)	(.3871)
F=MS(a-c)/MSE		1.02	1.09	3.38	3.44	2.98	0.17	0.18	0.41
		(.3166)	(.3187)	(.0705)	(.0682)	(.0875)	(.6770)	(.6708)	(.5250)

There are three things to observe in Output 8.7. First, error mean squares are approximately equal to 0.5 at all hours. This means we do not have to be concerned about heterogeneous variances in analyses over time. Second, the F-ratios indicate significant differences between drugs at hours 1 through 5, but not at hours 6 through 8. Third, the difference between drugs A and C is only modestly significant at hours 3 through 5, but not at other hours. However, these tests do not reveal trends of FEV1 response over times. Proceed with repeated analyses to examine trends over time.

8.3.2 Using the REPEATED Statement in PROC GLM to Perform Multivariate Analysis of Repeated-Measures Data

The first thing to know about the REPEATED statement is that it is used when the data set is in the *multivariate* mode. In the FEV1 example, use the data set FEV1MULT.

Run the statements

```
proc glm data=fev1mult;
  class drug;
  model fev11h fev12h fev13h fev14h fev15h fev16h fev17h fev18h
     = drug;
    repeated hour / printe;
run;
```

Note that the MODEL statement alone would simply compute ANOVA tables for each hour, as are summarized in Output 8.7. Now we discuss the syntax of the REPEATED statement. Following the key word *repeated* you see the word HOUR. This is *not* a variable in the SAS data set FEV1MULT. It is simply a word used to refer to the eight response variables FEV11H, ... , FEV18H. This is the basic REPEATED statement. It produces both multivariate and univariate tests that will be discussed later. The PRINTE option at the end of the REPEATED statement produces the sums of squares and cross products of residuals from fitting the model to the eight response variables, and correlation coefficients based on the sum of squares and cross products. The correlation coefficients are shown in Output 8.8.

Output 8.8
Correlations
between FEV1
Repeated
Measures

```
                                    FEV1 Data

Partial Correlation Coefficients from the Error SS&CP Matrix / Prob > |r|

  DF = 69     FEV11H     FEV12H     FEV13H     FEV14H     FEV15H     FEV16H     FEV17H     FEV18H

  FEV11H    1.000000   0.947354   0.939327   0.877290   0.849003   0.833356   0.747934   0.803328
              0.0001     0.0001     0.0001     0.0001     0.0001     0.0001     0.0001     0.0001

  FEV12H    0.947354   1.000000   0.953603   0.928445   0.904953   0.845329   0.786126   0.835204
              0.0001     0.0001     0.0001     0.0001     0.0001     0.0001     0.0001     0.0001

  FEV13H    0.939327   0.953603   1.000000   0.950646   0.907989   0.867392   0.810935   0.855238
              0.0001     0.0001     0.0001     0.0001     0.0001     0.0001     0.0001     0.0001

  FEV14H    0.877290   0.928445   0.950646   1.000000   0.905495   0.848955   0.810209   0.852875
              0.0001     0.0001     0.0001     0.0001     0.0001     0.0001     0.0001     0.0001

  FEV15H    0.849003   0.904953   0.907989   0.905495   1.000000   0.928608   0.864207   0.918064
              0.0001     0.0001     0.0001     0.0001     0.0001     0.0001     0.0001     0.0001

  FEV16H    0.833356   0.845329   0.867392   0.848955   0.928608   1.000000   0.899701   0.932298
              0.0001     0.0001     0.0001     0.0001     0.0001     0.0001     0.0001     0.0001

  FEV17H    0.747934   0.786126   0.810935   0.810209   0.864207   0.899701   1.000000   0.896999
              0.0001     0.0001     0.0001     0.0001     0.0001     0.0001     0.0001     0.0001

  FEV18H    0.803328   0.835204   0.855238   0.852875   0.918064   0.932298   0.896999   1.000000
              0.0001     0.0001     0.0001     0.0001     0.0001     0.0001     0.0001     0.0001


  Applied to Orthogonal Components:
  Test for Sphericity: Mauchly's Criterion = 0.0654899
  Chisquare Approximation = 181.13982 with 27 df    Prob > Chisquare = 0.0000
```

The correlation coefficients in Output 8.8, combined with the error mean squares in Output 8.7, tell much about the covariance structure of the data. The correlation between FEV11H and FEV12H through FEV18H generally decrease from 0.95 down to 0.80. Thus, the correlation coefficients generally decrease as length of the time interval increases. This is a common phenomenon of repeated measures. It usually means the covariance structure *does not* obey the H-F conditions, and, as a consequence, the univariate AVOVA analysis of repeated measures is flawed. Below the correlation coefficients in Output 8.8 you see results of a chi-square test based on the so-called Mauchly's criterion. This is a likelihood ratio statistic to test whether the model that imposes the H-F conditions on the covariance structure fits as well as the model that imposes no conditions at all. The highly significant chi-square value indicates the H-F conditions are too restrictive. Thus the univariate ANOVA tests in Section 8.2.1 are suspect.

Output 8.9 contains the results of multivariate tests applied to differences between the repeated measures.

Output 8.9
Multivariate
Tests for
HOUR and
DRUG*HOUR
Effects

```
        Manova Test Criteria and Exact F Statistics for the Hypothesis of no HOUR Effect
              H = Type III SS&CP Matrix for HOUR     E = Error SS&CP Matrix


                           S=1      M=2.5     N=30.5

          Statistic                    Value          F       Num DF   Den DF   Pr > F

          Wilks' Lambda            0.41804056      12.5290        7       63    0.0001
          Pillai's Trace           0.58195944      12.5290        7       63    0.0001
          Hotelling-Lawley Trace   1.39211241      12.5290        7       63    0.0001
          Roy's Greatest Root      1.39211241      12.5290        7       63    0.0001



       Manova Test Criteria and F Approximations for the Hypothesis of no HOUR*DRUG Effect
             H = Type III SS&CP Matrix for DRUG*HOUR    E = Error SS&CP Matrix


                            S=2       M=2      N=30.5

          Statistic                    Value          F       Num DF   Den DF   Pr > F

          Wilks' Lambda            0.51191319       3.5789       14      126    0.0001
          Pillai's Trace           0.55490980       3.5108       14      128    0.0001
          Hotelling-Lawley Trace   0.82292043       3.6444       14      124    0.0001
          Roy's Greatest Root      0.60834527       5.5620        7       64    0.0001


              NOTE: F Statistic for Roy's Greatest Root is an upper bound.
                 NOTE: F Statistic for Wilks' Lambda is exact.



        Manova Test Criteria and Exact F Statistics for the Hypothesis of no HOUR*a-c Effect
              H = Contrast SS&CP Matrix for HOUR*a-c    E = Error SS&CP Matrix


                           S=1      M=2.5     N=30.5

          Statistic                    Value          F       Num DF   Den DF   Pr > F

          Wilks' Lambda            0.81742760       2.0101        7       63    0.0676
          Pillai's Trace           0.18257240       2.0101        7       63    0.0676
          Hotelling-Lawley Trace   0.22334994       2.0101        7       63    0.0676
          Roy's Greatest Root      0.22334994       2.0101        7       63    0.0676
```

The multivariate tests in Output 8.9 show highly significant effects of HOUR and DRUG*HOUR. In this particular example there are highly pronounced changes over time and the changes are not the same for all drugs. Statistical significance of these effects is not really in question. In other examples the trends over time and differences between trends for treatment groups may be less pronounced. In such cases it often happens that the univariate ANOVA tests will indicate statistical significance but the multivariate tests will not. In fact, this phenomenon occurs in the test for interaction between the A-C contrast and HOUR. The univariate ANOVA test in Output 8.6 showed a *p*-value of 0.0157 for this interaction, but the multivariate test in Output 8.9 shows a *p*-value of 0.0676. This apparent contradiction might be an artifact of a failed assumption of the H-F covariance structure, leading to a Type I error in the univariate ANOVA. It might also be due to weakness in the multivariate tests because they do not exploit identifiable trends in the covariance structure.

The remaining results from the REPEATED statement are shown in Output 8.10.

Output 8.10
Univariate
Tests for
DRUG, HOUR
and
DRUG*HOUR
Effects, with
Adjusted p-
Values

```
Tests of Hypotheses for Between Subjects Effects

Source          DF          Type III SS         Mean Square    F Value     Pr > F

DRUG            2           25.78256701         12.89128351    3.60        0.0327

Error           69          247.41249063        3.58568827


Univariate Tests of Hypotheses for Within Subject Effects

Source: HOUR
                                                                    Adj  Pr > F
     DF        Type III SS      Mean Square     F Value   Pr > F    G - G    H - F
     7         17.17039931      2.45291419      38.86     0.0001    0.0001   0.0001

Source: DRUG*HOUR
                                                                    Adj  Pr > F
     DF        Type III SS      Mean Square     F Value   Pr > F    G - G    H - F
     14        6.28006632       0.44857617      7.11      0.0001    0.0001   0.0001

Source: Error(HOUR)

     DF        Type III SS        Mean Square
     483       30.49025938        0.06312683

                            Greenhouse-Geisser Epsilon = 0.4971
                                 Huynh-Feldt Epsilon = 0.5419

Contrast: HOUR*a-b
                                                                    Adj  Pr > F
     DF        Contrast SS      Mean Square     F Value   Pr > F    G - G    H - F
     7         1.10406432       0.15772347      2.50      0.0157    0.0514   0.0461
```

The entries in Output 8.10 duplicate the univariate ANOVA in Output 8.3. The F-test for DRUG in Output 8.10 under the heading "Tests of Hypotheses for between Subject Effects" is the same as that obtained using the TEST option on the RANDOM statement in Output 8.3. The validity of this test does not depend on the covariance structure of repeated measures because it is based on means for each subject averaged over the repeated measures. The tests for HOUR ($F = 38.86$) and DRUG*HOUR ($F = 7.11$) in Output 8.3 also are found in Output 8.10 under the heading "Univariate Tests of Hypotheses for Within Subject Effects." Both of these F-tests show highly significant p-values of 0.0001. The validity of these tests that involve the within-subjects effect HOUR depend on the covariance structure. Specifically, the covariance structure must obey the Huynh-Feldt conditions. In other words, the computed F-values for HOUR and DRUG*HOUR have F-distributions under the respective null hypotheses only if the H-F condition is met. But the results of the chi-square test in Output 8.8 showed that the H-F condition is not met. Thus the univariate ANOVA F-tests are suspect.

Output 8.10 contains two adjusted p-values, labeled G-G and H-F, for the F-tests for HOUR and DRUG*HOUR. These adjustments are obtained by referring the F-values to distributions with reduced degrees of freedom using a method due to Box (1954). The reduced degrees of freedom are equal to the ordinary degrees of freedom multiplied by a number called epsilon. But the value of epsilon depends on unknown parameters and must be estimated. Two estimates of epsilon are given, equal to 0.4971 and 0.5419. The first (G-G) is due to Greenhouse and Geisser (1959) and the second (H-F) is due to Huynh and Feldt (1976). The G-G adjusted p-value for DRUG*HOUR is obtained by referring $F = 7.11$ to an F-distribution with 14*.4971=6.96 numerator degrees of freedom and 483*.4971=240.1 denominator degrees of freedom. The adjusted p-value is still less than 0.0001. The other adjusted p-values are computed likewise, and all are less than 0.0001. Thus, in this example, the univariate F-values for HOUR and DRUG*HOUR are so large that they are highly significant even with the reduced degrees of freedom. This is not always the case. It commonly happens that the p-value for a univariate F for a within-subjects effect is "significant," but the adjusted *p*-value is not. This more or less occurs with interaction contrast A-C*HOUR, for which the univariate ANOVA gave *p*=.0157, but has G-G and H-F adjusted *p*-values of 0.0514 and 0.0461, respectively.

8.3.3 Univariate ANOVA of Contrasts of Repeated Measures

The SUMMARY option in the REPEATED statement provides univariate ANOVA results of contrast variables. Results appear in Output 8.11.

Output 8.11
Univariate
Tests for
Contrast
Variables

```
                  Analysis of Variance of Contrast Variables

         HOUR.N represents the contrast between the nth level of HOUR and the last

Contrast Variable: HOUR.1

Source          DF           Type III SS        Mean Square     F Value    Pr > F
MEAN            1            15.47533889        15.47533889      81.76     0.0001
DRUG            2             4.95388611         2.47694306      13.09     0.0001
Error           69           13.05957500         0.18926920

Contrast        DF           Contrast SS        Mean Square     F Value    Pr > F
a-c             1             0.05135208         0.05135208       0.27     0.6041

Contrast Variable: HOUR.2

Source          DF           Type III SS        Mean Square     F Value    Pr > F
MEAN            1            13.84256806        13.84256806      82.37     0.0001
DRUG            2             2.85541111         1.42770556       8.50     0.0005
Error           69           11.59632083         0.16806262

Contrast        DF           Contrast SS        Mean Square     F Value    Pr > F
a-c             1             0.07207500         0.07207500       0.43     0.5147

Contrast Variable: HOUR.3

Source          DF           Type III SS        Mean Square     F Value    Pr > F
MEAN            1             8.96761250         8.96761250      62.21     0.0001
DRUG            2             1.95842500         0.97921250       6.79     0.0020
Error           69            9.94626250         0.14414873

Contrast        DF           Contrast SS        Mean Square     F Value    Pr > F
a-c             1             0.69841875         0.69841875       4.85     0.0311
                               FEV1 Data
Contrast Variable: HOUR.4

Source          DF           Type III SS        Mean Square     F Value    Pr > F
MEAN            1             4.63093889         4.63093889      31.57     0.0001
DRUG            2             1.19181111         0.59590556       4.06     0.0215
Error           69           10.12265000         0.14670507

Contrast        DF           Contrast SS        Mean Square     F Value    Pr > F
a-c             1             0.72030000         0.72030000       4.91     0.0300
```

Output 8.11
(Continued)
Univariate
Tests for
Contrast
Variables

```
Contrast Variable: HOUR.5

Source             DF          Type III SS          Mean Square    F Value    Pr > F
MEAN                1            1.77347222           1.77347222      19.50    0.0001
DRUG                2            0.54738611           0.27369306       3.01    0.0559
Error              69            6.27594167           0.09095568

Contrast           DF          Contrast SS          Mean Square    F Value    Pr > F
a-c                 1            0.02296875           0.02296875       0.25    0.6169

Contrast Variable: HOUR.6
Source             DF          Type III SS          Mean Square    F Value    Pr > F
MEAN                1            0.62533472           0.62533472       9.28    0.0033
DRUG                2            0.02916944           0.01458472       0.22    0.8058
Error              69            4.64719583           0.06735066

Contrast           DF          Contrast SS          Mean Square    F Value    Pr > F
a-c                 1            0.02566875           0.02566875       0.38    0.5390
                                     FEV1 Data
Contrast Variable: HOUR.7

Source             DF          Type III SS          Mean Square    F Value    Pr > F

MEAN                1            0.00700139           0.00700139       0.07    0.7953
DRUG                2            0.08841944           0.04420972       0.43    0.6535
Error              69            7.12547917           0.10326781

Contrast           DF          Contrast SS          Mean Square    F Value    Pr > F
a-c                 1            0.02296875           0.02296875       0.22    0.6387
```

The contrasts are differences between each hour and hour 8. For example, HOUR.1 is the difference between hour 1 and hour 8. For each contrast variable there are tests labeled MEAN, DRUG, and A-C. These results are what you would obtain if you created the differences in a SAS data set and ran univariate analyses on the contrasts. Since they are univariate analyses, their validity does not depend on the covariance structure. See Littell et al. (1998) for more details.

Some care is required to correctly interpret the tests in Output 8.11. Consider the variable HOUR.1. The *F*-test for DRUG is highly significant with $p<0.0001$. This is evidence that the means of the variable HOUR.1 for the three drugs are different. But since the variable HOUR.1 is the difference between hours 1 and 8, the test for DRUG is really a test of whether the change from hour 1 to hour 8 differs between the drugs. Thus the test for DRUG is really a test for the interaction between drugs and the contrast hour 1 minus hour 8. Likewise, the test for A-C is a test for the interaction between the drug contrast A-C and the hour contrast 1-8. The test for MEAN is a test of whether the average value of HOUR.1 over the drugs differs from zero.

8.4 Mixed-Model Analysis of Repeated Measures

The first sections of this chapter reviewed methods for repeated-measures data that predate the availability of PROC MIXED. Data analysts essentially faced a choice between "split-plot in time" univariate ANOVA, presented in Section 8.2, contrasts, or multivariate analysis of variance (MANOVA), as presented in Section 8.3. These approaches all have serious drawbacks.

As noted earlier in this chapter, univariate ANOVA assumes that within-subjects errors are uncorrelated, or, equivalently, that correlation between observations on the same subject is constant, regardless of the time lapse between pairs of observations. This is often an unrealistic assumption. If you ignore time-dependent correlation, you risk underestimating standard errors of mean comparisons involving different times and overestimating test statistics for time and time-by-treatment effects, resulting in excessive Type I error rates. While adjustments such as the Greenhouse-Geisser and Huynh-Feldt corrections used in PROC GLM attempt to account for within-subjects correlation, these corrections are often inadequate and, in any event, do not address the impact of correlation on standard errors of differences that may be of interest to researchers.

MANOVA assumes a correlation structure that is far more general than most repeated-measures data require. The MANOVA covariance matrix assumes that the correlation between every pair of times of observation within subjects is unique. For example, the correlations between the first and second, second and third, third and fourth, and so on, observations are each unique parameters in MANOVA. Usually, a simpler model, such as a single correlation parameter for all observations that are adjacent in time, is sufficient to characterize the data. Thus, MANOVA typically wastes a great deal of information, which in turn adversely affects efficiency and power. Moreover, if an observation is lost at a single time for a given subject, MANOVA must delete the *entire* subject. Thus, a great deal of valid data may be thrown out, a draconian approach to missing data.

The mixed-model procedures of PROC MIXED allow a flexible approach to modeling correlated errors. By using a parsimonious covariance model that adequately accounts for within-subjects correlation, you can avoid the problems associated with univariate and multivariate ANOVA.

As noted in the introduction to this chapter, mixed-model analysis involves two stages: first, estimate the covariance structure, then second, assess treatment and time effects using generalized least squares with the estimated covariance. Littell, Pendergast, and Natajan (2000) break these stages further into a four-step procedure for mixed-model analysis:

Step 1: Model the mean structure, usually by specification of the fixed effects.

Step 2: Specify the covariance structure, between subjects as well as within subjects.

Step 3: Fit the mean model accounting for the covariance structure; MIXED uses generalized least squares to do this.

Step 4: Make statistical inference based on the results of Step 3; this step may include making the means model more parsimonious.

As Littell et al. (2000) point out, other authors, for example, Diggle (1988) and Wolfinger (1993), recommend similar model-fitting and inference processes. The following sections show you how to use PROC MIXED to implement these fours steps. Sections 8.4.1 through 8.4.4 apply Steps 1 through 4, respectively, to the univariate data (FEV1UNI) from Output 8.2.

8.4.1 The Fixed-Effects Model and Related Considerations

The model equation for mixed-model analysis is identical to the univariate ANOVA model described in Section 8.2. Model (8.2) is a mixed model for the FEV1 data.

As discussed earlier, the FEV1 data set contains a baseline observation, BASEFEV1, on each subject. Often, you can get a more accurate assessment of treatment effects by accounting for baseline measurements in the model using analysis of covariance. A model accounting for baseline observations is

$$y_{ijk} = \mu + \alpha_i + \beta X_{ij} + b_{ij} + \gamma_k + (\alpha\gamma)_{ik} + e_{ijk} \tag{8.3}$$

where X_{ij} is the baseline observation on the *ij*th subject, β is the regression coefficient, and all other terms are as defined previously for model (8.2).

The analyses presented in Sections 8.2 and 8.3 do not lend themselves to using baseline data. In general, for any data with repeated measures or split-plot features, analysis of covariance with covariates observed on the *larger* units—for example, subjects or whole-plot experimental units—is difficult and awkward, if not impossible, to implement with PROC GLM, or any other procedure not explicitly written to use mixed-model methods. However, baseline covariate models are relatively easy to compute with PROC MIXED.

In the mixed model, the between-subjects error, b_{ij} and the within-subjects error, e_{ijk}, are considered random effects. The b_{ij}'s and e_{ijk}'s are assumed to be independent of one another. For each set of effects, the covariance structure can be quite general.

Typically, the between-subjects errors are assumed i.i.d. $N(0, \sigma_B^2)$. Thus, the vector of random model effects, $\mathbf{b}' = \left[b_{11}, b_{12}, ..., b_{a,n_a} \right] \sim \text{MVN}(\mathbf{0}, \mathbf{I}\sigma_B^2)$. The covariance for the within-subjects effects is more complex. Denote $\mathbf{e}'_{ij} = \left[e_{ij1}, e_{ij2}, ..., e_{ijK} \right]$ as the vector of within-subjects errors for the ijth subject. In general, $\mathbf{e}_{ij} \sim \text{MVN}(\mathbf{0}, \Sigma)$, where Σ is a K×K covariance matrix. The full vector of within-subjects errors, $\mathbf{e}' = \left[e'_{11}, e'_{12}, ..., e'_{a,n_a} \right]$ is distributed $\text{MVN}(\mathbf{0}, \mathbf{R})$, where R is a block diagonal matrix, with one block per subject, and each block equal to Σ. Thus $\mathbf{R} = I_{n_\bullet} \otimes \Sigma$, where $n_\bullet = \sum_i n_i$, is the total number of subjects.

PROC MIXED allows you to choose from many **covariance models**, in other words, forms of Σ. The simplest model is the **independent** covariance model, where the within-subjects error correlation is zero, and hence $\Sigma = \mathbf{I}\sigma_s^2$. The most complex is the **unstructured** covariance model, where within-subjects errors for each pair of times have their own unique correlation, and

$$
\text{hence } \Sigma = \begin{bmatrix} \sigma_1^2 & \sigma_{12} & \sigma_{13} & . & \sigma_{1,K} \\ & \sigma_2^2 & \sigma_{23} & . & \sigma_{2,K} \\ & & \sigma_3^2 & . & \sigma_{3,K} \\ & & & . & . \\ & & & & \sigma_K^2 \end{bmatrix}
$$

. In some applications, the within-subjects correlation is negligible. For example, in some agronomic and large animal nutrition trials, repeated measurements may occur at long enough intervals, for example, monthly, so that correlation is effectively zero relative to other variation. In such cases, the independent structure is acceptable. However, this should be checked before analyzing the data assuming uncorrelated errors.

In most experiments, *some* correlation is present. However, correlation is usually not as complex as the **unstructured** model. The simplest correlation model is **compound symmetry**, referred to as **CS** in PROC MIXED syntax. The CS model is written

$$
\Sigma = \sigma^2 \begin{bmatrix} 1 & \rho & \rho & . & \rho \\ & 1 & \rho & . & \rho \\ & & 1 & . & . \\ & & & . & \\ & & & & 1 \end{bmatrix}
$$

It assumes that correlation is constant regardless of the distance between pairs of repeated measurements. Note that the split-plot model, $y_{ijk} = \mu + \alpha_i + b_{ij} + \gamma_k + (\alpha\gamma)_{ik} + e_{ijk}$, with independent errors, and the model $y_{ijk} = \mu + \alpha_i + \gamma_k + (\alpha\gamma)_{ik} + e_{ijk}$, dropping b_{ij} and assuming compound symmetry among the e_{ijk} for the ijth subject, are equivalent expressions of the same model. Under the independent errors model, $\text{var}(y_{ijk}) = \sigma_B^2 + \sigma_S^2$ and $\text{cov}(y_{ijk}, y_{ijk'}) = \sigma_B^2$ for $k \neq k'$, identical to the CS model with $\sigma^2 = \sigma_B^2 + \sigma_S^2$ and $\rho = \dfrac{\sigma_B^2}{\sigma_B^2 + \sigma_S^2}$.

Typically, correlation between observations is a function of their distance in time: adjacent observations tend to be more highly correlated than observations farther apart in time. Several models may adequately describe such correlation. The simplest is **the first-order autoregressive**, or **AR(1)**, model. For the AR(1),

$$\Sigma = \sigma^2 \begin{bmatrix} 1 & \rho & \rho^2 & . & \rho^{K-1} \\ & 1 & \rho & . & \rho^{K-2} \\ & & 1 & . & . \\ & & & . & \rho \\ & & & & 1 \end{bmatrix}$$

The AR(1) model assumes that $e_{ijk} = \rho e_{ij,k-1} + s_{ijk}$, where $s_{ijk} \sim N(0, \sigma_S^2)$. It follows that $\sigma^2 = \dfrac{\sigma_S^2}{1 - \rho^2}$.

This helps explain why independent errors models tend to underestimate within-subjects variance when correlation among the errors is non-negligible.

For AR(1), correlation between adjacent within-subjects errors is ρ, regardless of whether the pair of observations is the 1st and 2nd, 2nd and 3rd, or (K–1)st and Kth. With the unstructured model, each pair has its own correlation. The correlation is ρ^2 for any pair of errors 2 units apart, for example, the 1st and 3rd. In general, errors d units apart have correlation ρ^d. Note that the AR(1) model requires estimates of just two parameters, σ^2 and ρ, whereas unstructured models require estimating $K + \dfrac{K(K-1)}{2}$ parameters.

The **Toeplitz** model is similar to the AR(1) model in the sense that pairs of within-subjects errors separated by a common distance share the same correlation. Errors d units apart have correlation ρ_d. However, there is no known function relating ρ_d to d. Thus, for the Toeplitz model,

$$\Sigma = \sigma^2 \begin{bmatrix} 1 & \rho_1 & \rho_2 & . & \rho_{K-1} \\ & 1 & \rho_1 & . & \rho_{K-2} \\ & & 1 & . & . \\ & & & . & \rho_1 \\ & & & & 1 \end{bmatrix}$$

The Toeplitz model is less restrictive than AR(1), but it requires estimating K parameters (σ^2, $\rho_1, ..., \rho_{K-1}$) instead of just two.

The AR(1) and Toeplitz models make sense when observations are equally spaced and the correlation structure does not change appreciably over time. A more general model that preserves the main features of these models, but allows for unequal spacing and change over time is the **first-order ante dependence** model, or **ANTE(1)**. The model structure is

$$
\Sigma = \begin{bmatrix}
\sigma_1^2 & \sigma_1\sigma_2\rho_1 & \sigma_1\sigma_3\rho_1\rho_2 & \cdot & \sigma_1\sigma_K\rho_1\rho_2\cdots\rho_{K-1} \\
 & \sigma_2^2 & \sigma_2\sigma_3\rho_2 & \cdot & \sigma_2\sigma_K\rho_2\rho_3\cdots\rho_{K-1} \\
 & & \sigma_3^2 & \cdot & \cdot \\
 & & & \cdot & \sigma_{K-1}\sigma_K\rho_{K-1} \\
 & & & & \sigma_K^2
\end{bmatrix}
$$

You can see that the ANTE(1) model permits the variance among observations to change over time. Correlation between pairs of observations is the product of the correlations between adjacent times between observations, so that correlation may change over time. The ANTE(1) model requires estimating $2K-1$ parameters.

There are several other structures, including the first-order autoregressive model and Toeplitz models that allow for unequal variances over time, ARH(1) and TOEPH, respectively. See SAS OnlineDoc, Version 8 for a complete listing of the covariance structures available in PROC MIXED.

Section 8.4.2 presents methods for selecting an appropriate covariance model—that is, a model that adequately accounts for within-subjects correlation but does not require estimating an excessive number of covariance parameters. As with any model selection activity, evaluating covariance structure should not be a purely statistical exercise. You should first rule out covariance structures that clearly make no sense in the context of a given data set. For example, AR(1) or TOEP models are generally inappropriate if the times of observation are not equally spaced either chronologically or in terms of some meaningful criterion such as biological stage of development.

8.4.2 Selecting an Appropriate Covariance Model

To draw accurate conclusions from repeated-measures data, you must use an appropriate model of within-subjects correlation. If you ignore important correlation by using a model that is too *simple*, you risk increasing Type I error rate and underestimating standard errors. If the model is too *complex* you sacrifice power and efficiency. Guerin and Stroup (2000) documented the effects of various covariance modeling decisions using PROC MIXED for repeated-measures data. Their work supports the idea that repeated-measures analysis is robust as long as the covariance model used is approximately correct, but inference is severely compromised by blatantly poor choice of a covariance model. This is the philosophy underlying this section.

This section illustrates two types of tools you can use with PROC MIXED to help you select a covariance model. First are graphical tools to help visualize patterns of correlation between observations at different times. Second are **information criteria** that measure the relative fit of competing covariance models. As noted at the end of the last section, these methods work best when you first rule out covariance structures that are obviously inconsistent with the characteristics of the data you are analyzing.

You can visualize the correlation structure by plotting changes in covariance and correlation among residuals on the same subject at different times over distance between times of observation. Distance between times is called **lag**. Although you can use either PROC GLM or PROC MIXED to create the data sets needed for these plots, PROC MIXED is easier. To see why, consider PROC GLM first. With PROC GLM, you can use using the following SAS statements to obtain residuals from the multivariate form of the data given in Output 8.1:

```
proc glm data=fev1mult;
 class drug;
 model fev11h fev12h fev13h fev14h fev15h fev16h
       fev17h fev18h=drug;
  output out=resid residual=e1-e8;
```

This program outputs the residuals from the means model. The variable E1 denotes residuals at time 1, E2 at time 2, and so forth. You can use the output data set RESID to construct scatter plots among the residuals for all pairs of times, as shown by Littell et al. (2000). Once you compute the residuals, you can use PROC CORR to compute correlation and the sum of squares and cross-products among the residuals for all pairs of times. Although PROC CORR has an option to compute covariance, for repeated measures it is unable to determine the correct denominator degrees of freedom, so the covariance is computed incorrectly.

You can obtain correlation and covariance among residuals more simply and in more readily usable form for plotting with the following PROC MIXED statements:

```
proc mixed data=fev1uni;
   class drug patient hour;
   model fev1=drug|hour;
   repeated /type=un subject=patient(drug) sscp rcorr;
   ods output covparms=cov;
   ods output rcorr=corr;
```

Note that PROC MIXED uses the univariate form of the FEV1 data. The MODEL follows from the fixed-effects of model (8.2) given in Section 8.2. The REPEATED statement determines the form of the covariance among the e_{ijk}. TYPE=UN specifies an unstructured correlation model, as described in Section 8.4.1. SUBJECT=PATIENT(DRUG) specifies that errors are correlated *within* each PATIENT(DRUG)—that is, observations on different PATIENT(DRUG) levels are independent, but observations on the same PATIENT(DRUG) are not. In mixed-model covariance matrix terms, TYPE= specifies the form of Σ and SUBJECT= specifies the block in the block diagonal structure of R, that is, R= $I_{n\bullet} \otimes \Sigma$. The SSCP option causes the $\sigma_{kk'}=\mathrm{cov}(e_{ijk}, e_{ijk'})$ to be computed directly from the corrected sums of squares and cross products matrix rather than the default REML procedure. The RCORR option causes the correlations to be computed as well as the covariance matrix. The two ODS statements create new SAS data sets containing the covariances and correlations, respectively.

Use the following SAS statements to create the first plot for visualizing an appropriate covariance model:

```
data times;
 do time1=1 to 8;
  do time2=1 to time1;
   dist=time1-time2;
   output;
  end;
 end;

data covplot;
 merge times cov;
proc print;

axis1 value=(font=swiss2 h=2) label=(angle=90 f=swiss
   h=2    'Covariance of Between Subj Effects');
axis2 value=(font=swiss h=2 )label=(f=swiss h=2 'Distance');
legend1 value=(font=swiss h=2  ) label=(f=swiss h=2 'From Time');
```

```
symbol1 color=black interpol=join
          line=1 value=square;
symbol2 color=black interpol=join
        line=2 value=circle;
symbol3 color=black interpol=join
          line=20 value=triangle;
symbol4 color=black interpol=join
          line=3 value=plus;
symbol5 color=black interpol=join
        line=4 value=star;
symbol6 color=black interpol=join
          line=5 value=dot;
symbol7 color=black interpol=join
        line=6 value=_;
symbol8 color=black interpol=join
          line=10 value==;

proc gplot data=covplot;
  plot estimate*dist=time2/vaxis=axis1 haxis=axis2
    legend=legend1;
```

First, you create a data set called TIMES containing the pairs of observation times and the distance between them. You then merge the data set TIMES with the covariance data set COV created by the first ODS statement in the PROC MIXED program given above. The resulting data set COVPARMS appears in Output 8.12. The GPLOT procedure with its associated AXIS, LEGEND, and SYMBOL definitions creates a plot of covariance between pairs of repeated measures by distance, shown in Figure 8.2.

Output 8.12
Data Set
Containing
Pairs of Times
and Estimates

Obs	time1	time2	dist	CovParm	Subject	Estimate	adjcov	adjcorr
1	1	1	0	UN(1,1)	PATIENT(DRUG)	0.4541	0.08893	0.24350
2	2	1	1	UN(2,1)	PATIENT(DRUG)	0.4587	0.09352	0.25607
3	2	2	0	UN(2,2)	PATIENT(DRUG)	0.5163	0.15108	0.41370
4	3	1	2	UN(3,1)	PATIENT(DRUG)	0.4441	0.07894	0.21617
5	3	2	1	UN(3,2)	PATIENT(DRUG)	0.4808	0.11556	0.31644
6	3	3	0	UN(3,3)	PATIENT(DRUG)	0.4923	0.12711	0.34806
7	4	1	3	UN(4,1)	PATIENT(DRUG)	0.4154	0.05023	0.13754
8	4	2	2	UN(4,2)	PATIENT(DRUG)	0.4688	0.10358	0.28362
9	4	3	1	UN(4,3)	PATIENT(DRUG)	0.4687	0.10351	0.28344
10	4	4	0	UN(4,4)	PATIENT(DRUG)	0.4938	0.12858	0.35208
11	5	1	4	UN(5,1)	PATIENT(DRUG)	0.4349	0.06974	0.19098
12	5	2	3	UN(5,2)	PATIENT(DRUG)	0.4943	0.12912	0.35355
13	5	3	2	UN(5,3)	PATIENT(DRUG)	0.4843	0.11912	0.32619
14	5	4	1	UN(5,4)	PATIENT(DRUG)	0.4837	0.11851	0.32452
15	5	5	0	UN(5,5)	PATIENT(DRUG)	0.5779	0.21273	0.58249
16	6	1	5	UN(6,1)	PATIENT(DRUG)	0.3934	0.02816	0.07711
17	6	2	4	UN(6,2)	PATIENT(DRUG)	0.4254	0.06024	0.16496
18	6	3	3	UN(6,3)	PATIENT(DRUG)	0.4263	0.06109	0.16728
19	6	4	2	UN(6,4)	PATIENT(DRUG)	0.4179	0.05265	0.14418
20	6	5	1	UN(6,5)	PATIENT(DRUG)	0.4945	0.12927	0.35397
21	6	6	0	UN(6,6)	PATIENT(DRUG)	0.4906	0.12542	0.34343
22	7	1	6	UN(7,1)	PATIENT(DRUG)	0.3562	-0.00901	-0.02467
23	7	2	5	UN(7,2)	PATIENT(DRUG)	0.3992	0.03398	0.09305
24	7	3	4	UN(7,3)	PATIENT(DRUG)	0.4021	0.03690	0.10105
25	7	4	3	UN(7,4)	PATIENT(DRUG)	0.4023	0.03714	0.10171
26	7	5	2	UN(7,5)	PATIENT(DRUG)	0.4643	0.09909	0.27133
27	7	6	1	UN(7,6)	PATIENT(DRUG)	0.4454	0.08015	0.21948
28	7	7	0	UN(7,7)	PATIENT(DRUG)	0.4994	0.13422	0.36753
29	8	1	7	UN(8,1)	PATIENT(DRUG)	0.3840	0.01878	0.05143
30	8	2	6	UN(8,2)	PATIENT(DRUG)	0.4257	0.06046	0.16557
31	8	3	5	UN(8,3)	PATIENT(DRUG)	0.4256	0.06044	0.16549
32	8	4	4	UN(8,4)	PATIENT(DRUG)	0.4251	0.05989	0.16400
33	8	5	3	UN(8,5)	PATIENT(DRUG)	0.4950	0.12984	0.35553
34	8	6	2	UN(8,6)	PATIENT(DRUG)	0.4632	0.09799	0.26832
35	8	7	1	UN(8,7)	PATIENT(DRUG)	0.4496	0.08443	0.23119
36	8	8	0	UN(8,8)	PATIENT(DRUG)	0.5031	0.13791	0.37763

You can see that data set COVPLOT contains the time pairs and distance between them created in the data set TIMES and the covariance information created in the ODS step of PROC MIXED. Notice that because of the way the ODS statement constructs the data set COV, the variable TIME2 is actually the observation in the pair that is taken first. In the plot shown in Figure 8.2, notice that there are seven profiles, each corresponding to the time of the first observation in the pair. Since the variable TIME2 is the first observation in the pair, this explains why you use TIME2 in the PLOT statement of PROC GPLOT.

Figure 8.2 *Plot of Covariance as a Function of Distance in Time between Pairs of Observations*

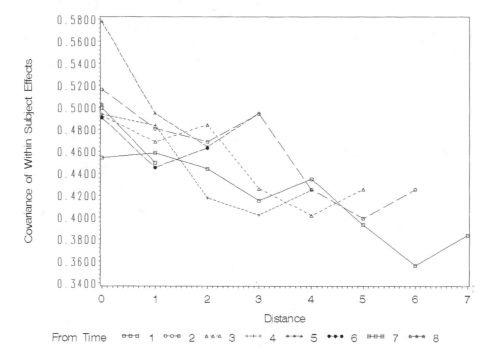

The values plotted at distance=0 are the variances among the observations at each of the eight times. These range from roughly 0.45 to just less than 0.60. This, and the fact that there is no trend of increasing or decreasing variance with time of observation, suggests that a covariance model with constant variance over time is probably adequate. You will see how to confirm this more formally when we discuss **model-fitting criteria** below.

Figure 8.2 shows that for the FEV1 data, as the distance between pairs of observations increases, covariance tends to decrease. Also, the pattern of decreasing covariance with distance is roughly the same for all reference times, although seeing this takes a bit of practice. Start with the profile labeled "From Time 1," which gives the HOUR of the first observation of a given pair of repeated measures. The square symbol on the plot tracks the covariance between pairs of repeated measurements whose first observation occurs at HOUR=1. The position of this plot at distance 0 plots the sample variance of observations taken at hour 1. The plot at distance 1 gives the covariance between hours 1 and 2, distance 2 plots covariance for hours 1 and 3, and so forth. Following the profile across the distances shows how the covariance decreases as distance increases for pairs of observations whose first time of observation is HOUR=1. You can see that there is a general pattern of decrease from roughly 0.50 at distance 1 to a little less than 0.40. If you follow the plots labeled "From Time" 2, 3, and so forth, the overall pattern is very similar. The covariance among adjacent observations is consistently between 0.45 and 0.50, with the HOUR of the first element of the pair making little difference. The covariance at distance 2 is a bit lower, averaging around 0.45, and does not appear to depend on the hour of the first element of the pair.

You can draw two main conclusions from this plot:

1. AR(1) or Toeplitz covariance models are probably most appropriate for the data. They allow covariance to decrease with increasing distance and they assume that correlation is strictly a function of distance and does not depend on the time of the first observation in the pair. Compound symmetry appears to be inappropriate because the covariance does not remain constant for all distances; unstructured, ante dependence, or heterogeneous variance models also are probably inappropriate because there is no visual evidence of changes in variance or covariance-distance relationships over time.

2. The between-subjects variance component, σ_B^2 is nonzero. In fact, it is roughly between 0.35 and 0.40. Note that all of the plots of covariance decline for distances 1, 2, and 3, and then seem to flatten out. This is exactly what should happen if the between-subjects variance component is approximately equal to the plotted covariance at the larger distances and there is AR(1) correlation among the observations within each subject. At the larger distances the AR(1) correlation ρ^d, should approach zero. The nonzero covariance plotted for larger distances results from the intraclass correlation $\dfrac{\sigma_B^2}{\sigma_B^2 + \sigma_S^2}$.

Figure 8.3 shows idealized plots for three covariance structures: compound symmetry, AR(1) only, and AR(1)+random between-subjects effects.

Figure 8.3 *Idealized Plot of Covariance by Distance for Three Covariance Models*

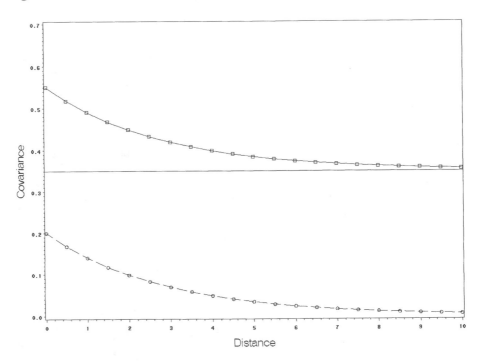

The flat line across all distances at an approximate covariance of 0.35 shows the plot you would expect to see with compound symmetry. The plot with a covariance decreasing from approximately 0.20 to 0 as distance increases is what you would expect to see with pure AR(1) covariance. The plot that decreases from approximately 0.55 to 0.35 is the expected plot from the AR(1)+random between-subjects effects model. The numbers on the covariance axis follow from the approximate magnitudes of the observed covariances in Output 8.12 and Figure 8.2. You can see that the plots of observed covariances in Figure 8.2 correspond most closely to the AR(1)+random between-subjects effects model.

You can remove the between-subjects component and plot the remaining within-subjects covariance. Use the SAS statements

```
data covplot;
  merge times cov;
  adjcov=estimate-0.3652;
  adjcorr=adjcov/0.3652;
```

Note that these statements simply add the ADJCOV and ADJCORR formula to the statements used previously to create the COVPLOT data set. You use the minimum covariance among pairs of repeated measures—specifically $\hat{\sigma}_{71} = 0.3652$, UN(7,1) in Output 8.12. This is a reasonable lower bound on σ_B^2 from the covariance plot. ADJCOV simply removes σ_B^2 from the covariance. ADJCORR assumes that an upper bound of the correlation between a pair of repeated measures, $\dfrac{\mathrm{cov}(e_{ijk}, e_{ijk'})}{\sqrt{\mathrm{var}(e_{ijk})\,\mathrm{var}(e_{ijk'})}} \simeq \dfrac{UN(k,k')}{\hat{\sigma}_B^2}$. ADJCOV and ADJCORR appear in Output 8.12. You can revise the PROC GPLOT statements used to create Figure 8.2 to plot these values by distance. Figure 8.4 shows the adjusted correlation. The adjusted covariance is not shown.

Figure 8.4 *Plot of Approximate Adjusted Correlation by Distance*

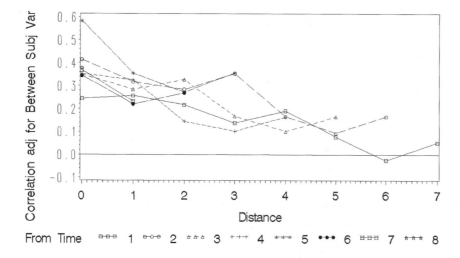

These plots reveal essentially the same pattern as in Figure 8.2. The values at distance 0 for ADJCORR in Output 8.15 are meaningless. Starting with distance 1, the correlation is between 0.44 and 0.50 and decreases to essentially zero by distance 6. Again, these values seem consistent with either the AR(1) or Toeplitz model. Because AR(1) requires fewer parameters, it is the more parsimonious model and thus it may be the better choice.

You can also plot the unadjusted correlation by distance using the data set CORR created by the second ODS statement in the PROC MIXED program at the beginning of this section. The form of the data set is not as convenient, but the following statements allow you to create the plot, which appears in Figure 8.5. The AXIS2, LEGEND, and SYMBOL statements for the PROC GPLOT statements are identical to those used for Figure 8.2. Only the AXIS1 LABEL is changed.

```
data corrlplot;
 set corr;
  if row=1 then do;
    dist=1; corr=col2; output;
    dist=2; corr=col3; output;
    dist=3; corr=col4; output;
    dist=4; corr=col5; output;
    dist=5; corr=col6; output;
    dist=6; corr=col7; output;
    dist=7; corr=col8; output;
  end;
  if row=2 then do;
    dist=1; corr=col3; output;
    dist=2; corr=col4; output;
    dist=3; corr=col5; output;
    dist=4; corr=col6; output;
    dist=5; corr=col7; output;
    dist=6; corr=col8; output;
  end;
 if row=3 then do;
    dist=1; corr=col4; output;
    dist=2; corr=col5; output;
    dist=3; corr=col6; output;
    dist=4; corr=col7; output;
    dist=5; corr=col8; output;
  end;
  if row=4 then do;
    dist=1; corr=col5; output;
    dist=2; corr=col6; output;
    dist=3; corr=col7; output;
    dist=4; corr=col8; output;
  end;
  if row=5 then do;
    dist=1; corr=col6; output;
    dist=2; corr=col7; output;
    dist=3; corr=col8; output;
  end;
  if row=6 then do; output;
    dist=1; corr=col7; output;
    dist=2; corr=col8; output;
  end;
  if row=7 then do;
    dist=1; corr=col8; output;
  end;


axis1 value=(font=swiss2 h=2) label=(angle=90 f=swiss
    h=2 'Correlation between Pairs of Observations');

proc gplot data=corrlplot;
 plot corr*dist=row/vaxis=axis1 haxis=axis2
    legend=legend1;
```

You can see from Figure 8.5 that correlation decreases from roughly 0.90-0.95 at distance 1 to approximately 0.80 for a pair of observations whose distance exceeds 5 units in time. From Output 8.12, the adjusted variance after removing σ_B^2 was between 0.15 and 0.20. This provides a crude idea of what you can expect the within-subjects variance σ_S^2 to be. The intraclass correlation could then be very crudely approximated by $\dfrac{0.35}{0.35+0.15} = 0.7$. This is fairly close to the value approached by the correlation at larger distances in the plot in Figure 8.4. As with the covariance plot of Figure 8.2, when the correlation plot appears to asymptotically approach a value appreciably greater than 0 for the larger distances, this suggests a correlation process among the repeated measures over and above a substantial between-subjects error component. A flat correlation plot over distance would suggest compound symmetry. Here, the plot is not flat for the shorter distances, suggesting a within-subjects correlation model for which correlation is a decreasing function of distance. This is very similar to the conclusion you can draw from the covariance plots discussed earlier. One disadvantage of the correlation plot is that at distance 0, correlation always equals 1 regardless of the variance. This means you cannot use the correlation plot to diagnose possible heterogeneous variance over time, as you can with the covariance plots.

Figure 8.5 *Plot of Correlation between Repeated Measurements by Distance between Times of Observation*

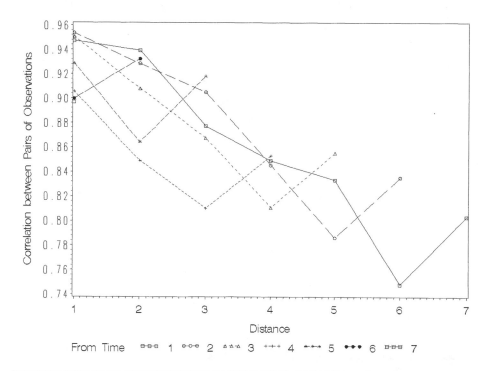

8.4.3 Reassessing the Covariance Structure with a Means Model Accounting for Baseline Measurement

Recall at the beginning of Section 8.4.1 we discussed the possible use of the baseline covariate BASEFEV1. You can refit the means model to include the covariate, output the resulting covariance and correlation matrices, and plot them using the same approach used for Figures 8.2, 8.4, and 8.5. The revised PROC MIXED statements for the baseline covariate model are

```
proc mixed data=fev1uni;
 class drug hour patient;
 model fev1=drug|hour basefev1;
 repeated / type=un sscp subject=patient(drug) rcorr;
 ods output covparms=cov;
 ods output rcorr=corr;
```

Figure 8.6 shows the plot of the covariance by distance.

Figure 8.6 *Plot of Covariance as a Function of Distance in Time between Pairs of Observations*

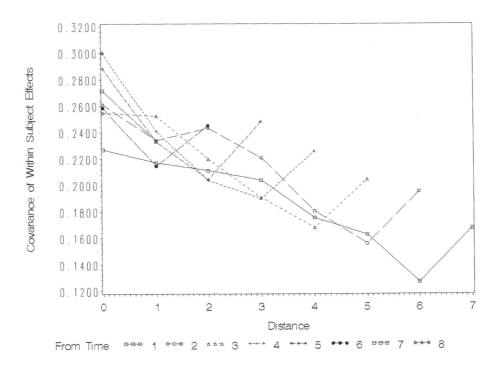

The pattern of Figure 8.6 is similar to Figure 8.2 except that using BASEFEV1 as a covariate substantially reduces the variance—from roughly 0.50 to less than 0.30. As distance increases, the covariance appears to reach an asymptote between 0.15 and 0.20. Again, this suggests a between-subjects variance between 0.15 and 0.20, and an additional within-subjects covariance model with covariance a decreasing function of distance, but independent of time.

8.4.4 Information Criteria to Compare Covariance Models

The default output from PROC MIXED includes values under the heading "Fit Statistics." These include −2 × the Residual (or REML) Log Likelihood (labeled "−2 Res Log Likelihood"), and three **information criteria**. In theory, the greater the residual log likelihood, the better the fit of the model. However, somewhat analogously to R^2 in multiple regression, you can always improve the log likelihood by adding parameters to the point of absurdity.

Information criteria attach penalties to the log likelihood. The information criterion equals the residual log likelihood *minus* some function of the number of parameters in the covariance model, or, alternatively, −2 × the residual log likelihood *plus* −2 × the function of the number of covariance model parameters. The latter form is used beginning with Release 8.2 of PROC MIXED. For example, the penalty for the compound symmetry model is a function of 2, because there are two covariance parameters, σ^2 and ρ. The penalty for an unstructured model is a function of $K + \dfrac{K(K-1)}{2}$. Hence, the residual log likelihood for an unstructured model is *always* greater than the residual log likelihood for compound symmetry, but the penalty is always greater as well. Unless the improvement in the residual log likelihood exceeds the size of the penalty, the simpler model yields the higher information criterion and is thus the preferred covariance model.

Two commonly used information criteria are Akaike's (1974) and Schwarz's (1978). A more recent information criterion, also included as a default starting with Release 8.2 of PROC MIXED, is a finite-population corrected Akaike criterion developed by Burnham and Anderson (1998). The Akaike Information Criterion is often referred to by the acronym AIC. The finite-population corrected AIC has the acronym AICC. Two commonly used acronyms for Schwarz's Bayesian Information Criterion are BIC and SBC—hereafter this text uses the latter. Refer to SAS OnlineDoc, Version 8 for a complete explanation of how these terms are computed.

The basic idea for repeated-measures analysis is that among the models for within-subjects covariance that are considered plausible in the context of a particular study (for example, biologically or physically reasonable), the model that *minimizes* the AIC, AICC, or SBC—in its $-2 \times$ (REML log likelihood penalty) form—is the preferred model. When AIC, AICC, or SBC are close, the simpler model is generally considered preferable in the interest of using a parsimonious model.

Keselman et al. (1998) compared AIC and SBC for their ability to select "the right" covariance model. Their study used Release 6.12 of SAS, which was determined to compute SBC incorrectly. This was corrected in Release 8.0. Guerin and Stroup (2000) compared the information criteria using Release 8.0 for their ability to select "the right" model *and* for the impact of choosing "the wrong" model on Type I error rate. They found that AIC tends to choose more complex models than SBC. They found that choosing a model that is too simple affects Type I error control more adversely than choosing a model that is too complex. When Type I error control is the highest priority, AIC is the model-fit criterion of choice. However, if loss of power is relatively more serious, SBC may be preferable. AICC was not available at the time of the Guerin and Stroup study; a reasonable inference from their study is that its performance is similar to AIC, but somewhat less likely to choose a more complex model. Thus, loss of power is less than with AIC, but still greater than with SBC.

You can compare candidate covariance models by running PROC MIXED with the same fixed-effects model, varying the RANDOM and REPEATED statements to obtain the AIC, AICC, and SBC for all candidate models. For the FEV1 data, the models that appear to warrant consideration based on the covariance plots are the AR(1) and the Toeplitz models, both in conjunction with a between-subjects random effect. For the AR(1) model, the needed SAS statements are

```
proc mixed;
  class drug hour patient;
  model fev1=drug|hour basefev1;
  random patient(drug);
  repeated / type=ar(1) subject=patient(drug);
```

The parameter estimates and fit statistics appear in Output 8.13.

Output 8.13
Covariance
Parameter
Estimates and
Fit Statistics for
FEV1 Data
Using the AR(1)
Model

```
                   Covariance Parameter Estimates
        Cov Parm              Subject           Estimate

        PATIENT(DRUG)                             0.1848
        AR(1)                PATIENT(DRUG)        0.5401
        Residual                                  0.08309

                          Fit Statistics

           -2 Res Log Likelihood         247.0
           AIC (smaller is better)       253.0
           AICC (smaller is better)      253.1
           BIC (smaller is better)       259.9
```

Note the close correspondence between the REML estimates of between-subjects variance (the PATIENT(DRUG) variance, 0.1848) and AR(1) correlation, $\hat{\rho}=0.5401$, and the rough approximations anticipated from the covariance plot in Figure 8.2. You can compare the estimates and fit statistics to the Toeplitz model by changing the TYPE= option. Replace the AR(1) REPEATED statement with

```
repeated / type=toep subject=patient(drug);
```

This yields the results in Output 8.14.

Output 8.14
Covariance
Parameter
Estimates and
Fit Statistics for
FEV1 Data
Using the
Toeplitz Model

```
                   Covariance Parameter Estimates

        Cov Parm              Subject           Estimate

        PATIENT(DRUG)                            6.38E-34
        TOEP(2)              PATIENT(DRUG)        0.2284
        TOEP(3)              PATIENT(DRUG)        0.2162
        TOEP(4)              PATIENT(DRUG)        0.2070
        TOEP(5)              PATIENT(DRUG)        0.1908
        TOEP(6)              PATIENT(DRUG)        0.1826
        TOEP(7)              PATIENT(DRUG)        0.1691
        TOEP(8)              PATIENT(DRUG)        0.1579
        Residual                                  0.2664

                          Fit Statistics

           -2 Res Log Likelihood         227.9
           AIC (smaller is better)       243.9
           AICC (smaller is better)      244.2
           BIC (smaller is better)       262.1
```

Note that the Toeplitz model adequately modeled the decreasing covariance with distance without a separate between-subjects variance component: In this case the PATIENT(DRUG) variance is effectively 0. The TOEP(d+1) parameters estimate correlation at distance d. For example, TOEP(2) estimates correlation between repeated measures d=1 unit apart in time. The AIC and AICC for the Toeplitz model are slightly better: the AIC is 243.9 vs. the AR(1) AIC of 253.0 and the AICC is 244.2 vs. 253.1 for the AR(1). However, the SBC is slightly worse: 262.1 vs. the AR(1) SBC of 259.9. You could use either model, but the AR(1) would generally be considered preferable because it is a simpler model.

For completeness you can also fit the compound symmetry and unstructured models.

Important note about compound symmetry and unstructured models: In the AR(1) and TOEP models above, you use both a RANDOM statement for the between-subjects error term, PATIENT(DRUG) and a REPEATED statement for the AR(1) or Toeplitz covariance among repeated measures within subjects. The latter measures covariance over and above that induced by between-subjects variation. For the AR(1) and Toeplitz models, these two sources of variation are distinct and clearly identifiable. However, this is not true for compound symmetry and unstructured covariance.

In Section 8.4.1, you saw that compound symmetry and the model with random between-subjects effect and independent errors are equivalent. Thus, the between-subjects variance component, σ_B^2, and compound symmetry covariance are not identifiable. This situation also holds, in more complex form, for unstructured covariance. Therefore, you should *not* use a RANDOM statement for the effect used as subject in the REPEATED statement. For example, the SAS statements

```
proc mixed;
  class drug hour patient;
  model fev1=drug|hour basefev1;
  random patient(drug);
  repeated / type=cs subject=patient(drug);
```

are *inappropriate* for the compound symmetry model. You should delete the RANDOM statement.

Note that you *should* attempt to use separate RANDOM and REPEATED statements for structures such as AR(1) and Toeplitz, particularly when covariance-by-distance plots such as Figures 8.2 and 8.6 show evidence of nonzero between subject variation. Guerin and Stroup (2000) showed that failure to model a separate between-subjects random effect can adversely affect inference on time and treatment × time effects.

The AIC, AICC, and SBC results are given below. Also computed are values for the AR(1) model deleting the RANDOM PATIENT(DRUG) statement.

Model	AIC	AICC	SBC
Compound Symmetry	351.3	351.3	355.8
AR(1) with random patient(drug)	253.0	253.1	259.9
AR(1) with no random patient(drug)	279.0	279.1	283.6
Toeplitz	243.9	244.2	262.1
Unstructured	220.3	225.5	302.2

You can see that the unstructured model actually has the best AIC and AICC. However, it is a distinctly poor choice according to the SBC. This illustrates the tendency of AIC, and to a lesser extent AICC, to "choose" more complex models. In this case, given the covariance plots, choosing the unstructured model on the basis of the AIC is probably "modeling overkill," unless there is some compelling medical or biological process that supports the unstructured model. Sections 8.4.5 and 8.4.6 show inference on the FEV1 data using the AR(1) + between-subjects PATIENT(DRUG) effect model.

8.4.5 PROC MIXED Analysis of FEV1 Data

Once you have selected the covariance model, you can proceed with the analysis of baseline covariate and the treatment effects just as you would in any other analysis of variance or, in this case, analysis of covariance. Start with the "Type 3 Tests of Fixed Effects" from the output you obtained when you fit the model with AR(1) covariance among the errors plus PATIENT(DRUG) between-subjects error. Output 8.15 shows the Type 3 tests.

Output 8.15
Type 3
Analysis-of-
Covariance
Tests for FEV1
Data

```
Type 3 Tests of Fixed Effects

                   Num     Den
     Effect         DF      DF    F Value    Pr > F

     DRUG            2      68       7.28     0.0014
     hour            7     483      17.10     <.0001
     DRUG*hour      14     483       3.94     <.0001
     BASEFEV1        1     483      75.93     <.0001
```

You can see that the two main results are

❑ there is very strong evidence of a relationship between the baseline covariate BASEFEV1 and the subsequent responses FEV1. The *p*-value is <0.0001.
❑ there is strong evidence of a DRUG*HOUR interaction ($p<0.0001$). That is, changes to the response variable FEV1 over time are not the same for all DRUG treatments.

As with any factorial structure, inference on the DRUG and HOUR main effects should not proceed until the DRUG*HOUR interaction is understood.

Important note about degrees of freedom, standard errors, and test statistics: Output 8.15 shows the default denominator degrees of freedom and *F*-values computed by PROC MIXED. The degree of freedom default is based on traditional analysis-of-variance assumptions, specifically an independent errors model. Denominator degrees of freedom are often substantially affected by more complex covariance structures, including those typical of repeated-measures analysis. Also, PROC MIXED computes so-called "naive" standard errors and test statistics—it uses estimated covariance parameters in formulas that assume these quantities are known. Kacker and Harville (1984) showed that using estimated covariance parameters in this way results in test statistics that are biased upward and standard errors that are biased downward, for all cases except independent errors models with balanced data. Kenward and Roger (1997) obtained a correction for standard errors and *F*-statistics and a generalized procedure to obtain degrees of freedom. The Kenward-Roger (KR) correction is applicable to most covariance structures available in PROC MIXED, including all of those used in repeated measures analysis. The KR correction was added as an option with PROC MIXED in SAS Version 8 and is highly recommended whenever MIXED is used for repeated measures. Guerin and Stroup (2000) compared Type I error rates for default versus KR-adjusted test statistics. Their results strongly supported Kenward and Roger's early work: Unless you use the adjustment, Type I error rates tend to be highly inflated, especially for more complex covariance structures.

The amended Type 3 statistics for fixed effects appear in Output 8.16. You obtain this output by using the option DDFM=KR in the MODEL statement. The full set of SAS statements is

```
proc mixed data=fev1uni;
  class drug hour patient;
  model fev1=drug|hour basefev1 / ddfm=kr;
  random patient(drug);
  repeated / type=ar(1) subject=patient(drug);
```

Output 8.16
Type 3
Analysis-of-
Covariance
Tests for FEV1
Data

```
Type 3 Tests of Fixed Effects

                    Num     Den
    Effect          DF      DF      F Value    Pr > F

    DRUG             2      68.8      7.30      0.0013
    hour             7      395      16.99      <.0001
    DRUG*hour       14      424       3.90      <.0001
    BASEFEV1         1       68      75.83      <.0001
```

You can see that the test statistics for HOUR and DRUG*HOUR, the statistics that should be most strongly affected, are somewhat lower. The degrees of freedom, especially for BASEFEV1, are also affected. The default showed BASEFEV1 with the within-subjects error denominator degrees of freedom, which makes no sense given that there is only one BASEFEV1 measurement per subject. The adjustment gives BASEFEV1 the between-subjects error degrees of freedom, which is more reasonable. The denominator degrees of freedom for HOUR and DRUG*HOUR also receive a downward adjustment to account for the fact that an estimated covariance matrix is used. In this case, the basic conclusions of a significant BASEFEV1 covariance effect and a significant DRUG*HOUR interaction are not changed. In many cases, however, conclusions with the KR adjustment will differ from those obtained using the default.

The next step is to explain the DRUG*HOUR interaction. To help visualize the interaction, you can plot the DRUG*HOUR LSMEANS over time for each treatment. First, output the least-squares means computed by PROC MIXED using ODS output. This procedure was introduced in the discussion of factorial ANOVA in Chapter 3. Add the following statements to the above SAS program:

```
lsmeans drug*hour;
ods output lsmeans=lsm;
```

This creates a new data set, called LSM, containing the least-squares means. You can then use the following statements to plot the means:

```
proc gplot data=lsm;
  plot estimate*hour=trt;
```

Figure 8.7 shows the GPLOT results.

Figure 8.7 *Plot of LS Means Adjusted for Baseline Covariate by Hour for Each Drug*

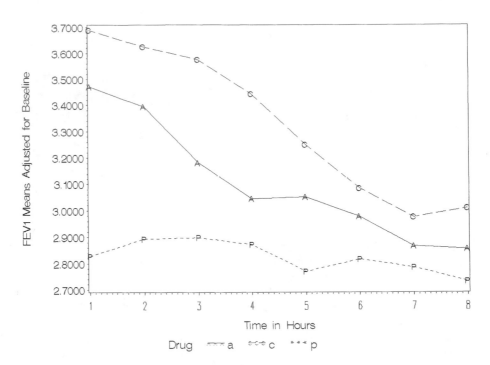

Inspecting Figure 8.7, you can see that after adjustment for the baseline covariate, the mean responses for the two drugs, A and C, are much greater than those for the placebo, P, for the first hours of measurement. This suggests that the two drugs do improve respiratory performance relative to the placebo initially after the patient uses them. Also, for the two drugs the responses generally decrease over time, whereas for the placebo, P, there is little change. The change is approximately linear over time with a negative slope whose magnitude appears to be greater for DRUG C than for DRUG A. This suggests fitting a linear regression over HOUR, possibly with a quadratic term to account for decreasing slope as HOUR increases, and testing its interaction with HOUR. Alternatively, depending on the objectives, you might want to test DRUG differences at specific hours during the experiment. Inference along these lines continues in the next section.

8.4.6 Inference on the Treatment and Time Effects of FEV1 Data Using PROC MIXED

As indicated by Output 8.16, the main task of inference on the treatment (DRUG) and time (HOUR) effects is to explain the DRUG*HOUR interaction in a manner consistent with the objectives of the research. The section presents two approaches, one based on comparisons among the DRUG*HOUR treatment combination means, and the other based on comparing regression of the response variable, FEV1, on HOUR for the three DRUG treatments. The mean comparison strategy is very similar to methods for factorial experiments presented in Chapter 3. The regression analysis is similar to procedures for comparing slopes presented in Chapter 7. Each approach is appropriate for some data sets and not for others. Which approach you use for a given data set depends on the objective. The purpose of this section is simply to show you how to implement and interpret each approach.

8.4.6.1 Comparisons of DRUG*HOUR Means

Interest usually focuses on three main *types* of tests.

❑ Estimates or tests of simple effects, either among DRUGs holding HOUR constant or vice versa.

❑ SLICE options to test the effects of DRUG at a given HOUR or HOUR for a given DRUG. The former is more common because the latter is usually addressed more directly by regression methods present later in this section.

❑ Contrasts defined on specific aspects of the DRUG*HOUR interaction.

The following SAS statements demonstrate examples of each of these types of tests. This is not an exhaustive analysis. You can think of different tests that might be of interest. Certainly, different data sets will call for other tests. However, the methods used to construct these examples can be adapted for most tests of potential interest.

```
lsmeans drug*hour/ diff slice=hour;
contrast 'hr=1 vs hr=8 x a vs c'
         drug*hour 1 0 0 0 0 0 0 -1
              -1 0 0 0 0 0 0 1   0;
contrast 'hr=1 vs hr=8 x p v trt'
         drug*hour 1 0 0 0 0 0 0 -1
              1 0 0 0 0 0 0 -1  -2 0 0 0 0 0 0 2;
```

You add these statements to the PROC MIXED statements given above in this section. The SLICE =HOUR option produces tests of the DRUG effect at each HOUR. Often, researchers who do repeated-measures experiments want to measure the change between the first and last time of measurement and want to know if this change is the same for all treatments. The CONTRAST statements perform such a comparison. The contrast HR=1 vs HR=8 × A vs C compares the change from the first to last HOUR for the two DRUGs, A and C. The contrast HR=1 vs HR=8 × P v TRT compares the first to last HOUR change in the PLACEBO to the average of the two DRUGs. The SLICE and CONTRAST results are shown in Output 8.17.

Output 8.17 PROC MIXED SLICE and CONTRAST Results for FEV1 Data

Contrasts

Label	Num DF	Den DF	F Value	Pr > F
hr=1 vs hr=8 x a vs c	1	160	0.31	0.5773
hr=1 vs hr=8 x p v trt	1	160	29.78	<.0001

Tests of Effect Slices

Effect	hour	Num DF	Den DF	F Value	Pr > F
DRUG*hour	1	2	107	18.03	<.0001
DRUG*hour	2	2	107	12.57	<.0001
DRUG*hour	3	2	107	10.38	<.0001
DRUG*hour	4	2	107	7.67	0.0008
DRUG*hour	5	2	107	5.21	0.0070
DRUG*hour	6	2	107	1.64	0.1998
DRUG*hour	7	2	107	0.82	0.4427
DRUG*hour	8	2	107	1.72	0.1839

You can see that the difference between HOUR 1 and HOUR 8 is significantly different for the PLACEBO than it is for the two DRUG treatments ($p<0.0001$), but there is no evidence that the change from beginning to end is different for the two DRUG treatments ($p=0.5773$). The SLICE results suggest significant differences among DRUG treatments for HOURs 1 through 5, but no statistically significant differences among DRUG for HOURs 6, 7, and 8. Note that these are two degree-of-freedom comparisons. You could partition each SLICE into two single degree-of-freedom comparisons, for example, A vs C and PLACEBO vs TRT, using CONTRAST statements. For HOUR=1, the contrasts would be

```
contrast 'A vs C at HOUR=1' drug 1 -1 0
    drug*hour 1 0 0 0 0 0 0 0  -1 0 ;
contrast 'Placebo vs trt at HOUR=1' drug 1 1 -2
    drug*hour 1 0 0 0 0 0 0 0  1 0 0 0 0 0 0 0  -2 0;
```

These contrast results are not shown.

The DIFF option shown in the LSMEANS statement above produces 24*23/2=276 treatment differences, an unmanageably large output. As shown in Chapter 3, you can reduce the size of the output to only the simple effects of interest using ODS statements. The following is an example. Add the following statements to the PROC MIXED program:

```
ods listing exclude lsmeans;
ods listing exclude diffs;
ods output diffs=dhdiff;
```

The ODS LISTING EXCLUDE statements prevent the LSMEANS and DIFFS to be printed when you run PROC MIXED. The SLICE results will be printed unless you add another ODS EXCLUDE statement specifically for them. Add the following statements to select only the simple effects and eliminate all other DIFFS:

```
data smpleff;
 set dhdiff;
 if drug=_drug or hour=_hour;

proc print data=smpleff;
```

Output 8.18 shows selected results.

Output 8.18 Simple Effect Differences from the PROC MIXED Analysis of FEVI Data

Obs	Effect	DRUG	hour	_DRUG	_hour	Estimate	StdErr	DF	tValue	Probt
1	DRUG*hour	a	1	a	2	0.07667	0.05678	436	1.35	0.1777
2	DRUG*hour	a	1	a	3	0.2896	0.07049	456	4.11	<.0001
3	DRUG*hour	a	1	a	4	0.4271	0.07684	334	5.56	<.0001
4	DRUG*hour	a	1	a	5	0.4200	0.07999	248	5.25	<.0001
5	DRUG*hour	a	1	a	6	0.4942	0.08159	200	6.06	<.0001
6	DRUG*hour	a	1	a	7	0.6050	0.08241	174	7.34	<.0001
7	DRUG*hour	a	1	a	8	0.6154	0.08282	160	7.43	<.0001
8	DRUG*hour	a	1	c	1	-0.2182	0.1494	107	-1.46	0.1471
9	DRUG*hour	a	1	p	1	0.6447	0.1495	107	4.31	<.0001
52	DRUG*hour	c	1	p	1	0.8629	0.1494	107	5.77	<.0001

Output 8.18 shows the differences between the FEV1 mean at HOUR 1 and each of the subsequent HOURs for DRUG A. The full output also contains comparisons between the mean for each pair of hours holding each DRUG constant. Output 8.18 also shows comparisons between each pair of DRUG treatments holding HOUR equal to 1. The full output contains similar comparisons for HOURs 2 through 8. You can see that with the AR(1) covariance structure, the standard error of a difference increases, and the degrees of freedom decrease, as the distance

between HOUR increases. For DRUG A, the change from HOUR 1 to HOUR 2 is not significant, but change from HOUR 1 to each subsequent HOUR is. Also, at HOUR 1, DRUG A and C are not significantly different, but both DRUG treatments are significantly different from the PLACEBO.

8.4.6.2 Comparisons Using Regression

From the plot in Figure 8.7, there appears to be a linear regression of FEV1 on HOUR with a negative slope. The slope appears to be different for each DRUG treatment. The slopes appear to be less negative as HOUR increases, indicating a possible quadratic trend. You can test these impressions. One approach is to write orthogonal polynomial contrasts for linear effects, quadratic effects, and so forth, and more importantly, their interactions with DRUG contrasts of interest. As you saw in Chapter 7, you can do this far less tediously using analysis-of-covariance methods. Create a new data set with variables H and H2 defined as follows:

```
data fev1regr;
 set fev1uni;
  h=hour;
  h2=hour*hour;
```

The following SAS statements allow you to see if

❑   there is a significant quadratic component to the regression.

❑   the slopes for the linear and, if applicable, quadratic regressions, are the same for DRUG treatments.

❑   there is any lack of fit resulting from regression of a higher order than quadratic.

```
proc mixed;
 class drug hour patient;
 model fev1=basefev1 drug h h2 hour
     drug*h drug*h2 drug*hour/htype=1;
 random patient(drug);
 repeated / type=ar(1) subject=patient(drug);
```

You can see that this program is similar to analysis of covariance for qualitative × quantitative factorial treatment structures presented in Chapter 7. Here, there is an additional covariate, BASEFEV1. When you attempt to run this program with the REPEATED covariance error structure, you often get the following warning:

```
The Mixed Procedure
```

```
WARNING: Stopped because of infinite likelihood.
```

The REML algorithm cannot simultaneously compute the analysis-of-covariance MODEL and the correlated error structure in the REPEATED statement. You can get around this problem by using the covariance parameters already estimated from the mean comparison analysis and preventing MIXED from attempting to obtain new REML estimates. Use the following statements:

```
proc mixed noprofile;
 class drug hour patient;
 model fev1=basefev1 drug h h2 hour drug*h drug*h2
     drug*hour/htype=1;
 random patient(drug);
 repeated / type=ar(1) subject=patient(drug);
 parms (0.1848)(0.08309)  (0.5401) / noiter;
```

The PARMS statement gives the estimates of σ_R^2, σ^2, and ρ, obtained previously. You can determine the order the variance and covariance components appear in the PARMS statement by looking at the order in which they are printed in the standard MIXED analysis, for example, in Output 8.13. The NOPROFILE and NOITER commands, *used together*, cause MIXED to use these estimates literally as is. **If you use NOITER without the NOPROFILE option in the PROC MIXED statement, this procedure will not work.** Note also that you cannot use the DDFM=KR option with this particular analysis. Output 8.19 shows the results.

Output 8.19
Regression Analysis to Measure Lack of Fit, Order of Polynomial Regression, and Equality of Slopes for the Mixed-Model Analysis of FEV1 Data with AR(1) Correlated Error Structure

```
              Covariance Parameter Estimates

   Cov Parm              Subject            Estimate

   PATIENT(DRUG)                             0.1848
   Variance            PATIENT(DRUG)         0.08309
   AR(1)               PATIENT(DRUG)         0.5401

              Type 1 Tests of Fixed Effects

                     Num     Den
   Effect             DF      DF     F Value    Pr > F

   BASEFEV1            1     483      76.51     <.0001
   DRUG               2      68       7.26     0.0014
   h                  1     483     118.46     <.0001
   h2                 1     483       0.43     0.5142
   hour               5     483       1.27     0.2739
   h*DRUG             2     483      17.78     <.0001
   h2*DRUG            2     483       2.20     0.1115
   DRUG*hour          10    483       1.54     0.1227
```

Notice that the order of the "Covariance Parameter Estimates" on this analysis is different from the previous results in Output 8.13. The NOPROFILE option causes the reordering. The BASEFEV1 denominator degrees of freedom are clearly wrong. You could correct this by using the DDF option with the MODEL statement to manually set degrees of freedom. This is not done here because inference with this analysis is not concerned with BASEFEV1. These are the important elements:

❑ Neither DRUG*HOUR nor HOUR is significant, indicating no lack of fit from the quadratic model, that is, no 3rd or higher order regression effects.

❑ Neither H2*DRUG nor H2 is statistically significant. The H2*DRUG term tests the equality of quadratic regressions, if any, for the DRUG treatments. Since there is no evidence of unequal slope, we test the quadratic main effect, H2, which is also not significant. Taken together, these terms indicate that there is no evidence of quadratic regression effects.

❑ H*DRUG is statistically significant. There is evidence of unequal linear regressions of FEV1 on HOUR for the three DRUG treatments.

You can then fit separate regressions of FEV1 on HOUR for each DRUG using the following statements:

```
proc mixed data=fev1regr;
  class drug patient;
  model fev1=basefev1 drug h(drug) /noint solution
      ddfm=kr htype=1;
  random patient(drug);
  repeated / type=ar(1) subject=patient(drug);
```

This program is similar to examples shown in Chapter 7. Output 8.20 shows relevant results.

Output 8.20
Regression Analysis to Measure Lack of Fit, Order of Polynomial Regression, and Equality of Slopes for the Mixed-Model Analysis of FEV1 Data with AR(1) Correlated Error Structure

```
                        Solution for Fixed Effects

                                  Standard
   Effect      DRUG     Estimate    Error      DF    t Value    Pr > |t|

   BASEFEV1              0.8947     0.1026      68      8.72     <.0001
   DRUG        a         1.1487     0.2940     72.5     3.91     0.0002
   DRUG        c         1.4407     0.2917     72.5     4.94     <.0001
   DRUG        p         0.5146     0.2909     72.6     1.77     0.0811
   h(DRUG)     a        -0.08887    0.01164    113     -7.63     <.0001
   h(DRUG)     c        -0.1057     0.01164    113     -9.08     <.0001
   h(DRUG)     p        -0.01583    0.01164    113     -1.36     0.1766
```

You could add CONTRAST statements to the above program to see if there is evidence that regressions for DRUG A and C are different. You could also test PLACEBO versus the treated drugs but in context this difference has already been established. The parameter estimates shown above give you the linear regression equations adjusted for BASEFEV1. For example, for DRUG A, the regression is $1.15 - 0.089*HOUR$. Consistent with the plot in Output 8.22, the slopes for DRUG A and C are significantly different from 0 ($p<0.0001$), whereas the slope for the PLACEBO is not ($p=0.1766$). The contrasts for comparing the DRUG A vs C regressions are

```
contrast 'drug a vs c intercept' drug 1 -1 0;
contrast 'drug a vs c slope' h(drug) 1 -1 0;
```

The results are shown in Output 8.21.

Output 8.21
Test of Equality of Regression for DRUG A versus C

```
                             Contrasts

                          Num     Den
        Label              DF      DF     F Value    Pr > F

        drug a vs c intercept     1      111      3.69     0.0572
        drug a vs c slope         1      113      1.04     0.3094
```

There is some evidence that the intercepts are different ($p=0.0572$), but no evidence of a difference in slopes ($p=0.3094$). The statistical conclusions cannot be interpreted practically, however. They have meaning only between HOUR=1 and HOUR=8. The intercepts are only extrapolations to HOUR=0, and have no real meaning. Also, the true FEV1 curves would return approximately the baseline level. Thus, the slopes of the true curves for DRUG A and DRUG C are logically different, even though not statistically different.

Chapter 9 Multivariate Linear Models

9.1 Introduction

Although the methods presented in earlier chapters are sufficient for studying one dependent variable at a time, there are many situations in which several dependent variables are studied simultaneously. For example, in monitoring the growth of animals, a researcher might measure the length, weight, and girth of animals receiving different treatments. The goal of the experiment would be to see if the treatments had an effect on the growth of the animals. One way to determine this would be to use analysis of variance or regression methods to analyze the effects of the treatment on the length, weight, and girth of the animals. There are two problems with this approach. First, trying to interpret results produced by separate univariate (one variable at a time) analyses of each of the variables can be unwieldy. In the example with three dependent variables, this may not seem like a problem. But if you have ten or twenty dependent variables, the task could be substantial. Second, and more importantly, when many variables are studied simultaneously, they are almost always correlated—that is, the value of each variable may be related to the values of others. This is true for measurements such as height and weight, or responses to similar questions on a questionnaire. In cases like these, considering the univariate analyses separately would not take into account information contained in the data due to the correlation. Moreover, this approach could mislead a naïve researcher into believing that a factor has a very significant effect, when in fact it does not. On the other hand, a significant effect that only becomes apparent when all the dependent variables are studied simultaneously may not be discovered from the univariate analyses alone. In most multivariate data applications, you should usually examine the results of the multivariate tests first, then examine the univariate analyses cautiously if significant results do not appear in the multivariate analysis.

Although some of the details of a multivariate analysis differ from those of a univariate analysis, the two are similar in many ways. Experimental factors of interest are related to the dependent variables by a linear model, and functions of sums of squares are computed to test hypotheses about these factors. In general, if you have designed an experiment with only one dependent variable, the extension of the analysis to the multivariate case can be carried out in a very straightforward manner.

We present examples of several types of multivariate analyses. Basic theory for the methods is presented in Section 9.7, "Statistical Background."

9.2   A One-Way Multivariate Analysis of Variance

Test scores from two exams taken by students with three different teachers are analyzed in Output 9.1.

Output 9.1
Two Exam
Scores for
Students in
Three
Teachers'
Classes

Obs	teach	score1	score2
1	JAY	69	75
2	JAY	69	70
3	JAY	71	73
4	JAY	78	82
5	JAY	79	81
6	JAY	73	75
7	PAT	69	70
8	PAT	68	74
9	PAT	75	80
10	PAT	78	85
11	PAT	68	68
12	PAT	63	68
13	PAT	72	74
14	PAT	63	66
15	PAT	71	76
16	PAT	72	78
17	PAT	71	73
18	PAT	70	73
19	PAT	56	59
20	PAT	77	79
22	ROBIN	64	65
23	ROBIN	74	74
24	ROBIN	72	75
25	ROBIN	82	84
26	ROBIN	69	68
27	ROBIN	76	76
28	ROBIN	68	65
29	ROBIN	78	79
30	ROBIN	70	71
31	ROBIN	60	61

We first perform an analysis of variance to compare the teacher means for each variable, SCORE1 and SCORE2. Run the SAS statements for these analyses:

```
proc glm;
   class teach;
   model score1 score2=teach;
```

Results are shown in Output 9.2. There is no difference among teachers when considering a univariate analysis of either SCORE1 or SCORE2.

```
                              The GLM Procedure

Dependent Variable: score1

                                   Sum of
        Source           DF        Squares      Mean Square    F Value    Pr > F

        Model             2      60.6050831      30.3025415      0.91     0.4143

        Error            28     932.8787879      33.3170996
        Corrected Total  30     993.4838710


              R-Square      Coeff Var      Root MSE      score1 Mean

              0.061003      8.144515       5.772097       70.87097


        Source           DF     Type III SS     Mean Square    F Value    Pr > F

        teach             2     60.60508309     30.30254154      0.91     0.4143
```

```
                              The GLM Procedure

Dependent Variable: score2

                                   Sum of
        Source           DF        Squares      Mean Square    F Value    Pr > F

        Model             2      49.735861       24.867930       0.56     0.5776

        Error            28    1243.941558       44.426484

        Corrected Total  30    1293.677419


              R-Square      Coeff Var      Root MSE      score2 Mean

              0.038445      9.062496       6.665320       73.54839


        Source           DF     Type III SS     Mean Square    F Value    Pr > F

        teach             2     49.73586091     24.86793046      0.56     0.5776
```

The objective now is to see if a model using both scores shows a difference among teachers. The following SAS statements are used for this analysis:

```
proc glm;
     class teach;
     model score1 score2=teach;
     manova h=teach / printh printe;
```

The first three statements produce the usual univariate analyses of the two scores as shown in Output 9.2. The MANOVA statement produces results in Output 9.3.

*Output 9.3
Results of
One-Way
Multivariate
Analysis:
The
MANOVA
Statement*

```
                            The GLM Procedure
                      Multivariate Analysis of Variance

                          E = Error SSCP Matrix


       ❶                      score1              score2

             score1      932.87878788         1018.6818182
             score2     1018.6818182          1243.9415584



   ❷  Partial Correlation Coefficients from the Error SSCP Matrix / Prob > |r|

                  DF = 28          score1           score2

                  score1         1.000000         0.945640
                                                  <.0001


                  score2         0.945640         1.000000
                                 <.0001



              ❸     H = Type III SSCP Matrix for teach

                               score1              score2

             score1       60.605083089         31.511730205
             score2       31.511730205         49.735860913



           Characteristic Roots and Vectors of: E Inverse * H, where
                      H = Type III SSCP Matrix for teach
                      E = Error SSCP Matrix


       ❹   Characteristic              Characteristic Vector  V'EV=1
              Root      Percent          score1           score2

           0.43098027     91.86       -0.10044686        0.08416103
           0.03821194      8.14        0.00675930        0.02275380


 MANOVA Test Criteria and F Approximations for the Hypothesis of No Overall teach Effect
       ❺                   H = Type III SSCP Matrix for teach
                           E = Error SSCP Matrix

                        S=2     M=-0.5      N=12.5

        Statistic                     Value    F Value   Num DF   Den DF   Pr > F

        Wilks' Lambda               0.67310116    2.95      4        54    0.0279
        Pillai's Trace              0.33798387    2.85      4        56    0.0322
        Hotelling-Lawley Trace      0.46919220    3.13      4      31.389  0.0281
        Roy's Greatest Root         0.43098027    6.03      2        28    0.0066

           NOTE: F Statistic for Roy's Greatest Root is an upper bound.
           NOTE: F Statistic for Wilks' Lambda is exact.
```

The PRINTH and PRINTE options cause the printing of the hypothesis and error matrices, respectively. In addition, the PRINTE option produces a matrix of partial correlation coefficients derived from the error SSCP matrix. This correlation matrix represents the correlations of the dependent variables corrected for all the independent factors in the MODEL statement.

The results in Output 9.3 are described below. The callout numbers have been added to the output to key the following descriptions:

❶ The elements of the error matrix. The diagonal elements of this matrix represent the error sums of squares from the corresponding univariate analyses (see Output 9.2).

❷ The associated partial correlation matrix. In this example it appears that SCORE1 and SCORE2 are highly correlated ($r=0.945640$).

❸ The elements of the hypothesis matrix, **H**. Again, the diagonal elements correspond to the hypothesis sums of squares from the corresponding univariate analysis.

❹ The characteristic roots and vectors of $\mathbf{E}^{-1}\mathbf{H}$. The elements of the characteristic vector describe a linear combination of the analysis variables that produces the largest possible univariate F-ratio.

❺ The four test statistics previously discussed. The values of S, M, and N, printed above the table of statistics, provide information that is used in constructing the F-approximations for the criteria. (For more information, see Morrison (1976).) All four tests give similar results, although this is not always the case. Note that the p-values for the "Hypothesis of No Overall TEACH Effect" are much lower for the multivariate tests than any of the univariate tests would indicate. This is an example of how viewing a set of variables together can help you detect differences that you would not detect by looking at the individual variables.

9.3 Hotelling's T² Test

Consider a common situation in multivariate analysis. You have several different measurements taken on each of several subjects, and you want to know if the means of the different variables are all the same. For example, you may have used different recording devices to measure the same phenomenon, or you may have observed subjects under a variety of conditions and administered a test under each of the conditions. If you had only two means to compare, you could use the familiar t-test, but it is important to use an analysis that takes into account the correlations among the dependent variables, just as in the previous examples, even though there are no independent factors in the model—that is, no terms on the right side of the MODEL statement. In this situation you could use Hotelling's T^2 test. As an example, consider the following data taken from Morrison (1976). Weight gains in rats given a special diet were measured at one (GAIN1), two (GAIN2), three (GAIN3), and four (GAIN4) weeks after administration of the diet. The question of interest is whether the rats' weight gains stayed constant over the course of the experiment; in other words, were the mean weight gains of the rats the same at each of the four weeks? Output 9.4 shows the data.

Output 9.4
Data for
Hotelling's T²
Test

Obs	gain1	gain2	gain3	gain4
1	29	28	25	33
2	33	30	23	31
3	25	34	33	41
4	18	33	29	35
5	25	23	17	30
6	24	32	29	22
7	20	23	16	31
8	28	21	18	24
9	18	23	22	28
10	25	28	29	30

Note that the following MODEL statement fits a model with only an intercept:

```
model gain1 gain2 gain3 gain4 = ;
```

Following this statement with a MANOVA statement to test the intercept tests the hypothesis that the means of the four weight gains are all 0. It does *not* test the hypothesis of interest—that the four means are equal. To test the hypothesis that the four means are equal, the dependent variables must be transformed in such a way that their transformed means being 0 will imply that the original means are equal for all four variables. It turns out that there are many different transformations that have this effect. (This problem is discussed in detail in Chapter 8, "Repeated-Measures Analysis.") One simple transformation that achieves this goal is subtracting one of the variables from each of the other variables; in this example, the first gain could be subtracted from each of the other gains. You should understand that the only way all the means could be equal to each other is if each of these differences is 0 and that if all the means are equal, then all the differences must be 0. In this way, Hotelling's T^2 test for equality of means can be performed using the MANOVA statement.

One way of producing the transformed variables necessary to perform Hotelling's T^2 test is to produce new variables in a DATA step and to perform an analysis on the new variables. A quicker and more efficient way, however, is to use the M= option in the MANOVA statement. Using the M= option, you can perform an analysis on a set of variables that is a linear transformation of the original variables as listed on the left side of the equal sign in the MODEL statement.

The following SAS statements perform the appropriate analysis:

```
proc glm data=wtgain;
     model gain1 gain2 gain3 gain4 = / nouni;
     manova h=intercept
            m=gain2-gain1, gain3-gain1, gain4-gain1
            mnames=diff2 diff3 diff4 / summary;
```

The NOUNI option in the MODEL statement suppresses the individual analyses of the gain variables. This is done because the multivariate hypothesis of equality of the four means is the hypothesis of interest. The SUMMARY option in the MANOVA statement produces analysis-of-variance tables of the transformed variables. The MNAMES= option provides labels for these transformed variables; if omitted, the procedure uses the names MVAR1, MVAR2, and so on. The results appear in Output 9.5.

*Output 9.5
Analysis of
Transformed
Variables:
Hotelling's T2
Test*

```
                          The GLM Procedure

              Number of observations      10
                     The ANOVA Procedure
                Multivariate Analysis of Variance

              M Matrix Describing Transformed Variables

                    gain1            gain2            gain3            gain4
       diff2          -1                1                0                0
       diff3          -1                0                1                0
       diff4          -1                0                0                1


          Characteristic Roots and Vectors of: E Inverse * H, where
                   H = Type III SSCP Matrix for Intercept
                        E = Error SSCP Matrix

             Variables have been transformed by the M Matrix

       Characteristic             Characteristic Vector  V'EV=1
              Root     Percent         diff2            diff3            diff4

         2.97211676    100.00      -0.11825538       0.11174598      -0.02428445
         0.00000000      0.00      -0.11610475       0.05331366       0.06160662
         0.00000000      0.00       0.00519921       0.03899407       0.00000000


MANOVA Test Criteria and Exact F Statistics for the Hypothesis of No Overall
Intercept Effect
           on the Variables Defined by the M Matrix Transformation
                   H = Type III SSCP Matrix for Intercept
                        E = Error SSCP Matrix

                   S=1       M=0.5      N=2.5

Statistic                       Value    F Value    Num DF    Den DF    Pr > F

Wilks' Lambda                0.25175494     6.93        3         7      0.0167
Pillai's Trace               0.74824506     6.93        3         7      0.0167
Hotelling-Lawley Trace       2.97211676     6.93        3         7      0.0167
Roy's Greatest Root          2.97211676     6.93        3         7      0.0167

Dependent Variable: diff2

Source              DF     Type III SS    Mean Square    F Value    Pr > F

Intercept            1      90.0000000     90.0000000       2.10     0.1814
Error                9     386.0000000     42.8888889

Dependent Variable: diff3

Source              DF     Type III SS    Mean Square    F Value    Pr > F

Intercept            1       1.6000000      1.6000000       0.03     0.8735
Error                9     536.4000000     59.6000000


Dependent Variable: diff4

Source              DF     Type III SS    Mean Square    F Value    Pr > F

Intercept            1     360.0000000    360.0000000       6.53     0.0309
Error                9     496.0000000     55.1111111
```

Although PROC ANOVA does not produce the Hotelling T^2 statistic, it produces the correct *F*-test and probability level. Note that all the multivariate test statistics result in the same *F*-values and that they are labeled as "Exact F Statistics." These tests are always exact when the hypothesis being tested has only 1 degree of freedom, such as the hypothesis of no INTERCEPT effect in this example. In order to calculate the actual value of the T^2 statistic, the following formula can be used:

$$T^2 = (\text{nobs} - 1)\big((1 / \lambda) - 1\big)$$

In the data set, nobs is the number of observations and λ is the value of Wilks's Criterion that is printed by PROC ANOVA. (See Section 9.7, "Statistical Background.") In this case, the calculation leads to $T^2 = (10 - 1)(1/0.2517 - 1) = 26.757$.

9.4   A Two-Factor Factorial

The total weight of a mature cotton boll can be divided into three parts: the weight of the seeds, the weight of the lint, and the weight of the bract. Lint and seed constitute the economic yield of cotton.

In the following data, the differences in the three components of the cotton bolls due to two varieties (VARIETY) and two plant spacings (SPACING) are studied. Five plants are chosen at random from each of the four treatment combinations. Two bolls are picked from each plant, and the weights of the seeds, lint, and bract are recorded. The most appropriate error term for testing VARIETY, SPACING, and the interaction of the two is the variation among plants. In univariate analyses, the TEST statement is used to specify this alternative error term; for multivariate analyses, the error term is specified in the MANOVA statement or statements. (Each MANOVA statement can have, at most, one error term specified.)

The following SAS statements are used:

```
proc glm;
   class variety spacing plant;
   model seed lint bract=variety spacing variety*spacing
                         plant(variety spacing)/ss3;
   test h=variety|spacing e=plant(variety spacing);
   means variety|spacing;
   manova h=variety|spacing e=plant(variety spacing);
```

The data used in this analysis appear in Output 9.6.

Output 9.6
*Data for
Two-Way
Multivariate
Analysis*

Obs	variety	spacing	plant	seed	lint	bract
1	213	30	3	3.1	1.7	2.0
2	213	30	3	1.5	1.7	1.4
3	213	30	5	3.0	1.9	1.8
4	213	30	5	1.4	0.9	1.3
5	213	30	6	2.3	1.7	1.5
6	213	30	6	2.2	2.0	1.4
7	213	30	8	0.4	0.9	1.2
8	213	30	8	1.7	1.6	1.3
9	213	30	9	1.8	1.2	1.0
10	213	30	9	1.2	0.8	1.0
11	213	40	0	2.0	1.0	1.9
12	213	40	0	1.5	1.5	1.7
13	213	40	1	1.8	1.1	2.1
14	213	40	1	1.0	1.3	1.1
15	213	40	2	1.3	1.1	1.3
16	213	40	2	2.9	1.9	1.7
17	213	40	3	2.8	1.2	1.3
18	213	40	3	1.8	1.2	1.2
19	213	40	4	3.2	1.8	2.0
20	213	40	4	3.2	1.6	1.9
21	37	30	1	3.2	2.6	1.4
22	37	30	1	2.8	2.1	1.2
23	37	30	2	3.6	2.4	1.5
24	37	30	2	0.9	0.8	0.8
25	37	30	3	4.0	3.1	1.8
26	37	30	3	4.0	2.9	1.5
27	37	30	5	3.7	2.7	1.6
28	37	30	5	2.6	1.5	1.3
29	37	30	8	2.8	2.2	1.2
30	37	30	8	2.9	2.3	1.2
31	37	40	1	4.1	2.9	2.0
32	37	40	1	3.4	2.0	1.6
33	37	40	3	3.7	2.3	2.0
34	37	40	3	3.2	2.2	1.8
35	37	40	4	3.4	2.7	1.5
36	37	40	4	2.9	2.1	1.2
37	37	40	5	2.5	1.4	0.8
38	37	40	5	3.6	2.4	1.6
39	37	40	6	3.1	2.3	1.4
40	37	40	6	2.5	1.5	1.5

Note that the TEST statement is used to obtain the appropriate error terms. The results of the three univariate analyses appear in Output 9.7.

Dependent Variable: seed

Source	DF	Sum of Squares	Mean Square	F Value	Pr > F
Model	19	24.06500000	1.26657895	2.22	0.0425
Error	20	11.43000000	0.57150000		
Corrected Total	39	35.49500000			

R-Square	Coeff Var	Root MSE	seed Mean
0.677983	29.35830	0.755976	2.575000

Source	DF	Type III SS	Mean Square	F Value	Pr > F
variety	1	12.99600000	12.99600000	22.74	0.0001
spacing	1	0.57600000	0.57600000	1.01	0.3274
variety*spacing	1	0.02500000	0.02500000	0.04	0.8364
plant(variet*spacin)	16	10.46800000	0.65425000	1.14	0.3823

Tests of Hypotheses Using the Type III MS for plant(variet*spacin) as an Error Term

Source	DF	Type III SS	Mean Square	F Value	Pr > F
variety	1	12.99600000	12.99600000	19.86	0.0004
spacing	1	0.57600000	0.57600000	0.88	0.3620
variety*spacing	1	0.02500000	0.02500000	0.04	0.8475

Dependent Variable: lint

Source	DF	Sum of Squares	Mean Square	F Value	Pr > F
Model	19	10.62875000	0.55940789	2.28	0.0377
Error	20	4.91500000	0.24575000		
Corrected Total	39	15.54375000			

R-Square	Coeff Var	Root MSE	lint Mean
0.683796	27.35072	0.495732	1.812500

Source	DF	Type III SS	Mean Square	F Value	Pr > F
variety	1	6.64225000	6.64225000	27.03	<.0001
spacing	1	0.05625000	0.05625000	0.23	0.6375
variety*spacing	1	0.00025000	0.00025000	0.00	0.9749
plant(variet*spacin)	16	3.93000000	0.24562500	1.00	0.4934

Tests of Hypotheses Using the Type III MS for plant(variet*spacin) as an Error Term

Source	DF	Type III SS	Mean Square	F Value	Pr > F
variety	1	6.64225000	6.64225000	27.04	<.0001
spacing	1	0.05625000	0.05625000	0.23	0.6387
variety*spacing	1	0.00025000	0.00025000	0.00	0.9749

Output 9.7
(Continued)
Results of
Two-Way
Multivariate
Analysis:
Univariate
Analyses

```
Dependent Variable: bract

                                     Sum of
        Source              DF       Squares      Mean Square    F Value    Pr > F

        Model               19     2.70500000     0.14236842       1.63     0.1442

        Error               20     1.75000000     0.08750000

        Corrected Total     39     4.45500000


              R-Square       Coeff Var       Root MSE      bract Mean

              0.607183       20.05451        0.295804        1.475000


   Source                   DF     Type III SS    Mean Square    F Value    Pr > F

   variety                   1     0.03600000     0.03600000       0.41     0.5285
   spacing                   1     0.44100000     0.44100000       5.04     0.0362
   variety*spacing           1     0.00400000     0.00400000       0.05     0.8329
   plant(variet*spacin)     16     2.22400000     0.13900000       1.59     0.1626


   Tests of Hypotheses Using the Type III MS for plant(variet*spacin) as an Error Term

   Source                   DF     Type III SS    Mean Square    F Value    Pr > F

   variety                   1     0.03600000     0.03600000       0.26     0.6178
   spacing                   1     0.44100000     0.44100000       3.17     0.0939
   variety*spacing           1     0.00400000     0.00400000       0.03     0.8674
```

VARIETY has a statistically significant effect on SEED and LINT; no effects are statistically significant for BRACT.

The means for all levels and combinations of levels of VARIETY and SPACING produced by the MEANS statement appear in Output 9.8.

Output 9.8
Results of
Two-Way
Multivariate
Analysis: The
MEANS
Statement

```
                                  The GLM Procedure

Level of      --------seed----------   ------------lint----------   -----------bract------
variety  N       Mean       Std Dev        Mean       Std Dev         Mean        Std Dev

37       20   3.14500000  0.72799291    2.22000000  0.57087191    1.44500000   0.32843328
213      20   2.00500000  0.80881655    1.40500000  0.37763112    1.50500000   0.35314378

Level of      --------seed----------   ------------lint----------   -----------bract------
spacing  N       Mean       Std Dev        Mean       Std Dev         Mean        Std Dev

30       20   2.45500000  1.04351380    1.85000000  0.70150215    1.37000000   0.29037181
40       20   2.69500000  0.86540225    1.77500000  0.56835404    1.58000000   0.35629674

Level of  Level of      -------------seed------------   -------------lint--------
variety   spacing    N        Mean        Std Dev            Mean        Std Dev

37        30        10     3.05000000   0.91439111       2.26000000   0.68182761
37        40        10     3.24000000   0.51251016       2.18000000   0.46856756
213       30        10     1.86000000   0.82219219       1.44000000   0.44771022
213       40        10     2.15000000   0.81137743       1.37000000   0.31287200

            Level of   Level of       ------------bract------------
            variety    spacing    N         Mean        Std Dev

            37         30        10     1.35000000   0.27588242
            37         40        10     1.54000000   0.36270588
            213        30        10     1.39000000   0.31780497
            213        40        10     1.62000000   0.36453928
```

The results of the multivariate analyses appear in Output 9.9.

Output 9.9
Results of
Two-Way
Multivariate
Analysis

```
                          The GLM Procedure
                   Multivariate Analysis of Variance

            Characteristic Roots and Vectors of: E Inverse * H, where
                     H = Type III SSCP Matrix for variety
                   E = Type III SSCP Matrix for plant(variet*spacin)


        Characteristic              Characteristic Vector  V'EV=1
              Root     Percent           seed            lint           bract


        3.43919116     100.00        0.13061027       0.48969379     -0.64083380
        0.00000000       0.00       -0.52400501       0.74035630      0.10041133
        0.00000000       0.00        0.01704640       0.02228579      0.62659680



MANOVA Test Criteria and Exact F Statistics for the Hypothesis of No Overall variety Effect
                     H = Type III SSCP Matrix for variety
                   E = Type III SSCP Matrix for plant(variet*spacin)


                         S=1     M=0.5      N=6


        Statistic                    Value    F Value   Num DF   Den DF   Pr > F

        Wilks' Lambda             0.22526626    16.05       3       14    <.0001
        Pillai's Trace            0.77473374    16.05       3       14    <.0001
        Hotelling-Lawley Trace    3.43919116    16.05       3       14    <.0001
        Roy's Greatest Root       3.43919116    16.05       3       14    <.0001
```

```
                          The GLM Procedure
                   Multivariate Analysis of Variance

            Characteristic Roots and Vectors of: E Inverse * H, where
                     H = Type III SSCP Matrix for spacing
                   E = Type III SSCP Matrix for plant(variet*spacin)

        Characteristic              Characteristic Vector  V'EV=1
              Root     Percent           seed            lint           bract


        0.63209472     100.00        0.27027458      -0.73247531      0.62673044
        0.00000000       0.00        0.24001974       0.22618207     -0.19352896
        0.00000000       0.00       -0.40158816       0.44804654      0.61897451



MANOVA Test Criteria and Exact F Statistics for the Hypothesis of No Overall spacing Effect
                     H = Type III SSCP Matrix for spacing
                   E = Type III SSCP Matrix for plant(variet*spacin)


                         S=1     M=0.5      N=6


        Statistic                    Value    F Value   Num DF   Den DF   Pr > F

        Wilks' Lambda             0.61270954     2.95       3       14    0.0692
        Pillai's Trace            0.38729046     2.95       3       14    0.0692
        Hotelling-Lawley Trace    0.63209472     2.95       3       14    0.0692
        Roy's Greatest Root       0.63209472     2.95       3       14    0.0692
```

*Output 9.9
(Continued)
Results of
Two-Way
Multivariate
Analysis*

```
                          The GLM Procedure
                    Multivariate Analysis of Variance

               Characteristic Roots and Vectors of: E Inverse * H, where
                    H = Type III SSCP Matrix for variety*spacing
                    E = Type III SSCP Matrix for plant(variet*spacin)

      Characteristic                 Characteristic Vector  V'EV=1
             Root     Percent              seed              lint            bract

        0.00616711    100.00          0.42143581        -0.67443149        0.35670210
        0.00000000      0.00         -0.33315217         0.01818488        0.82833421
        0.00000000      0.00         -0.05772656         0.57726561        0.00000000


  MANOVA Test Criteria and Exact F Statistics for the Hypothesis of No Overall variety*spacing Effect
                    H = Type III SSCP Matrix for variety*spacing
                    E = Type III SSCP Matrix for plant(variet*spacin)


                            S=1      M=0.5      N=6

       Statistic                     Value    F Value    Num DF    Den DF    Pr > F

       Wilks' Lambda              0.99387069     0.03         3        14    0.9931
       Pillai's Trace             0.00612931     0.03         3        14    0.9931
       Hotelling-Lawley Trace     0.00616711     0.03         3        14    0.9931
       Roy's Greatest Root        0.00616711     0.03         3        14    0.9931
```

The only highly significant effect is VARIETY, and all four multivariate statistics are the same because there is only one hypothesis degree of freedom.

9.5 Multivariate Analysis of Covariance

This section illustrates multivariate analysis of covariance using the data on orange sales presented in Output 7.6. Sales of two types of oranges are related to experimentally determined prices (PRICE) as well as stores (STORE) and days of the week (DAY). The analysis is expanded here to consider the simultaneous multivariate relationship of the price to both types of oranges.

The following SAS statements are used for the analysis:

```
proc glm;
   class store day;
   model q1 q2=store day p1 p2 / nouni;
   manova h=store day p1 p2 / printh printe;
```

Note that PROC ANOVA is not appropriate in this situation because of the presence of the covariates. The NOUNI option in the MODEL statement suppresses printing of the univariate analyses that are already shown in Chapter 7, "Analysis of Covariance." Results of the multivariate analysis appear in Output 9.10.

Output 9.10
Results of
Multivariate
Analysis of
Covariance

```
                        The GLM Procedure
                 Multivariate Analysis of Variance

                     E = Error SSCP Matrix

                            q1                 q2

          q1     408.30824182        74.603217758
          q2     74.603217758        706.94116552


    Partial Correlation Coefficients from the Error SSCP Matrix / Prob > |r|

          DF = 23                 q1                 q2

          q1              1.000000           0.138858
                                             0.5176


          q2              0.138858           1.000000
                          0.5176
```

```
                 H = Type III SSCP Matrix for store

                            q1                 q2

          q1     223.83267344        93.801152319
          q2     93.801152319        155.09933793


        Characteristic Roots and Vectors of: E Inverse * H, where
                 H = Type III SSCP Matrix for store
                      E = Error SSCP Matrix

        Characteristic                Characteristic Vector  V'EV=1
             Root       Percent            q1                 q2

         0.57363829      78.23        0.04622384         0.00941459
         0.15960322      21.77       -0.01899048         0.03679295


   MANOVA Test Criteria and F Approximations for the Hypothesis of No Overall store Effect
                     H = Type III SSCP Matrix for store
                          E = Error SSCP Matrix

                       S=2    M=1    N=10

   Statistic                     Value    F Value    Num DF    Den DF    Pr > F

   Wilks' Lambda              0.54800645     1.54        10        44    0.1564
   Pillai's Trace             0.50216601     1.54        10        46    0.1553
   Hotelling-Lawley Trace     0.73324151     1.57        10    30.372    0.1634
   Roy's Greatest Root        0.57363829     2.64         5        23    0.0501

          NOTE: F Statistic for Roy's Greatest Root is an upper bound.
              NOTE: F Statistic for Wilks' Lambda is exact.
```

Output 9.10
(Continued)
Results of
Multivariate
Analysis of
Covariance

```
                         H = Type III SSCP Matrix for day


                              q1                    q2

             q1        433.09686996         461.05064188
             q2        461.05064188         614.4088834


            Characteristic Roots and Vectors of: E Inverse * H, where
                      H = Type III SSCP Matrix for day
                      E = Error SSCP Matrix

            Characteristic                 Characteristic Vector  V'EV=1
                 Root       Percent             q1              q2


             1.60708776     93.18          0.03517603       0.02300242
             0.11766546      6.82         -0.03549548       0.03021993


    MANOVA Test Criteria and F Approximations for the Hypothesis of No Overall day Effect
                      H = Type III SSCP Matrix for day
                      E = Error SSCP Matrix


                        S=2     M=1     N=10

    Statistic                    Value    F Value   Num DF   Den DF    Pr > F

    Wilks' Lambda             0.34318834    3.11       10        44     0.0044
    Pillai's Trace            0.72170813    2.60       10        46     0.0137
    Hotelling-Lawley Trace    1.72475321    3.69       10    30.372     0.0026
    Roy's Greatest Root       1.60708776    7.39        5        23     0.0003

        NOTE: F Statistic for Roy's Greatest Root is an upper bound.
           NOTE: F Statistic for Wilks' Lambda is exact.
```

```
                         H = Type III SSCP Matrix for p1


                              q1                    q2

             q1        538.16885116        -212.5196287
             q2       -212.5196287          83.922717744


            Characteristic Roots and Vectors of: E Inverse * H, where
                      H = Type III SSCP Matrix for p1
                      E = Error SSCP Matrix

            Characteristic                 Characteristic Vector  V'EV=1
                 Root       Percent             q1              q2

             1.57701930    100.00          0.04805513      -0.01539025
             0.00000000      0.00          0.01371082       0.03472026


                         The GLM Procedure
                    Multivariate Analysis of Variance


    MANOVA Test Criteria and Exact F Statistics for the Hypothesis of No Overall p1 Effect
                      H = Type III SSCP Matrix for p1
                      E = Error SSCP Matrix

                        S=1     M=0     N=10

    Statistic                    Value    F Value   Num DF   Den DF    Pr > F

    Wilks' Lambda             0.38804521   17.35        2        22    <.0001
    Pillai's Trace            0.61195479   17.35        2        22    <.0001
    Hotelling-Lawley Trace    1.57701930   17.35        2        22    <.0001
    Roy's Greatest Root       1.57701930   17.35        2        22    <.0001
```

Output 9.10
(Continued)
Results of
Multivariate
Analysis of
Covariance

```
                          H = Type III SSCP Matrix for p2

                                   q1                   q2

                     q1      39.542251923        -183.5850939
                     q2     -183.5850939          852.34110489


                Characteristic Roots and Vectors of: E Inverse * H, where
                            H = Type III SSCP Matrix for p2
                            E = Error SSCP Matrix

               Characteristic                Characteristic Vector  V'EV=1
                    Root       Percent              q1                q2


              1.42489030       100.00        -0.01960102         0.03666503
              0.00000000         0.00         0.04596827         0.00990107


         MANOVA Test Criteria and Exact F Statistics for the Hypothesis of No Overall p2 Effect
                            H = Type III SSCP Matrix for p2
                            E = Error SSCP Matrix

                              S=1      M=0      N=10

         Statistic                      Value    F Value   Num DF   Den DF   Pr > F

         Wilks' Lambda               0.41238979    15.67       2       22    <.0001
         Pillai's Trace              0.58761021    15.67       2       22    <.0001
         Hotelling-Lawley Trace      1.42489030    15.67       2       22    <.0001
         Roy's Greatest Root         1.42489030    15.67       2       22    <.0001
```

For the STORE effect, none of the statistics produce significant results. This is not totally consistent with univariate results where sales of the first type of oranges are nearly significant at the 5% level.

The DAY effect is quite significant according to all statistics, although there is a considerable difference in the level of significance (Pr > F) among the statistics. Also, there is one dominant eigenvalue, indicating that the trend in sales over days is roughly parallel for the two types of oranges. This can be verified by printing and plotting the least-squares means.

The results for the PRICE effects are relatively straightforward. Even though both P1 and P2 are significant for only one dependent variable in the univariate analyses, the multivariate analysis indicates that their effects are substantial enough to be significant overall.

9.6 Contrasts in Multivariate Analyses

Chapter 3, "Analysis of Variance for Balanced Data," discusses how specialized questions concerning certain levels of factors in univariate analyses can be answered by using the CONTRAST statement to define a hypothesis to be tested. This same technique can be useful in multivariate analyses of variance. The GLM procedure prints output for CONTRAST statements as part of its multivariate analysis. As an example, consider the study described in Section 9.5, "Multivariate Analysis of Covariance." Assume you want to know if Saturday sales differ from weekday sales, averaged across the two types of oranges. Because the levels for DAY are coded as 1 through 6 corresponding to Monday through Saturday, you need to construct a contrast that compares the average of the first five levels of DAY to the sixth. The following SAS statement is required:

```
contrast 'SAT. vs. WEEKDAYS' day .2 .2 .2 .2 .2 -1;
```

The label SAT. vs. WEEKDAYS appears in the output to identify the contrast. If this CONTRAST statement is appended to the program of the previous section, preceding the MANOVA statement, then Output 9.11 is produced. This statement must precede the MANOVA statement so that PROC GLM will know that the multivariate test for the contrast is wanted.

Output 9.11
Results of
Multivariate
Analysis: The
CONTRAST
Statement

```
                          The GLM Procedure
                    Multivariate Analysis of Variance

            H = Contrast SSCP Matrix for sat vs weekdays

                             q1                    q2

            q1      9.3680815712          7.8469238548
            q2      7.8469238548          6.5727666348


        Characteristic Roots and Vectors of: E Inverse * H, where
            H = Contrast SSCP Matrix for sat vs weekdays
                    E = Error SSCP Matrix

        Characteristic              Characteristic Vector  V'EV=1
                Root     Percent           q1                q2

            0.02873910   100.00        0.04110205        0.01705466
            0.00000000     0.00       -0.02842364        0.03393368


            MANOVA Test Criteria and Exact F Statistics for the
            Hypothesis of No Overall sat vs weekdays Effect
                H = Contrast SSCP Matrix for sat vs weekdays
                        E = Error SSCP Matrix

                      S=1     M=0     N=10

        Statistic                Value   F Value   Num DF   Den DF   Pr > F

        Wilks' Lambda         0.97206377   0.32      2        22     0.7322
        Pillai's Trace        0.02793623   0.32      2        22     0.7322
        Hotelling-Lawley Trace 0.02873910  0.32      2        22     0.7322
        Roy's Greatest Root   0.02873910   0.32      2        22     0.7322
```

The results of the multivariate tests indicate no significant overall difference between the Saturday and weekday sales for the two types of oranges (Pr > F=0.7322).

9.7 Statistical Background

The multivariate linear model can be written as

$$\mathbf{Y} = \mathbf{XB} + \mathbf{U}$$

where

Y   is an $n \times k$ matrix of observed values of k dependent variables or responses. Each column corresponds to a specific dependent variable and each row to an observation.

X   is an $n \times m$ matrix of n observations on the m independent variables (which may contain dummy variables).

B   is an mXk matrix of regression coefficients or parameters. Each column of **B** is a vector of coefficients corresponding to each of k dependent variables, and each row contains the coefficients associated with each of m independent variables.

U   is the nXk matrix of the n random errors, with columns corresponding to the dependent variables.

The matrix of estimated coefficients is

$$\hat{\mathbf{B}} = \left(\mathbf{X'X}\right)^{-1}\mathbf{X'Y}$$

Each column of $\hat{\mathbf{B}}$ is the vector of estimated coefficients that would be obtained by estimating coefficients for each response variable separately.

The partitioning of sums of squares is parallel to that developed in previous chapters except that the partitions consist of kXk matrices of sums of squares and crossproducts:

Sums of Squares	SSCP Matrix
TOTAL	$\mathbf{Y'Y}$
MODEL	$\hat{\mathbf{B}}'\mathbf{X'Y}$
ERROR	$\mathbf{Y'Y} - \hat{\mathbf{B}}'\mathbf{X'Y}$

In the univariate analysis of variance, the F-statistic is the statistic of choice in most cases for testing hypotheses about the factors being considered. Recall that this statistic is derived by taking the ratio of two sums of squares, one derived from the hypothesis being tested and the other derived from an appropriate error term. In multivariate linear models, these sums of squares are replaced by matrices of sums of squares and crossproducts. These matrices are represented by **H** for the hypothesis, corresponding to the numerator sum of squares, and **E** for the error matrix, corresponding to the denominator sum of squares. Since division of matrices is not possible, $\mathbf{E}^{-1}\mathbf{H}$ is the matrix that is the basis for test statistics for multivariate hypotheses. Four different functions of this matrix are used as test statistics and are available in the GLM, ANOVA, and other multivariate procedures in SAS. Each of these statistics is a function of the **characteristic roots** (also known as eigenvalues) of the matrix $\mathbf{E}^{-1}\mathbf{H}$. In the formulas below, λ_i represents the characteristic roots.

Corresponding to each characteristic root is a **characteristic vector**, or eigenvector, that represents a linear combination of the dependent variables being analyzed. A function of the characteristic root, $\lambda/(1+\lambda)$, is the value of R^2 that would be obtained if the linear combination of dependent variables represented by the corresponding characteristic vector were used as the dependent variable in a univariate analysis of the same model. For this reason, that function of the characteristic root is sometimes called the **canonical correlation**. In the formulas below, r_i^2 represents the canonical correlations.

Hotelling-Lawley Trace

$$= \text{Tr}\,(\mathbf{E}^{-1}\mathbf{H})$$

$$= \sum r_i^2 / (1 - r_i^2)$$

$$= \sum \lambda_i$$

Pillai's Trace

$$= \text{Tr}(\mathbf{H}(\mathbf{H} + \mathbf{E})^{-1})$$

$$= \sum r_i^2$$

$$= \sum \lambda_i / (1 + \lambda_i)$$

Wilks' Lambda

$$= |E| / |H + E|$$

$$= \prod (1 - r_i^2)$$

$$= \prod (1 / (1 + \lambda_i))$$

Roy's Greatest Root

$$= \max \lambda_i$$

Not one of these criteria has been identified as being universally superior to the others, although there are hypothesized situations where one criterion may outperform the others. Because we generally do not know the exact form of the alternative hypotheses being studied, the decision of which test criterion to use often becomes a matter of personal choice. Wilks' criterion is derived from a likelihood-ratio approach and appeals to some statisticians on those grounds.

Chapter 10 Generalized Linear Models

10.1 Introduction

The models considered in the previous chapters assumed a normal distribution. Many studies, however, use observations that cannot reasonably be assumed to be normally distributed. For example, a quality improvement engineer may be interested in whether a manufactured item is defective or not, or a medical researcher may want to know if an experimental treatment cures or fails to cure patients. Responses with only two possible outcomes are called Bernoulli random variables, and observations on them are called Bernoulli trials. Usually, the goal of studies that observe Bernoulli responses is to estimate the probability of a particular outcome—for example, defective items or cured patients—and see how treatments or other predictor variables affect the probability. Probabilities are estimated by sample proportions, that is, the number of times the outcome of interest occurs divided by the number of independent Bernoulli trials observed. Observations of this type have a binomial distribution. The binomial distribution is covered in most introductory statistics classes and is one of the best-known non-normal probability distributions.

Here are some other common examples of observations with non-normal distributions. Researchers often focus on counts—for example, numbers of birds, insects, or subatomic particles. Depending on the context in which the data are collected, distributions such as the Poisson or negative binomial are typically used to model random count data. In studies involving the reliability of a component, or survival of a patient, researchers want to know the time until an event occurs. Time-to-event variables are commonly assumed to have exponential or gamma distributions. In many studies, the main objective is to discover how treatments affect variation, rather than average response. In these studies, the variance is often used as the response variable. Variance estimates are typically proportional to random variables with chi-square distributions.

The methods described in previous chapters are unsuitable for non-normal response variables in general. This is because standard linear models methods assume normality and homogeneity of variance, but Bernoulli, count, time-to-failure, and other non-normal response variables often seriously violate these assumptions. Homogeneity of variance, for example, is usually violated because in most common non-normal distributions, the variance is a function of the mean. Consequently, changes in the independent variables that affect the mean of the response also affect its variance. Concerning normality, true normal distributions probably do not exist in nature and some departure from normality is inevitable. What *is* important is that the distribution of errors at least be reasonably symmetric. However, several of the most important non-normal distributions are skewed, a particularly serious deviation from normality.

While the methods described in previous chapters may not be suitable for non-normal data, the *approach* is still worthwhile. That is, researchers and data analysts find regression and analysis-of-variance models to be extremely convenient and useful tools. Therefore, statistical methods that allow regression and ANOVA to be adapted to non-normal data are very desirable. Traditionally, the main tool to adapt non-normal data to regression and ANOVA models has been transformations. More recently, *generalized linear models* have provided a systematic approach to adapting linear model methods.

Why generalized linear models rather than transformations? Most introductory statistical methods texts suggest various transformations for non-normal data. Standard transformations include $\sin^{-1}\sqrt{\text{pct}}$ for binomial data, where "pct" is the percent or sample proportion, $\sqrt{\text{count}}$ or $\sqrt{\text{count}+1}$ for count data, $\log(y)$ or $\log(y+1)$, for response variables, denoted by y, whose mean and variance are proportional. There are several problems with transformations. First, most commonly used transformations are intended to stabilize the variance, but they typically fail to address the problem of skewness. Consequently, transformations are often ineffective. Second, transformations express the data on scales that are unfamiliar to those who use the results of the analysis, often leaving data analysts with awkward problems when presenting their results. For example, when binomial data are analyzed using $\sin^{-1}\sqrt{\text{pct}}$, results such as estimates of treatment means usually need to be transformed back to the percent scale. When the standard errors are also transformed, a statistically significant difference on the $\sin^{-1}\sqrt{\text{pct}}$ scale may appear to be nonsignificant on the percent scale (or vice versa). The apparent contradiction can be difficult, even embarrassing, to explain.

Generalized linear models adapt linear model methods for use with non-normal data. Although special purpose models were in use prior to 1972, their connection was not well understood until Nelder and Wedderburn (1972) presented a unifying theory and first used the term "generalized linear model" in print. Their work provided an overall framework for models that predated their article, and formed the basis for subsequent development of new applications.

The main features of the generalized linear model are the **link function** and the **variance function**. The variance function describes the relationship between the mean and the variance of the distribution of the response variable. The variance function allows the generalized linear model to account for the unique features of the distribution. The link function is a mathematical model of the expected value of a random variable to which a linear model can be fit. Most non-normal random variables are not linear with respect to their expected value, unlike the normal random variable, which *is*. Therefore, while fitting a linear model directly to the mean works well with normally distributed data, you can usually do better with non-normal data by fitting the linear model indirectly, via some function of the mean. This is the link function. Because link functions are typically non-linear, generalized linear models can be viewed as a special case of non-linear models. However, because *the main benefit of generalized linear models is the ability to use linear regression and ANOVA methods on non-normal data,* they are best viewed as an extension of linear models.

Generalized linear models include such diverse applications that they are a topic for an entire textbook. Indeed, certain specialized applications of generalized linear models, such as categorical data analysis and survival analysis, are in themselves topics for entire texts. This chapter has a more limited aim. Its purpose is to give a brief introduction to how the basic linear regression models from Chapter 2, the ANOVA models from Chapter 3, the analysis-of-covariance models from Chapter 7, and the repeated-measures analysis from Chapter 8 are adapted to common kinds of non-normal data (Bernoulli response and random count data) using generalized linear models, and how to use SAS to compute basic analyses. Readers wanting more depth are referred to several more specialized texts.

PROC GENMOD is SAS' procedure for generalized linear models. PROC GENMOD serves roughly the same role for generalized linear models that PROC GLM serves for fixed-effects linear models for normally distributed observations. SAS also has several procedures, such as PROC CATMOD, PROC LIFEREG, PROC LOGISTIC, and PROC PROBIT, for an in-depth analysis of certain highly specialized applications. In the interests of staying focused on basic overall concepts, this chapter uses PROC GENMOD and does not discuss these other procedures. For an in-depth discussion of SAS for specialized applications, see Allison (1999) and Cantor (1997), for survival analysis, and Allison (1999), Stokes et al. (2000), and Zelterman (2002) for categorical data analysis.

Although PROC GENMOD can analyze certain forms of repeated measures and split-plot data, it has no facility for random-model effects. Generalized linear models with random effects, called generalized linear mixed-models (GLMMs), do exist and there is SAS software available. The non-linear mixed-model procedure NLMIXED can compute analysis for certain GLMMs. A macro, GLIMMIX, provides supplementary computations in conjunction with PROC MIXED to facilitate the analysis of a wide variety of GLMMs. Readers can refer to Chapter 11 in Littell et al. (1996), which covers examples of generalized linear mixed models and their analysis using the GLIMMIX macro.

A special note about nomenclature: Since Nelder and Wedderburn published their article, GLM has become the standard acronym for generalized linear models. This is confusing, especially for SAS users, because PROC GLM is the SAS procedure for fixed effects, normal errors linear models. However, PROC GLM does *not* analyze generalized linear models!

Prior to 1972, GLM was the accepted acronym for the general linear model—that is, what would now be called the normal errors linear model. Linear model specialists agree that what was once known as the general linear model is not at all general by contemporary standards, and hence the word "general" is usually dropped. The acronym LM, for linear model, is gaining acceptance for the normal errors linear model.

To summarize, PROC GLM analyzes LMs, that is, normal errors linear models, once known as general linear models. PROC GENMOD analyzes GLMs, that is, generalized linear models.

However, because this text uses PROC GLM extensively, referring to generalized linear models as GLMs risks hopeless confusion. In this chapter, when an acronym is needed for generalized linear model, we will use GzLM.

10.2 The Logistic and Probit Regression Models

The logistic regression model is useful when you want to fit a linear regression model to a binary response variable. You have several levels of an independent, or predictor variable, X. Denote these levels $X_1, X_2,...,X_m$. At the ith level of X, you have N_i ($i=1, 2,...,m$) observations, each of which is an independent Bernoulli trial. Of the N_i observations, y_i are classified as "the outcome of interest"—or "success"—and the remaining N_i-y_i have "the other" classification, that is, "failure."

At the ith level of X, y_i has a binomial distribution, or, more formally, y_i~Binomial(N_i, π_i), where N_i is the number of trials and π_i is the probability of a success on a given trial. The object of logistic regression is to estimate or test for changes in π_i associated with changes in X_i, specifically by modeling these changes via regression.

10.2.1 Logistic Regression: The *Challenger* Shuttle O-Ring Data Example

Following the 1986 *Challenger* space shuttle disaster, investigators focused on a suspected association between O-ring failure and low temperature at launch. Data documenting the presence or absence of primary O-ring thermal distress in the 23 shuttle launches preceding the *Challenger* mission appeared in Dalal et al. (1989) and were reproduced in Agresti (1996). Output 10.1 shows the raw data. Temperature at launch (TEMP) is the X variable. At each TEMP, TD denotes the number of launches in which thermal distress occurred and TOTAL gives the number of launches. TOTAL is the N variable, TD is the y variable, and the variable NO_TD is equal to $N-y$.

Output 10.1
Challenger
O-Ring Thermal
Distress Data

Obs	temp	td	no_td	total
1	53	1	0	1
2	57	1	0	1
3	58	1	0	1
4	63	1	0	1
5	66	0	1	1
6	67	0	3	3
7	68	0	1	1
8	69	0	1	1
9	70	2	2	4
10	72	0	1	1
11	73	0	1	1
12	75	1	1	2
13	76	0	2	2
14	78	0	1	1
15	79	0	1	1
16	81	0	1	1

Inspection of the data in Output 10.1 reveals that the incidence of thermal distress, indicated by the frequency of TD versus NO_TD, appears to be greater at low temperatures. Therefore, it is of interest to fit a model for which π, the probability of thermal distress, decreases as temperature increases. However, fitting a model directly to π, such as $\hat{\pi}_i = \beta_0 + \beta_1 X_i$, where X_i denotes the temperature at the ith launch, is not necessarily a reasonable approach. This is partly because, for theoretical reasons explained in Section 10.6, "Background Theory," the binomial random variable is not linear with respect to π. It is also partly because fitted values of π from this model are not bounded by 0 or 1, allowing the possibility of nonsense estimates of π.

A better approach is to fit the linear regression model to a function of π that is bounded by 0 and 1 and with which the binomial random variable at least theoretically has a linear relationship. Two such functions are the **logit**, defined as $\text{logit}(\pi_i) = \log\left(\dfrac{\pi_i}{1 - \pi_i}\right)$, and the **probit**, defined as $\text{probit}(\pi_i) = \Phi^{-1}(\pi_i)$, where $\Phi^{-1}(\cdot)$ is the inverse of the cumulative density function of the standard normal distribution—that is, the value on a standard normal table corresponding to a probability of π_i.

The logit and probit are both examples of **link functions**. The link function is a fundamental component of generalized linear models, because it specifies the relationship between the mean of the response variable and the linear model. Note that the mean of the sample proportion, y_i/N_i, is π_i. For reasons explained in more detail in Section 10.6, the logit is the most natural link function for binomial data. Models using the logit are called "logistic" models; in this case we are interested in a logistic *regression* model because we want to regress a binomial random variable on temperature.

The simplest logistic regression model for these data is $\text{logit}(\pi_i) = \beta_0 + \beta_1 X_i$. You can fit the logistic regression model by using PROC GENMOD and the following SAS program statements:

```
proc genmod;
model td/total=temp / link=logit dist=binomial type1;
```

From the SAS statements, you can see that GENMOD has a number of features in common with PROC GLM and PROC MIXED, but a number of unique features as well. As with GLM and MIXED, the MODEL statement has the general form of ⟨response variable⟩=⟨independent variable(s)⟩. The independent variables can be direct regression variables, or they can be CLASS variables, which you use in GENMOD to create the generalized linear model analog of analysis of variance. As with GLM and MIXED, GENMOD treats independent variables as direct regression variables by default and as ANOVA variables only if they appear first in a CLASS statement. Examples that use the CLASS statement appear later in this chapter.

For binomial **response variables** the syntax differs from other SAS linear model procedures. You specify the response variable as the ratio of the number of outcomes of interest (the y variable—in this case TD) divided by the number of observations per level of X (the N variable—in this case TOTAL). The binomial is unique in this respect. For other distributions, shown in examples later in this chapter, the form of the response variable is the same as other linear model procedures in SAS.

To complete the MODEL statement, you also specify the **distribution** of the response variable, the **link** function, and **other options**. The DIST option specifies the distribution. If you do not specify a distribution, GENMOD uses either the binomial distribution (if the response variable is a ratio, as above) or the normal distribution (for all other response variables) as the default. Several distributions are available in GENMOD. Consult SAS OnlineDoc, Version 8 for a complete list. Alternatively, you can provide your own distribution or quasi-likelihood, if none of the distributions provided with GENMOD are suitable. Section 10.4.5 presents an example of a user-specified distribution. The LINK option specifies the link function. If you do not specify a link function, GENMOD will use the **canonical link**—that is, the link that follows naturally from the probability distribution (see Section 10.6) that you select. In this example, the logit link is the default because the ratio response variable implies the binomial distribution and the logit is its canonical link. Thus, neither the DIST=BINOMIAL nor the LINK=LOGIT statements are actually needed for this example. However, it is good practice to include the DIST and LINK options even when they are not strictly necessary, if only for the sake of clarity.

The TYPE1 option yields likelihood ratio test statistics for hypotheses based on Type I estimable functions, as described in Chapter 6. You can also compute tests based on Type III estimable functions by using the TYPE3 option. For Type 3 tests, you can use likelihood ratio statistics, the default, or you can use the WALD option to compute Wald statistics. Section 10.6 gives explanations of likelihood ratio and Wald test statistics. Several other options are also available. This chapter illustrates several of these options where appropriate.

Output 10.2 shows the results generated by PROC GENMOD.

Output 10.2
Basic
GENMOD
Output for
Challenger
O-Ring
Logistic
Regression

```
                    The GENMOD Procedure

                     Model Information

          Data Set                    WORK.O_RING
          Distribution                   Binomial
          Link Function                     Logit
          Response Variable (Events)          td
          Response Variable (Trials)       total
          Observations Used                   16
          Number Of Events                     7
          Number Of Trials                    23

            Criteria For Assessing Goodness Of Fit

      Criterion               DF        Value      Value/DF

      Deviance                14       11.9974       0.8570
      Scaled Deviance         14       11.9974       0.8570
      Pearson Chi-Square      14       11.1303       0.7950
      Scaled Pearson X2       14       11.1303       0.7950
      Log Likelihood                  -10.1576

  Algorithm converged.

                  Analysis Of Parameter Estimates

                              Standard    Wald 95%         Chi-
  Parameter  DF  Estimate     Error    Confidence Limits   Square  Pr > ChiSq

  Intercept   1   15.0429     7.3786    0.5810   29.5048    4.16     0.0415
  temp        1   -0.2322     0.1082   -0.4443   -0.0200    4.60     0.0320
  Scale       0    1.0000     0.0000    1.0000    1.0000

  NOTE: The scale parameter was held fixed.

              LR Statistics For Type 1 Analysis

                                         Chi-
          Source        Deviance    DF   Square   Pr > ChiSq

          Intercept     28.2672
          temp          20.3152      1    7.95      0.0048
```

The beginning of the output contains some basic information about the data set. You can use this output to make sure that the data were read as intended, that the correct response variable was analyzed, that the right distribution and link were used, and so forth. The first substantive output is the "Criteria for Assessing Goodness of Fit." You can use the **deviance**, defined in Section 10.6, to check the fit of the model by comparing the computed deviance to a χ^2 distribution with 14 DF. In this case, the deviance is 11.9974 whereas the table value of $\chi^2_{(14)}$ at α=0.25 is 17.12, indicating no evidence of lack of fit.

The Pearson Chi-Square provides an alternative way to check goodness of fit. Like the deviance, the Pearson χ^2 also has an approximate $\chi^2_{(14)}$ distribution. Its computed value is 11.1303, similar to the deviance and also suggesting no evidence of lack of fit. The "Scaled Deviance" and "Scaled Pearson Chi-Square" are not of interest in this example. They are relevant when there is evidence of lack of fit resulting from overdispersion. Section 10.4.3 presents an example.

The "Analysis of Parameter Estimates" gives the estimates of the regression parameters as well as their standard errors and confidence limits. Here, the estimated intercept is $\hat{\beta}_0$ =15.0429 with a standard error of 7.3786. The estimated slope is $\hat{\beta}_1$ = –0.2322 with a standard error of 0.1082.

The "Chi-Square" statistics and associated *p*-values (Pr > ChiSq) given in the "Analysis of Parameter Estimates" table are Wald statistics for testing null hypotheses of zero intercept and slope. For example, the Wald χ^2 statistic to test H_0: β_1=0 is 4.60 and the *p*-value is 0.0320. You can also test the hypothesis of zero slope using the likelihood ratio statistic generated by the TYPE1 option and printed under "LR Statistics for Type I Analysis." The likelihood ratio χ^2 is 7.95 and its *p*-value is 0.0048. The fact that the likelihood ratio statistic is larger than the corresponding Wald statistic in this case is coincidental. In general, no pattern exists, and there is no compelling evidence in the literature to indicate that either statistic is preferable.

10.2.2 Using the Inverse Link to Get the Predicted Probability

From the output, you can see that 15.049–0.2322*TEMP is the estimated regression equation. The regression equation allows you to compute the predicted **logit** for a desired temperature. For example, at 50°, the predicted logit is 15.0429–0.2322*50=3.4329.

Typically, the logit is not of direct interest. On the other hand, the predicted probability *is* of interest—in this case the probability of O-ring thermal distress occurring at a given temperature. You use the **inverse link** function to convert the logit to a probability. In this example, the logit link function, $\eta = \log \frac{\pi}{1-\pi}$, hence, the inverse link is $\pi = \frac{e^{\eta}}{1+e^{\eta}}$.

For 50°, using $\hat{\eta} = 3.4329$ as calculated above, the predicted probability is therefore

$\hat{\pi} = \frac{e^{3.4329}}{1 + e^{3.4329}} = 0.9687$. That is, according to the logistic regression estimated from the

data, the probability of observing primary O-ring thermal distress at 50° is 0.9687.

You can convert the standard error from the link function scale to the inverse link scale using the **Delta Rule**. The general form of the Delta Rule for generalized linear models is

$\text{var}[h(\hat{\eta})]$ is approximately equal to $\left(\dfrac{\partial h(\eta)}{\partial \eta}\right)^2 \text{var}(\hat{\eta})$. For the logit link, some algebra

yields $\dfrac{\partial h(\eta)}{\partial \eta} = \pi(1-\pi)$ and hence the standard error of $\hat{\pi} = \hat{\pi}(1-\hat{\pi}) \times \text{s.e.}(\hat{\eta})$. You can use

GENMOD to compute s.e.$(\hat{\eta})$, as well as $\hat{\eta}$ and related statistics, using the ESTIMATE statement. The syntax and placement of the ESTIMATE statement are similar to PROC GLM and PROC MIXED. Here are the statements to compute $\hat{\eta}$ for several temperatures of interest. Output 10.3 shows the results.

```
estimate 'logit at 50 deg' intercept 1 temp 50;
estimate 'logit at 60 deg' intercept 1 temp 60;
estimate 'logit at 64.7 deg' intercept 1 temp 64.7;
estimate 'logit at 64.8 deg' intercept 1 temp 64.8;
estimate 'logit at 70 deg' intercept 1 temp 70;
estimate 'logit at 80 deg' intercept 1 temp 80;
```

Output 10.3
Estimated
Logits for
Various
Temperatures of
Interest

```
                        Contrast Estimate Results

                              Standard                              Chi-
    Label            Estimate   Error    Alpha  Confidence Limits  Square

    logit at 50 deg    3.4348   2.0232    0.05   -0.5307   7.4002    2.88
    logit at 60 deg    1.1131   1.0259    0.05   -0.8975   3.1238    1.18
    logit at 64.7 deg  0.0220   0.6576    0.05   -1.2669   1.3109    0.00
    logit at 64.8 deg -0.0012   0.6518    0.05   -1.2788   1.2764    0.00
    logit at 70 deg   -1.2085   0.5953    0.05   -2.3752  -0.0418    4.12
    logit at 80 deg   -3.5301   1.4140    0.05   -6.3014  -0.7588    6.23
```

The column "Estimate" gives you the estimated logit. For LOGIT AT 50 DEG, $\hat{\eta}$ at $50°$, the computed value is 3.4348, rather than the hand-calculated $\hat{\eta} = 3.4329$ given above. This reflects a rounding error: SAS computations involve much greater precision. From the output, you can see that the standard error of $\hat{\eta}$ at $50°$ is 2.0232. Using the Delta Rule, the standard error for $\hat{\pi}$ is $\hat{\pi}(1-\hat{\pi}) \times \text{s.e.}(\hat{\eta}) = 0.9687 \times (1-0.9687) \times 2.0232 = 0.0613$.

In addition to $\hat{\eta}$ and s.e. $(\hat{\eta})$, Output 10.3 also gives upper and lower 95% confidence limits for the predicted logit. You can use the inverse link to convert these to confidence limits for the predicted probability. For example, at $50°$, the lower confidence limit for η

is -0.5307. Applying the inverse link, the lower confidence limit for π is $\dfrac{e^{-0.5307}}{1+e^{-0.5307}} =$

0.3704. A similar computation using the upper confidence limit for η, 7.4002, yields the upper confidence limit for π, 0.9994. It is better to use the upper and lower limits for η and convert them using the inverse link rather than using the standard error of $\hat{\pi}$ computed from the Delta Rule. The standard error results in a symmetric interval, that is, $\hat{\pi} \pm t \times \text{s.e.}(\pi)$, which is not, in general, a sensible confidence interval. The confidence interval should be asymmetric reflecting the non-linear nature of the link function.

You can compute $\hat{\pi}$, its standard error and confidence interval, using the ODS output statement in GENMOD followed by program statements to implement the inverse link and Delta Rule. First, you insert the following ODS statement after the ESTIMATE statements in the GENMOD procedure:

```
ods output estimates=logit;
```

Then use the following statements:

```
data prob_hat;
  set logit;
    phat=exp(estimate)/(1+exp(estimate));
    se_phat=phat*(1-phat)*stderr;
    prb_LcL=exp(LowerCL)/(1+exp(LowerCL));
    prb_UcL=exp(UpperCL)/(1+exp(UpperCL));
proc print data=prob_hat;
run;
```

The statements produce Output 10.4.

Output 10.4
PROC PRINT
of a Data Set
Containing
s.e.( ), and
Upper and
Lower
Confidence
Limits

Obs	Label	Estimate	StdErr	Alpha	LowerCL	UpperCL
1	logit at 50 deg	3.4348	2.0232	0.05	-0.5307	7.4002
2	logit at 60 deg	1.1131	1.0259	0.05	-0.8975	3.1238
3	logit at 64.7 deg	0.0220	0.6576	0.05	-1.2669	1.3109
4	logit at 64.8 deg	-0.0012	0.6518	0.05	-1.2788	1.2764
5	logit at 70 deg	-1.2085	0.5953	0.05	-2.3752	-0.0418
6	logit at 80 deg	-3.5301	1.4140	0.05	-6.3014	-0.7588

Obs	ChiSq	Prob ChiSq	phat	se_phat	prb_LcL	prb_UcL
1	2.88	0.0896	0.96877	0.06121	0.37036	0.99939
2	1.18	0.2779	0.75271	0.19095	0.28956	0.95786
3	0.00	0.9733	0.50549	0.16439	0.21978	0.78766
4	0.00	0.9985	0.49969	0.16296	0.21775	0.78183
5	4.12	0.0423	0.22997	0.10541	0.08509	0.48955
6	6.23	0.0125	0.02847	0.03911	0.00183	0.31891

The variables PHAT, SE_PHAT, PRB_LCL, and PRB_UCL give $\hat{\pi}$, its standard error, and the confidence limits.

Output 10.3 and Output 10.4 also give chi-square statistics. You can use these to test H_0: $\eta=0$ for a given temperature. In categorical data analysis, $\frac{\pi}{1=\pi}$ is defined as the **odds** and hence $\hat{\eta}$ estimates the log of the odds for a given temperature. An odds of 1, and hence a log odds of 0, means that an event is equally likely to occur or not occur. In Output 10.4, those temperatures whose χ^2 and associated *p*-values (ProbChiSq) result in a failure to reject H_0 are temperatures for which there is insufficient evidence to contradict the hypothesis that there is a 50-50 chance of thermal distress occurring at that temperature. Whether this hypothesis is useful depends on the context. In many cases, the confidence limits of $\hat{\pi}$ may be important. What is striking in the O-ring data is that the upper confidence limit for the likelihood of O-ring thermal distress is fairly high (considering the consequences of O-ring failure), even at 80°. When the *Challenger* was launched, it was 31°.

One final note regarding the odds. The estimated slope $\hat{\beta}_1 = -0.2322$ is interpreted as the log odds **ratio** per one-unit change in *X*. Thus $e^{\hat{\beta}_1} = e^{-0.2322} = 0.793$ is the ratio defined as $\frac{\text{odds at a given temperature}}{\text{odds at temperature} - 1}$. An odds ratio < 1 indicates the odds of thermal distress decrease as temperature increases.

10.2.3 Alternative Logistic Regression Analysis Using 0-1 Data

In the previous section, there was one row in the data set for each temperature level with a variable for *N*, the number of observations per level, and one for *y*, the number of outcomes with the characteristic of interest. You can also enter binomial data with one row per observation, with each observation classified by which of the two possible outcomes was observed. Output 10.5 shows the O-ring data entered in this way.

Output 10.5
O-Ring Data Entered by Observation Rather Than by Temperature Level

Obs	launch	temp	td
1	1	66	0
2	2	70	1
3	3	69	0
4	4	68	0
5	5	67	0
6	6	72	0
7	7	73	0
8	8	70	0
9	9	57	1
10	10	63	1
11	11	70	1
12	12	78	0
13	13	67	0
14	14	53	1
15	15	67	0
16	16	75	0
17	17	70	0
18	18	81	0
19	19	76	0
20	20	79	0
21	21	75	1
22	22	76	0
23	23	58	1

There are three variables for each observation—an identification for the shuttle launch (LAUNCH), the temperature at the time of launch (TEMP) and an indicator for whether or not there was thermal distress (TD=0 means no distress, TD=1 means there was distress).

You can estimate the logistic regression model using the 0-1 data with the following GENMOD statements:

```
proc genmod;
  model td=temp  /dist=binomial link=logit type1;
```

These statements differ from the GENMOD program used in the previous section to obtain Output 10.2. First, the sample proportion *y*/*N*, used as the response variable to compute Output 10.2, is replaced here by TD, the 0-1 variable. Also, because TD is not a ratio response variable, you must specify DIST=BINOMIAL, or GENMOD will use the normal distribution. As before, the LINK=LOGIT statement is not necessary because the logit link is the default for the binomial distribution, but it *is* good form. The results appear in Output 10.6.

Output 10.6
Results of the PROC GENMOD Analysis of the 0-1 Form of O-Ring Data

```
                        The GENMOD Procedure

                        Model Information

            Data Set                WORK.TBL_5_10
            Distribution                 Binomial
            Link Function                   Logit
            Dependent Variable                 td
            Observations Used                  23
            Probability Modeled     Pr( td = 1 )

                        Response Profile

                Ordered     Ordered
                Level       Value       Count

                    1         0            16
                    2         1             7

            Criteria For Assessing Goodness Of Fit

        Criterion                DF       Value      Value/DF

        Deviance                 21     20.3152       0.9674
        Scaled Deviance          21     20.3152       0.9674
        Pearson Chi-Square       21     23.1691       1.1033
        Scaled Pearson X2        21     23.1691       1.1033
        Log Likelihood                 -10.1576

Algorithm converged.

                Analysis Of Parameter Estimates

                        Standard      Wald 95%         Chi-
    Parameter   DF  Estimate   Error  Confidence Limits  Square  Pr > ChiSq

    Intercept   1   15.0429   7.3786   0.5810   29.5048   4.16      0.0415
    temp        1   -0.2322   0.1082  -0.4443   -0.0200   4.60      0.0320
    Scale       0    1.0000   0.0000   1.0000    1.0000

NOTE: The scale parameter was held fixed.

                LR Statistics For Type 1 Analysis

                                            Chi-
        Source          Deviance      DF    Square    Pr > ChiSq

        Intercept       28.2672
        temp            20.3152       1      7.95       0.0048
```

Compared to Output 10.2, the "Model Information" is in somewhat different form, reflecting the difference between using individual outcomes of each Bernoulli response rather than the sample proportion for each temperature level. The goodness-of-fit statistics, deviance and Pearson χ^2, are also different because the response variable and hence the log likelihood are not the same. Using the data in Output 10.1, there were $N=16$ observations, that is, 16 sample proportions, one per temperature level, and hence the deviance had $N-p=16-2=14$ DF, where p corresponds to the 2 model degrees of freedom for β_0 and β_1. Using the data in Output 10.5, there are $N=23$ distinct observations, and hence $N-p = 23-2=21$ degrees of freedom for the lack-of-fit statistics. The deviance and Pearson χ^2 are the only statistics affected by whether you use sample proportion data or 0-1 data.

The "Analysis of Parameter Estimates" and likelihood ratio test statistics for the Type I test of H_0: $\beta_1=0$ are identical to those computed using the sample proportion data. You can also compute estimated logit for various temperatures using the same ESTIMATE statements shown previously in Section 10.2.2. The output is identical to that shown in Output 10.3. Therefore, when you apply the inverse link and Delta Rule, you use the same program statements and get the same results as those presented in Output 10.4.

10.2.4 An Alternative Link: Probit Regression

As mentioned in Section 10.2.1, the probit link is another function suitable for fitting regression and ANOVA models to binomial data. The probit model assumes that the observed Bernoulli "success" or "failure" results from an underlying, but not directly observable, normally distributed random variable. Figure 10.1 illustrates the hypothesized model.

Figure 10.1 *Illustration of a Model Underlying a Probit Link*

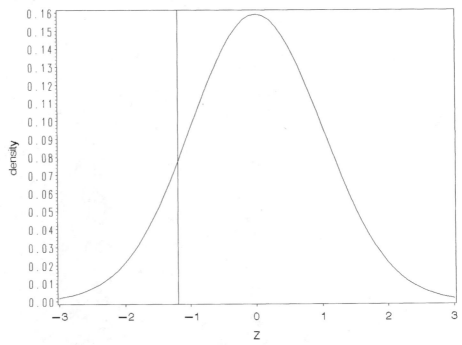

Denote the underlying, unobservable random variable by Z and suppose that Z is associated with a predictor variable X according to the linear regression equation, $Z = \beta_0 + \beta_1 X$. Remember, you cannot observe Z; all you can observe is the consequences of Z. If Z is below a certain level, you observe a success. Otherwise, you observe a failure. The regression of Z on X models how the failure-success boundary changes with X. Figure 10.1 depicts a case for which the boundary, denoted Z_x, for a given X is equal to -1.2. Thus, the area under the normal curve below $Z_x = -1.2$ is the probability of a success for the corresponding X. As X changes, the boundary value Z_x changes, thereby altering the probability of a success.

Formally, the standard normal cumulative distribution function, that is, the area under the curve less than Z, is denoted $\Phi(z) = \int_{-\infty}^{z} \frac{1}{\sqrt{2\pi}} e^{-\frac{X^2}{2}} dx$. Thus, the probit linear regression model can be written $\pi = \Phi(\beta_0 + \beta_1 X)$. Note that this give the model in the form of the inverse link. You can write the probit model in terms of the link function as probit$(\pi) = \Phi^{-1}(\pi) = \beta_0 + \beta_1 X$, where $\Phi^{-1}(\pi)$ means the Z value such that the area under the curve less than Z is π.

You can fit the probit regression model to the O-ring data using the following SAS statements:

```
proc genmod data=agr_135;
  model td/total=temp/ link=probit type1;
```

Note the use of the LINK=PROBIT option but no DIST option. Because of the ratio response variable, the binomial distribution is assumed by default, but a LINK statement is required because the PROBIT link is not the default. The results appear in Output 10.7.

Output 10.7
GENMOD
Results Fitting
the PROBIT
Link to O-Ring
Data

```
                   Criteria For Assessing Goodness Of Fit

           Criterion              DF         Value        Value/DF

           Deviance               14        12.0600        0.8614
           Scaled Deviance        14        12.0600        0.8614
           Pearson Chi-Square     14        10.9763        0.7840
           Scaled Pearson X2      14        10.9763        0.7840

                     Analysis Of Parameter Estimates

                          Standard      Wald 95%          Chi-
    Parameter  DF  Estimate   Error   Confidence Limits  Square  Pr > ChiSq

    Intercept   1    8.7750  4.0286   0.8790  16.6709     4.74      0.0294
    temp        1   -0.1351  0.0584  -0.2495  -0.0206     5.35      0.0207

                    LR Statistics For Type 1 Analysis

                                            Chi-
           Source        Deviance    DF    Square    Pr > ChiSq

           Intercept      19.9494
           temp           12.0600     1     7.89        0.0050
```

The results are not strikingly different from the results of the logistic regression. The deviance is 12.060 (versus 11.997 for the logit link) and the *p*-value for the likelihood ratio test of H_0: $\beta_1=0$ is 0.0050 (versus 0.0320 using the logit link). The estimate of β_1 is different, reflecting a different scale for the probit versus the logit. However, the sign and conclusion regarding the effect of temperature on thermal distress is the same.

You can use the ESTIMATE statements as shown in Output 8.3 to obtain predicted probits for various temperatures. You use the inverse link, Φ(estimate), to convert predicted probits to predicted probabilities. The SAS function to evaluate Φ(estimate) is PROBNORM. You use the following SAS statements to obtain the probit model analog to Output 10.4:

```
estimate 'probit at 50 deg' intercept 1 temp 50;
estimate 'probit at 60 deg' intercept 1 temp 60;
estimate 'probit at 64.7 deg' intercept 1 temp 64.7;
estimate 'probit at 64.8 deg' intercept 1 temp 64.8;
estimate 'probit at 70 deg' intercept 1 temp 70;
estimate 'probit at 80 deg' intercept 1 temp 80;
ods output estimates=probit;
  run;
```

```
data prob_hat;
  set probit;
    phat=probnorm(estimate);
    pi=3.14159;
    invsqrt=1/(sqrt(2*pi));
    se_phat=invsqrt*exp(-0.5*(estimate**2))*stderr;
    prb_LcL=probnorm(LowerCL);
    prb_UcL=probnorm(UpperCL);
  proc print data=prob_hat;
```

The results appear in Output 10.8. Note the form of the Delta Rule for the probit model to obtain the approximate standard error of $\hat{\pi}$. This follows from the fact that the approximate standard error of $\hat{\pi}_1$, using the Delta Rule, is $\dfrac{\partial \Phi(\eta)}{\partial \eta} \times$ s.e.(η). The

derivative $\dfrac{\partial \Phi(\eta)}{\partial \eta} = \dfrac{\partial}{\partial \eta} \int_{-\infty}^{\eta} \dfrac{1}{\sqrt{2\pi}} e^{-\frac{X^2}{2}} dx = \dfrac{1}{\sqrt{2\pi}} e^{-\frac{\eta^2}{2}}$.

Output 10.8
Predicted
Probits and
Probabilities
Obtained
from the
PROBNORM
Inverse Link
and the Probit
Form of the
Delta Rule

Obs	Label	Estimate	StdErr	Alpha	LowerCL	UpperCL
1	probit at 50 deg	2.0201	1.1413	0.05	-0.2167	4.2570
2	probit at 60 deg	0.6692	0.6024	0.05	-0.5115	1.8498
3	probit at 64.7 deg	0.0342	0.3960	0.05	-0.7420	0.8104
4	probit at 64.8 deg	0.0207	0.3925	0.05	-0.7487	0.7901
5	probit at 70 deg	-0.6818	0.3244	0.05	-1.3175	-0.0461
6	probit at 80 deg	-2.0328	0.7277	0.05	-3.4590	-0.6066

Obs	ChiSq	Prob ChiSq	phat	pi	invsqrt	se_phat	prb_LcL	prb_UcL
1	3.13	0.0767	0.97832	3.14159	0.39894	0.05917	0.41421	0.99999
2	1.23	0.2666	0.74831	3.14159	0.39894	0.19211	0.30450	0.96783
3	0.01	0.9312	0.51365	3.14159	0.39894	0.15790	0.22905	0.79115
4	0.00	0.9579	0.50826	3.14159	0.39894	0.15657	0.22703	0.78526
5	4.42	0.0356	0.24768	3.14159	0.39894	0.10257	0.09383	0.48163
6	7.80	0.0052	0.02104	3.14159	0.39894	0.03678	0.00027	0.27207

Comparing Output 10.8 to the analogous output for the logistic model in Output 10.4, the estimated probabilities, approximate standard errors, and lower and upper confidence limits are similar, though not equal, for the two models. For example, for the logit model, at 50°, the predicted probability of thermal distress was 0.969 with an approximate standard error of 0.061, whereas for the probit model the predicted probability (PHAT) is 0.978, with an approximate standard error of 0.059. Other "discrepancies" are similarly small; you reach essentially the same conclusions about the O-ring data with either link function.

In general, the logit and probit models produce similar results. In fact, the logit and probit are very similar functions of π, so the fact that they produce similar results is not surprising. There are no compelling statistical reasons to choose one over the other. In some studies, you use the logistic model because its interpretation in terms of odds-ratios fits the subject matter. In other disciplines, the probit model of the mean has a theoretical basis, so the probit is preferred.

10.3 Binomial Models for Analysis of Variance and Analysis of Covariance

Logit and probit link functions are not limited to regression models. You can use them in conjunction with any of the classification models—one-way, two-way, and so forth—used in analysis of variance with normally distributed data. This allows you to perform the equivalent of analysis of variance for non-normal data. You can also fit the logit and probit links to analysis-of-covariance models. This section presents two examples, one for analysis of variance, the other for analysis of covariance. The analysis of variance example uses both the logit link (Section 10.3.1) and the probit link (Section 10.3.2). Section 10.3.3 presents analysis of covariance with a logit link.

10.3.1 Logistic ANOVA

Output 10.9 shows data from a clinical trial comparing two treatments, an experimental drug versus a control.

Output 10.9
Binomial
Clinical Trial
Data

Obs	clinic	trt	fav	unfav	nij
1	1	drug	11	25	36
2	1	cntl	10	27	37
3	2	drug	16	4	20
4	2	cntl	22	10	32
5	3	drug	14	5	19
6	3	cntl	7	12	19
7	4	drug	2	14	16
8	4	cntl	1	16	17
9	5	drug	6	11	17
10	5	cntl	0	12	12
11	6	drug	1	10	11
12	6	cntl	0	10	10
13	7	drug	1	4	5
14	7	cntl	1	8	9
15	8	drug	4	2	6
16	8	cntl	6	1	7

The study was conducted at eight clinics. At each clinic, patients were assigned at random to either the experimental drug or the control. Denote n_{ij} as the number of patients (NIJ in the SAS data set) at the jth clinic assigned to the ith treatment. At the end of the trial, patients either had favorable or unfavorable responses to treatment. For each treatment-clinic combination, the number of patients with favorable (FAV) and unfavorable (UNFAV) responses was recorded.

Beitler and Landis (1985) discussed these data assuming that CLINIC constituted a random effect. Littell et al. (1996) used the data as an example of a generalized linear mixed model, again assuming random clinic effects. This section presents a simpler analysis, using PROC GENMOD, assuming that the design can be viewed as a randomized-blocks design with CLINIC assumed to be a fixed-blocks effect.

Assuming the RCBD with fixed blocks, a model for the data is

$$logit(\pi_{ij}) = \mu + \tau_i + \beta_j,$$

where

π_{ij} is the probability of a favorable response for the *i*th treatment at the *j*th clinic ($i = 1, 2; j = 1, 2,...,8$).

μ is the intercept.

τ_i is the effect of the *i*th treatment.

β_j is the effect of the *j*th clinic.

Note that the response variable is FAV, the number of patients having favorable responses, or, alternatively, the sample proportion, FAV/NIJ. However, generalized linear models are given in terms of the link function, rather than the response variable, on the left-hand side of the model equation.

You can run the analysis using the SAS statements

```
proc genmod;
 class clinic trt;
 model fav/nij=clinic trt/dist=binomial link=logit type3;
run;
```

Output 10.10 shows the results of interest for this analysis.

Output 10.10
Logistic
ANOVA for
Binomial
Clinical Trial
Data

```
                      The GENMOD Procedure

                      Model Information

            Data Set                        WORK.A
            Distribution                    Binomial
            Link Function                   Logit
            Response Variable (Events)      fav
            Response Variable (Trials)      nij
            Observations Used               16
            Number Of Events                102
            Number Of Trials                273

                  Class Level Information

            Class      Levels    Values

            clinic        8       1 2 3 4 5 6 7 8
            trt           2       cntl drug

              Criteria For Assessing Goodness Of Fit

       Criterion             DF       Value      Value/DF

       Deviance               7      9.7463       1.3923
       Scaled Deviance        7      9.7463       1.3923
       Pearson Chi-Square     7      8.0256       1.1465
       Scaled Pearson X2      7      8.0256       1.1465
       Log Likelihood              -138.5100

  Algorithm converged.

             LR Statistics For Type 3 Analysis

                              Chi-
           Source       DF    Square    Pr > ChiSq

           clinic        7     81.21      <.0001
           trt           1      6.67      0.0098
```

The first part of the output gives the model and class level information. You use this to verify that GENMOD is interpreting the data and your model instructions correctly.

The important items in the output are the "Criteria for Assessing Goodness of Fit" and the "LR Statistics for Type 3 Analysis." You use the goodness-of-fit criteria as they were used for logistic regression. At $\alpha=0.25$, the table $\chi^2_{(7)}=9.04$ and at $\alpha=0.10$, $\chi^2_{(7)}=12.02$. Neither the deviance, 9.7463, nor the Pearson χ^2, 8.0256, gives evidence of significant lack of fit. The Type 3 likelihood ratio statistics are equivalent to the analysis of variance results for hypotheses based on Type 3 estimable functions for data with normally distributed errors. The LR χ^2 statistic for CLINIC is 81.21 with 7 degrees of freedom. As in the case of conventional ANOVA, formal testing of hypotheses about blocking criteria is questionable, but you *can* interpret the large χ^2 value as evidence of clinic differences and hence the effectiveness of blocking by CLINIC. The main conclusion follows from the LR test for treatment. Based on the χ^2 of 6.67 and its *p*-value of 0.0098, you reject H_0: $\tau_1=\tau_2$ and thus conclude that there is a statistically significant treatment effect.

You can compute LS means just as you can with conventional ANOVA. After the model statement in the GENMOD statements above, add the command:

```
lsmeans trt;
```

This produces the results given in Output 10.11. The estimate and standard error are given on the logit scale. For example, for CNTL, the control treatment, the LS mean is -1.2554 with a standard error of 0.2692. This is the estimated logit that results from computing $\hat{\mu}+\hat{\tau}_1+\overline{\beta}_\bullet$. The chi-square values are the Wald statistics to test H_0: LSMEAN (TRT i) = 0. On the logit scale, LSMEAN=0 implies that the log odds for the ith treatment is 0 or the odds equal 1.

Output 10.11
Least-Squares
Means for
Binomial
Clinical Trial
Data

```
                         Least Squares Means

                            Standard          Chi-
    Effect   trt   Estimate    Error    DF   Square   Pr > ChiSq

    trt      cntl   -1.2554   0.2692    1    21.74     <.0001
    trt      drug   -0.4784   0.2592    1     3.41     0.065
```

If you want the equivalent of the least-squares mean for the probability of a favorable outcome, you need to use the inverse link. Section 8.3.1 showed you how to use the inverse link to get the probability, the Delta Rule for the standard error, and how to obtain upper and lower confidence limits. For the control treatment, the estimated probability is $\frac{\exp^{-1.2554}}{1+\exp^{-1.2554}}=0.22$ and the standard error of $\hat{\pi}_1$, the "LS Mean of treatment 1," is approximately $0.22*(1-0.22)*(0.2692)=0.046$. As shown in Section 8.3.1, you can use ODS statements to output the logit-scale LS means and apply the inverse link and Delta Rule to convert them to the probability scale. You add the following statement to GENMOD after the LSMEANS statement:

```
ods output lsmeans=lsm;
```

and then run the following:

```
data prob_hat;
 set lsm;
  phat=exp(estimate)/(1+exp(estimate));
  se_phat=phat*(1-phat)*stderr;
proc print data=prob_hat;
run;
```

Output 10.12 shows the results. PHAT is $\hat{\pi}_i$, the LS mean probability for the *i*th treatment.

Output 10.12
LS Means
Converted to
Probability
Scale for
Binomial
Clinical Trial
Data

Obs	Effect	trt	Estimate	StdErr	DF	ChiSq	ChiSq	phat	se_phat
1	trt	cntl	-1.2554	0.2692	1	21.74	<.0001	0.22177	0.046467
2	trt	drug	-0.4784	0.2592	1	3.41	0.0650	0.38262	0.061238

GENMOD allows you to use the ESTIMATE and CONTRAST statements. You include these statements after the MODEL statement. They can go either before or after the LSMEANS statement, if you are using both. Here are typical examples:

```
estimate 'lsm - cntl' intercept 1 trt 1 0;
 estimate 'lsm - drug' intercept 1 trt 0 1;
  estimate 'diff' trt 1 -1;
  contrast 'diff' trt 1 -1;
```

You can see that the syntax is identical to other SAS PROCs, such as GLM and MIXED. Output 10.13 shows the results.

Output 10.13
ESTIMATE
and
CONTRAST
Results for
Binomial
Clinical Trial
Data

Contrast Estimate Results

Label	Estimate	Standard Error	Alpha	Confidence Limits		Chi-Square	Pr > ChiSq
lsm - cntl	-1.2554	0.2692	0.05	-1.7830	-0.7277	21.74	<.0001
lsm - drug	-0.4784	0.2592	0.05	-0.9865	0.0297	3.41	0.0650
diff	-0.7769	0.3067	0.05	-1.3780	-0.1758	6.42	0.0113

Contrast Results

Contrast	DF	Chi-Square	Pr > ChiSq	Type
diff	1	6.67	0.0098	LR

The estimates LSM-CNTL and LSM-DRUG are the treatment least-squares means; the statistics are identical to those discussed earlier for the LS means. The estimate of DIFF is $\hat{\tau}_1 - \hat{\tau}_2$ and the associated chi-square is the Wald statistic for testing H_0: $\tau_1 - \tau_2 = 0$. The CONTRAST statement computes a test for H_0: $\tau_1 - \tau_2 = 0$, but it uses the likelihood ratio statistic. You can see that the two statistics are similar and would thus result in similar conclusions. As mentioned in Section 8.3, there is no clear evidence in favor of the Wald or likelihood ratio statistic relative to the other.

Caution: You *cannot* convert a difference to the probability scale using the inverse link—it results in nonsense. For example, on the logit scale, $\hat{\tau}_1 - \hat{\tau}_2 = -0.7769$, thus the inverse link is $\dfrac{\exp^{-0.7769}}{1+\exp^{-0.7769}} = 0.32$. However, from the previous discussion of the LS means, the difference between $\hat{\pi}_1$ and $\hat{\pi}_2$ was approximately 0.16. Because the logit is a non-linear function, differences are not preserved when the function or its inverse is applied.

You *can* interpret $\hat{\tau}_1 - \hat{\tau}_2$ as the log of the odds-*ratio* between the treatments. An odds-ratio of 1, and hence a log odds-ratio of 0 mean that the odds, for example, of favorable versus unfavorable response, are the same for the two treatments. A significant negative log odds means that the odds of a favorable response are lower for the control than they are for the experimental drug, evidence that the experimental drug is effective. You can also see this reflected in the estimates of $\hat{\pi}_1$ and $\hat{\pi}_2$ given earlier.

An alternative model for these data includes a term for the treatment-clinic interaction. The model is

$$\text{logit}(\pi_{ij}) = \mu + \tau_i + \beta_j + (\tau\beta)_{ij},$$

where $(\tau\beta)_{ij}$ is the treatment-clinic interaction for the *i*th treatment and *j*th clinic and all other terms are as previously defined. The SAS statements for the model are

```
proc genmod data=a;
  class clinic trt;
  model fav/nij=clinic trt clinic*trt/dist=binomial
link=logit type3;
```

These statements are identical to the SAS program for the model without $(\tau\beta)_{ij}$ except for the addition of the CLINIC*TRT term to the model. Output 10.14 gives the goodness-of-fit statistics and the ANOVA analog, that is, the Type 3 likelihood ratio statistics.

Output 10.14 Goodness of Fit and Likelihood Ratio Type 3 ANOVA of a Model with an Interaction for Binomial Clinical Trial Data

```
                 Criteria For Assessing Goodness Of Fit

     Criterion              DF          Value        Value/DF

     Deviance               0          0.0000            .
     Pearson Chi-Square     0          0.0000            .

              LR Statistics For Type 3 Analysis

                                  Chi-
         Source        DF        Square     Pr > ChiSq

         clinic         7        83.14        <.0001
         trt            1         5.86        0.0155
         clinic*trt     7         9.75        0.2034
```

For the model with interaction, there are 0 degrees of freedom for the goodness-of-fit statistics. This is a **saturated model** because there are 16 treatment clinic combinations, and hence 15 total degrees of freedom, all of which are used in the model. Also, the CLINIC*TRT likelihood ratio χ^2 statistic has the same degrees of freedom and the same value, 9.75, as the deviance in the no-interaction model. In fact, the likelihood ratio χ^2 for CLINIC*TRT and the deviance in the no-interaction model are the same statistic, expressed in different ways. In general, the deviance in ANOVA-type generalized linear models *is* the sum of all effects—typically interaction effects—left out of the model. In principle, this is similar to using interaction effects assumed to be zero in place of MS(ERROR) in standard ANOVA models.

10.3.2 The Analysis-of-Variance Model with a Probit Link

Just as you saw with regression in Section 10.2, you can fit an analysis-of-variance model to binomial data using a logit or probit link. In many cases, a mathematical model that follows from underlying theory in a given discipline provides the link function. **Probit analysis,** data analysis using a generalized linear model for binomial data with a **probit link**, is a common example of a subject matter theory-driven approach. For example, quantitative genetics has a well-developed theory framed in terms of the normal distribution. The probit link allows this theory to be adapted easily to Bernoulli response variables.

The basic premise of the probit link is that there exists an unobservable, normally distributed process. The observable consequence of the process is a Bernoulli random variable. For example, in the Beitler and Landis (1985) binomial clinic trial data, imagine that there is a continuous random variable roughly defined as the amount of disease stress to which the patient is subjected. The actual disease stress random variable cannot be measured. If the disease stress is sufficient, we observe an unfavorable outcome. Otherwise, we observe a favorable outcome. Only the favorable or unfavorable Bernoulli response variable is observable.

A probit model for the data shown in Output 10.9 is

$$\text{probit}(\pi_{ij}) = \Phi^{-1}(\pi_{ij}) = \mu + \tau_i + \beta_j$$

where μ, τ_i, and β_j are the intercept, treatment effect, and block effect as defined in Section 8.4.1. You use the following statements for the analysis:

```
proc genmod data=a;
  class clinic trt;
  model fav/nij=clinic trt/dist=binomial link=probit type3;
  lsmeans trt;
  estimate 'lsm - cntl' intercept 1 trt 1 0;
  estimate 'lsm - drug' intercept 1 trt 0 1;
  estimate 'diff' trt 1 -1;
  contrast 'diff' trt 1 -1;
```

The only difference between the GENMOD statement above and the logistic ANOVA in Section 10.3.1 is the substitution of LINK=PROBIT in the above MODEL statement in place of LINK=LOGIT. Note that if you leave out the LINK= statement, GENMOD uses the logit link by default, because it is the canonical link for the binomial distribution. If you want to use the probit link, you must use the LINK=PROBIT option. Output 10.15 shows the results.

Output 10.15
Analysis of
Binomial
Clinical Trial
Data Using
the Probit
Link

```
                    Criteria For Assessing Goodness Of Fit

          Criterion                 DF          Value        Value/DF

          Deviance                   7          9.6331        1.3762
          Pearson Chi-Square         7          8.0048        1.1435
```

```
                       Analysis Of Parameter Estimates
```

Parameter		DF	Estimate	Standard Error	Wald 95% Confidence Limits		Chi-Square
Intercept		1	0.9792	0.3934	0.2082	1.7503	6.20
clinic	1	1	-1.3158	0.4120	-2.1233	-0.5082	10.20
clinic	2	1	-0.0678	0.4254	-0.9014	0.7659	0.03
clinic	3	1	-0.6120	0.4339	-1.4625	0.2385	1.99
clinic	4	1	-2.1118	0.4922	-3.0765	-1.1472	18.41
clinic	5	1	-1.6482	0.4695	-2.5685	-0.7280	12.32
clinic	6	1	-2.4909	0.6171	-3.7004	-1.2813	16.29
clinic	7	1	-1.7758	0.5666	-2.8862	-0.6653	9.82
clinic	8	0	0.0000	0.0000	0.0000	0.0000	.
trt	cntl	1	-0.4587	0.1778	-0.8072	-0.1101	6.65
trt	drug	0	0.0000	0.0000	0.0000	0.0000	.

```
                     LR Statistics For Type 3 Analysis
                                      Chi-
              Source         DF      Square     Pr > ChiSq

              clinic          7      81.33        <.0001
              trt             1       6.78        0.0092
```

```
                          Least Squares Means
```

Effect	trt	Estimate	Standard Error	DF	Chi-Square	Pr > ChiSq
trt	cntl	-0.7322	0.1481	1	24.45	<.0001
trt	drug	-0.2735	0.1407	1	3.78	0.0518

```
                       Contrast Estimate Results
```

Label	Estimate	Standard Error	Alpha	Confidence Limits		Chi-Square	Pr > ChiSq
lsm - cntl	-0.7322	0.1481	0.05	-1.0224	-0.4420	24.45	<.0001
lsm - drug	-0.2735	0.1407	0.05	-0.5493	0.0022	3.78	0.0518
diff	-0.4587	0.1778	0.05	-0.8072	-0.1101	6.65	0.0099

```
                            Contrast Results
```

Contrast	DF	Chi-Square	Pr > ChiSq	Type
diff	1	6.78	0.0092	LR

There are minor differences between the results, but the overall conclusions are similar. The deviance for the logit model was 9.7463; here it is 9.6331. The likelihood ratio χ^2 for the test of no treatment effect, H_0: $\tau_1 = \tau_2$, was 6.67 for the logit model; here it is 6.78. The major difference is in the estimates of the intercept μ, the treatment effects τ_i, the clinic effects β_j, and the results estimates and least-squares means. These are estimated on the probit scale, and hence the estimates and standard errors are different.

As with the logistic model, in most practical situations, you will want to use the inverse link to convert estimates from the probit scale to the probability scale, and use the Delta Rule to convert the standard error. As with the logistic model, you can use the ODS statement to output LSMEANS or ESTIMATES and then use program statements to compute the inverse link and Delta Rule. For the LSMEANS, you follow these steps:

1. Insert `ods output lsmeans=lsm;` into the GENMOD program.
2. Run these program statements:

```
data prob_hat;
  set lsm;
    phat=probnorm(estimate);
    pi=3.14159;
    invsqrt=1/(sqrt(2*pi));
    se_phat=invsqrt*exp(-0.5*(estimate**2))*stderr;
proc print data=prob_hat;
run;
```

You can also add upper and lower confidence limits, as shown in Section 10.3.1. The statements are identical except you use the PROBNORM inverse link. Output 10.16 shows the result for $\hat{\pi}_1$ and its standard error.

Output 10.16
Probability of Favorable Outcome—Estimate and Standard Error from a Probit Link Model

Obs	Effect	trt	Estimate	StdErr	DF	ChiSq
1	trt	cntl	-0.7322	0.1481	1	24.45
2	trt	drug	-0.2735	0.1407	1	3.78

Obs	Prob ChiSq	phat	pi	invsqrt	se_phat
1	<.0001	0.23202	3.14159	0.39894	0.045181
2	0.0518	0.39222	3.14159	0.39894	0.054060

You can see that the estimated probabilities (PHAT), 0.23 and 0.39 for control and drug treatments, respectively, are slightly different from the 0.22 and 0.38 obtained using the logit link. Also, the standard errors (SE_PHAT) are somewhat different from those obtained from the logit model. However, the differences are slight, and the overall conclusions you would reach from these two models are very similar.

The same comments made at the end of Section 10.2 about logistic versus probit regression also apply to logit versus probit analysis-of-variance models. They typically produce similar results. The main reason to choose one link over the other often has to do with the subject matter of the study. Some disciplines characterize binomial data mainly by using odds ratios; because logistic models have natural interpretations in terms of odds ratios, they are an obvious choice. Other areas, such as quantitative genetics, have a well-developed theory expressed in terms of the normal distribution. For these disciplines, the probit model has obvious advantages.

10.3.3 Logistic Analysis of Covariance

Output 10.17 shows data from a bioassay involving two drugs, standard (STD) and treated (TRT) injected in varying dosages to 20 mice per treatment-dose combination. The response variable of interest is the number of mice out of the 20 that are ALIVE versus the number DEAD. These data appear in Koch et al. (1975) and are discussed in Freeman's (1987) text.

Output 10.17
Treatment-
Dose Bioassay
Data

Obs	drug	dosage	alive	dead	total
1	std	1	19	1	20
2	std	2	15	5	20
3	std	4	11	9	20
4	std	8	4	16	20
5	std	16	1	19	20
6	trt	1	16	4	20
7	trt	2	12	8	20
8	trt	4	5	15	20
9	trt	8	2	18	20

As discussed in Freeman's text, the researchers wanted to know how DRUG and DOSAGE affected the probability of a dead mouse. Their objectives suggested analysis of covariance to answer the following specific questions:

❑   Was there a dosage effect?

❑   Was the dosage effect the same for the two DRUG groups?

❑   Was there a significant difference between the DRUG groups, adjusted for constant dosage?

You can use PROC GENMOD to compute logistic analysis of covariance to answer these questions. Following Freeman, this example uses $X=\log_2(\text{DOSAGE})$ as the covariate. A generalized linear model for these data is

$$\text{logit}(\pi_{ij})=\mu+\tau_i + (\beta+\delta_i)X$$

where

π_{ij} is the probability of a dead mouse given the ith treatment (DRUG group) and the jth dosage.

μ is the intercept.

τ_i is the effect of the ith treatment.

β is the slope for the regression of X on the logit.

δ_i is the effect of the ith treatment on the slope.

Alternatively, you can give the above model as

$$\text{logit}(\pi_{ij}) = \beta_{0_i} + \beta_{1_i} X$$

where

$\beta_{0_i} = \mu + \tau_i$ is the intercept for the *i*th treatment.

$\beta_{1_i} = \beta + \delta_i$ is the slope for the *i*th treatment.

The first form of the model allows you an easy test for equal slopes, that is, H_0: all $\delta_i=0$. The second form of the model allows easier characterization of the regression equation for each DRUG group.

You can compute the analysis for the $\mu+\tau_i + (\beta+\delta_i)X$ form of the model using these statements:

```
proc genmod;
 class drug;
 model dead/total=drug x drug*x
       /dist=bin link=logit type1;
run;
```

In the model, DRUG corresponds to τ_i, X corresponds to β, and X*DRUG corresponds to δ_i. Output 10.18 gives the relevant results.

Output 10.18
Logistic
Analysis of
Covariance to
Test for
Unequal
Slopes

```
              Criteria For Assessing Goodness Of Fit

     Criterion              DF          Value        Value/DF

     Deviance                5          0.7464        0.1493
     Scaled Deviance         5          0.7464        0.1493
     Pearson Chi-Square      5          0.7541        0.1508
     Scaled Pearson X2       5          0.7541        0.1508

              LR Statistics For Type 1 Analysis

                                         Chi-
     Source        Deviance      DF      Square    Pr > ChiSq

     Intercept     80.6276
     drug          79.9301        1       0.70       0.4036
     x              0.8580        1      79.07       <.0001
     x*drug         0.7464        1       0.11       0.7383
```

You can see from the goodness-of-fit statistics that there is no evidence of lack of fit. The likelihood ratio χ^2 for X*DRUG tests H_0:$\delta_i=0$, the hypothesis of equal slopes for the drug groups.

You can also test the hypothesis of equal slopes and at the same time estimate the regressions for each DRUG group using the $\beta_{0_i} + \beta_{1_i} X$ form of the model. Here are the required SAS statements:

```
proc genmod;
 class drug;
 model alive/total=drug x(drug)/dist=bin link=logit noint;
 contrast 'equal slope?' x(drug) 1 -1;
```

As in other SAS linear model programs, the NOINT option in the MODEL statement suppresses the intercept, so that DRUG in the model corresponds to β_{0_i} and X(DRUG) corresponds to β_{1_i}. Thus the CONTRAST statement computes the test statistic for H_0: $\beta_{1_1} = \beta_{1_2}$, that is, the hypothesis of equal slopes for the two DRUG groups. Output 10.19 shows the relevant results.

Output 10.19
Logistic
Analysis of
Covariance:
Estimates of
Regression
Equations for
Each Group

```
                    Analysis of Parameter Estimates

                              Standard   Wald 95% Confidence    Chi-
    Parameter       DF  Estimate  Error       Limits          Square

    drug      std    1   2.7345   0.5732   1.6111    3.8579     22.76
    drug      trt    1   1.4960   0.4633   0.5880    2.4040     10.43
    x(drug)   std    1  -1.3673   0.2529  -1.8630   -0.8715     29.22
    x(drug)   trt    1  -1.2404   0.2827  -1.7945   -0.6863     19.25

                     Analysis Of Parameter
                          Estimates

              Parameter           Pr > ChiSq

              drug      std        <.0001
              drug      trt        0.0012
              x(drug)   std        <.0001
              x(drug)   trt        <.0001

                     Contrast Results

                              Chi-
        Contrast        DF    Square   Pr > ChiSq   Type

        equal slope?     1    0.11       0.7383     LR
```

You can see that the likelihood ratio χ^2 in the "Contrast Results" for the test of equal slopes is identical to the previous result for X*DRUG. The two are alternative ways to compute the same statistic. The deviance and Pearson χ^2 goodness-of-fit statistics are also the same (not shown for the latter model). From the output, you can see that the estimated regression equation for the DRUG STD is 2.7345–1.3673*X, and for the DRUG TRT it is 1.4960–1.2404*X. You could determine these regression coefficients from the program using the former model that produced Output 10.18 by using appropriately defined ESTIMATE statements.

Because there is no evidence of unequal slopes, the next step is to fit the equal slopes analysis-of-covariance model

$$\text{logit}(\pi_{ij}) = \mu + \tau_i + \beta X$$

where the terms in the model are defined as given above. Use the following program to compute the analysis:

```
proc genmod;
  class drug;
  model alive/total=drug x /dist=bin link=logit type1 type3;
```

Output 10.20 shows the results.

Output 10.20
Equal Slopes
Analysis of
Covariance for
Treatment-
Dose Bioassay
Data

```
                       Analysis Of Parameter Estimates

                                    Standard   Wald 95% Confidence        Chi-
Parameter          DF    Estimate    Error        Limits                 Square

Intercept          1     1.5923     0.3707     0.8658    2.3187          18.45
drug     std       1     1.0337     0.4031     0.2437    1.8237           6.58
drug     trt       0     0.0000     0.0000     0.0000    0.0000            .
x                  1    -1.3130     0.1875    -1.6805   -0.9455          49.02


                  LR Statistics For Type 1 Analysis

                                            Chi-
         Source          Deviance     DF    Square    Pr > ChiSq

         Intercept        80.6276
         drug             79.9301      1      0.70       0.4036
         x                 0.8580      1     79.07       <.0001


                  LR Statistics For Type 3 Analysis

                                      Chi-
            Source          DF        Square    Pr > ChiSq

            drug             1         7.04       0.0080
            x                1        79.07       <.0001
```

From the "Analysis of Parameter Estimates" you can see that the estimated regression coefficient is $\hat{\beta} = 1.313$, with a standard error of 0.1875. The regression equation for the *i*th treatment is $\mu + \tau_i + \beta X$. For example, for the DRUG STD, the estimated regression equation is 1.5923+1.0337−1.313*X, or 2.626−1.313*X. For the DRUG TRT, the regression equation is 1.5923−1.313*X.

As with standard analysis of covariance, as presented in Chapter 7, the Type 1 analysis compares the DRUG STD with TRT at their respective mean *X* levels, rather than at a common value. The DRUG STD had a DOSAGE level of 16 (*X*=4) whereas TRT did not. Hence the mean of *X* for STD is 2, whereas the mean of *X* for TRT is only 1.5. Formally, the Type 1 likelihood ratio χ^2 tests $H_0{:}\tau_{STD}{-}\tau_{TRT}{+}\beta(\overline{X}_{STD} - \overline{X}_{TRT})$, that is, $H_0{:}\tau_{STD}{-}\tau_{TRT}{+}\beta*0.5$. Because the slope β is positive, the observed performance of STD is shifted upward relative to TRT resulting in a χ^2 of 0.70 (*p*=0.4036), making it appear that there is no statistically significant DRUG effect.

The Type 3 analysis compares STD with TRT at a common value of *X*—that is, $H_0{:}\tau_{STD}{-}\tau_{TRT}$, resulting in a likelihood ratio χ^2 of 7.04 (*p*=0.008). The DRUG STD effect of 1.0337 in the "Analysis of Parameter Estimates" estimates the magnitude of the DRUG effect for a common *X*. The chi-square of 6.58 is the Wald χ^2 to test the DRUG effect.

You can estimate least-squares means by adding the statement

```
lsmeans drug/e;
```

to the GENMOD statements given above. The E option prints the coefficients of the LS means, allowing you to see the value of *X* used to compute them. Output 10.21 gives the result.

Output 10.21
Least-Squares
Means for
Common Slope
Analysis of
Treatment-
Dose Bioassay
Data

```
                Coefficients for drug Least Squares Means

        Label        Row      Prm1      Prm2      Prm3      Prm4

        drug          1        1         1         0       1.7778
        drug          2        1         0         1       1.7778


                        Least Squares Means

                              Standard            Chi-
    Effect    drug  Estimate    Error     DF    Square   Pr > ChiSq

    drug      std    0.2918    0.2686      1     1.18      0.2774
    drug      trt   -0.7419    0.2884      1     6.62      0.0101

                Coefficients for drug Least Squares Means

        Label        Row      Prm1      Prm2      Prm3      Prm4

        drug          1        1         1         0       1.7778
        drug          2        1         0         1       1.7778


                        Least Squares Means

                              Standard            Chi-
    Effect    drug  Estimate    Error     DF    Square   Pr > ChiSq

    drug      std    0.2918    0.2686      1     1.18      0.2774
    drug      trt   -0.7419    0.2884      1     6.62      0.0101
```

You can see that the LS means are computed at the overall $\bar{X}=1.778$. The estimates and standard errors are computed on the logit scale. You can use ODS output and program statements for the inverse link and Delta Rule, as shown in previous examples, to express the estimates and standard errors on the probability scale.

If you want to estimate the performance of the DRUG for X different from the default overall $\bar{X}$ used in the LSMEANS statement, you can use ESTIMATE statements. For example, for the DRUG STD at dosages of 1 and 16, that is, $X=0$ and $X=4$, you use the following statements. Output 10.22 shows the results.

```
estimate 'STD at dose=1' intercept 1 drug 1 0 x 0;
estimate 'STD at dose=16' intercept 1 drug 1 0 x 4;
```

Output 10.22
Estimates of
DRUG STD
at Different
Dosages

```
                        Contrast Estimate Results

                           Standard                              Chi-
    Label         Estimate   Error    Alpha  Confidence Limits  Square

    diff           1.0337    0.4031   0.05    0.2437   1.8237    6.58
    STD at dose=1  2.6260    0.4594   0.05    1.7255   3.5265   32.67
    STD at dose=16 -2.6260   0.4594   0.05   -3.5265  -1.7255   32.67
```

The estimate gives the logit for each dosage. As with the LS means, you can convert these to probabilities using the inverse link. Output 10.22 also gives the results for the statement

```
estimate 'diff' trt 1 -1;
```

You can see that the estimate, standard error, and χ^2 values for DIFF are identical to those of DRUG STD in the "Analysis of Parameter Estimates."

An alternative common-intercepts model uses β_{1_i} instead of $\mu+\tau_i$. Use the following SAS statements to compute the model. Output 10.23 shows the results.

```
proc genmod;
  class drug;
  model alive/total=drug x /dist=bin link=logit noint;
  contrast 'trt effect' drug 1 -1;
  estimate 'STD at dose=1'  drug 1 0 x 0;
  estimate 'STD at dose=16' drug 1 0 x 4;
```

The only difference between these statements and those shown earlier involves the intercept. In the MODEL statement, these statements use the NOINT (no intercept) option. NOINT suppresses μ so that DRUG estimates β_{1_i} instead of τ_i. In the ESTIMATE statements, the options remove references to INTERCEPT. Thus, DRUG 1 0 multiplies β_{1_i} instead of τ_i by a coefficient of 1. The CONTRAST statement obtains a likelihood ratio test of the difference between β_{1_1} and β_{1_2}, which is equivalent to testing $H_0:\tau_1-\tau_2$. You can use the ESTIMATE statement as shown in the previous analysis, if you prefer.

Output 10.23
Alternative
Common
Slopes
Analysis of
Covariance for
Treatment-
Dose Bioassay
Data

```
                     Analysis Of Parameter Estimates

                                  Standard   Wald 95% Confidence      Chi-
     Parameter       DF   Estimate   Error         Limits           Square

     drug      std    1     2.6260   0.4594    1.7255    3.5265       32.67
     drug      trt    1     1.5923   0.3707    0.8658    2.3187       18.45
     x                1    -1.3130   0.1875   -1.6805   -0.9455       49.02

                     Contrast Estimate Results

                            Standard                                  Chi-
     Label         Estimate    Error    Alpha   Confidence Limits    Square

     STD at dose=1   2.6260   0.4594     0.05    1.7255    3.5265      32.67
     STD at dose=16 -2.6260   0.4594     0.05   -3.5265   -1.7255      32.67

                        Contrast Results

                              Chi-
          Contrast       DF   Square   Pr > ChiSq    Type

          trt effect      1    7.04       0.0080     LR
```

The main advantage of computing the analysis this way is that the estimates of the intercepts, β_{1_i}, for each treatment are already computed, and do not need to be calculated from μ and τ_i. Also, you have the standard errors of the $\hat{\beta}_{1_i}$ in this output, whereas in the previous output there was no way to determine them. This output is more convenient when you need to report the regression equations for each DRUG. All other aspects of the analysis are identical to the results given in Outputs 10.20 through 10.22.

10.4 Count Data and Overdispersion

The examples in the previous sections of this chapter involved categorical data. Generalized linear models are useful for other types of non-normal data as well. Many studies, for example, use counts as the primary response variable. This section presents a typical example.

Historically, statistical theory has placed counting processes in a framework that implies a Poisson distribution. However, recent work, particularly in biological settings such as agriculture, ecology, and environmental science, suggests that the Poisson distribution is often inappropriate for counts. The Poisson assumes that the mean and variance are equal. In biological settings, the birds, insects, plants, and so forth being counted tend to be more aggregated—that is, clustered together—than you would expect under the Poisson model. Thus, the variance tends to be larger than the mean, in many cases *much* larger. **Overdispersion** is said to occur when the variance is larger than expected under a given model.

There are two approaches to overdispersion in GzLMs. First, you can use the GzLM assuming a Poisson distribution, but adjust standard errors and test statistics, as suggested by McCullagh and Nelder (1989). Alternatively, you can use a different distribution. For example, agricultural and ecological count data are often accurately characterized by the negative binomial distribution (Young and Young 1998).

This section uses an example of insect count to illustrate the main features of count models. Section 10.4.1 shows the standard Poisson model with no adjustments. Section 10.4.2 shows some basic model-checking procedures you can do with PROC GENMOD. Section 10.4.3 shows you how to do the overdispersion correction suggested by McCullagh and Nelder (1989). Section 10.4.4 presents an alternative model assuming the negative binomial and shows you how to fit it using PROC GENMOD.

10.4.1 An Insect Count Example

Output 10.24 contains data from an insect control experiment. The treatment design consisted of an untreated control and a 3×3 factorial for a total of ten treatments. The experiment was conducted as a randomized-complete-blocks design with four blocks. The response variable was insect count. TRT 0 is the control. The variable CTL_TRT is coded 0 for control and 1 otherwise. The two treatment factors, A and B, have three levels each (1, 2, and 3) and are both coded 0 for the control.

Output 10.24
Insect Count
Data

Obs	BLOCK	TRT	CTL_TRT	A	B	COUNT
1	1	0	0	0	0	16
2	1	1	1	1	1	6
3	1	2	1	1	2	2
4	1	3	1	1	3	4
5	1	4	1	2	1	5
6	1	5	1	2	2	3
7	1	6	1	2	3	1
8	1	7	1	3	1	1
9	1	8	1	3	2	3
10	1	9	1	3	3	1
11	2	0	0	0	0	25
12	2	1	1	1	1	9
13	2	2	1	1	2	6
14	2	3	1	1	3	3
15	2	4	1	2	1	3
16	2	5	1	2	2	4
17	2	6	1	2	3	5
18	2	7	1	3	1	2
19	2	8	1	3	2	2
20	2	9	1	3	3	0
21	3	0	0	0	0	5
22	3	1	1	1	1	2
23	3	2	1	1	2	14
24	3	3	1	1	3	5
25	3	4	1	2	1	3
26	3	5	1	2	2	6
27	3	6	1	2	3	17
28	3	7	1	3	1	2
29	3	8	1	3	2	3
30	3	9	1	3	3	2
31	4	0	0	0	0	9
32	4	1	1	1	1	22
33	4	2	1	1	2	4
34	4	3	1	1	3	5
35	4	4	1	2	1	2
36	4	5	1	2	2	3
37	4	6	1	2	3	1
38	4	7	1	3	1	3
39	4	8	1	3	2	4
40	4	9	1	3	3	9

As mentioned above, historically, the Poisson distribution has been assumed for count data. Using the Poisson canonical link, a GLM for these data is

$$\log(\lambda_{ij}) = \mu + \gamma_i + \tau_j$$

where

λ_{ij} is the mean count for the *i*th block (*i* = 1, 2, 3, 4) and *j*th treatment (*j* = 0, 1,...,9).

μ is the intercept.

γ_i is the *i*th block effect.

τ_j is the *j*th treatment effect.

The treatment effect can be expanded to account for control versus treated and the factorial treatment design. You partition it as

$$\tau_{jkl} = \delta_j + \alpha_k + \beta_l + (\alpha\beta)_{kl}$$

where

δ_j is the contrast between control and treated (j = 0, 1).

α_k is the main effect of the *k*th level of factor A (k = 1, 2, 3).

β_l is the main effect of the *l*th level of factor B (k = 1, 2, 3).

$(\alpha\beta)_{kl}$ is the *kl*th A×B interaction effect.

Use the following SAS statements to compute the analysis:

```
proc genmod data=a;
  class block ctl_trt a b;
  model count=block ctl_trt a b a*b/dist=poisson type1
 type3;
```

The MODEL statement allows you to partition the treatment effects according to the model for τ_{jkl} given above. The LINK= option is omitted, but the DIST=POISSON statement causes the canonical link for the Poisson—the log link—to be fitted by default. Output 10.25 shows the results of the Type 1 and Type 3 likelihood ratio tests. These results are discussed below. The goodness-of-fit statistics are shown in Section 10.4.2.

Output 10.25
The First
Version of the
GENMOD
Analysis of the
Poisson Model
for Insect
Count Data

```
                   LR Statistics For Type 1 Analysis

                                        Chi-
        Source          Deviance     DF  Square   Pr > ChiSq

        Intercept       175.0974
        CTL_TRT         135.1948      1   39.90     <.0001
        BLOCK           130.4411      3    4.75     0.1907
        A               107.5545      2   22.89     <.0001
        B               107.0454      2    0.51     0.7753
        A*B              93.9652      4   13.08     0.0109


                   LR Statistics For Type 3 Analysis

                                   Chi-
           Source          DF      Square    Pr > ChiSq

           CTL_TRT          0       0.00          .
           BLOCK            3       4.75       0.1907
           A                2      19.40       <.0001
           B                2       0.13       0.9394
           A*B              4      13.08       0.0109
```

You could put only BLOCK and TRT in the CLASS and MODEL statements and use the CONTRAST statements to compute tests for the (CTL) versus (TRT), the main effects of A and B, and the A×B interaction. The SAS program is then

```
proc genmod data=a;
  class block trt;
  model count=block trt/dist=poisson type1 type3;
  contrast 'ctl vs trt' trt 9 -1 -1 -1 -1 -1 -1 -1 -1 -1;
```

```
contrast 'a'     trt 0  1 1 1   0 0 0   -1 -1 -1,
                 trt 0  0 0 0   1 1 1   -1 -1 -1;
contrast 'b'     trt 0  1 0 -1  1 0 -1   1 0 -1,
                 trt 0  0 1 -1  0 1 -1   0 1 -1;
contrast 'a x b' trt 0  1 0 -1  0 0  0  -1 0 1,
                 trt 0  0 0  0  1 0 -1  -1 0 1,
                           trt 0  0 1 -1  0 0  0   0 -1 1,
                           trt 0  0 0  0  0 1 -1   0 -1 1;
```

Using A, B, and A*B in the MODEL statement as shown in the first program above is clearly more convenient than writing the equivalent contrasts. The results of the likelihood ratio tests for the contrasts in the second program appear in Output 10.26.

Output 10.26
Texts of
Contrasts, the
Second
Version of the
GENMOD
Analysis of the
Poisson Model
for Insect Data

Contrast	DF	Chi-Square	Pr > ChiSq	Type
ctl vs trt	1	45.86	<.0001	LR
a	2	19.40	<.0001	LR
b	2	0.13	0.9394	LR
a x b	4	13.08	0.0109	LR

You can see that the TYPE3 likelihood ratio χ^2 statistics for A, B, and A*B in Output 10.25 are the same as you get when you use TRT in the model and write contrasts, shown in Output 10.26. The χ^2 statistic for A*B is 13.08 with a *p*-value of 0.0109, strong evidence of an A×B interaction.

The situation is not quite so simple for CTL_TRT. In Output 10.25, the Type I likelihood ratio χ^2 statistic is 39.90 with a *p*-value <0.0001. This statistic tests $H_0: \delta_0 = \delta_1$, using the partitioned form of the model for τ_{jkl} given above. The TYPE3 χ^2 statistic for CTL_TRT is 0 with 0 DF. This is because there are 0 levels of A and B, and the CTL_TRT effect is confounded with the 0 versus other levels of A and B comparisons. Hence, there are 0 degrees of freedom left for CTL_TRT once adjusted for A and B. Therefore, if you use the approach shown in Output 10.25, you must use the TYPE1 test for CTL_TRT, rather than the Type 3, and CTL_TRT must appear *before* A, B, and A*B in the MODEL statement. In Output 10.26, the control versus treated CONTRAST tests $H_0: \tau_0 = \frac{1}{9} \sum_{j=1}^{9} \tau_j$.

This results in a χ^2 statistic of 45.86, somewhat different from the test statistic obtained in Output 10.25. The discrepancy in the test statistics results from the non-linearity of the link function. The test statistics for CTL_TRT in the first SAS program and the control versus treated CONTRAST defined on TRT in the second form of the model would be identical under conventional, normal errors linear models—for example, using PROC GLM or PROC MIXED. In this example, the two approaches lead to identical conclusions about the difference between the control and the other treatments. We know of no definitive statistical reason to favor either approach over the other. This and related issues are the subject of ongoing research in generalized linear models. The first set of SAS statements is clearly more convenient, but you should check the CONTRAST as well to be sure the results are not in conflict.

10.4.2 Model Checking

Both GENMOD programs in Section 10.4.1 produce the goodness-of-fit statistics shown in Output 10.27.

Output 10.27
Goodness-of-
Fit Statistics
for Poisson
GLM of Insect
Count Data

```
            Criteria For Assessing Goodness Of Fit

   Criterion              DF         Value      Value/DF

   Deviance               27       93.9652        3.4802
   Pearson Chi-Square      27       94.6398        3.5052
```

Both the deviance and the Pearson χ^2 are greater than you would expect under the Poisson generalized linear model given in Section 10.4.1. This casts doubt on whether the Poisson model is appropriate and therefore whether the conclusions reached in the last section are valid.

There are a number of possible reasons for poor fit. Some of these reasons are familiar from conventional normal errors linear models. For example, the right-hand side of the model equation ($\mu + \gamma_i + \tau_j$) might be inadequate, in this case indicating a possible treatment×block interaction. Some of the reasons are unique to the generalized linear model, because of its added flexibility. The choice of probability distribution, or link function, or variance function, may be inappropriate. This section surveys model-checking procedures for GLMs you can do using PROC GENMOD. For more details, see generalized linear model texts such as McCullagh and Nelder (1989). Certain SAS procedures, for example, PROC LOGISTIC, have far more extensive model-checking facilities for certain specialized GLMs. These are not covered in this book.

McCullagh and Nelder suggest several plots to assist in identifying obvious problems with the model, including these:

- **Plot standardized residuals against the predicted mean.** This serves the same purpose as similar plots in conventional linear models. Unequal scatter indicates violation of the homogeneity of variance assumption. More generally, unequal scatter may suggest a poor choice of variance function. Systematic pattern may indicate a poor choice of $X\beta$, for example, fitting a linear regression to a quadratic pattern, or it may indicate a poor choice of link function or probability distribution.

- **Plot y*, for example, $\hat{\eta} + \dfrac{y - \hat{\lambda}}{\hat{\lambda}}$ for the Poisson, against $\hat{\eta}$, the estimated link function.** This plot is unique to GLMs. The plot should be linear. Departure from linear suggests a poor choice of link function.

You can get these plots by using the GENMOD and PLOT procedures in SAS. For the insect count data and the Poisson model from Section 10.4.1, the SAS statements are

```
proc genmod data=a;
 class BLOCK trt;
 model count=BLOCK trt/dist=poisson type1 ObStats;
 ODS OUTPUT ObStats=check;
 title 'compute model checking statistics';
```

```
data plot;
 set check;
 adjlamda=2*sqrt(pred);
 ystar=xbeta+(count-pred)/pred;
 absres=abs(resdev);

options ps=28;
proc plot;
 plot resdev*(pred xbeta);
 plot (resdev reschi)*adjlamda;
 plot ystar*xbeta;
 plot absres*adjlamda;
```

The GENMOD statement is the same as you used to fit the Poisson model in Section 10.4.1, with the addition of an OBSTATS option to the MODEL statement and an ODS statement to output the OBSTATS to a new data set. The OBSTATS option computes several statistics analogous to predicted values and residuals for model checking in conventional linear models. The pertinent statistics are

STATISTIC	Definition
XBETA	estimate of link function, that is, $\hat{\eta} = X\hat{\beta}$
PRED	predicted value on COUNT scale, that is, $\hat{\lambda}_{ij} = h(\hat{\eta}_{ij}) = \exp(\hat{\eta}_{ij})$
RESDEV	residual deviance = sign(residual)*deviance of y_{ij} = $\text{sign(resid)} \times [\ell(\lambda_{ij}; y_{ij}) - \ell(\lambda_{ij}; \hat{\lambda}_{ij})]$
RESCHI	similar to residual deviance but uses component of Pearson χ^2 from ijth obs

The DATA PLOT step computes three additional functions of statistics in the OBSTATS output useful for model checking. McCullagh and Nelder (1989) suggest adjusting predicted values, in this case $\hat{\lambda}$, to a constant information scale before plotting it against residual, because unadjusted predicted values may result in a misleading plot. For the Poisson, the adjustment is $2\sqrt{\hat{\lambda}}$, ADJLAMDA in the SAS program. YSTAR is the linear predictor, $y* = \hat{\eta} + \dfrac{y - \hat{\lambda}}{\hat{\lambda}}$, to be plotted against $\hat{\eta}$ to check the link function. Finally, plotting the absolute value of the residuals as well as the residuals is often suggested, because the absolute values may reveal features the signed residuals miss. As an example, ABSRES is the absolute value of the residual deviance. You could also plot the absolute value of the residual χ^2. The plots generated by the PROC PLOT step are examples. Though not shown here, you could also do several plots using the residual χ^2 and plot absolute values of residuals against, say, XBETA as well as ADJLAMDA.

The example plots are shown below.

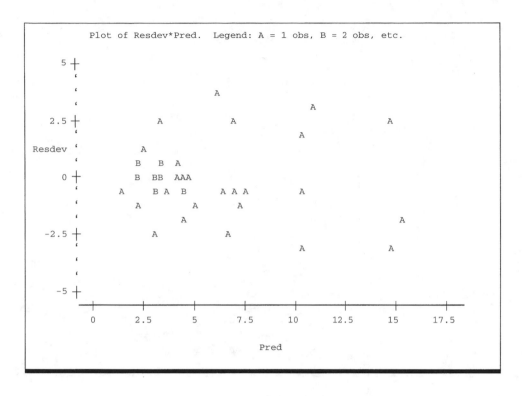

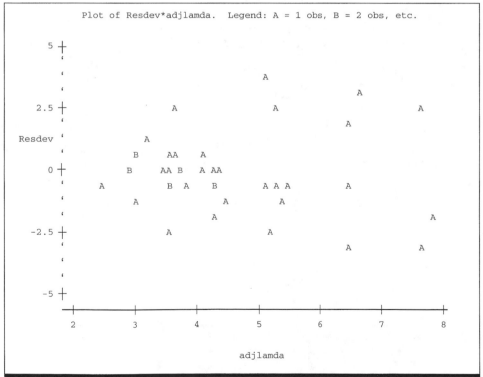

The plot of ADJLAMDA spreads the plot more evenly over the horizontal axis and reveals more detail for the lower predicted counts. While there is no overt visual evidence of unequal scatter or systematic pattern, there is a hint of lower variance among the residuals for the lowest predicted counts, especially on the ADJLAMDA plot. The plots of residual deviance against XBETA and residual χ^2 against ADJLAMDA reveal similar information:

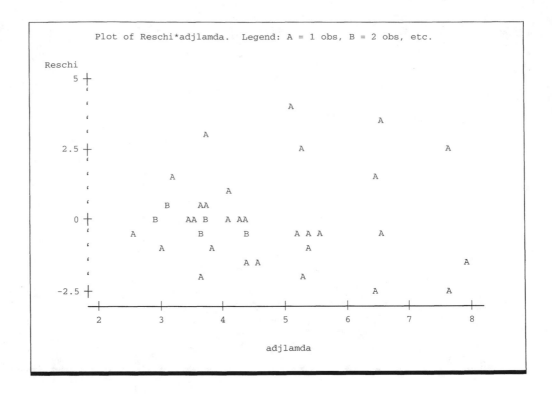

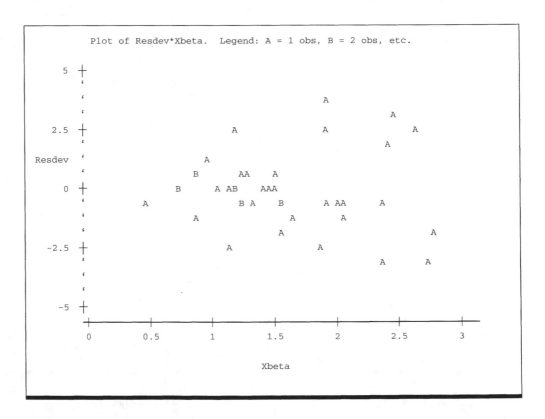

The plot of y^* versus $\hat{\eta}$ is next. While there is scatter, there is no overt departure from linearity and hence no obvious evidence of a poor choice of link function.

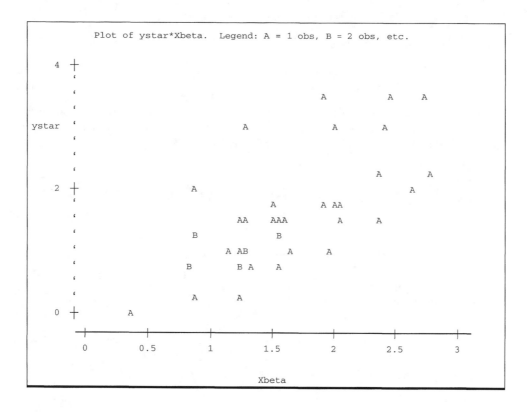

Finally, the plot of absolute value of residual deviance versus ADJLAMDA reinforces the information contained in the previous residual plots.

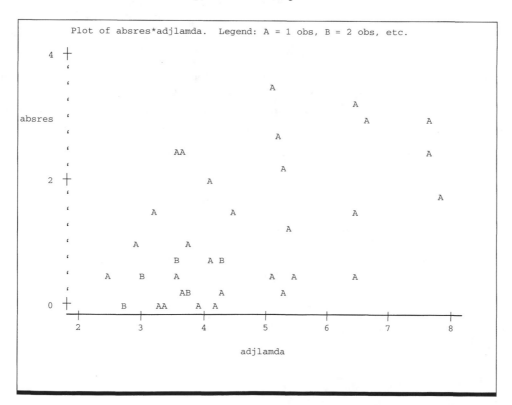

There are two main results of this model-checking exercise:

1. The goodness-of-fit statistics, deviance, and Pearson χ^2 are larger than expected.

2. There is slight visual evidence that scatter among the residuals is not constant, but is lower for lower predicted counts.

Taken together, these symptoms indicate that the Poisson model may be inappropriate and, in particular, the assumed variance structure may not be correct. The next section discusses the likely reason as well as a commonly used adjustment.

10.4.3 Correction for Overdispersion

The Poisson model assumes that the mean and variance are equal. As mentioned previously, in biological counting processes the variance is typically greater than the mean. When the variance is larger than expected under a given model, the condition is called **overdispersion**. Unaccounted for overdispersion tends to cause standard errors to be underestimated and test statistics to be overestimated, resulting in excessive Type I error rates.

With generalized linear models, you have two ways to account for overdispersion—fit the Poisson model but adjust the standard errors and test statistics—or choose a different distribution for the model. This section shows you how to adjust the Poisson model. Section 10.4.4 gives an example of using PROC GENMOD to fit an alternative distribution.

You can model overdispersion by letting the actual variance equal the assumed variance multiplied by an additional scale parameter that adjusts for the discrepancy between assumed and actual. For the Poisson distribution, the assumed variance is the mean, λ, so the adjusted variance is $\phi\lambda$. Overdispersion occurs when $\phi>1$. In GzLM literature, ϕ is typically referred to as the **overdispersion parameter**.

As a theoretical note, when you assume $E(y)=\lambda$ and $var(y)=\phi\lambda$, you no longer have a true probability distribution for the random variable, y, but you *do* have a valid *quasi-likelihood* (see Section 10.6.4). You can show that for estimable $K'\beta$ from the resulting GLM, standard errors of $K'\hat{\beta}$ are the standard errors you get from the Poisson model multiplied by $\sqrt{\phi}$, and the Wald and likelihood ratio χ^2 statistics are equal to the corresponding χ^2 statistics divided by ϕ.

Because the overdispersion parameter is unknown, you must estimate it. McCullagh (1983) gives $\hat{\phi} = \dfrac{(y-\hat{\mu})'V_{\mu}^{-1}(y-\hat{\mu})}{N-p} = \dfrac{\text{Pearson } \chi^2}{N-p}$, where $N–p$ is the degrees of freedom for lack of fit and V_{μ} is a diagonal matrix of variance functions defined in Section 10.6.2. McCullagh and Nelder (1989) also suggest using the deviance, that is,

$$\hat{\phi} = \frac{Deviance}{N-p} .$$

You can compute the Poisson model corrected for overdispersion by using PROC GENMOD. For the insect count data, use the following SAS statements:

```
proc genmod data=a;
  class BLOCK CTL_TRT a b;
  model count=BLOCK CTL_TRT a b a*b/dist=poisson dscale
type1 type3;
```

This is the same set of statements used in Section 10.4.1 except for the addition of the DSCALE option in the MODEL statement. DSCALE causes all standard errors and test statistics to be corrected for a scale parameter estimated using the deviance, that is,

$\hat{\phi} = \dfrac{Deviance}{N-p}$. Alternatively, you can substitute PSCALE for DSCALE to get the scale

parameter estimated by using the Pearson χ^2. Output 10.28 shows the results:

Output 10.28
The Poisson
Analysis of
Insect Count
Data
Corrected for
Overdispersion

```
              Criteria For Assessing Goodness Of Fit

     Criterion                DF         Value       Value/DF

     Deviance                 27        93.9652       3.4802
     Scaled Deviance          27        27.0000       1.0000
     Pearson Chi-Square       27        94.6398       3.5052
     Scaled Pearson X2        27        27.1938       1.0072
```

```
              LR Statistics For Type 1 Analysis
                                                  Chi-
Source       Deviance  Num DF  Den DF  F Value  Pr > F   Square  Pr > ChiSq

Intercept    175.0974
BLOCK        170.3437     3       27     0.46   0.7157    1.37    0.7135
CTL_TRT      130.4411     1       27    11.47   0.0022   11.47    0.0007
A            107.5545     2       27     3.29   0.0527    6.58    0.0373
B            107.0454     2       27     0.07   0.9296    0.15    0.9295
A*B           93.9652     4       27     0.94   0.4561    3.76    0.4397
```

```
              LR Statistics For Type 3 Analysis
                                           Chi-
Source       Num DF  Den DF  F Value  Pr > F   Square  Pr > ChiSq

BLOCK           3      27      0.46   0.7157    1.37    0.7135
CTL_TRT         0      27        .       .      0.00      .
A               2      27      2.79   0.0794    5.57    0.0616
B               2      27      0.02   0.9822    0.04    0.9822
A*B             4      27      0.94   0.4561    3.76    0.4397
```

The scaled deviance is the deviance divided by the estimated overdispersion parameter. Here, the overdispersion parameter is estimated by using the deviance, that is,

$\hat{\phi} = \dfrac{93.9652}{27} = 3.48$. The scaled deviance is therefore equal to 1. In this context, the

scaled goodness-of-fit statistics have no real interpretation except to indicate whether subsequent adjustment of standard errors and test statistics used a scale parameter estimate based on the deviance or the Pearson χ^2. If you use the latter, the scaled Pearson χ^2 would equal 1.

In the table of likelihood ratio statistics, there are two sets of test statistics. The right-hand two columns give an adjusted χ^2, equal to the original χ^2 computed in Section 10.4.1, divided by $\hat{\phi}$, and its associated *p*-value. For example, in Section 10.4.1, the χ^2 statistic for A*B was 13.08. Dividing by $\hat{\phi} = 3.48$ yields 3.76. The table also gives an

F-value and its associated degrees of freedom and *p*-value. The *F*-statistic accounts for the fact that $\hat{\phi}$ is an estimate. The adjusted χ^2 results from dividing either a Wald or a likelihood ratio χ^2 by either deviance/DF or Pearson χ^2/DF. In a normal errors linear model, the Wald and likelihood ratio χ^2 are equivalent to the sum of squares for the hypothesis (SSH). Thus, in conventional ANOVA terms, the adjusted χ^2 is equivalent to $\dfrac{\text{SSH}}{\text{MSE}}$. Dividing the adjusted χ^2 by the degrees of freedom for the hypothesis yields

$$\frac{\text{SSH}}{\text{MSE}}/\text{DF(hypothesis)} = \frac{\text{MS(hypothesis)}}{\text{MSE}},$$ which is an *F*-statistic. For example, there are 4

DF for the A×B interaction; dividing the adjusted χ^2 by 4 yields $\dfrac{3.76}{4} = 0.94$. Analogous

to ANOVA *F*-tests, the numerator degrees of freedom are the DF for the hypothesis and the denominator degrees of freedom are the DF used to estimate ϕ.

In the analysis of the Poisson model in Section 10.4.1, there was very strong evidence of a CTL_TRT effect—that is, a difference between the untreated control and the mean of the AB treatment combinations. There was also strong evidence of an A×B interaction. When the analysis is corrected for overdispersion, evidence of the CTL_TRT effect remains (*F*=11.47, *p*=0.0022), but the A×B interaction no longer is statistically significant. Instead, there is some evidence of an A main effect (*F*=2.79, *p*= 0.0794), but no other differences among the treated groups.

To complete the analysis, you need the estimated mean counts for control and treated, and for the levels of factor A. While the model using CTL_TRT A B A*B is convenient for obtaining test statistics, it results in non-estimable LS means. The only alternative is to define the LS means using ESTIMATE statements defined on the model that substitutes TRT for CTL_TRT A B A*B. The SAS statements follow. For completeness, the program includes the CONTRAST from Section 10.4.1.

```
proc genmod data=a;
 class block trt;
 model count=block trt/dist=poisson pscale type1 type3;
 contrast 'ctl vs trt' trt 9 -1 -1 -1 -1 -1 -1 -1 -1 -1;
 contrast 'a' trt 0  1 1 1  0 0 0  -1 -1 -1,
          trt 0  0 0 0  1 1 1  -1 -1 -1;
 contrast 'b' trt 0  1 0 -1  1 0 -1  1 0 -1,
          trt 0  0 1 -1  0 1 -1  0 1 -1;
 contrast 'a x b' trt 0  1 0 -1  0 0  0  -1 0 1,
             trt 0  0 0  0  1 0 -1  -1 0 1,
                    trt 0  0 1 -1  0 0  0  0 -1 1,
                    trt 0  0 0  0  0 1 -1  0 -1 1;
 estimate 'ctl lsmean' intercept 1 trt 1 0/exp;
 estimate 'treated lsm' intercept 9 trt 0 1 1 1 1 1 1 1 1 1/
      divisor=9 exp;
 estimate 'A=1 lsmean' intercept 3 trt 0 1 1 1 0/divisor=3 exp;
 estimate 'A=2 lsmean' intercept 3 trt 0 0 0 0 1 1 1
      0/divisor=3 exp;
 estimate 'A=3 lsmean' intercept 3 trt 0 0 0 0 0 0 0 1 1 1 0/
      divisor=3 exp;
```

The results appear in Output 10.29. The PSCALE option appears in the above MODEL statement, so the Pearson χ^2 is used to estimate the overdispersion parameter. You could just as well use DSCALE. The EXP option in the ESTIMATE statements computes exp(estimate), as well as its standard error and upper and lower confidence limits. For the log link, the inverse link is $\lambda=exp(\eta)$, so the EXP option computes the inverse link to obtain the estimate of the actual counts, applies the Delta Rule to get the standard error, and so forth. The log is the only link function for which there is an option in the ESTIMATE statement to compute the inverse link. Otherwise, you must output the LS means using an ODS statement in the WRITE statement to compute the inverse link as shown for the binomial data in Sections 10.2 and 10.3.

Output 10.29
CONTRAST
and ESTIMATE
Results for
Insect Count
Analysis Using
the Poisson
Model
Corrected for
Overdispersion

```
                     Criteria For Assessing Goodness Of Fit

            Criterion                DF         Value        Value/DF

            Deviance                 27        93.9652        3.4802
            Scaled Deviance          27        26.8076        0.9929
            Pearson Chi-Square       27        94.6398        3.5052
            Scaled Pearson X2        27        27.0000        1.0000

                            Contrast Results

                                                 Chi-
     Contrast    Num DF  Den DF  F Value  Pr > F  Square  Pr > ChiSq  Type

     ctl vs trt     1      27     13.08   0.0012  13.08     0.0003    LR
     a              2      27      2.77   0.0807   5.53     0.0629    LR
     b              2      27      0.02   0.9823   0.04     0.9823    LR
     a x b          4      27      0.93   0.4597   3.73     0.4435    LR

                        Contrast Estimate Results

                           Standard                                  Chi-
     Label         Estimate   Error  Alpha  Confidence Limits       Square

     ctl lsmean      2.6096  0.2532   0.05   2.1133    3.1059       106.21
     Exp(ctl lsmean)13.5940  3.4422   0.05   8.2758   22.3296
     treated lsm     1.4156  0.1618   0.05   1.0986    1.7326        76.59
     Exp(treated lsm)4.1189  0.6663   0.05   2.9999    5.6555
     A=1 lsmean      1.8539  0.2197   0.05   1.4233    2.2845        71.22
     Exp(A=1 lsmean) 6.3848  1.4026   0.05   4.1510    9.8207
     A=2 lsmean      1.4408  0.2663   0.05   0.9189    1.9628        29.27
     Exp(A=2 lsmean) 4.2242  1.1249   0.05   2.5065    7.1189
     A=3 lsmean      0.9520  0.3376   0.05   0.2903    1.6137         7.95
     Exp(A=3 lsmean) 2.5910  0.8747   0.05   1.3369    5.0216
```

The adjusted statistics for A B and A*B are somewhat different from Output 10.28 because overdispersion was estimated using the Pearson χ^2 instead of the deviance. Here, $\hat{\phi}$ is 3.5052. The test for CTL vs TRT also differs from the result for the CTL_TRT effect in Output 10.28 because of the difference in parameterization discussed in Section 10.4.1. Nonetheless, like Output 10.28, after correcting for overdispersion, there is strong evidence of a difference between the untreated control and treated, weaker evidence of a main effect of A, and no statistically significant evidence of an A×B interaction or a main effect of B.

The table of "Contrast Estimate Results" gives the estimates in pairs of lines. The first line is the literal $K'\hat{\beta}$ on the log link scale. The next line, generically titled "Exp(estimate)" is exp($k'\hat{\beta}$), that is, the estimate on the count scale. For example, the untreated control's LS mean is 2.6096 with a standard error of 0.2532; the LS mean count is 13.5940 with a standard error of 3.4422. The treated LS mean count is only 4.1189; the statistically significant CTL vs TRT difference translates to a reduction in mean count from roughly 13.6 to 4.1. Level 3 of factor A has the lowest LS mean count, 2.5910. You could write additional CONTRAST statements to further partition the A main effect.

There were two main points of this section. First, the DSCALE or PSCALE options in GENMOD allow you to adjust for overdispersion. Second, adjusting for overdispersion can substantially affect the conclusions you draw from the analysis. Without correction for overdispersion, you would focus on a highly significant A×B interaction within the treated group. With correction for overdispersion, your emphasis shifts to interpreting the main effect of factor A.

While the overdispersion correction presented in this section is simple, many statisticians criticize it for being *simplistic* as well. They contend that the real problem is that the assumed distribution is inappropriate, and advocate using a more realistic model. Section 10.4.4 presents an example of using GENMOD to fit a user-supplied distribution, when none of the distributions supported by GENMOD are considered appropriate.

10.4.4 Fitting a Negative Binomial Model

For biological count data, recent work indicates that overdispersion is the rule, not the exception, and that distributions where the variance exceeds the mean are usually more appropriate than the Poisson as the basis for generalized linear models. One such distribution is the negative binomial.

The negative binomial appears in statistical theory texts as the distribution of the probability that it takes exactly N Bernoulli trials for exactly Y successes to occur. If π denotes the probability of a success, the $P(N=n)=\dfrac{(N-1)!}{(y)!(N-y-1)!}\pi^y(1-\pi)^{N-y}$. A more useful form of the negative binomial results from letting $N=y+k$ and $\pi=\dfrac{\lambda}{\lambda+k}$, resulting in the probability distribution function $\dfrac{(y+k-1)!}{y!(k-1)!}\left(\dfrac{\lambda}{\lambda+k}\right)^y\left(\dfrac{k}{\lambda+k}\right)^k$ and a log likelihood of $y\log\dfrac{\lambda}{\lambda+k}-k\log\dfrac{\lambda}{\lambda+k}+\log\dfrac{(y+k-1)!}{y!(k-1)!}$. While the negative binomial does not belong to the exponential family, $y\log\dfrac{\lambda}{\lambda+k}-k\log\dfrac{\lambda}{\lambda+k}$ is a quasi-likelihood with $E(y)=\lambda$ and var$(y)=\lambda+\dfrac{\lambda^2}{k}$. The parameter μ and k are the mean and the aggregation parameter, respectively. As k becomes large, var(y) approaches λ—that is, the limiting distribution of negative binomial as k goes to infinity is the Poisson. For small k, however, the variance can be considerably larger than the mean, depending on how large λ is relative to k. A small k is interpreted to mean the data are highly aggregated, that is, the data are more clustered, or "clumped," than would be expected in a Poisson population.

The natural parameter for the negative binomial is $\log \dfrac{\lambda}{\lambda + k}$ and thus it can be used as the canonical link function for a generalized linear model if you can assume a known k. Setting k=1 gives you the geometric distribution. Young and Young (1998) have found the geometric distribution to be common in experiments involving insect counts, so it is a good place to begin. You can also estimate k and then use it in the generalized linear model estimating equations as if it was known. A simpler approach uses the log link,

$\log(\lambda)$, as the link function, while using the variance function $\lambda + \dfrac{\lambda^2}{k}$ of the negative binomial.

A generalized linear model for the insect count data using the negative binomial with a canonical link is

$$\eta_{ij} = \log \frac{\lambda_{ij}}{\lambda_{ij} + k} = \mu + \beta_i + \tau_j$$

The alternative model using the log link is

$$\eta_{ij} = \log\left(\lambda_{ij}\right) = \mu + \beta_i + \tau_j$$

where the terms of both models share the same definitions given in previous sections. As shown earlier, you can partition τ_j into its components. You can fit the latter model using the DIST=NEGBIN option in PROC GENMOD. However, the canonical link shown in the former model is not one of the links supported by PROC GENMOD—that is, it is not available in the LINK option of the MODEL statement. You can, however, use program statements in GENMOD to supply your own distribution or quasi-likelihood. This section shows you how to work with both link functions.

10.4.5 Using PROC GENMOD to Fit the Negative Binomial with a Log Link

Use the following SAS statements to estimate the model:

```
proc genmod data=a;
  class BLOCK CTL_TRT a b;
  model count=BLOCK CTL_TRT a b a*b/dist=negbin type1 type3;
```

The DIST=NEGBIN option estimates the parameters for the negative binomial distribution. Output 10.30 shows the results.

Output 10.30
An Analysis of
Count Data
Using Negative
Binomial
Distribution
and a Log Link

```
                        Model Information

          Data Set                        WORK.A
          Distribution          Negative Binomial
          Link Function                      Log
          Dependent Variable               COUNT
          Observations Used                   40

          Criteria For Assessing Goodness Of Fit

     Criterion              DF        Value      Value/DF

     Deviance               27      36.7953        1.3628
     Pearson Chi-Square     27      33.8797        1.2548

               Analysis Of Parameter Estimates

                                 Standard   Wald 95% Confidence
Parameter         DF   Estimate    Error          Limits

Dispersion         1     0.2383    0.0899     0.1137    0.4992

NOTE: The negative binomial dispersion parameter was estimated by maximum
      likelihood.
               LR Statistics For Type 1 Analysis

                         2*Log               Chi-
          Source       Likelihood    DF     Square    Pr > ChiSq

          Intercept    399.7286
          BLOCK        400.9381       3       1.21       0.7507
          CTL_TRT      411.5677       1      10.63       0.0011
          A            420.5913       2       9.02       0.0110
          B            420.6331       2       0.04       0.9794
          A*B          425.4540       4       4.82       0.3062

               LR Statistics For Type 3 Analysis

                                     Chi-
          Source          DF       Square    Pr > ChiSq

          BLOCK            3         3.12       0.3733
          CTL_TRT          0         0.00         .
          A                2         8.96       0.0113
          B                2         0.03       0.9849
          A*B              4         4.82       0.3062
```

Several items in Output 10.30 are worth noting. First, under "Model Information," you can see that the log link function was used. The VALUE/DF ratios for the deviance and Pearson χ^2 are both greater than 1, but not enough to suggest lack of fit. The dispersion parameter is estimated obtained via maximum likelihood. The estimate given, 0.2383, is actually an estimate of $1/k$, as the negative binomial was set up earlier in this section. Thus, $\hat{k} = \dfrac{1}{0.2383} = 4.196$. The test statistics for CTL_TRT, A, B, and A*B are interpreted as in earlier examples in this section. The results are similar, but not identical, to the overdispersion-corrected tests in Section 10.4.3.

You can use the alternative analysis with TRT identifying treatments and the CTL_TRT, A, B, and A*B effects broken out by the CONTRAST statements. The SAS statements are

```
proc genmod data=a;
 class block trt;
 model count=block trt/dist=negbin type1 type3 wald;
 contrast 'ctl vs trt' trt 9 -1 -1 -1 -1 -1 -1 -1 -1 -1/wald;
 contrast 'a' trt 0  1 1 1  0 0 0  -1 -1 -1,
          trt 0  0 0 0  1 1 1  -1 -1 -1/wald;
 contrast 'b' trt 0  1 0 -1  1 0 -1  1 0 -1,
          trt 0  0 1 -1  0 1 -1  0 1 -1/wald;
 contrast 'a x b' trt 0  1 0 -1  0 0  0  -1 0 1,
              trt 0  0 0  0  1 0 -1  -1 0 1,
                      trt 0  0 1 -1  0 0  0  0 -1 1,
                      trt 0  0 0  0  0 1 -1  0 -1 1/wald;
 estimate 'ctl lsmean' intercept 1 trt 1 0/exp;
 estimate 'treated lsm'intercept 9 trt 0 1 1 1 1 1 1 1 1 1
      /divisor=9 exp;
 estimate 'A=1 lsmean' intercept 3 trt 0 1 1 1 0/divisor=3 exp;
 estimate 'A=2 lsmean' intercept 3 trt 0 0 0 0 1 1 1
      0/divisor=3 exp;
 estimate 'A=3 lsmean' intercept 3 trt 0 0 0 0 0 0 0 1 1 1
      /divisor=3 exp;
 ods output estimates=lsm;
```

Output 10.31 shows the results.

Output 10.31 CONTRAST Results for a Negative Binomial Model of Count Data with a Log Link

```
                           Contrast Results

                                  Chi-
            Contrast         DF   Square    Pr > ChiSq   Type

            ctl vs trt        1   15.99      <.0001      LR
            a                 2    8.96       0.0113     LR
            b                 2    0.03       0.9849     LR
            a x b             4    4.82       0.3062     LR
```

You can see that the χ^2 statistics for A, B, and A×B are identical to those in Output 10.30. CTL_TRT on the other hand, differs: the χ^2 statistic obtained using the CTL_TRT class variable was 10.63; here it is 15.99. This discrepancy is similar to what was observed in Sections 10.4.1 and 10.4.3.

10.4.6 Fitting the Negative Binomial with a Canonical Link

You need to write program statements to define the canonical link. Specifically, you need to provide the link and inverse link functions. The following SAS statements fit the desired model:

```
proc genmod data=a;
 k=1/0.2383;
 mu=_mean_;
 eta=_xbeta_;
 fwdlink link=log(mu/(mu+k));
 invlink ilink=k*exp(eta)/(1-exp(eta));
 class block trt;
 model count=block trt/dist=negbin type1 type3 wald;
 contrast 'ctl vs trt' trt 9 -1 -1 -1 -1 -1 -1 -1 -1 -1;
 contrast 'a' trt 0  1 1 1  0 0 0  -1 -1 -1,
          trt 0  0 0 0  1 1 1  -1 -1 -1;
 contrast 'b' trt 0  1 0 -1  1 0 -1  1 0 -1,
          trt 0  0 1 -1  0 1 -1  0 1 -1;
 contrast 'a x b' trt 0  1 0 -1  0 0 0  -1 0 1,
              trt 0  0 0 0  1 0 -1  -1 0 1,
                    trt 0  0 1 -1  0 0 0  0 -1 1,
                    trt 0  0 0 0  0 1 -1  0 -1 1;
```

The statements FWDLINK and INVLINK define the link and inverse link, respectively. You need to provide the estimate of k in order to define these two functions. MU and ETA are not required; they are conveniences. _MEAN_ and _XBETA_ are internal SAS names for the estimates of λ and η, respectively. They are awkward. MU and ETA are simply more manageable names for subsequent program statements.

Alternatively, you could use the CLASS statements for CTL_TRT, A, and B, that is, use the statements

```
proc genmod data=a;
 k=1/0.2383;
 mu=_mean_;
 eta=_xbeta_;
 fwdlink link=log(mu/(mu+k));
 invlink ilink=k*exp(eta)/(1-exp(eta));
 class block ctl_trt a b;
 model count=BLOCK CTL_TRT a b a*b/dist=negbin type1 type3;
```

The "Criteria for Assessing Goodness of Fit" and parameter estimates for both models are identical. Output 10.32 shows these results. The tests of treatment effects are different. Output 10.33 shows the treatment effect results.

Output 10.32
Goodness of Fit and Dispersion Estimate for a Negative Binomial with a Canonical Link

```
            Criteria For Assessing Goodness Of Fit

      Criterion              DF         Value      Value/DF

      Deviance               27       37.5857       1.3921
      Pearson Chi-Square     27       36.0722       1.3360

Algorithm converged.

                  Analysis Of Parameter Estimates

                              Standard   Wald 95% Confidence
Parameter       DF   Estimate    Error        Limits

Dispersion       1     0.2456   0.0933    0.1167    0.5171
```

The deviance and Pearson χ^2 values are very close to those obtained using the log link (Output 10.30). The dispersion estimate is also slightly changed by using the canonical link. Here, $\hat{k} = \frac{1}{0.2456} = 4.072$.

Output 10.33
Treatment
Effect Tests
Using a
Negative
Binomial with a
Canonical Link

```
Run 1: TRT in model, effects defined by CONTRAST

                        Contrast Results

                               Chi-
        Contrast        DF     Square    Pr > ChiSq    Type

        ctl vs trt       1     15.22      <.0001       LR
        a                2      8.05       0.0179      LR
        b                2      0.29       0.8656      LR
        a x b            4      5.13       0.2739      LR


Run 2: CTL_TRT, A, and B in CLASS statement

              LR Statistics For Type 1 Analysis

                        2*Log              Chi-
        Source         Likelihood    DF    Square    Pr > ChiSq

        Intercept       399.7286
        BLOCK           400.9381      3      1.21      0.7507
        CTL_TRT         410.0197      1      9.08      0.0026
        A               418.7535      2      8.73      0.0127
        B               418.9195      2      0.17      0.9203
        A*B             424.0528      4      5.13      0.2739


              LR Statistics For Type 3 Analysis

                               Chi-
        Source          DF     Square    Pr > ChiSq

        BLOCK            3      1.72       0.6325
        CTL_TRT          0      0.00        .
        A                2      8.05       0.0179
        B                2      0.29       0.8656
        A*B              4      5.13       0.2739
```

These results follow the same pattern: The Run 1 CONTRAST results and Run 2 Type 3 likelihood ratio statistics are the same for A, B, and A×B, whereas the CTL_TRT results vary somewhat. All results are similar to those obtained using the log link.

10.4.7 Advanced Application: A User-Supplied Program to Fit the Negative Binomial with a Canonical Link

To try other distributions, including values of the aggregation parameter k, you need to write program statements to provide the variance function and deviance as well as the link and inverse link. The following examples illustrate how to do this. Suppose, for example, you want to fit the geometric distribution, that is, the negative binomial with $k=1$. For the insect count data and the negative binomial, the SAS GENMOD statements are

```
proc genmod;
    count=_resp_;
    y=count;
    if y=0 then y=0.1;
    mu=_mean_;
    eta=_xbeta_;
    K=1;
    FWDLINK LINK=LOG(MU/(MU+K));
    INVLINK ILINK=K*EXP(ETA)/(1-EXP(ETA));
    lly=y*log(y/(y+k))-k*log((k+y)/k);
    llm=y*log(mu/(mu+k))-k*log((k+mu)/k);
    d=2*(lly-llm);
    VARIANCE VAR=MU+(MU*MU/K);
    DEVIANCE DEV=D;
    class block ctl_trt a b;
    model y=block ctl_trt/type3 wald;
    model y=block ctl_trt a b a*b/ type3 wald;
```

The required program statements appear in uppercase. The statement $K=1$ sets the aggregation parameter. FWDLINK LINK= gives the link function for your model; you supply the expression after the equal sign. Similarly, INVLINK ILINK= supplies the inverse link function. VARIANCE VAR= and DEVIANCE DEV= give the forms of the variance function and deviance, respectively. The statements immediately preceding VARIANCE VAR= are for programming convenience. LLY defines the log likelihood (or quasi-likelihood) evaluated at the observations, and LLM defines the log likelihood or quasi-likelihood evaluated using the estimated natural parameter from the model. Then D=2*(LLY–LLM) is the formula for the deviance. You could write the full formula immediately following DEV=, but separating the task into steps may help you reduce the chances of a hard-to-recognize mistake. GENMOD uses the internal SAS names _RESP_ for the response variable, _MEAN_ for the expected value of the observation based on the model, and _XBETA_ for the current value of $\hat{\eta}=X\hat{\beta}$. You can use the SAS names in your program, but many users find it easier to use their own names.

Note: The response variable for these data is COUNT. For one observation (block 2, treatment 9), COUNT=0. The deviance is 2*(LLY–LLM), but for this observation, LLY cannot be determined since $\log \dfrac{y}{y+k} = \log(0)$. This produces the following output for the goodness-of-fit statistics:

Output 10.34
Goodness-of-Fit Statistics for a Negative Binomial Model with a User-Supplied Program and Uncorrected Zero Counts

```
                  Criteria For Assessing Goodness Of Fit

       Criterion                 DF          Value        Value/DF

       Deviance                  27         0.0000         0.0000
       Scaled Deviance           27         0.0000         0.0000
       Pearson Chi-Square        27         8.1026         0.3001
       Scaled Pearson X2         27         8.1026         0.3001
       Log Likelihood                  -1.79769E308

  ERROR:  Error in computing deviance function.
```

Therefore, for observations where COUNT=0, you must add a small positive constant in order to avoid this problem, hence the statements Y=COUNT followed by IF Y=0 THEN Y=0.1. This prevents the zero counts from causing an error when computing LLY.

Finally, user-supplied distributions may introduce difficulty computing certain likelihood ratio statistics. The best alternative is to use Wald statistics. You can compute Wald statistics for Type III hypotheses with GENMOD. Therefore, to fully test the model, you must run it twice, once with CTL_TRT only in the model, adjusted for BLOCK, then with the full model. Run 1 tests control versus treated. Run 2 tests the factorial effects within the treated group. Output 10.35 shows the results.

Output 10.35
Deviance and Wald Statistics for a Geometric Model Fit to Insect Count Data

```
                  Criteria For Assessing Goodness Of Fit

       Criterion                 DF          Value        Value/DF

       Deviance                  27        14.4938         0.5368
       Pearson Chi-Square        27        13.2924         0.4923

                  Wald Statistics For Type 3 Analysis
Run 1
                                    Chi-
             Source        DF      Square    Pr > ChiSq

             BLOCK          3       0.56        0.9052
             CTL_TRT        1       6.62        0.0101
Run 2
                                    Chi-
             Source        DF      Square    Pr > ChiSq

             BLOCK          3       0.51        0.9177
             CTL_TRT        0       0.00          .
             A             2       2.89        0.2361
             B             2       0.17        0.9173
             A*B           4       1.80        0.7733
```

These results suggest there is a statistically significant difference between the control and treated groups (Wald χ^2=6.62, p=0.0101), but no evidence of any statistically significant A or B effects within the treated group. These results, however, may be excessively conservative. The deviance/DF ratio is only 0.5368, considerably less than 1. This may suggest **underdispersion**, that is, the variance may be *less* than you would expect with the geometric model. Possibly the geometric model assumes too low a value for the aggregation parameter.

You can try different values of k using the program statements given above. Setting k=2.5 yields the results in Output 10.36.

Output 10.36
Deviance and
Wald Statistics
for a Negative
Binomial
Model, k=2.5,
Fit to Insect
Count Data

```
                     Criteria For Assessing Goodness Of Fit

          Criterion             DF           Value         Value/DF

          Deviance              27          27.6607         1.0245
          Pearson Chi-Square    27          26.5857         0.9847

                    Wald Statistics For Type 3 Analysis
Run 1
                                          Chi-
                  Source          DF     Square    Pr > ChiSq

                  BLOCK            3       1.18       0.7580
                  CTL_TRT          1      13.94       0.0002
Run 2
                  Source          DF     Square    Pr > ChiSq

                  BLOCK            3       1.08       0.7818
                  CTL_TRT          0       0.00        .
                  A               2       5.81       0.0547
                  B               2       0.26       0.8781
                  A*B             4       3.66       0.4546
```

The results are more consistent with the Poisson model corrected for overdispersion discussed in Section 10.4.3.

You can also fit the model with block and treatment and use ESTIMATE statements to compute LS means, similar to what was done in previous sections. For example, the following statements allow you to do this for the model with $k=2.5$:

```
PROC GENMOD;
 k=2.5;
 count=_resp_;
 y=count;
 if y=0 then y=0.1;
 mu=_mean_;
 eta=_xbeta_;
 fwdlink link=log(mu/(mu+k));
 invlink ilink=k*exp(eta)/(1-exp(eta));
 lly=y*log(y/(y+k))-k*log((k+y)/k);
 llm=y*log(mu/(mu+k))-k*log((k+mu)/k);
 d=2*(lly-llm);
 variance var=mu+(mu*mu/k);
 deviance dev=d;
 CLASS BLOCK TRT;
 MODEL y=trt block;
 contrast 'ctl vs trt' trt 9 -1 -1 -1 -1 -1 -1 -1 -1 -1/wald;
 contrast 'a' trt 0  1 1 1  0 0 0  -1 -1 -1,
              trt 0  0 0 0  1 1 1  -1 -1 -1/wald;
 contrast 'b' trt 0  1 0 -1  1 0 -1  1 0 -1,
              trt 0  0 1 -1  0 1 -1  0 1 -1/wald;
 contrast 'a x b' trt 0  1 0 -1  0 0  0  -1 0 1,
                  trt 0  0 0  0  1 0 -1  -1 0 1,
                       trt 0  0 1 -1  0 0  0  0 -1 1,
                            trt 0  0 0  0  0 1 -1  0 -1 1/wald;
 contrast 'ctl vs trt' trt 9 -1 -1 -1 -1 -1 -1 -1 -1 -1;
 contrast 'a' trt 0  1 1 1  0 0 0  -1 -1 -1,
              trt 0  0 0 0  1 1 1  -1 -1 -1;
 contrast 'b' trt 0  1 0 -1  1 0 -1  1 0 -1,
              trt 0  0 1 -1  0 1 -1  0 1 -1;
```

```
contrast 'a x b' trt 0  1 0 -1  0 0  0  -1 0 1,
                 trt 0  0 0  0  1 0 -1  -1 0 1,
                     trt 0  0 1 -1  0 0  0  0 -1 1,
                     trt 0  0 0  0  0 1 -1  0 -1 1;
    estimate 'ctl lsmean' intercept 1 trt 1 0;
  estimate 'treated lsm'intercept 9 trt 0 1 1 1 1 1 1 1 1 1
/divisor=9;
    estimate 'A=1 lsmean' intercept 3 trt 0 1 1 1 0/divisor=3;
    estimate 'A=2 lsmean' intercept 3 trt 0 0 0 0 1 1 1
0/divisor=3;
    estimate 'A=3 lsmean' intercept 3 trt 0 0 0 0 0 0 0 1 1 1 0
/divisor=3;
    ods output estimates=lsm;
```

These statements are identical to the program shown above except MODEL COUNT=BLOCK CTL_TRT A B A*B is replaced by MODEL COUNT=BLOCK TRT and the CONTRAST and ESTIMATE statements shown. Notice that the program includes the ODS OUTPUT statement for the estimates of the LS means. Output 10.37 shows the results for the goodness-of-fit statistics and the contrasts. The LS means are discussed below.

Output 10.37
An Alternative Analysis of a Negative Binomial Model Using Contrasts

Criteria For Assessing Goodness Of Fit

Criterion	DF	Value	Value/DF
Deviance	27	27.6607	1.0245
Pearson Chi-Square	27	26.5857	0.9847

Contrast Results

Contrast	DF	Chi-Square	Pr > ChiSq	Type
ctl vs trt	1	17.76	<.0001	Wald
a	2	5.81	0.0547	Wald
b	2	0.26	0.8781	Wald
a x b	4	3.66	0.4546	Wald
ctl vs trt	1	13.46	0.0002	LR
a	2	6.41	0.0406	LR
b	2	0.28	0.8703	LR
a x b	4	4.05	0.3993	LR

The deviance and Pearson χ^2 are similar but not identical to the previous model. The difference results from the fact that with the non-linearity of the link function, the change in the parameterization of the model induces slight differences in many of the statistics. This model did allow the likelihood ratio statistics to be computed; they are given along with the Wald statistics. The numbers are somewhat different, but the overall conclusions are similar: There is very strong evidence of a difference between the control and treated groups, fairly strong evidence of an A main effect, and no evidence of either an A×B interaction or a B main effect.

The ESTIMATE statements given above compute LS means on the link function scale. As with previous examples, you can use an ODS statement to output the results. Then you can use the following program statements to compute the inverse link and the Delta Rule to get LS means and their standard errors on the COUNT scale.

```
data c_lsm;
 set lsm;
 k=2.5;
 counthat=k*exp(estimate)/(1-exp(estimate));
 deriv=k*exp(estimate)/((1-exp(estimate))**2);
 se_count=deriv*stderr;

proc print data=c_lsm;
run;
```

Output 10.38 shows the results. You can adapt these statements to obtain predicted counts from the negative binomial models fit with the maximum likelihood estimate of k earlier in this section. For example, you could substitute $k=4.072$ above and use the LS means for the negative binomial model, canonical link, and maximum likelihood estimate of k.

Output 10.38
LS Means and Standard Errors for Negative Binomial Model of Insect Count Data

Obs	Label	Estimate	StdErr	Alpha	LowerCL	UpperCL
1	ctl lsmean	-0.1752	0.0537	0.05	-0.2804	-0.0700
2	treated lsm	-0.4975	0.0561	0.05	-0.6075	-0.3875
3	A=1 lsmean	-0.3437	0.0650	0.05	-0.4710	-0.2163
4	A=2 lsmean	-0.4718	0.0880	0.05	-0.6442	-0.2993
5	A=3 lsmean	-0.6770	0.1269	0.05	-0.9257	-0.4283

Obs	ChiSq	Prob ChiSq	k	counthat	deriv	se_count
1	10.65	0.0011	2.5	13.0587	81.2700	4.36218
2	78.61	<.0001	2.5	3.8785	9.8956	0.55525
3	27.98	<.0001	2.5	6.0955	20.9576	1.36169
4	28.74	<.0001	2.5	4.1473	11.0272	0.97043
5	28.46	<.0001	2.5	2.5827	5.2508	0.66637

ESTIMATE and STDERR give the estimates on the link function scale. COUNTHAT and SE_COUNT are the LS means on the COUNT scale. These results are similar to those obtained using the Poisson model corrected for overdispersion. Recall, for example, that the control LS mean was 13.59 with a standard error of 3.44, whereas for the negative binomial model the control LS mean is 13.06 with a standard error of 4.36. Although the overall conclusions about which treatment effects are significant are reasonably consistent, the estimates of the means and hence the magnitude of the treatment effects are different.

This is an ongoing area of research. At this time, no compelling evidence favors either approach over the other. However, Young et al. (1999) showed that Type I error control in tests using GLMs are severely affected by model misspecification. They did not address the question of how best to estimate treatment means and treatment effects when the treatments *are* different. Further research will certainly shed more light on how best to use GLMs. The main point of this section is that your choice of model can greatly affect your conclusions. This section presented a number of approaches that are possible and how to use tools available with PROC GENMOD to check these approaches for problems.

10.5 Generalized Linear Models with Repeated Measures—Generalized Estimating Equations

Chapter 8 discussed repeated measures with standard linear models. The main distinguishing feature of repeated-measures data is the possible correlation among observations observed at different times on the same subject. PROC MIXED allows you to fit a variety of correlation models when the data have a normal distribution. When there are no random-model effects, the MIXED procedure uses generalized least squares (GLS) to estimate model parameters. PROC GENMOD uses generalized estimating equations (GEEs), a generalized linear model analog of GL developed by Liang and Zeger (1986) and Zeger et al. (1988). Section 10.6 presents background theory for the GEE method. The section discusses a repeated-measures example and shows you how to use PROC GENMOD'S GEE option to do repeated-measured analysis of generalized linear models.

10.5.1 A Poisson Repeated-Measures Example

Output 10.39 shows data from a study evaluating a new treatment for epilepsy. These data appeared in Leppik et al. (1985) and were subsequently discussed by Thall and Vail (1990), and Breslow and Clayton (1993). The variable ID identifies each patient in the study. The treatments are TRT=0, a placebo, and TRT=1, an anti-epileptic drug. The response variable is the number of seizures over a two-week interval. For the eight weeks prior to placing the participants on treatment, the number of seizures was counted for each patient in order to form a baseline (BASE) measurement. Also, the patients' AGE, in years, was thought to be a potentially important covariate. The number of seizures was recorded for each of four time intervals and appears in the data set as Y1 through Y4 for the first through fourth observation interval, respectively.

Output 10.39
Epilepsy
Seizure
Repeated-
Measures
Data

Obs	id	trt	base	age	y1	y2	y3	y4
1	104	0	11	31	5	3	3	3
2	106	0	11	30	3	5	3	3
3	107	0	6	25	2	4	0	5
4	114	0	8	36	4	4	1	4
5	116	0	66	22	7	18	9	21
6	118	0	27	29	5	2	8	7
7	123	0	12	31	6	4	0	2
8	126	0	52	42	40	20	23	12
9	130	0	23	37	5	6	6	5
10	135	0	10	28	14	13	6	0
11	141	0	52	36	26	12	6	22
12	145	0	33	24	12	6	8	4
13	201	0	18	23	4	4	6	2
14	202	0	42	36	7	9	12	14
15	205	0	87	26	16	24	10	9
16	206	0	50	26	11	0	0	5
17	210	0	18	28	0	0	3	3
18	213	0	111	31	37	29	28	29
19	215	0	18	32	3	5	2	5
20	217	0	20	21	3	0	6	7
21	219	0	12	29	3	4	3	4
22	220	0	9	21	3	4	3	4
23	222	0	17	32	2	3	3	5
24	226	0	28	25	8	12	2	8
25	227	0	55	30	18	24	76	25
26	230	0	9	40	2	1	2	1
27	234	0	10	19	3	1	4	2
28	238	0	47	22	13	15	13	12
29	101	1	76	18	11	14	9	8
30	102	1	38	32	8	7	9	4
31	103	1	19	20	0	4	3	0
32	108	1	10	30	3	6	1	3
33	110	1	19	18	2	6	7	4
34	111	1	24	24	4	3	1	3
35	112	1	31	30	22	17	19	16
36	113	1	14	35	5	4	7	4
37	117	1	11	27	2	4	0	4
38	121	1	67	20	3	7	7	7
39	122	1	41	22	4	18	2	5
40	124	1	7	28	2	1	1	0
41	128	1	22	23	0	2	4	0
42	129	1	13	40	5	4	0	3
43	137	1	46	33	11	14	25	15
44	139	1	36	21	10	5	3	8
45	143	1	38	35	19	7	6	7
46	147	1	7	25	1	1	2	3
47	203	1	36	26	6	10	8	8
48	204	1	11	25	2	1	0	0
49	207	1	151	22	102	65	72	63
50	208	1	22	32	4	3	2	4
51	209	1	41	25	8	6	5	7
52	211	1	32	35	1	3	1	5
53	214	1	56	21	18	11	28	13
54	218	1	24	41	6	3	4	0
55	221	1	16	32	3	5	4	3
56	225	1	22	26	1	23	19	8
57	228	1	25	21	2	3	0	1
58	232	1	13	36	0	0	0	0
59	236	1	12	37	1	4	3	2

The data have both repeated measures and generalized linear model features. Repeated measures result from the fact that each subject was observed four times. Because the observations are counts, it is reasonable to fit a model with a link function and an assumed probability distribution appropriate for a random count variable.

The analyses reported by Thall and Vail (1990) used transformed covariates for baseline and age. The baseline covariate was log (BASE/4), the 4 accounting for the baseline period being four times as long as the two-week observation periods during the study. The age covariate was log(AGE). These variables appear at LOG_BASE and LOG_AGE, respectively, in the SAS programs and output shown below.

Following Thall and Vail, a generalized linear model for these data is

$$\log(\lambda_{ij}) = \mu + \alpha_i + \tau_j + (\alpha\tau)_{ij} + \beta_{1i}*(\log\_base) + \beta_2*(\log\_age)$$

where

λ_{ij} is the mean count for treatment i ($i = 0, 1$) and time j ($j = 1, 2, 3, 4$).

μ is the intercept.

α_i is the effect of the ith treatment.

τ_j is the effect of the jth time period.

$(\alpha\tau)_{ij}$ is the ijth TIME×TREATMENT interaction effect.

β_{1i} is the regression coefficient for LOG_BASE for the ith treatment.

β_2 is the regression coefficient for LOG_AGE.

Alternatively, you can write $\beta_{1i} = \beta_1 + \delta_i$, equivalent to specifying a common slope plus a TREATMENT-BY-LOG_base interaction term. Note that Thall and Vail fit a separate regression of LOG_BASE on count for each treatment, but a common regression for LOG_AGE. If you did not have the advantage of their work, you would fit a model with separate regressions over LOG_AGE for each treatment as well and test the hypothesis of equal regressions.

10.5.2 Using PROC GENMOD to Compute a GEE Analysis of Repeated Measures

You can fit the model using PROC GENMOD. First, you must modify the data set so that there is one observation per time, rather than all four observations over time on a single data line as shown in Output 10.39. The one-line-per-time requirement is identical to the format PROC MIXED uses for repeated-measures data. You can use the following SAS statements to convert the data set. Set LEPPIK is the original data set shown in Output 10.39; SEIZURE is the modified form suitable for PROC GENMOD.

```
data seizure;
  set Leppik;
    time = 1; y = y1; output;
    time = 2; y = y2; output;
    time = 3; y = y3; output;
    time = 4; y = y4; output;
```

Then use the following GENMOD statements:

```
proc genmod;
 class id trt time;
 model y=trt time trt*time log_base trt*log_base log_age/
      dist=poisson link=log type1 type3;
 repeated subject=id / type=exch corrw;
```

The REPEATED statement implements the generalized estimating equations (GEEs) for the repeated measures over the four times. GEE uses a "working correlation matrix" to account for the correlation among the repeated measurements within subjects. Refer to Section 10.6 for more about working correlation matrices. The SUBJECT= and TYPE= commands are required in the REPEATED statement. SUBJECT= identifies the unit on which the repeated measurements were taken. In this case, they are on each patient, identified by the variable ID. The SUBJECT=ID command creates a block-diagonal working correlation matrix, one 4×4 block per ID. TYPE= defines the type of correlation model, similar to the TYPE command in the REPEATED statement in PROC MIXED. Available types include IND (independent), EXCH (exchangeable, or, equivalently, CS for compound symmetry), AR (first-order autoregressive, similar to AR(1) in PROC MIXED) and UN (unstructured). Following Thall and Vail, this example uses TYPE=EXCH.

The CORRW option in the REPEATED statement causes the estimated working correlation to be included in the output. In the MODEL statement, LOG_BASE and TRT*LOG_BASE account for the separate regressions over LOG_BASE for each treatment. Alternatively, you could use LOG_BASE(TRT). Relating these to the MODEL statement given above, the latter models β_{1i}, whereas the former models $\beta_1 + \delta_i$.

Output 10.40 shows selected results.

Output 10.40
GEE Results for
Epilepsy Seizure
Data: Test or
Treatment×Time
Interaction

```
                         GEE Model Information

              Correlation Structure              Exchangeable
              Subject Effect                    id (59 levels)
              Number of Clusters                          59
              Correlation Matrix Dimension                 4
              Maximum Cluster Size                         4
              Minimum Cluster Size                         4


                      Working Correlation Matrix

                      Col1        Col2        Col3        Col4

            Row1    1.0000      0.3569      0.3569      0.3569
            Row2    0.3569      1.0000      0.3569      0.3569
            Row3    0.3569      0.3569      1.0000      0.3569
            Row4    0.3569      0.3569      0.3569      1.0000


                 Score Statistics For Type 3 GEE Analysis

                                     Chi-
              Source          DF     Square    Pr > ChiSq

              trt              1      5.82        0.0159
              time             3      5.03        0.1695
              trt*time         3      1.53        0.6751
              log_base         1      6.34        0.0118
              log_base*trt     1      3.61        0.0576
              log_age          1      6.48        0.0109
```

The "GEE Model Information" tells you how many subjects (clusters) and the dimension of each block of the working correlation matrix (four, that is, the number of time periods). The estimated working correlation matrix appears next. In this case, the observations at any two time periods for the same subject have a correlation of 0.3569. The test for TRT*TIME has a χ^2 statistic of 1.53, with a *p*-value of 0.6751. Thall and Vail reported their analyses without a treatment×time interaction term, which this result clearly justifies.

The SAS statements for the revised model without a TRT*TIME interaction are

```
proc genmod;
 class id trt time;
 model y=trt time log_base(trt)log_age/
     dist=poisson link=log type1 type3;
 repeated subject=id / type=exch corrw;
 lsmeans trt / e;

 estimate 'lsm trt 0' intercept 1 trt 1 0 time 0.25 0.25
0.25 0.25
         log_base(trt) 1.7679547 0 log_age 3.3197835/exp;
 estimate 'lsm at t=4' intercept 1 trt 1 0 time 0 0 0 1
         log_base(trt) 1.7679547 0 log_age 3.3197835/exp;
 estimate 'lsm at t<4' intercept 3 trt 3 0 time 1 1 1 0
         log_base(trt) 5.3038641 0
         log_age 9.9593505/ divisor=3 exp;
 contrast 'log_b slopes =' log_base(trt) 1 -1;
 contrast 'visit 4 vs others' time 1 1 1 -3;
 contrast 'among visit 1-3' time 1 -1 0 0, time 1 1 -2 0;
```

Several additional statements appear. This program uses the alternative form of the regressions over LOG_BASE for each treatment, LOG_BASE(TRT), based on the parameterization β_{1i}. The associated contrast, LOG_B SLOPES = tests H_0: $\beta_{1_0} = \beta_{1_1}$, the same hypothesis tested by LOG_BASE*TRT in Output 10.40. The LSMEANS statement computes least-square means on the log link scale, but does so at the means of LOG_BASE, LOG_AGE, and TIME. The E option prints the coefficients used to compute LS means so you can see if they are really what you want. The first ESTIMATE statement duplicates the coefficients of LS mean for TRT=0, the placebo, but adds the EXP option, which applies the inverse link and Delta Rule, as also shown in Section 10.2. The second and third ESTIMATE statements compute the expected means at time 4 and the average of times 1 through 3, respectively. Thall and Vail defined a variable to model the difference between time 4 and the others. The CONTRASTs shown here, time 4 versus the others and among the remaining time periods, accomplish the same thing. The results for the working correlation matrix, parameter estimates, and Type 3 ANOVA are shown in Output 10.41. Output 10.42 contains the LS mean, estimate, and contrast results.

Output 10.41
GEE Analysis
of Epilepsy
Seizure Data: A
Model without a
Treatment×Time
Interaction,
Working
Correlation
Matrix,
Parameter
Estimates,
and Type 3
Analysis

```
                        Working Correlation Matrix

                 Col1         Col2         Col3         Col4

      Row1     1.0000       0.3552       0.3552       0.3552
      Row2     0.3552       1.0000       0.3552       0.3552
      Row3     0.3552       0.3552       1.0000       0.3552
      Row4     0.3552       0.3552       0.3552       1.0000

             Analysis Of GEE Parameter Estimates
             Empirical Standard Error Estimates

                            Standard   95% Confidence
    Parameter      Estimate   Error       Limits          Z  Pr > |Z|

    Intercept       -4.3522   1.0788  -6.4667  -2.2377  -4.03   <.0001
    trt         0    1.3551   0.4289   0.5145   2.1956   3.16   0.0016
    trt         1    0.0000   0.0000   0.0000   0.0000    .       .
    time        1    0.2030   0.0987   0.0096   0.3964   2.06   0.0397
    time        2    0.1344   0.0762  -0.0149   0.2837   1.76   0.0776
    time        3    0.1445   0.1228  -0.0963   0.3852   1.18   0.2395
    time        4    0.0000   0.0000   0.0000   0.0000    .       .
    log_base(trt) 0  0.9500   0.0986   0.7567   1.1432   9.64   <.0001
    log_base(trt) 1  1.5202   0.1423   1.2413   1.7992  10.68   <.0001
    log_age          0.9194   0.2773   0.3759   1.4630   3.32   0.0009

             Score Statistics For Type 3 GEE Analysis

                                     Chi-
             Source          DF     Square    Pr > ChiSq

             trt              1      5.81       0.0159
             time             3      4.71       0.1941
             log_base(trt)    2      9.94       0.0070
             log_age          1      6.47       0.0110
```

Dropping TRT*TIME has little effect on the estimated working correlation: Here, $\hat{\rho} = 0.3552$. The LOG_BASE(TRT) output under "Analysis of GEE Parameter Estimates" gives you $\hat{\beta}_{1_0} = 0.95$ and $\hat{\beta}_{1_1} = 1.52$, respectively. The estimate for LOG_AGE, 0.9194, is $\hat{\beta}_2$. From the "Score Statistics for Type III GEE Analysis" the χ^2 statistic that tests H_0: $\alpha_0 = \alpha_1$ (equal treatments) is 5.81 ($p = 0.0159$). **Caution:** The Type 3 TRT χ^2 statistic tests H_0: $\alpha_0 = \alpha_1$ at the mean of LOG_BASE. Because there are different slopes over LOG_BASE for different treatments, the size and statistical significance of the treatment effect will vary with LOG_BASE. If testing the treatment effect at a different LOG_BASE is more consistent with your objectives, you need to write a CONTRAST statement specifically for that purpose.

Output 10.42
GEE Analysis of Epilepsy Seizure Data: A Model without a Treatment×Time Interaction LS Means, Estimate, and CONTRAST Results

```
                 Coefficients for trt Least Squares Means

    Label      Row    Prm1    Prm2    Prm3    Prm4    Prm5    Prm6    Prm7
                      Prm8    Prm9    Prm10

    trt         1       1       1       0    0.25    0.25    0.25    0.25
                      1.768       0  3.3198
    trt         2       1       0       1    0.25    0.25    0.25    0.25
                          0   1.768  3.3198

                         Least Squares Means

                                 Standard           Chi-
      Effect    trt   Estimate      Error    DF   Square    Pr > ChiSq

       trt       0     1.8552     0.1047      1   313.92       <.0001
       trt       1     1.5084     0.1480      1   103.94       <.0001


                       Contrast Estimate Results

                              Standard                                  Chi-
    Label            Estimate    Error   Alpha   Confidence Limits    Square

    lsm trt 0          1.8552   0.1047    0.05     1.6500   2.0605    313.92
    Exp(lsm trt 0)     6.3932   0.6694    0.05     5.2070   7.8496

    lsm at t=4         1.7348   0.0944    0.05     1.5498   1.9197    337.91
    Exp(lsm at t=4)    5.6676   0.5349    0.05     4.7105   6.8191

    lsm at t<4         1.8954   0.1128    0.05     1.6743   2.1165    282.28
    Exp(lsm at t<4)    6.6552   0.7508    0.05     5.3350   8.3020


                     Contrast Results for GEE Analysis

                                     Chi-
        Contrast            DF     Square   Pr > ChiSq   Type

        log_b slopes =       1       3.64     0.0565     Score
        visit 4 vs others    1       4.16     0.0414     Score
        among visit 1-3      2       0.35     0.8408     Score
```

The coefficients for the LS means show the coefficients SAS uses by default. The order of the coefficients, from PRM1 through, in this case, PRM10, follows from the order in which the parameter estimates were given earlier. That is, PRM1 is the coefficient for INTERCEPT, PRM2 is for TRT 0, PRM3 is for TRT 1, and so forth. You can see that the TRT least-squares means are computed averaged over the four time periods and at the mean LOG_BASE and LOG_AGE. If you subtract the coefficients of the TRT 1 LS Mean from TRT 0, you get the difference tested by the Type III χ^2 statistic discussed above: $\alpha_0-\alpha_1+(\beta_{10}-\beta_{11})*$LOG_BASE.

The CONTRAST results show moderate evidence of unequal slopes for the regression over LOG_BASE for each treatment (χ^2=3.64, p=0.0565). They confirm the Thall and Vail (1990) result that time 4 has a different expected number of seizures from the other time periods (χ^2=4.16, p=0.0414), but no statistically significant evidence of differences among the first three periods (χ^2=0.35, p=0.8408). The expected number of counts, for example, for the placebo (TRT 0) is 6.66 with a standard error of 0.75, using the EXP(LSM AT T< 4) line of the "Contrast Estimate Results." Recall that the LS mean is computed on the log link scale; the EXP line applies the inverse link, giving you the estimate on the original count (number of seizures) scale. The expected number of seizures for the placebo at time 4 is 5.67 with a standard error of 0.35. Thus, the VISIT 4 VS OTHERS effect results from an overall reduction in the number of seizures in the fourth period.

This example showed the GEE method applied to Poisson data. You can apply GEE to other GLMs, for example, with binomial or gamma distributions as well. As with normal data using the REPEATED option in PROC MIXED, accounting for correlation among repeated measures can substantially affect your conclusions. Failing to account for repeated measures risks seriously misrepresenting the data. One caution: The GEE option in PROC GENMOD does not account for random-model effects. For more sophisticated analyses where your model needs to have both random-model effects and account for within-subjects correlation, you should not use PROC GENMOD. Generalized linear mixed-model programs such as the GLIMMIX macro or, in some cases, PROC NLMIXED, are better suited for such models. See Littell et al. (1996) for an introduction to GLIMMIX.

10.6 Background Theory

Nelder and Wedderburn (1972) presented the basic theory for **generalized linear models**, hereafter referred to in this section as GzLMs. As a point of information, people working in the area refer to generalized linear models as "GLMs." To avoid confusion with PROC GLM—which does *not* compute generalized linear models—we use the acronym GzLM. Nelder and Wedderburn's work applied linear model methods to response variables whose distributions belong to the exponential family. Two major extensions followed. **Quasi-likelihood** theory, developed by Wedderburn (1974) and discussed in detail by McCullagh (1983), allowed GzLMs to be used with a much broader class of response variables than the exponential family. Zeger et al. (1988) developed **generalized estimating equations** (GEEs), to permit GzLMs to be used with non-normal repeated-measures data. This section provides a brief overview of the main ideas.

To understand the basic idea of GzLMs, it is helpful to review the normal errors linear model from a slightly different perspective. In previous chapters, linear models were presented in the form $y = \beta_0 + \Sigma_{i=1}^{p} \beta_i X_i + e$, where e is i.i.d. $N(0, \sigma^2)$. This means that the mean of y, $E(y) = \mu$, is equal to $\beta_0 + \Sigma_{i=1}^{p} \beta_i X_i$. In other words, $\beta_0 + \Sigma_{i=1}^{p} \beta_i X_i$ models μ, the expected value of the response variable, y. The basic idea of the GLM is that for response variables with other than the normal distribution, it is often more reasonable to use $\beta_0 + \Sigma_{i=1}^{p} \beta_i X_i$ to model a function of μ, rather than μ itself. The binomial and Poisson distributions illustrate the main ideas.

For the binomial distribution, suppose the possible outcomes of each Bernoulli trial are coded 0 or 1. For example, if you flip a coin, 0 and 1 could represent tails and heads, respectively. The probability that you obtain y 1's out of N independent Bernoulli trials (for example, y heads out of N coin flips) is given by the formula

$P(Y=y) = \binom{N}{y} \pi^y (1-\pi)^{N-y}$, where π is the probability of a 1 on any given Bernoulli

trial. The expected value of the sample proportion, $p = \hat{\pi} = \dfrac{y}{N}$ is $E(p) = \pi$ and its variance is var$(p)=\dfrac{\pi(1-\pi)}{N}$. The **log likelihood** of the sample proportion, obtained by taking the log of $P(Y=y)$ and expressing it in terms of p, is

$$\ell(p) = \frac{p\,\log(\frac{\pi}{1-\pi})+\log(1-\pi)}{1/N} + \log\binom{N}{y} \ .$$

For the Poisson distribution, the probability of exactly Y occurrences of a discrete count is $P(Y=y)=\dfrac{y^{\lambda}e^{-\lambda}}{y!}$. For the Poisson, the mean and variance are equal. Specifically, the expected number of counts, $E(y)=\mathrm{var}(y)=\lambda$. The log likelihood of the count is $\ell(y) = y\,\log(\lambda)-\lambda-\log(y!)$.

The log likelihood, mean, and variance of these distributions, as well as the normal and other members of the exponential family, share a common form. The general form of the log likelihood is $\ell(\theta,\phi;y) = \dfrac{y\theta-b(\theta)}{\phi}+c(y,\phi)$, where θ is the **natural parameter** and ϕ is the scale parameter. The natural parameter is often written $\theta(\mu)$ because it is a function of the expected value, μ. For the binomial, the natural parameter is

$\theta = \log\dfrac{\pi}{1-\pi}$, and for the Poisson, $\theta = \log(\lambda)$. You can show that $\mu = \dfrac{\partial b(\theta)}{\partial\theta}$ and

$\mathrm{var}(y) = \phi\dfrac{\partial^2 b(\theta)}{\partial\theta^2}$. The second derivative, $\dfrac{\partial^2 b(\theta)}{\partial\theta^2}$, is usually denoted $V(\mu)$ and is called

the **variance function**. Thus, var$(y)=\phi V(\mu)$, illustrating the dependence of the variance on the scale parameter, through ϕ, and the mean, through $V(\mu)$. For the normal, $\phi=\sigma^2$ and $V(\mu)=1$, which shows the independence of the mean and the variance. For the binomial, $\phi=1/N$ and $V(\mu)=\pi(1-\pi)$. For the Poisson, $\phi=1$ and $V(\mu)=\lambda$. The binomial and Poisson both illustrate cases where the variance is a function of the mean. This dependence of the mean and variance is a key feature of GzLMs.

10.6.1 The Generalized Linear Model Defined

The GzLM models $g(\mu)=\beta_0+\sum_{t=1}^{P}\beta_i X_i$ or, alternatively, $\mu=h(\beta_0+\sum_{t=1}^{P}\beta_i X_i)$. Nelder and Wedderburn (1972) called $g(\mu)$ the **link function** because it "links" the linear model, $\beta_0+\sum_{t=1}^{P}\beta_i X_i$ to the mean, μ. Following Nelder and Wedderburn, the character η is often used in place of $g(\mu)$ to denote the link function. The function $h(\eta)$ is thus called the **inverse link function**.

The general form of the log likelihood provides one common rationale (but certainly not the only one!) for selecting a suitable link function for a given distribution. Notice that the observations, denoted by the random variable y, are linear in the natural parameter, θ. Therefore, it would make sense to fit a linear model to θ rather than directly to the expected value, μ. Link functions having the form $\eta=\theta$ are called **canonical** link functions. Two common types of GzLMs are **log-linear models,** where the observations are assumed to have a Poisson distribution and the canonical link function $\log(\lambda)$ is used, and **logistic models**, where the observations are assumed to have a binomial distribution and the canonical link $\log \dfrac{\pi}{1-\pi}$ is used. In logistic models, the link function,

$\log \dfrac{\pi}{1-\pi}$ is referred to as the **logit** of π. The normal errors linear model is a special
case of the canonical link, because for the normal distribution, $\theta = \mu$. The link function,
$\eta = \mu$ is called the **identity link**. In many cases the choice of link is motivated by a
theoretical model of the mean and its response to the regression or treatment variables.
The **probit** link, discussed in Sections 10.2 and 10.3 is an example.

10.6.2 How the GzLM's Parameters Are Estimated

The model parameters, β_0, β_1,..., β_p can be estimated using **maximum likelihood**, that is,
finding the values of the β_i's that maximize the likelihood, or equivalently, the log
likelihood. For GzLMs, maximum likelihood estimation results in a generalized form of
the normal equations. The normal equations for standard linear models were described in
Chapter 6. Recalling the matrix description from Chapter 6, $\mathbf{y}$ is an $n \times 1$ vector of
observations. Let μ denote the $n \times 1$ vector of expected values of the elements of $\mathbf{y}$, that
is, $E(\mathbf{y}) = \mu$, and V be an $n \times n$ diagonal matrix whose elements are the variances of the
elements of $\mathbf{y}$, that is, $V = \mathrm{diag}[\mathrm{var}(y)]$. Since $\mathrm{var}(y) = \phi V(\mu)$, then $V = \mathrm{diag}[\phi V(\mu)]$.

Under the matrix setup, the GzLM is $g(\mu) = \eta = X\beta$, where X is the $n \times p$ matrix of
constants for the linear model as defined in Chapter 5, and β is the $p \times 1$ vector of model
parameters. To estimate β, you solve the GzLM **estimating equations**,

$X'WX\beta = X'Wy^*$, where $W = DV^{-1}D$, D is an $n \times n$ diagonal matrix whose elements are

the derivatives of the elements of η with respect to μ, that is, $D = \mathrm{diag}[\dfrac{\partial \eta}{\partial \mu}]$, and

$\mathbf{y}^* = \hat{\eta} + D^{-1}(\mathbf{y} - \hat{\mu})$. Because several elements of the GzLM estimating equations, notably
D, $\hat{\eta}$ and $\hat{\mu}$ depend on estimates of β, the equations must be solved iteratively. For the
normal errors model, $\eta = \mu$ and $V = I\sigma^2$, and therefore $D = I$ and $\mathbf{y}^* = \mathbf{y}$. Thus, the
GzLM estimating equations reduce to $X'X\beta = X'y$, the normal equations given in
Chapter 6.

10.6.3 Standard Errors and Test Statistics

Several results are useful when working with GzLMs. Readers can refer to texts such as
McCullagh and Nelder (1989), Dobson (1990), and Lindsey (1997) for additional detail.
Important results include the following:

❑ For estimable functions $K'\beta$, the variance of the estimator, $K'\hat{\beta}$, is $K'(X'WX)^-K$.

❑ It follows that for vector $\mathbf{k}$, the approximate standard error of $\mathbf{k}'\hat{\beta}$ is

$\sqrt{\mathbf{k}'(X'WX)^-\mathbf{k}}$.

❑ For hypothesis testing, you can use either the Wald statistic or the likelihood ratio statistic.

❑ The **Wald statistic** for testing H_0: $K'\beta = K'\beta_0$ is computed by using the formula $(K'\hat{\beta} - K'\beta_0)' [K'(X'WX)^- K]^{-1}(K'\hat{\beta} - K'\beta_0)$. The Wald statistic has an approximate χ^2 distribution with degrees of freedom equal to rank(K). In the normal errors case, the Wald statistic reduces to the sum of squares for testing $K'\beta = \boldsymbol{K}'\beta_0$.

❑ When W depends on an unknown scale parameter, the Wald statistic computed using the estimate of ϕ divided by rank(K) has an approximate F-distribution. The ANOVA F-test is a well-known example of Wald/rank(K), but this chapter illustrates other examples using GzLMs for non-normal data.

❑ The **deviance** is defined as $2[\ell(\theta(\mathbf{y});\mathbf{y}) - \ell(\theta(X\hat{\beta});\mathbf{y}]$, where $\ell(\theta;\mathbf{y})$ is the log likelihood, with $\theta(\mathbf{y})$ the value of θ determined from the data and $\theta(X\hat{\beta})$ determined from the estimate of β under the model. The deviance has an approximate χ^2 distribution with N–p degrees of freedom, where N=the number of observations and p is the rank of X.

❑ For distributions that do not depend on an unknown scale parameter, such as the binomial and Poisson, the deviance tests **lack of fit** of the GzLM. This chapter presents several examples using the deviance for this purpose.

❑ **Likelihood ratio statistics** can be computed from the difference between deviances as follows. Partition the parameter vector β so that $\beta = \begin{matrix} \beta_1 \\ \beta_2 \end{matrix}$. The resulting model is $X\beta = X_1\beta_1 + X_2\beta_2$, rank($X_1$)=$p_1$, rank($X$)–rank($X_1$)=$p_2$, and p=p_1+p_2. The likelihood ratio statistic to test H_0: $\beta_2 = 0$, that is, $X\beta$ reduces to $X_1\beta_1$ is the difference between the deviance for the model $X\beta$ and the deviance for the model $X_1\beta_1$. The resulting likelihood ratio statistic has an approximate χ^2 distribution whose degrees of freedom equal the difference between the degrees of freedom of the deviances, in this case DF= $(N$–$p_1)$–$(N$–$p)$=p–p_1=p_2. For example, in a one-way ANOVA whose full model is $\mu + \tau_i$, the parameter vector $\beta' = [\mu, \tau_1, ..., \tau_t]$ can be partitioned into $\beta_1 = \mu$ and $\beta_2' = [\tau_1, ..., \tau_t]$. The difference between the deviance for the full model and the deviance for the reduced model μ tests H_0: all τ_i=0.

10.6.4 Quasi-Likelihood

The previous sections presented GzLMs as a tool for fitting linear models to non-normal response variables whose probability distributions belong to the exponential family. However, there are occasions when you want to fit a linear model to a response variable whose distribution does not belong to the exponential family and may not even be known. You can do this using quasi-likelihood theory.

Recall the basic form of the log likelihood for the exponential family, $\ell(\lambda, \phi; y) = \dfrac{y\theta - b(\theta)}{\phi} + c(\phi, y)$. The GzLM estimating equations follow from information contained in $\dfrac{y\theta - b(\theta)}{\phi}$ only, specifically the mean, $\mu = \dfrac{\partial b(\theta)}{\partial \theta}$ and the variance, $\phi V(\mu)$.

The function $\dfrac{y\theta - b(\theta)}{\phi}$ is a special case of quasi-likelihood. More generally, any

function whose general form is $Q(\mu;y) = \displaystyle\int_{y}^{\mu} \dfrac{y - t}{\phi V(t)} dt$ is defined as a quasi-likelihood

function. This implies that $E(y) = \mu$ and $\mathrm{var}(y) = \phi V(\mu)$. Hence, generalized linear models can be fit to any response variable provided the following two conditions hold:

❑ You can specify a function relating the linear model $X\beta$ to $E(y)$. The inverse link $\mu = h(X\beta)$ or the link $\eta = g(\mu) = X\beta$ set up the GzLM.

❑ You can specify a variance in the form $\mathrm{var}(y) = \phi V(\mu)$.

McCullagh and Nelder (1989) review several commonly used quasi-likelihood forms. Section 10.4.4 discussed an example of a quasi-likelihood analysis of count data using PROC GENMOD.

10.6.5 Repeated Measures and Generalized Estimating Equations

The previous sections discussed fitting generalized linear models to non-normally but *independently* distributed response variables. However, non-normal response variables are often observed in repeated-measures experiments. The resulting longitudinal data raise the same issues of correlation between observations within subjects discussed in Chapter 8. Zeger et al. (1988) presented the method of generalized estimating equations (GEEs), essentially an extension of the repeated-measures methods used by PROC MIXED from normally distributed data to non-normal data and generalized linear models.

The basic idea of GEEs is to build a correlation structure into V, the variance-covariance matrix of the observation vector $\mathbf{y}$ described in Section 10.6.2. Recall that V was defined as $V(\mathbf{y}) = \mathrm{diag}[\phi V(\mu)]$. This variance-covariance structure assumes that the observations are independent. GEEs account for possible serial correlation by modifying V as $V(\mathbf{y}) = V_{\mu}^{1/2} R V_{\mu}^{1/2}$, where $V_{\mu}^{1/2} = \mathrm{diag}[\sqrt{\phi V(\mu)}$ and R is a **working correlation matrix**.

The working correlation matrix is not a true correlation matrix, but it approximates the true correlation by using structures similar to covariance models you would use, for example, in PROC MIXED for normally distributed data. Once V is modified to include the working correlation matrix, the GzLM estimating equations described in Section 10.6.2 are used to estimate β. Inference with GEEs typically replace the model-based $\mathrm{var}(K'\beta) = K'(X'\hat{W}X)^{-} K$ by the robust, or empirical estimate,

$K'[(X'V^{-1}X)^{-} X'V^{-1}]\hat{V}[V^{-1}X(X'V^{-1}X)^{-}]K$ K where $V = V_{\mu}^{1/2}RV_{\mu}^{1/2}$ as defined above and

$\hat{V} = (\mathbf{y} - \hat{\mu})(y - \hat{\mu})'$. Provided the sample size is adequate, the empirical estimate has the advantage that when it is used, correct inference does not depend on getting the working correlation matrix right. With very small samples, however, test statistics computed with the empirical estimate can be inflated, sometimes wildly.

Diggle et al. (1994) provide additional background on GEEs. Section 10.5.1 presented an example with correlated Poisson data and their analysis using the GEE option available in PROC GENMOD.

11.1 Introduction

As already noted, the GLM and MIXED procedures can be used to analyze a multitude of data structures. In this chapter several applications are presented that utilize tools discussed in the previous chapters. Some of these applications involve statistical topics that are not discussed in great detail in this book. References are given to provide the necessary background information.

11.2 Confounding in a Factorial Experiment

Experiments use confounding in two forms. The first are factorial treatments designs in which all factorial combinations appear in the experiment, but they appear in incomplete blocks containing only a subset of the factor combinations. Thus, within a given block, one or more treatment effects are confounded with block effects. The second are **fractional factorial experiments** in which only a subset of the factor combinations appear in the experiment. Thus, some of the factorial effects are not estimable, but are **aliased** with other effects, meaning that the same estimable function estimates both effects. Confounding is covered in most textbooks on the design of experiments (for example, Hicks and Turner 2000).

11.2.1 Confounding with Blocks

The first example for this topic is a 2^3 factorial with factors labeled A, B, and C in blocks of size four. There are three replications with interactions ABC, AC, and BC, confounded with blocks in replications 1, 2, and 3, respectively. These factors are thus partially confounded with blocks. The data appear in Output 11.1.

Output 11.1
Data for a
Two-Cube
Factorial in
Blocks of Size
Four

Obs	rep	blk	a	b	c	y
1	1	1	1	1	1	3.99
2	1	1	1	0	0	1.14
3	1	1	0	1	0	1.52
4	1	1	0	0	1	3.33
5	1	2	1	1	0	2.06
6	1	2	1	0	1	5.58
7	1	2	0	1	1	2.06
8	1	2	0	0	0	-0.17
9	2	1	1	1	1	3.77
10	2	1	1	0	1	6.69
11	2	1	0	1	0	2.17
12	2	1	0	0	0	-0.01
13	2	2	1	1	0	2.43
14	2	2	0	1	1	1.22
15	2	2	1	0	0	0.37
16	2	2	0	0	1	2.06
17	3	1	1	1	1	4.53
18	3	1	0	1	1	1.90
19	3	1	1	0	0	1.62
20	3	1	0	0	0	-0.70
21	3	2	1	1	0	1.56
22	3	2	1	0	1	5.99
23	3	2	0	1	0	1.44
24	3	2	0	0	1	2.42

Contrasts corresponding to confounded effects can be estimated only from those replications in which they are not confounded. In this example, they are estimated from only two-thirds of the data; thus their standard errors should be larger by a factor of $\sqrt{3/2}$.

The analysis using PROC GLM is straightforward. You can generate contrasts in the DATA step instead of specifying classes for treatments and using CONTRAST statements, as the following code shows:

```
data confound;
   input rep blk a b c y;
      ca= -(a=0) + (a=1);
      cb= -(b=0) + (b=1);
      cc= -(c=0) + (c=1);
   datalines;

      data
```

By sorting the data and running the analysis by REP, you can use the ALIASING option in PROC GLM to print out the confounding pattern. Use the following statements:

```
proc sort;
    by rep;
proc glm;
    by rep;
    class blk;
    model y=blk ca|cb|cc/solution aliasing;
```

The results appear in Output 11.2.

Output 11.2
Aliasing
Output
Showing a
Confounding
Pattern for a 2^3
Factorial in
Blocks of Size
Four

```
----------------------------------- rep=1 -----------------------------------

          Parameter           Expected Value

          Intercept           Intercept + [blk 2] - ca*cb*cc
          blk        1        [blk 1] - [blk 2] + 2*ca*cb*cc
          blk        2
          ca                  ca
          cb                  cb
          ca*cb               ca*cb
          cc                  cc
          ca*cc               ca*cc
          cb*cc               cb*cc
          ca*cb*cc
----------------------------------- rep=2 -----------------------------------

          Parameter           Expected Value

          Intercept           Intercept + [blk 2] - ca*cc
          blk        1        [blk 1] - [blk 2] + 2*ca*cc
          blk        2
          ca                  ca
          cb                  cb
          ca*cb               ca*cb
          cc                  cc
          ca*cc
          cb*cc               cb*cc
          ca*cb*cc            ca*cb*cc


----------------------------------- rep=3 -----------------------------------

          Parameter           Expected Value

          Intercept           Intercept + [blk 2] - cb*cc
          blk        1        [blk 1] - [blk 2] + 2*cb*cc
          blk        2
          ca                  ca
          cb                  cb
          ca*cb               ca*cb
          cc                  cc
          ca*cc               ca*cc
          cb*cc
          ca*cb*cc            ca*cb*cc
```

The contents of Output 11.2 appear immediately after the parameter estimates generated by the SOLUTION option in the MODEL statement. For REP=1, you can see that the three-way interaction CA*CB*CC has a blank under "Expected Value" but the INTERCEPT and BLK 1 effects estimate their usual estimable functions plus the CA*CB*CC effect. This indicates that the ABC interaction effect is confounded with block in REP=1. Similarly, the output indicates that the AC interaction is confounded with block in REP=2, and the BC interaction is confounded with block in REP=2. Although in this example the ALIASING option merely confirms the confounding pattern stated in the introduction, it can be very useful in data sets where the confounding pattern is not obvious and needs to be investigated.

For a complete analysis of the data, combined over all replications, use the following SAS statements:

```
proc glm;
   classes rep blk;
   model y=rep blk(rep) ca|cb|cc/ solution;
```

The results appear in Output 11.3.

Output 11.3
ANOVA for a
Two-Cube
Factorial in
Blocks of Size
Four

```
                              The GLM Procedure
                          Class Level Information
                  Class         Levels    Values
                  rep                3     1 2 3
                  blk                2     1 2

                       Number of observations     24
                          The GLM Procedure
```

Dependent Variable: y

Source	DF	Sum of Squares	Mean Square	F Value	Pr > F
Model	12	81.74957500	6.81246458	33.60	<.0001
Error	11	2.23018750	0.20274432		
Corrected Total	23	83.97976250			

R-Square	Coeff Var	Root MSE	y Mean
0.973444	18.96878	0.450271	2.373750

Source	DF	Type I SS	Mean Square	F Value	Pr > F
rep	2	0.05092500	0.02546250	0.13	0.8832
blk(rep)	3	7.43221250	2.47740417	12.22	0.0008
ca	1	21.07500417	21.07500417	103.95	<.0001
cb	1	0.00453750	0.00453750	0.02	0.8838
ca*cb	1	1.72270417	1.72270417	8.50	0.0141
cc	1	37.77550417	37.77550417	186.32	<.0001
ca*cc	1	2.31800625	2.31800625	11.43	0.0061
cb*cc	1	11.34005625	11.34005625	55.93	<.0001
ca*cb*cc	1	0.03062500	0.03062500	0.15	0.7049

Source	DF	Type III SS	Mean Square	F Value	Pr > F
rep	2	0.05092500	0.02546250	0.13	0.8832
blk(rep)	3	1.66755417	0.55585139	2.74	0.0938
ca	1	21.07500417	21.07500417	103.95	<.0001
cb	1	0.00453750	0.00453750	0.02	0.8838
ca*cb	1	1.72270417	1.72270417	8.50	0.0141
cc	1	37.77550417	37.77550417	186.32	<.0001
ca*cc	1	2.31800625	2.31800625	11.43	0.0061
cb*cc	1	11.34005625	11.34005625	55.93	<.0001
ca*cb*cc	1	0.03062500	0.03062500	0.15	0.7049

Output 11.3 (Continued) ANOVA for a Two-Cube Factorial in Blocks of Size Four

Parameter			Estimate		Standard Error	t Value	Pr > |t|
Intercept			2.010625000	B	0.25170936	7.99	<.0001
rep	1		0.328125000	B	0.35597078	0.92	0.3764
rep	2		-0.110000000	B	0.35597078	-0.31	0.7631
rep	3		0.000000000	B	.	.	.
blk(rep)	1	1	0.200000000	B	0.38994646	0.51	0.6182
blk(rep)	2	1	0.000000000	B	.	.	.
blk(rep)	1	2	0.873750000	B	0.38994646	2.24	0.0466
blk(rep)	2	2	0.000000000	B	.	.	.
blk(rep)	1	3	0.668750000	B	0.38994646	1.71	0.1143
blk(rep)	2	3	0.000000000	B	.	.	.
ca			0.937083333		0.09191126	10.20	<.0001
cb			0.013750000		0.09191126	0.15	0.8838
ca*cb			-0.267916667		0.09191126	-2.91	0.0141
cc			1.254583333		0.09191126	13.65	<.0001
ca*cc			0.380625000		0.11256785	3.38	0.0061
cb*cc			-0.841875000		0.11256785	-7.48	<.0001
ca*cb*cc			-0.043750000		0.11256785	-0.39	0.7049

NOTE: The X'X matrix has been found to be singular, and a generalized inverse was used to solve the normal equations. Terms whose estimates are followed by the letter 'B' are not uniquely estimable.

The standard errors of the coefficients of the confounded effects (ABC, AC, and BC) are indeed larger by $\sqrt{3/2}$ than the coefficients of the effects not confounded. You can verify that the sums of squares of the confounded effects, based on data from the replications in which they are not confounded, are identical to the sums of squares in Output 11.3.

The estimable functions option can be used to indicate the nature of the confounding. Requesting the Type I functions for effects in the same order as in the MODEL statement above gives the effects for BLK(REP) unadjusted for the factorial effects and reveals how the blocks are related to the factorial effects.

Output 11.4 Estimable Functions for a Two-Cube Factorial in Blocks of Size Four

Type I Estimable Functions

Effect			Coefficients blk(rep)
Intercept			0
rep	1		0
rep	2		0
rep	3		0
blk(rep)	1	1	L5
blk(rep)	2	1	-L5
blk(rep)	1	2	L7
blk(rep)	2	2	-L7
blk(rep)	1	3	L9
blk(rep)	2	3	-L9
ca			0
cb			0
ca*cb			0
cc			0
ca*cc			2L7
cb*cc			2L9
ca*cb*cc			2L5

Output 11.4 gives the nonzero coefficients of BLK(REP). The coefficient L5 appears on the terms for BLK in REP 1 and also the CA*CB*CC interaction term. This happens because CA*CB*CC is confounded with BLK in REP 1. This is apparent from the data set shown in Output 11.1. The product CA*CB*CC is equal to 1 for all observations in REP 1 of BLK 1 and CA*CB*CC= –1 for REP 2 of BLK 1. In some data sets, the confounding pattern is not so obvious. Using the coefficients for estimable functions in conjunction with the output from the ALIASING option shown above, you can discover the confounding pattern.

11.2.2 A Fractional Factorial Example

The second example is a ½ fraction of a 2^4 factorial experiment. The defining contrast is ABCD. The data appear in Output 11.5.

Output 11.5
Data for a ½ Fraction of a 2^4 Factorial Experiment

```
        Obs    a    b    c    d      y      ca    cb    cc    cd

         1     0    0    0    0    2.29     -1    -1    -1    -1
         2     0    0    1    1    1.51     -1    -1     1     1
         3     0    1    0    1    1.49     -1     1    -1     1
         4     0    1    1    0    3.43     -1     1     1    -1
         5     1    0    0    1    3.78      1    -1    -1     1
         6     1    0    1    0    2.08      1    -1     1    -1
         7     1    1    0    0    3.30      1     1    -1    -1
         8     1    1    1    1    3.63      1     1     1     1
```

The data in Output 11.5 include the factor levels in their original form (A, B, C, and D) and in contrast (–1,1) form (CA, CB, CC, and CD).

You can compute the analysis with the aliasing pattern by using PROC GLM statements similar to those used in the previous example:

```
proc glm;
     model y=ca|cb|cc|cd/solution aliasing;
```

Output 11.6 shows the results.

Output 11.6
PROC GLM Analysis of Data from a ½ Fraction of a 2^4 Factorial Experiment

Source	DF	Squares	Mean Square	F Value	Pr > F
Model	7	6.35588750	0.90798393	.	.
Error	0	0.00000000	.		
Corrected Total	7	6.35588750			

	R-Square	Coeff Var	Root MSE	y Mean
	1.000000	.	.	2.688750

Source	DF	Type I SS	Mean Square	F Value	Pr > F
ca	1	2.07061250	2.07061250	.	.
cb	1	0.59951250	0.59951250	.	.
ca*cb	1	0.00031250	0.00031250	.	.
cc	1	0.00551250	0.00551250	.	.
ca*cc	1	0.80011250	0.80011250	.	.
cb*cc	1	2.82031250	2.82031250	.	.
ca*cb*cc	1	0.05951250	0.05951250	.	.
cd	0	0.00000000	.	.	.
ca*cd	0	0.00000000	.	.	.
cb*cd	0	0.00000000	.	.	.
ca*cb*cd	0	0.00000000	.	.	.
cc*cd	0	0.00000000	.	.	.
ca*cc*cd	0	0.00000000	.	.	.
cb*cc*cd	0	0.00000000	.	.	.
ca*cb*cc*cd	0	0.00000000	.	.	.

*Output 11.6
(Continued)
PROC GLM
Analysis of
Data from a ½
Fraction of a
2⁴ Factorial
Experiment*

Source	DF	Type III SS	Mean Square	F Value	Pr > F
ca	0	0	.	.	.
cb	0	0	.	.	.
ca*cb	0	0	.	.	.
cc	0	0	.	.	.
ca*cc	0	0	.	.	.
cb*cc	0	0	.	.	.
ca*cb*cc	0	0	.	.	.
cd	0	0	.	.	.
ca*cd	0	0	.	.	.
cb*cd	0	0	.	.	.
ca*cb*cd	0	0	.	.	.
cc*cd	0	0	.	.	.
ca*cc*cd	0	0	.	.	.
cb*cc*cd	0	0	.	.	.
ca*cb*cc*cd	0	0	.	.	.

Parameter	Estimate	Standard Error	t Value	Pr > |t|
Intercept	2.688750000 B	.	.	.
ca	0.508750000 B	.	.	.
cb	0.273750000 B	.	.	.
ca*cb	-0.006250000 B	.	.	.
cc	-0.026250000 B	.	.	.
ca*cc	-0.316250000 B	.	.	.
cb*cc	0.593750000 B	.	.	.
ca*cb*cc	-0.086250000 B	.	.	.
cd	0.000000000 B	.	.	.
ca*cd	0.000000000 B	.	.	.
cb*cd	0.000000000 B	.	.	.
ca*cb*cd	0.000000000 B	.	.	.
cc*cd	0.000000000 B	.	.	.
ca*cc*cd	0.000000000 B	.	.	.
cb*cc*cd	0.000000000 B	.	.	.
ca*cb*cc*cd	0.000000000 B	.	.	.

Parameter	Expected Value
Intercept	Intercept + ca*cb*cc*cd
ca	ca + cb*cc*cd
cb	cb + ca*cc*cd
ca*cb	ca*cb + cc*cd
cc	cc + ca*cb*cd
ca*cc	ca*cc + cb*cd
cb*cc	cb*cc + ca*cd
ca*cb*cc	ca*cb*cc + cd
cd	
ca*cd	
cb*cd	
ca*cb*cd	
cc*cd	
ca*cc*cd	
cb*cc*cd	
ca*cb*cc*cd	

From Output 11.6, you can see that because there are only eight observations, only the first seven parameters in the model plus the intercept can be estimated. Also, each estimate is confounded—aliased—with one other factorial effect. The tables of "Parameter" and "Expected Value" at the end of the printout give the aliases. For example, the estimate of the intercept is aliased with the ABCD interaction, indicated on the printout by the fact that the expected value of the intercept is INTERCEPT + CA*CB*CC*CD. Similarly, the output indicates that the expected value of the parameter CA is CA+CB*CC*CD, that is, the main effect of A is aliased with the BCD interaction. You can apply analogous interpretations to the remaining parameters. You can see that this aliasing pattern agrees with the pattern you would derive from standard fractional factorial methods. In this case, which uses a very basic design, the ALIASING option merely restates information someone familiar with fractional factorial design would already know. However, for nonstandard incomplete factorial designs, for instance those you could generate with PROC OPTEX, the ALIASING option can provide useful information that usually is not obvious.

There are three important additional points about the analysis in Output 11.6. First, the default order of effects from the CA|CB|CC|CD syntax used in the MODEL statement causes all estimates involving factors A, B, and C to be estimated first, before any effects involving D appear in the model. This is not very realistic. Normally, you would not use a fractional factorial design unless you expect higher-order interaction effects to be negligible. For example, the output gives an estimate of the ABC interaction, which is aliased with the main effect of D. In practice, you would assume this to be an estimate of the main effect of D. That is, you would use this design only if you could assume that the ABC interaction is essentially zero. Also, because all the two-factor interactions are aliased with other two-factor interactions, you must be sure which you can assume to be negligible, and not alias two potentially important effects.

The second point is that there are no degrees of freedom and hence no *F*-values or *p*-values given in Output 11.6. The model used to compute the analysis is saturated. There are various strategies to get around this. A common approach is to assume that all interactions are zero and compute a main-effects-only model using the three degrees of freedom for the two-way interactions to estimate experimental error. You can do this by using the following statements:

```
proc glm;
    model y=ca cb cc cd/solution aliasing;
```

The results appear in Output 11.7. However, you can easily question whether the results in Output 11.7 are valid, because in Output 11.6, the largest single source of variation was the BC (aliased with AD) interaction. For these data, at least, the assumption that all interaction effects are zero is questionable. If there is a non-negligible BC (or AD) interaction, then the MS(ERROR) in Output 11.7 overestimates σ^2 and hence the *F*-values are too low. An alternative strategy, not shown here, uses half-normal plots to estimate σ^2 and construct approximate tests for the model effects. See Milliken and Johnson (1989, Chapter 4) for an explanation of how to implement half-normal plot analysis using SAS. Under the half-normal plot method, the main effects of A and the BC (or AD) interaction are statistically significant. You would need sufficient understanding of the data to decide whether the interaction is a BC or an AD interaction.

Output 11.7 Main-Effects-Only Analysis of Fractional Factorial Data

Source	DF	Sum of Squares	Mean Square	F Value	Pr > F
Model	4	2.73515000	0.68378750	0.57	0.7075
Error	3	3.62073750	1.20691250		
Corrected Total	7	6.35588750			

R-Square	Coeff Var	Root MSE	y Mean
0.430333	40.85898	1.098596	2.688750

Source	DF	Type I SS	Mean Square	F Value	Pr > F
ca	1	2.07061250	2.07061250	1.72	0.2815
cb	1	0.59951250	0.59951250	0.50	0.5317
cc	1	0.00551250	0.00551250	0.00	0.9504
cd	1	0.05951250	0.05951250	0.05	0.8385

Source	DF	Type III SS	Mean Square	F Value	Pr > F
ca	1	2.07061250	2.07061250	1.72	0.2815
cb	1	0.59951250	0.59951250	0.50	0.5317
cc	1	0.00551250	0.00551250	0.00	0.9504
cd	1	0.05951250	0.05951250	0.05	0.8385

Output 11.7
(Continued)
Main-Effects-
Only Analysis
of Fractional
Factorial Data

Parameter	Estimate	Standard Error	t Value	Pr > |t|	Expected Value
Intercept	2.688750000	0.38841223	6.92	0.0062	Intercept
ca	0.508750000	0.38841223	1.31	0.2815	ca
cb	0.273750000	0.38841223	0.70	0.5317	cb
cc	-0.026250000	0.38841223	-0.07	0.9504	cc
cd	-0.086250000	0.38841223	-0.22	0.8385	cd

The final point concerns the use of the (–1,1) contrasts CA through CD instead of the original (0,1) coding of A through D. If you use the variables A through D in the model, the ALIASING option assesses the aliasing pattern based on the estimable functions that follow from the (0,1) coding. These do not correspond to the standard aliasing pattern for fractional factorial experiments, and can be difficult to interpret. For example, these SAS statements yield the results shown in Output 11.8:

```
proc glm;
    model y=a|b|c|d/ aliasing;
```

Output 11.8
An Analysis of
Fractional
Factorial Data
Using 0-1
Coding

Source	DF	Type I SS	Mean Square	F Value	Pr > F
a	1	2.07061250	2.07061250	.	.
b	1	0.59951250	0.59951250	.	.
a*b	1	0.00031250	0.00031250	.	.
c	1	0.00551250	0.00551250	.	.
a*c	1	0.80011250	0.80011250	.	.
b*c	1	2.82031250	2.82031250	.	.
a*b*c	1	0.05951250	0.05951250	.	.
d	0	0.00000000	.	.	.
a*d	0	0.00000000	.	.	.
b*d	0	0.00000000	.	.	.
a*b*d	0	0.00000000	.	.	.
c*d	0	0.00000000	.	.	.
a*c*d	0	0.00000000	.	.	.
b*c*d	0	0.00000000	.	.	.
a*b*c*d	0	0.00000000	.	.	.

Parameter	Estimate		Standard Error	t Value	Pr > |t|
Intercept	2.290000000		.	.	.
a	1.490000000	B	.	.	.
b	-0.800000000	B	.	.	.
a*b	0.320000000	B	.	.	.
c	-0.780000000	B	.	.	.
a*c	-0.920000000	B	.	.	.
b*c	2.720000000	B	.	.	.
a*b*c	-0.690000000	B	.	.	.
d	0.000000000	B	.	.	.
a*d	0.000000000	B	.	.	.
b*d	0.000000000	B	.	.	.
a*b*d	0.000000000	B	.	.	.
c*d	0.000000000	B	.	.	.
a*c*d	0.000000000	B	.	.	.
b*c*d	0.000000000	B	.	.	.
a*b*c*d	0.000000000	B	.	.	.

Output 11.8
(Continued)
An Analysis of
Fractional
Factorial Data
Using 0-1
Coding

```
     Expected Value

     Intercept
     a + d + a*d
     b + d + b*d
     a*b - 2*d - a*d - b*d
     c + d + c*d
     a*c - 2*d - a*d - c*d
     b*c - 2*d - b*d - c*d
     a*b*c + 4*d + 2*a*d + 2*b*d + a*b*d + 2*c*d + a*c*d + b*c*d + a*b*c*d

NOTE: The X'X matrix has been found to be singular, and a generalized inverse was used to
solve the normal equations.  Terms whose estimates are followed by the letter 'B' are not
uniquely estimable.
```

You can see that the sums of squares are the same as those computed from the contrast coding (–1,1). The parameter estimates are different, as you would expect, because the different coding changes the intercept and hence the other coefficients. The aliasing pattern shown in the "Expected Value" of the parameter estimates is also quite different. This reflects the fact that the (0,1) coding results in a different set of estimable functions. As shown in the theory section of Chapter 6, GLM determines estimable functions from the nonzero rows of the $(X´X)^-(X´X)$ matrix. The contrast coding results in estimable functions in standard form for assessing aliasing patterns in incomplete factorials. On the other hand, the (0,1) coding results in a different, and unfamiliar, form.

11.3 A Balanced Incomplete-Blocks Design

Incomplete-blocks designs are used whenever there are not enough experimental units in blocks to accommodate all treatments. Perhaps the best-known incomplete-blocks design is the so-called balanced incomplete-blocks (BIB) design. This design in not balanced in the sense that we have used the word in previous chapters, because, in fact, not all treatments are assigned in all blocks. Instead, balance in the context of incomplete-blocks designs has the specific definition that all treatments appear in the same number of blocks, and all pairs of treatments appear together in the same number of blocks. These requirements result in certain conditions on the numbers of blocks, treatments, and numbers of treatments per block. For the BIB design with four treatments in blocks of size two, six blocks (three replications) are required (Cochran and Cox 1957). The data appear in Output 11.9. The design is shown below:

BIB Design Example Data
(Numbers in parentheses indicate treatment number)

			Block		
1	2	3	4	5	6
2.7(1)	7.1(3)	7.1(1)	8.8(2)	9.7(1)	13.0(2)
2.7(2)	8.6(4)	9.7(3)	15.1(4)	17.4(4)	16.6(3)

Output 11.9
Data for a
Balanced
Incomplete-
Blocks Design

```
            Obs   blk   trt    y

             1     1     1    1.2
             2     1     2    2.7
             3     2     3    7.1
             4     2     4    8.6
             5     3     1    7.1
             6     3     3    9.7
             7     4     2    8.8
             8     4     4   15.1
             9     5     1    9.7
            10     5     4   17.4
            11     6     2   13.0
            12     6     3   16.6
```

Consider the following statements:

```
proc glm;
   class blk trt;
   model y=trt blk / e1 ss3;
   means trt blk;
   lsmeans trt / stderr pdiff cl;
run;
```

The analysis-of-variance portion appears in Output 11.10.

Output 11.10
ANOVA for a
Balanced
Incomplete-
Blocks Design

The GLM Procedure

Source	DF	Sum of Squares	Mean Square	F Value	Pr > F
Model	8	281.1275000	35.1409375	40.82	0.0056
Error	3	2.5825000	0.8608333		
Corrected Total	11	283.7100000			

Source	DF	Type I SS	Mean Square	F Value	Pr > F
trt	3	102.2566667	34.0855556	39.60	0.0065
blk	5	178.8708333	35.7741667	41.56	0.0057

Source	DF	Type III SS	Mean Square	F Value	Pr > F
trt	3	59.0175000	19.6725000	22.85	0.0144
blk	5	178.8708333	35.7741667	41.56	0.0057

Least Squares Means

trt	y LSMEAN	Standard Error	95% Confidence Limits	
1	6.8000000	0.6281310	4.801007	8.798993
2	7.6500000	0.6281310	5.651007	9.648993
3	10.9250000	0.6281310	8.926007	12.923993
4	13.6250000	0.6281310	11.626007	15.623993

i	j	Difference Between Means	95% Confidence Limits for LSMean(i)-LSMean(j)	
1	2	-0.850000	-3.802709	2.102709
1	3	-4.125000	-7.077709	-1.172291
1	4	-6.825000	-9.777709	-3.872291
2	3	-3.275000	-6.227709	-0.322291
2	4	-5.975000	-8.927709	-3.022291
3	4	-2.700000	-5.652709	0.252709

The Type I sum of squares is the **unadjusted** treatment sum of squares, based on the ordinary treatment means. Therefore, the unadjusted treatment sum of squares contains both treatment differences and block differences. The Type III treatment sum of squares is adjusted for blocks. This means that block effects have been removed from the sum of squares. Thus, the **adjusted** treatment mean square measures only differences between treatment means and random error. These concepts are revealed in the estimable functions. Table 11.1 shows the Type I estimable functions.

Table 11.1 *Type I Estimable Functions for Treatments*

Effect		Symbolic Expression	Coefficients for TRT1 Effect
TRT	1	L2	+.75
	2	L3	−.25
	3	L4	−.25
	4	−L2 − L3 −L4	−.25
BLK	1	.333L2 + .333L3	.167
	2	−.333L2 − .333L3	−.167
	3	−.333L2 − .333L4	.167
	4	−.333L2 − .333L4	−.167
	5	−.333L3 − .333L4	.167
	6	.333L3 + .333L4	−.167

The Type I estimable function for treatments (TRT) is of some interest. Consider the contrast

$$TRT1 - 1/4(TRT1 + TRT2 + TRT3 + TRT4)$$

This is often called the **effect** of treatment 1, or the difference between the treatment 1 mean from the mean of *all* treatments. Simplification gives

$$3/4(TRT1) - 1/4(TRT2) - 1/4(TRT3) - 1/4(TRT4)$$

This expression is obtained by defining

$$L2 = 3/4$$
$$L3 = -1/4$$
$$L4 = -1/4$$

and results in the coefficients that appear in the right-hand column of Table 11.1. You can see that the Type I (unadjusted) estimate of the TRT 1 effect is also a contrast between blocks 1, 3, and 5, which contain treatment 1, and blocks 2, 4, and 6, which do not.

The least-squares means (see Output 11.10) have been "adjusted" for block effects. The corresponding estimable functions (not reproduced here) show that the LS means contain equal representation of block parameters even though individual treatments do not appear in all the blocks. Differences between LS means provide the so-called **intra-block** comparisons of treatments. There is information about differences between the treatment means contained in the block means that is not used in the intra-block comparisons. This is called the **inter-block** information.

Expected mean squares from the RANDOM statement reveal the presence of block effects in the Type I mean squares, but not in the Type III mean squares, as shown in Output 11.11. The Type I EMS for TRT contains VAR(BLK), but the Type III EMS does not.

Output 11.11
Expected Mean
Squares for a
Balanced
Incomplete-
Blocks Design

```
                          The GLM Procedure

     Source                Type I Expected Mean Square

     trt                   Var(Error) + 0.6667 Var(blk) + Q(trt)

     blk                   Var(Error) + 1.6 Var(blk)


     Source                Type III Expected Mean Square

     trt                   Var(Error) + Q(trt)

     blk                   Var(Error) + 1.6 Var(blk)
```

The MIXED procedure can be used to obtain the *combined* inter- and intra-block information about differences between treatment means. Run the following statements:

```
proc mixed data=bibd;
   class blk trt;
   model y=trt / ddfm=satterth;
   random blk;
   lsmeans trt / pdiff cl;
run;
```

The results appear in Output 11.12.

Output 11.12
A Mixed-Model
Analysis of a
Balanced
Incomplete-
Blocks Design

```
                      The Mixed Procedure

                   Covariance Parameter
                        Estimates

                   Cov Parm     Estimate

                   blk          17.8543
                   Residual      0.8518


                 Type 3 Tests of Fixed Effects

                        Num    Den
              Effect     DF     DF     F Value   Pr > F

              trt         3    3.13     23.46    0.0121


                      Least Squares Means

                         Standard
Effect  trt  Estimate     Error    DF   t Value  Pr > |t|  Alpha   Lower    Upper

trt      1    6.7724     1.8337   5.96    3.69    0.0103    0.05   2.2773  11.2674
trt      2    7.6678     1.8337   5.96    4.18    0.0059    0.05   3.1728  12.1629
trt      3   10.9322     1.8337   5.96    5.96    0.0010    0.05   6.4371  15.4273
trt      4   13.6276     1.8337   5.96    7.43    0.0003    0.05   9.1325  18.1227
```

Output 11.12
(Continued)
A Mixed-Model
Analysis of a
Balanced
Incomplete-
Blocks Design

Differences of Least Squares Means

Effect	trt	_trt	Estimate	Standard Error	DF	t Value	Pr > |t|	Alpha	Lower	Upper
trt	1	2	-0.8955	0.9176	3.13	-0.98	0.3983	0.05	-3.7462	1.9552
trt	1	3	-4.1598	0.9176	3.13	-4.53	0.0183	0.05	-7.0105	-1.3092
trt	1	4	-6.8552	0.9176	3.13	-7.47	0.0043	0.05	-9.7059	-4.0045
trt	2	3	-3.2643	0.9176	3.13	-3.56	0.0353	0.05	-6.1150	-0.4137
trt	2	4	-5.9597	0.9176	3.13	-6.49	0.0065	0.05	-8.8104	-3.1091
trt	3	4	-2.6954	0.9176	3.13	-2.94	0.0574	0.05	-5.5461	0.1553

You can see the distinction between the intra-block and the combined inter- and intra-block comparisons of treatments by comparing results in output 11.10 and Output 11.12. First of all, the TRT LSMEANS are slightly different in the two output tables. Also, the confidence for the difference between TRT 1 and TRT 2 in Output 11.10 is (–3.802709, 2.102709), whereas the confidence interval in Output 11.12 is (–3.7462, 1.9552). The confidence interval using the combined information in Output 11.12 is slightly narrower. However, this can be misleading. The standard error in Output 11.12 does not take into account the variation induced by estimating the variance-covariance matrix to obtain the *estimated* GLS estimates of differences between treatment means. If you use DDFM=KENWARDROGER in the MODEL statement you will get a better assessment of the true error of estimation.

11.4 A Crossover Design with Residual Effects

Crossover designs are used in animal nutrition and pharmaceutical studies to compare two or more treatments (diets or drugs). The treatments are administered sequentially to each subject over a set of time periods. This enables the comparison of treatments on a within-subjects basis. However, there is a possibility that the response obtained after a particular time period might be influenced by the treatment assigned not only in that period but also in previous periods. If so, then the response contains **residual effects** from the previous periods. Some authors call these "carry-over" effects. Certain crossover designs permit the residual effects to be estimated, and thus to be effectively removed from estimates of treatment means and comparisons of means.

Cochran and Cox (1957) present two 3×3 Latin squares as a design for estimating the residual effects on milk yields of treatment from the preceding period. The treatment allocation is shown in the table below. The columns of the two squares contain the six possible sequences.

Square		1			2		
Cow		I	II	III	IV	V	VI
	1	A	B	C	A	B	C
Period	2	B	C	A	C	A	B
	3	C	A	B	B	C	A

Output 11.13 contains data from a study that was conducted to compare the effects on heart rate of three treatments; a *test* drug, a *standard* drug, and a *placebo*. Treatments were assigned in the six possible sequences to four patients each. The treatment design for the data in Output 11.13 is equivalent to the Cochran and Cox design in the table above with sequences A-F in Output 11.13 corresponding to Cows I-VI in the table, respectively.

Heart rate was measured one hour following the administration of treatment in each of three visits. The visits are labeled 2, 3, and 4, because visit 1 was a preliminary visit for baseline data. Thus, in the general terminology of crossover designs, period 1 is visit 2, period 2 is visit 3, and period 3 is visit 4. Baseline heart rate was measured, but it is not used in the illustrative analysis.

A model for the data is

$$y_{ijk} = \mu + \alpha_i + d_j + \beta_k + \tau_{l(ik)} + \rho_{m(ik)} + e_{ijk}$$

where α_i is the effect of sequence i, d_j is the random effect of patient j, β_k is the effect of visit k, $\tau_{l(ik)}$ is the *direct* effect of treatment l, $\rho_{m(ik)}$ is the *residual* effect of treatment m, and e_{ijk} is a random effect associated with patient j in visit k. The subscript $l(jk)$ on the treatment direct effect indicates that the treatment (l) is a function of the visit (k) and sequence (i). The same is true of the treatment residual effect subscript $m(ik)$.

When using PROC GLM to analyze data from a crossover design, the sequence, patient, period, and direct treatment effects can be incorporated into the model with the dummy variables that result from using a CLASS statement. However, it is more convenient to use explicitly created covariates in the model for the residual effects. In the data set for the heartrate data, we create covariates for the standard and test drug residual effects named RESIDS and RESIDT, respectively. Their values in the first period (visit 2) are zero because there is no period prior to the first period that would contribute a residual effect. In periods 2 and 3 (visits 3 and 4), the values of RESIDS and RESIDT are 0 or ±1 depending on the treatment in the preceding visit. This particular coding provides estimates of the residual effects corresponding to those prescribed by Cochran and Cox (1957). For example, patient number 2 is in sequence F (test, placebo, standard). The values of RESIDS and RESIDT are both 0 in the first period (visit 2). Patient 2 received the test drug in period 1, so in period 2 (visit 3), the covariates have values RESIDS=0 and REISIDT=1. This specifies that the residual effect ρ_T for test is contained in the observation on patient 2 in period 2. In period 3 (visit 4), the covariates both have values of –1. This coding specifies a sum-to-zero constraint on the residual effects. Thus, the residual effect ρ_P of the placebo satisfies the equation $\rho_P = -\rho_T - \rho_S$, and hence the residual effect ρ_P of the placebo can be represented with –1 times the residual effects of test and standard.

Output 11.13
Date for Crossover Design with Residual Effects

PATIENT	SEQUENCE	VISIT	BASEHR	HR	DRUG	RESIDT	RESIDS
1	B	2	86	86	placebo	0	0
1	B	3	86	106	test	-1	-1
1	B	4	62	79	standard	1	0
2	F	2	48	66	test	0	0
2	F	3	58	56	placebo	1	0
2	F	4	74	79	standard	-1	-1
3	B	2	78	84	placebo	0	0
3	B	3	78	76	test	-1	-1
3	B	4	82	91	standard	1	0
4	D	2	66	79	standard	0	0
4	D	3	72	100	test	0	1
4	D	4	90	82	placebo	1	0
5	C	2	74	74	test	0	0
5	C	3	90	71	standard	1	0
5	C	4	66	62	placebo	0	1
6	B	2	62	64	placebo	0	0
6	B	3	74	90	test	-1	-1
6	B	4	58	85	standard	1	0
7	A	2	94	75	standard	0	0
7	A	3	72	82	placebo	0	1
7	A	4	100	102	test	-1	-1

Output 11.13
(Continued)
Date for
Crossover
Design with
Residual
Effects

8	A	2	54	63	standard	0	0
8	A	3	54	58	placebo	0	1
8	A	4	66	62	test	-1	-1
9	D	2	82	91	standard	0	0
9	D	3	96	86	test	0	1
9	D	4	78	88	placebo	1	0
10	C	2	86	82	test	0	0
10	C	3	70	71	standard	1	0
10	C	4	58	62	placebo	0	1
11	F	2	82	80	test	0	0
11	F	3	80	78	placebo	1	0
11	F	4	72	75	standard	-1	-1
12	E	2	96	90	placebo	0	0
12	E	3	92	93	standard	-1	-1
12	E	4	82	88	test	0	1
13	D	2	78	87	standard	0	0
13	D	3	72	80	test	0	1
13	D	4	76	78	placebo	1	0
14	F	2	98	86	test	0	0
14	F	3	86	86	placebo	1	0
14	F	4	70	79	standard	-1	-1
15	A	2	86	71	standard	0	0
15	A	3	66	70	placebo	0	1
15	A	4	74	90	test	-1	-1
16	E	2	86	86	placebo	0	0
16	E	3	90	103	standard	-1	-1
16	E	4	82	86	test	0	1
17	A	2	66	83	standard	0	0
17	A	3	82	86	placebo	0	1
17	A	4	86	102	test	-1	-1
18	F	2	66	82	test	0	0
18	F	3	78	80	placebo	1	0
18	F	4	74	95	standard	-1	-1
19	E	2	74	80	placebo	0	0
19	E	3	78	79	standard	-1	-1
19	E	4	70	74	test	0	1
20	B	2	66	70	placebo	0	0
20	B	3	74	62	test	-1	-1
20	B	4	62	67	standard	1	0
21	C	2	82	90	test	0	0
21	C	3	90	103	standard	1	0
21	C	4	76	82	placebo	0	1
22	C	2	82	82	test	0	0
22	C	3	66	83	standard	1	0
22	C	4	90	82	placebo	0	1
23	E	2	82	66	placebo	0	0
23	E	3	74	87	standard	-1	-1
23	E	4	82	82	test	0	1
24	D	2	72	75	standard	0	0
24	D	3	82	86	test	0	1
24	D	4	74	82	placebo	1	0

The following SAS statements can be used to construct an analysis of variance and parameter estimates similar to those proposed by Cochran and Cox (1957):

```
proc glm data=hrtrate;
    class sequence patient visit drug;
    model hr = sequence patient(sequence) visit drug
        resids residt / solution;
    random patient(sequence)
run;
```

ANOVA results appear in Output 11.14.

Output 11.14
ANOVA for a
Crossover
Design

```
                                      Sum of
Source                    DF         Squares      Mean Square    F Value    Pr > F

Model                     29      6408.694444      220.989464       3.91    <.0001
Error                     42      2372.583333       56.490079
Corrected Total           71      8781.277778

             R-Square      Coeff Var      Root MSE       HR Mean
             0.729813       9.301326      7.515988      80.80556

Source                    DF        Type I SS      Mean Square    F Value    Pr > F

SEQUENCE                   5       508.944444      101.788889       1.80    0.1333
PATIENT(SEQUENCE)         18      4692.333333      260.685185       4.61    <.0001
VISIT                      2       146.777778       73.388889       1.30    0.2835
DRUG                       2       668.777778      334.388889       5.92    0.0054
resids                     1       391.020833      391.020833       6.92    0.0119
residt                     1         0.840278        0.840278       0.01    0.9035


Source                    DF      Type III SS      Mean Square    F Value    Pr > F

SEQUENCE                   5       701.183333      140.236667       2.48    0.0466
PATIENT(SEQUENCE)         18      4692.333333      260.685185       4.61    <.0001
VISIT                      2       146.777778       73.388889       1.30    0.2835
DRUG                       2       343.950000      171.975000       3.04    0.0583
resids                     1       309.173611      309.173611       5.47    0.0241
residt                     1         0.840278        0.840278       0.01    0.9035
```

The desired ANOVA table is constructed as follows:

Source of Variation	DF	SS	
Sequence	5	508.94	(Type I)
Patient(Sequence)	18	4692.33	(Type I)
Visits	2	146.78	(Type III)
Direct effect of drugs (adjusted for residual effects)	2	343.95	(Type III)
Residual effects(adjusted)	2	391.86	(Type I SS RESIDS+Type I SS RESIDT)

Expected mean squares shown in Output 11.15 show that appropriate tests for VISIT, DRUG, and the carry-over effect covariates utilize residual means square as an error term. A test for SEQUENCE would use PATIENT(SEQUENCE) in the error term.

Output 11.15
Expected Mean
Squares for a
Crossover
Design

```
Source                   Type III Expected Mean Square

SEQUENCE                 Var(Error) + 2.76 Var(PATIENT(SEQUENCE)) + Q(SEQUENCE)

PATIENT(SEQUENCE)        Var(Error) + 3 Var(PATIENT(SEQUENCE))

VISIT                    Var(Error) + Q(VISIT)

DRUG                     Var(Error) + Q(DRUG)

resids                   Var(Error) + Q(resids)

residt                   Var(Error) + Q(residt)
```

The effect of SEQUENCE is clearly not significant, since the *F*-ratio would be less than 1 using either a Type I or a Type III mean square in the numerator. The Type III test for DRUG has a significance level *p*=0.0538. The Type III mean square for DRUG has been adjusted for the residual effects. The Type I mean square for DRUG is *not* adjusted for the residual effects, and an *F*-test based on it has a significance probability *p*=0.0054. Thus, results from tests for DRUG depend on whether residual effects have been removed or not. Estimates of the direct and residual effect parameters can be obtained from Output 11.16.

Output 11.16
Parameter
Estimates for
a Crossover
Design

Parameter		Estimate	Standard Error	t Value	Pr > |t|
Intercept		82.06250000 B	4.72870558	17.35	<.0001
SEQUENCE	A	6.20833333 B	6.23192824	1.00	0.3249
SEQUENCE	B	-19.33333333 B	6.23192824	-3.15	0.0030
SEQUENCE	C	-0.47916667 B	6.23192824	-0.08	0.9391
SEQUENCE	D	-1.81250000 B	6.23192824	-0.29	0.7726
SEQUENCE	E	-5.79166667 B	6.23192824	-0.93	0.3580
SEQUENCE	F	0.00000000 B	.	.	.
PATIENT(SEQUENCE)	7 A	-4.00000000 B	6.13677871	-0.65	0.5181
PATIENT(SEQUENCE)	8 A	-29.33333333 B	6.13677871	-4.78	<.0001
PATIENT(SEQUENCE)	15 A	-13.33333333 B	6.13677871	-2.17	0.0355
PATIENT(SEQUENCE)	17 A	0.00000000 B	.	.	.
...					
PATIENT(SEQUENCE)	2 F	-18.66666667 B	6.13677871	-3.04	0.0040
PATIENT(SEQUENCE)	11 F	-8.00000000 B	6.13677871	-1.30	0.1995
PATIENT(SEQUENCE)	14 F	-2.00000000 B	6.13677871	-0.33	0.7461
PATIENT(SEQUENCE)	18 F	0.00000000 B	.	.	.
VISIT	2	-2.58333333 B	2.16967892	-1.19	0.2405
VISIT	3	0.75000000 B	2.16967892	0.35	0.7313
VISIT	4	0.00000000 B	.	.	.
DRUG	standard	2.31250000 B	2.42577478	0.95	0.3459
DRUG	test	5.93750000 B	2.42577478	2.45	0.0186
DRUG	placebo	0.00000000 B	.	.	.
resids		-4.39583333	1.87899706	-2.34	0.0241
residt		0.22916667	1.87899706	0.12	0.9035

First of all, the residual effects presented by Cochran and Cox (1957) are obtained from the parameter estimates for RESIDS and RESIDT. The values are

STD: −4.396
TST: 0.229
PCB: − (−4.396 + 0.229) = 4.167

Notice that these estimates come from the sum-to-zero coding for the residual effect dummy variables.

The direct treatment *effects* reported by Cochran and Cox (1957) can be obtained from the TRTMENT parameter estimates according to the following equations:

STD: −0.4375 = 2.3125 − (1/3)(2.3125 + 5.9375 + 0.0000)
TST: 3.1875 = 5.9375 − (1/3)(2.3125 + 5.9375 + 0.0000)
PCB: −2.7500 = 0.000 − (1/3)(2.3125 + 5.9375 + 0.0000)

Thus, the direct effects can be obtained from the following ESTIMATE statements:

```
estimate 'DIRECT EFFECT OF STD'
         drug 2 -1 -1 / divisor=3;
estimate 'DIRECT EFFECT OF TST'
         drug -1 2 -1 / divisor=3;
estimate 'DIRECT EFFECT OF PCB'
         drug -1 -1 2 / divisor=3;
```

Results from these ESTIMATE statements appear in Output 11.17.

Output 11.17
Direct Effect Estimates

The GLM Procedure

Parameter	Estimate	Standard Error	t Value	Pr > |t|
DIRECT EFFECT OF STD	-0.43750000	1.40052172	-0.31	0.7563
DIRECT EFFECT OF TST	3.18750000	1.40052172	2.28	0.0280
DIRECT EFFECT OF PCB	-2.75000000	1.40052172	-1.96	0.0562

The direct effect *means* reported by Cochran and Cox (1957) are equal to the overall mean 80.8056 (printed as HR mean in Output 11.14) added to the direct effects. They are also equal to the GLM least-squares means, obtained from the following statement:

```
lsmeans drug / pdiff cl e;
```

The results appear in Output 11.18. You can see from the estimable functions that the LS means contain the INTERCEPT, and average across the SEQUENCE, PATIENT(SEQUENCE), and VISIT parameters. Thus, the *correct* standard error of these LS means would contain variance due to PATIENT(SEQUENCE). However, this variance is *not* contained in the standard error computed by PROC GLM for the LS means. (That is why we specified the STDERR option in the LSMEANS statement.) As a consequence, the confidence intervals for LS means displayed in Output 11.18 are not valid. However, the confidence intervals for the *differences* between LS means in Output 11.18 are valid because the INTERCEPT, SEQUENCE, PATIENT(SEQUENCE), and VISIT parameters would drop out of the differences.

Output 11.18
Least-Squares Means for a Crossover Design

Least Squares Means

Coefficients for DRUG Least Square Means

Effect		DRUG Level standard	test	placebo
Intercept		1	1	1
SEQUENCE	A	0.16666667	0.16666667	0.16666667
...				
SEQUENCE	F	0.16666667	0.16666667	0.16666667
PATIENT(SEQUENCE)	7 A	0.04166667	0.04166667	0.04166667
PATIENT(SEQUENCE)	8 A	0.04166667	0.04166667	0.04166667
PATIENT(SEQUENCE)	16 A	0.04166667	0.04166667	0.04166667
PATIENT(SEQUENCE)	18 A	0.04166667	0.04166667	0.04166667
...				
PATIENT(SEQUENCE)	12 F	0.04166667	0.04166667	0.04166667
PATIENT(SEQUENCE)	17 F	0.04166667	0.04166667	0.04166667
PATIENT(SEQUENCE)	20 F	0.04166667	0.04166667	0.04166667
PATIENT(SEQUENCE)	24 F	0.04166667	0.04166667	0.04166667
VISIT	2	0.33333333	0.33333333	0.33333333
VISIT	3	0.33333333	0.33333333	0.33333333
VISIT	4	0.33333333	0.33333333	0.33333333
DRUG	standard	1	0	0
DRUG	test	0	1	0
DRUG	placebo	0	0	1
resids		0	0	0
residt		0	0	0

*Output 11.18
(Continued)
Least-Squares
Means for a
Crossover
Design*

```
                  Least Squares Means for Effect DRUG

        DRUG           HR LSMEAN        95% Confidence Limits

        standard       80.368056        77.023853    83.712258
        test           83.993056        80.648853    87.337258
        placebo        78.055556        74.711353    81.399758


                        Difference
                        Between        95% Confidence Limits for
        i    j           Means          LSMean(i)-LSMean(j)

        1    2          -3.625000       -8.520412     1.270412
        1    3           2.312500       -2.582912     7.207912
        2    3           5.937500        1.042088    10.832912
```

PROC MIXED can be used to analyze the crossover design data. Run the following statements:

```
proc mixed data=hrtrate order=internal;
class sequence patient visit drug;
model hr=sequence visit drug resides residt/solution
ddfm=satterth;
random patient(sequence);
lsmeans drug / pdiff cl e;
run;
```

Edited results appear in Output 11.19.

*Output 11.19
Partial Mixed-
Model Results
for a
Crossover
Design*

```
                        The Mixed Procedure

                  Covariance Parameter Estimates

                  Cov Parm              Estimate

                  PATIENT(SEQUENCE)      68.0650
                  Residual               56.4901


                  Type 3 Tests of Fixed Effects

                          Num      Den
              Effect       DF       DF      F Value   Pr > F

              SEQUENCE      5      18.7       0.58     0.7165
              VISIT         2        42       1.30     0.2835
              DRUG          2        42       3.04     0.0583
              resids        1        42       5.47     0.0241
              residt        1        42       0.01     0.9035


                        Least Squares Means
```

Effect	DRUG	Estimate	Standard Error	DF	t Value	Pr > |t|	Alpha	Lower	Upper
DRUG	standard	80.3681	2.3626	38	34.02	<.0001	0.05	75.5852	85.1510
DRUG	test	83.9931	2.3626	38	35.55	<.0001	0.05	79.2102	88.7760
DRUG	placebo	78.0556	2.3626	38	33.04	<.0001	0.05	73.2727	82.8385

Output 11.19
(Continued)
Partial Mixed-
Model Results
for a
Crossover
Design

```
                        Differences of Least Squares Means

                                      Standard
   Effect    DRUG      _DRUG    Estimate    Error     DF   t Value   Pr > |t|   Alpha

   DRUG      standard  test      -3.6250    2.4258    42    -1.49     0.1426    0.05
   DRUG      standard  placebo    2.3125    2.4258    42     0.95     0.3459    0.05
   DRUG      test      placebo    5.9375    2.4258    42     2.45     0.0186    0.05


                        Differences of Least Squares Means

           Effect    DRUG      _DRUG        Lower       Upper

           DRUG      standard  test        -8.5204     1.2704
           DRUG      standard  placebo     -2.5829     7.2079
           DRUG      test      placebo      1.0421    10.8329
```

The test of significance for DRUG in "Type 3 Tests of Fixed Effects" in Output 11.19 is the same as the test from GLM in Output 11.15. Likewise, the least-squares means are equal in the two analyses. This illustrates that ordinary least-squares analyses, as performed by GLM, can be equivalent to generalized least-squares analyses, as performed by MIXED. The phenomenon occurs in this example because the within-patients effects are orthogonal to the between-patients effects. However, notice that the confidence intervals for differences between LS means are the same in Outputs 11.18 and 11.19, but the confidence intervals for the LS means themselves are wider in Output 11.19 than in Output 11.18 because PROC MIXED computes standard errors of LS means that incorporate the PATIENT(SEQUENCE) variance.

11.5 Models for Experiments with Qualitative and Quantitative Variables

The material in this section is related to the discussions of regression analysis in Chapter 2 and analysis of covariance in Chapter 7. This section concerns details of certain models that contain dummy variables generated from the CLASS statement, and also a continuous variable. These are several regression models in one equation. Of particular interest are cases for which the regressions have a common intercept. These types of models are frequently used, for example, in relative potency and relative bioavailability studies (Littell et al. 1997).

Many experiments involve both qualitative and quantitative factors. For example, the tensile strength (TS) of a monofilament fiber depends on the amount (AMT) of a chemical used in the manufacturing process. This chemical can be obtained from three different sources (SOURCE), with values A, B, or C. SOURCE is a qualitative variable and AMT is a quantitative variable. Measurements of TS were obtained from samples from different amounts and sources. The SAS data set named MONOFIL appears in Output 11.20.

Output 11.20
Data for an
Experiment
with
Qualitative
and
Quantitative
Variables

```
        Obs    SOURCE    AMT    TS

          1      A        1    11.5
          2      A        2    13.8
          3      A        3    14.4
          4      A        4    16.8
          5      A        5    18.7
          6      B        1    10.8
          7      B        2    12.3
          8      B        3    13.7
          9      B        4    14.2
         10      B        5    16.6
         11      C        1    13.1
         12      C        2    16.2
         13      C        3    19.0
         14      C        4    22.9
         15      C        5    26.5
```

A simple linear regression model for each source relates to TS and AMT:

$$TS = \alpha_A + \beta_A + \varepsilon \quad (SOURCE\ A)$$

$$TS = \alpha_B + \beta_B + \varepsilon \quad (SOURCE\ B)$$

$$TS = \alpha_C + \beta_C + \varepsilon \quad (SOURCE\ C)$$

The parameters α_A and β_A are the intercept and slope, respectively, for SOURCE=A.

The following statements produce the analysis of variance and parameter estimates in Output 11.21.

```
proc glm data=monofil;
class source;
model ts=source amt source*amt / solution;
run;
```

Output 11.21
A Model with
Main Effects
and
Interactions

```
                              The GLM Procedure

Dependent Variable: ts

                                     Sum of
Source                   DF          Squares     Mean Square   F Value   Pr > F

Model                     5      258.7273333     51.7454667    263.71    <.0001

Error                     9        1.7660000      0.1962222

Corrected Total          14      260.4933333


          R-Square     Coeff Var     Root MSE        ts Mean

          0.993221      2.762805     0.442970       16.03333


Source                   DF       Type I SS     Mean Square   F Value   Pr > F

source                    2      98.0013333     49.0006667    249.72    <.0001
amt                       1     138.2453333    138.2453333    704.53    <.0001
amt*source                2      22.4806667     11.2403333     57.28    <.0001


Source                   DF     Type III SS     Mean Square   F Value   Pr > F

source                    2       0.0702424      0.0351212      0.18    0.8390
amt                       1     138.2453333    138.2453333    704.53    <.0001
amt*source                2      22.4806667     11.2403333     57.28    <.0001


                                           Standard
      Parameter             Estimate          Error    t Value   Pr > |t|

      Intercept          9.490000000 B    0.46459062     20.43     <.0001
      source      A       0.330000000 B    0.65703036      0.50     0.6275
      source      B      -0.020000000 B    0.65703036     -0.03     0.9764
      source      C       0.000000000 B        .             .         .
      amt                 3.350000000 B    0.14007934     23.92     <.0001
      amt*source A       -1.610000000 B    0.19810211     -8.13     <.0001
      amt*source B       -2.000000000 B    0.19810211    -10.10     <.0001
      amt*source C        0.000000000 B        .             .         .

NOTE: The X'X matrix has been found to be singular, and a generalized inverse was used to
solve the normal equations. Terms whose estimates are followed by the letter 'B' are not
uniquely estimable.
```

These parameter estimates pertain to the integrated model

$$TS = \alpha_C + \alpha'_A D_A + \alpha'_B D_B + \beta_C AMT + \beta'_A D_A AMT + \beta'_B D_B AMT + \varepsilon$$

The parameters α' and β' are further defined as

$$\alpha'_A = \alpha_A - \alpha_C \quad \alpha'_B = \alpha_B - \alpha_C$$
$$\beta'_A = \beta_A - \beta_C \quad \beta'_B = \beta_B - \beta_C$$

The variable D_A is a dummy variable equal to 1 for SOURCE=A and equal to 0 otherwise, and D_B has a corresponding definition with respect to SOURCE=B. Thus, the regression models for the three nitrogen sources are

$$TS = (\alpha_C + \alpha'_A) + (\beta_C + \beta'_A)\ AMT + \varepsilon \quad (\text{SOURCE A})$$
$$TS = (\alpha_C + \alpha'_B) + (\beta_C + \beta'_B)\ AMT + \varepsilon \quad (\text{SOURCE B})$$
$$TS = \alpha_C + \beta_C AMT + \varepsilon \qquad\qquad\quad (\text{SOURCE C})$$

Therefore, the fitted equations are

$$
\begin{aligned}
TS &= 9.49 + 0.33 + (3.35 - 1.61)\ AMT \quad (\text{SOURCE A})\\
&= 9.82 + 1.74\ AMT\\
TS &= 9.49 - 0.02 + (3.35 - 2.00)\ AMT \quad (\text{SOURCE B})\\
&= 9.47 + 1.35\ AMT\\
TS &= 9.49 + 3.35\ AMT \qquad\qquad\qquad\quad (\text{SOURCE C})
\end{aligned}
$$

The GLM parameter estimates, in effect, treat the regression line for SOURCE=C as a reference line, and the parameters $\alpha'_A, \alpha'_B, \beta'_A$, and β'_B are parameters for lines A and B minus parameters for line C. The AMT source parameters β'_A and β'_B measure differences between the slopes for regression lines A and B, and line C, respectively. Thus, a test that these parameters are 0 is testing that the lines are parallel, that is, they have equal slopes. The appropriate statistic is the $F=57.28$ for the AMT*SOURCE effect, which has a significant probability $p=0.0001$.

Caution is advised in using the Type III F-test for SOURCE. It is a test of the equality of the intercepts $(H_0: \alpha_A = \alpha_B = \alpha_C)$, which probably has no practical interpretation because the intercepts are simply extrapolations of the lines to L=0. The Type I F-test, on the other hand, tests the equality of the midpoints of the regression lines $(H_0: \alpha_A + \beta_A(2) = \alpha_B + \beta_B(2) = \alpha_C + \beta_C(2))$.

You can compare two sources at a given amount with an ESTIMATE statement. Suppose you want to compare SOURCE=A with SOURCE=B using AMT=3.5. This difference is

$$
\begin{aligned}
&(\alpha_A + \beta_A(3.5)) - (\alpha_B + \beta_B(3.5))\\
&= ((\alpha_C + \alpha'_A) + (\beta_C + \beta'_C)\ 3.5)\\
&= ((\alpha_C + \alpha'_B) + (\beta_C + \beta'_C)\ 3.5)\\
&= \alpha'_A + \alpha'_B + (\beta'_A + \beta'_B)\ 3.5
\end{aligned}
$$

So the appropriate ESTIMATE statement is

```
estimate 'A vs B at AMT=3.5'
source 1 -1 0
   source*amt 3.5 -3.5 0;
```

The results appear in Output 11.22.

Output 11.22
The Difference between SOURCE=A and SOURCE=B at AMT=3.5

```
                              The GLM Procedure

Dependent Variable:   ts
                                          Standard
Parameter                   Estimate        Error        t Value      Pr > |t|

A vs B at AMT=3.5         1.71500000      0.29715316         5.77        0.0003
```

Suppose TS also is measured for AMT=0. This variation of the experiment is commonly mishandled by data analysts. Since AMT=0 means there is no chemical, the intercepts for the models are all equal, $\alpha_A = \alpha_B = \alpha_C$. Thus, a correct analysis should provide equal estimates of the intercepts. The regressions can be written simultaneously as

$$TS = \alpha + \gamma_A D_A AMT + \gamma_B D_B AMT + \gamma_C D_C AMT + \varepsilon$$

where D_A is a dummy variable equal to 1 for SOURCE=A and equal to 0 otherwise, and D_B and D_C have corresponding definitions with respect to SOURCE=B and SOURCE=C. Use PROC GLM to create D_A, D_B, and D_C by including the SOURCE variable in a CLASS statement.

Look at the data set MONOFIL2 printed in Output 11.23. The value C is arbitrarily assigned to SOURCE when AMT=0.

Output 11.23
Data with AMT=0

```
                    Qual and Quant Variables

            Obs    source        amt        ts

             1       A            1        11.5
             2       A            2        13.8
             3       A            3        14.4
             4       A            4        16.8
             5       A            5        18.7
             6       B            1        10.8
             7       B            2        12.3
             8       B            3        13.7
             9       B            4        14.2
            10       B            5        16.6
            11       C            1        13.1
            12       C            2        16.2
            13       C            3        19.0
            14       C            4        22.9
            15       C            5        26.5
            16       C            0        10.1
            17       C            0        10.2
            18       C            0         9.8
            19       C            0         9.9
            20       C            0        10.2
```

The following statements produce Output 11.24:

```
proc glm;
   class source;
   model ts=amt*source / solution
```

Output 11.24
Parameter
Estimates for
Data with
AMT=0

```
                          The GLM Procedure

Dependent Variable: ts

                                Sum of
Source                   DF     Squares    Mean Square   F Value   Pr > F

Model                     3   393.0051791   131.0017264    903.34   <.0001

Error                    16     2.3203209     0.1450201

Corrected Total          19   395.3255000

             R-Square    Coeff Var    Root MSE      ts Mean

             0.994131    2.619986     0.380815      14.53500


Source                   DF    Type I SS    Mean Square   F Value   Pr > F

amt*source                3   393.0051791   131.0017264    903.34   <.0001


Source                   DF    Type III SS   Mean Square   F Value   Pr > F

amt*source                3   393.0051791   131.0017264    903.34   <.0001

                                            Standard
     Parameter           Estimate             Error    t Value   Pr > |t|

     Intercept          9.882352941        0.13699380    72.14     <.0001
     amt*source A       1.722994652        0.06350310    27.13     <.0001
     amt*source B       1.237540107        0.06350310    19.49     <.0001
     amt*source C       3.242994652        0.06350310    51.07     <.0001
```

Parameter estimates in Output 11.24 yield the three prediction equations

TS = 9.88 + 1.72 AMT   (SOURCE A)
TS = 9.88 + 1.24 AMT   (SOURCE B)
TS = 9.88 + 3.24 AMT   (SOURCE C)

The relative effect of one source to another can be measured by the ratio of slopes of the regression parameters. For example, the strength of SOURCE B relative to SOURCE A is the ratio 1.24/1.72 = 0.72. This means that one unit of the chemical from SOURCE B has the same effect on tensile strength as .72 units of the chemical from SOURCE A.

Similar models are used in other types of applications. The potency of one drug relative to another in a drug study, or the bioavailability of one nutrient relative to another in a nutrition study, is measured in the same way.

11.6 A Lack-of-Fit Analysis

A lack-of-fit analysis can provide information on the adequacy of a model that does not include all possible terms. The basic principle is to compare the fits of "complete" and "reduced" models as described in Chapter 2. The sum of squares for the complete model can be obtained from a one-way analysis of variance that computes the sums of squares among all unique treatments. The error sum of squares for the full model is subtracted from the error sum of squares for the incompletely specified model to obtain the sum of squares for all terms not specified.

One of the most common applications of lack-of-fit analysis is testing the adequacy of a regression model. In this procedure, you want to determine if a fitted model accounts for essentially all of the variation in a response variable due to differences between the levels of a quantitative independent variable. For example, consider an experiment in which chickens were fed a form of dietary copper to relate the copper uptake in the liver to copper intake. The chickens were fed a basal diet of 11 ppm copper sulfate, plus a supplemental rate of 0, 150, 300, or 450 ppm. There were six chickens in each of the four treatment groups. The data for this experiment are shown in the SAS data set LIVCU printed in Output 11.25. The variable LOGLIVCU is the logarithm (base 10) of the copper in the livers of the chickens.

Output 11.25
Data for a
Lack-of-Fit
Analysis

```
            Liver Copper in Poultry Fed Sulfate or Lysine Source

            Obs     level     lackofit     loglivcu

             1         0          0         1.16761
             2         0          0         1.25789
             3         0          0         1.27312
             4         0          0         1.09688
             5         0          0         1.26881
             6         0          0         1.24391
             7       150        150         1.38957
             8       150        150         1.46716
             9       150        150         1.51402
            10       150        150         1.30969
            11       150        150         1.24596
            12       150        150         1.37160
            13       300        300         1.99269
            14       300        300         2.19897
            15       300        300         2.14038
            16       300        300         1.83695
            17       300        300         1.97164
            18       300        300         2.11470
            19       450        450         2.41911
            20       450        450         2.34434
            21       450        450         2.15644
            22       450        450         2.32868
            23       450        450         2.46058
            24       450        450         2.43342
```

The variable LACKOFIT was defined to be equal to LEVEL in the DATA step. Its purpose will become apparent.

We want to fit a linear regression of LOGLIVCU on LEVEL and determine if the linear equation adequately models the response of LOGLIVCU to LEVEL. There are three degrees of freedom for differences between treatments. The linear regression accounts for one of those. The other two account for lack of fit of the linear regression, plus random error. The challenge is to determine how much is lack of fit.

Within each treatment group there are six observations. Variation between these observations, within a group, measures random variation. This is sometimes called "pure error." Pooled across treatment groups, there are 20 DF for pure error. The analysis of variance is

Source	DF
LEVEL	1
Lack of Fit	2
Pure Error	20

This ANOVA can be obtained by the Type I sums of squares that are provided by the following statements:

```
proc glm;
    class lackofit;
    model loglivcu=level lackofit/ss1;
```

The term LEVEL in the MODEL statement to the right of the equal sign is the usual linear effect of LEVEL. The LACKOFIT variable measures the variation in LOGLIVCU due to treatment that is not accounted for by the linear regression. Specifying the variable LACKOFIT, which has four levels, in the CLASS statement causes the generation of four dummy variables. If LACKOFIT were the only variable in the MODEL statement, then it would account for 3 DF. One of these is confounded with the linear effect. Preceding LACKOFIT by LEVEL leaves only 2 DF for LACKOFIT in the Type I sums of squares. It is important to precede LACKOFIT by LEVEL, because if LACKOFIT preceded LEVEL, then all 3 DF would go to LACKOFIT and 0 DF would go to LEVEL. Note, however, that all other types of sums of squares for LEVEL would be zero.

Results of the preceding statements appear in Output 11.26.

Output 11.26
A Lack-of-Fit
Analysis

```
                              The GLM Procedure

Dependent Variable: loglivcu

                              Sum of
Source                DF      Squares      Mean Square   F Value   Pr > F

Model                 3       5.23096460   1.74365487    155.47    <.0001

Error                 20      0.22430992   0.01121550

Corrected Total       23      5.45527452


            R-Square    Coeff Var     Root MSE     loglivcu Mean

            0.958882    6.051018      0.105903        1.750172


Source                DF      Type I SS    Mean Square   F Value   Pr > F

level                 1       4.98592403   4.98592403    444.56    <.0001
lackofit              2       0.24504057   0.12252029     10.92    0.0006
```

The test for LACKOFIT is significant with $p=0.0006$. This indicates that the relation between LOGLIVCU and LEVEL is not linear. However, the analysis sheds no light on the true relationship.

You could test whether the relationship is quadratic with a similar analysis provided by the statements

```
proc glm;
    class lackofit;
    model loglivcu=level level*level lackofit/ss1;
```

Output is not shown. There would be 1 DF for LEVEL and 1 DF for LEVEL*LEVEL.

Now consider a three-factor factorial experiment with factors A, B, and C. Suppose you specify an incomplete model that omits the B*C and A*B*C interactions, and you want to test for lack of fit of this model. Run the following statements:

```
proc glm;
   class a b c;
   model y=a b a*b c a*c;
```

The difference between the error sum of squares that you obtain and the error sum of squares from a between-cell analysis of variance provides the additional sum of squares due to both B*C and A*B*C.

You can get the between-cell analysis of variance from PROC ANOVA. If the CLASS variables are integers, a single variable can be generated to represent all cell combinations. Assume, for example, that the values of three CLASS variables (A, B, C) consist of integers between 1 and 10. The following assignment statement placed in the DATA step provides the subscript:

```
group=100*a+10*b+c;
```

The following SAS statements provide the desired error sums of squares:

```
proc glm;
   class group;
   model y=group;
```

The CLASS variables may not conform to these specifications. Character values can be concatenated, but the resulting single classification variable may exceed eight characters. Alternately, the following statements can be used to compute the sum of squares for differences between the cells:

```
proc glm;
   class a b c;
   model y=a*b*c;
```

11.7 An Unbalanced Nested Structure

Nested structure concerns samples within samples, as discussed in Section 4.2, "Nested Classifications." An example is treatments applied to plants in pots. There are several pots per treatment and several plants per pot. The pots do not necessarily have the same number of plants, and there may be different numbers of pots per treatment. Another example of nested structure occurs in sample surveys in which households are sampled within blocks, blocks are sampled within precincts, precincts are sampled within cities, and so on. In many cases, PROC NESTED is adequate for such analyses. Some applications, however, require PROC GLM or PROC MIXED. This section addresses some basic issues of computing means with unbalanced data and random effects. Section 11.8 continues in addressing the issues in a more complex setting.

Data containing a nested structure often have both fixed and random components, which raises questions about proper error terms. Consider an experiment with *t* treatments (TRT) applied randomly to a number of pots (POT), each containing several plants (PLANT). The data appear in Output 11.27.

Output 11.27
Data from an Unbalanced Nested Classification

```
                        Unbalanced Nested Structure

             Obs      TRT      POT     PLANT      Y

              1        1        1        1        15
              2        1        1        2        13
              3        1        1        3        16
              4        1        2        1        17
              5        1        2        2        19
              6        1        3        1        12
              7        2        1        1        20
              8        2        1        2        21
              9        2        2        1        20
             10        2        2        2        23
             11        2        2        3        19
             12        2        2        4        19
             13        3        1        1        12
             14        3        1        2        13
             15        3        1        3        14
             16        3        2        1        11
             17        3        3        1        12
             18        3        3        2        13
             19        3        3        3        15
             20        3        3        4        11
             21        3        3        5         9
```

The model is

$$y_{ijk} = \mu + \lambda_i + \rho_{ij} + \varepsilon_{ijk}$$

where

y_{ijk} is the observed response in the kth PLANT of the jth POT in the ith TREATMENT.

μ is the overall mean response.

λ_i is the effect of the ith TREATMENT.

ρ_{ij} is the effect of the jth POT within the ith TREATMENT.

ε_{ijk} is the effect of the kth individual PLANT in the jth POT of the ith TREATMENT. This effect is usually considered to be the random error.

You are primarily interested in tests and estimates related to the TREATMENTs as well as the variation among POTs and PLANTs. The analysis is implemented by the following SAS statements:

```
proc glm;
   class trt pot;
   model y=trt pot(trt) / ss1 ss3;
   means trt pot(trt);
   lsmeans trt pot(trt);
```

Most items in these statements are similar to those from previous examples. The analysis of variance appears in Output 11.28.

Output 11.28
Types I and III
ANOVA for an
Unbalanced
Nested
Classification

```
                              The GLM Procedure

Dependent Variable: Y

                                    Sum of
Source                    DF       Squares      Mean Square    F Value    Pr > F

Model                      7     267.2261905     38.1751701      12.43    <.0001

Error                     13      39.9166667      3.0705128

Corrected Total           20     307.1428571


              R-Square      Coeff Var      Root MSE       Y Mean

              0.870039      11.35742       1.752288       15.42857


Source                    DF      Type I SS      Mean Square    F Value    Pr > F

TRT                        2     236.9206349     118.4603175      38.58    <.0001
POT(TRT)                   5      30.3055556       6.0611111       1.97    0.1499


Source                    DF     Type III SS     Mean Square    F Value    Pr > F

TRT                        2     200.1109726     100.0554863      32.59    <.0001
POT(TRT)                   5      30.3055556       6.0611111       1.97    0.1499
```

The analysis of variance has the same form as given in Output 4.4. Note, however, that there is a slight difference in the Type I and Type III sums of squares of TRT in Output 11.28, due to the unbalanced structure of the data. This difference is made clearer by noting the differences between the unadjusted means from the MEANS statement and the adjusted or least-squares means from the LSMEANS statement. Table 11.2 shows the differences.

The means and least-squares means are shown in Output 11.29.

Output 11.29
Means and
Least Squares
from the GLM
Procedure

```
                              The GLM Procedure

            Level of          --------------Y--------------
            TRT           N           Mean           Std Dev

             1            6       15.3333333        2.58198890
             2            6       20.3333333        1.50554531
             3            9       12.2222222        1.78730088


     Level of    Level of        --------------Y--------------
     POT         TRT         N           Mean           Std Dev

      1           1          3       14.6666667       1.52752523
      2           1          2       18.0000000       1.41421356
      3           1          1       12.0000000        .
      1           2          2       20.5000000       0.70710678
      2           2          4       20.2500000       1.89296945
      1           3          3       13.0000000       1.00000000
      2           3          1       11.0000000        .
      3           3          5       12.0000000       2.23606798
                        Unbalanced Nested Structure                     43
                                      10:05 Thursday, January 10, 2002
```

Output 11.29
(Continued)
Means and
Least Squares
from the GLM
Procedure

```
                        The GLM Procedure
                      Least Squares Means

            TRT            Y LSMEAN

             1            14.8888889
             2            20.3750000
             3            12.0000000


            POT    TRT        Y LSMEAN

             1      1       14.6666667
             2      1       18.0000000
             3      1       12.0000000
             1      2       20.5000000
             2      2       20.2500000
             1      3       13.0000000
             2      3       11.0000000
             3      3       12.0000000
```

Table 11.2 *Means and Least-Squares Means*

TRT	POT	N	Means	Least-Squares Means
1		6	15.333	14.899
2		6	20.333	20.375
3		9	12.222	12.000
1	1	3	14.667	14.667
	2	2	18.000	18.000
	3	1	12.000	12.000
2	1	2	20.500	20.500
	2	4	20.250	20.250
3	1	3	13.000	13.000
	2	1	11.000	11.000
	3	5	12.000	12.000

The values produced by the MEANS statement are the means of all observations in a TREATMENT. These are the weighted POT means, as shown in the following equation:

$$\text{mean}(i) = \bar{y}_{i..} = (1/n_{i.}) \Sigma_j n_{ij} \bar{y}_{ij.}$$

n_{ij} is the number of plants in POT j of TREATMENT i. On the other hand, the least-squares means are the unweighted POT means, as shown in the following equation:

$$\text{least-squares mean}(i) = (1/k_i) \Sigma_j \bar{y}_{ij.}$$

k_i is the number of pots in TREATMENT i.

Both of these types of means have specific uses. In sample surveys, particularly self-weighting samples, it is usually appropriate to use the ordinary means. For the present case, POTs would probably be considered a random effect (see Section 4.2.1, "Analysis of Variance for Nested

Classifications.") In this event, the variance of least-squares mean (*i*) is less than the variance of mean (*i*) if σ_ρ^2 is large relative to σ^2 and conversely, where $\sigma_\rho^2 = V\left(\rho_{ij}\right)$ and $\sigma^2 = V\left(\varepsilon_{ijk}\right)$.

11.8 An Analysis of Multi-Location Data

Multiple location studies, such as clinical trials conducted at several centers, or on-farm trials in agriculture, raise several linear model issues. These issues primarily involve mixed-model *inference* considerations introduced in Chapter 4, and linear model *unbalanced data* concepts discussed in Chapters 5 and 6. The analysis of multi-location data can be both confusing and controversial, partly because different kinds of multi-location studies call for different approaches and partly because there is disagreement within the statistics community on what methods are appropriate. This section presents an example multi-location data set and several alternative analyses using linear and mixed-model methods. The purpose of this section is not to prescribe, but simply to demonstrate the various linear model approaches and in the process frame the main linear model issues.

Output 11.30 contains data from a study to compare 3 treatments (TRT) conducted at 8 locations (LOC). At each location, a randomized complete-blocks design was used, but the number of blocks varied. Locations 1-4 used 3 blocks each, locations 5 and 6 used 6 blocks each, and locations 7 and 8 used 12 blocks each. In the interest of space, not all the data are shown in Output 11.30. However, Output 11.31 shows the response variable (*Y*) mean and number of observations (blocks) per treatment for each location.

Output 11.30
Multi-Location
Data

Obs	loc	blk	trt	y
1	1	1	1	46.6
2	1	1	2	46.4
3	1	1	3	44.4
4	1	2	1	43.7
5	1	2	2	43.6
6	1	2	3	31.4
7	1	3	1	37.9
8	1	3	2	39.5
9	1	3	3	48.2
10	2	1	1	34.0
				.
				.
				.
124	8	6	1	43.5
125	8	6	2	52.1
126	8	6	3	61.4
127	8	7	1	44.1
128	8	7	2	54.8
129	8	7	3	59.9
130	8	8	1	43.3
131	8	8	2	49.4
132	8	8	3	63.0
133	8	9	1	44.2
134	8	9	2	54.6
135	8	9	3	64.8
136	8	10	1	54.6
137	8	10	2	56.6
138	8	10	3	64.6
139	8	11	1	52.1
140	8	11	2	44.3
141	8	11	3	59.7
142	8	12	1	44.9
143	8	12	2	43.3
144	8	12	3	65.0

Output 11.31
Mean Response for Each Treatment by Location

Obs	loc	trt	_FREQ_	y_mean
1	1	1	3	42.7333
2	1	2	3	43.1667
3	1	3	3	41.3333
4	2	1	3	33.5333
5	2	2	3	37.0000
6	2	3	3	22.2333
7	3	1	3	36.6667
8	3	2	3	43.4000
9	3	3	3	47.9000
10	4	1	3	47.7000
11	4	2	3	52.3000
12	4	3	3	73.7000
13	5	1	6	41.8000
14	5	2	6	45.9500
15	5	3	6	47.0000
16	6	1	6	33.9667
17	6	2	6	38.1667
18	6	3	6	30.2333
19	7	1	12	38.6417
20	7	2	12	44.1833
21	7	3	12	51.8500
22	8	1	12	47.5417
23	8	2	12	50.6500
24	8	3	12	60.5500

The variable _FREQ_ in Output 11.31 refers to the number of blocks in a given location.

These are the main controversies for the analysis of multi-location data:

❑ Should the location×treatment interaction be included in the model, or should it be assumed to be zero?

❑ Should locations be considered fixed or random?

❑ Should means be weighted by the number of observations per location, or equivalently, should Type I or Type III SS be used if nonzero location×treatment interactions are assumed?

❑ If random locations and hence random location×treatment interaction effects are assumed, how should location-specific treatment effects, if they arise, be handled?

The following analyses illustrate different approaches to these questions. These illustrations suggest advantages and disadvantages for each approach.

11.8.1 An Analysis Assuming No Location×Treatment Interaction

This approach assumes the model $y_{ijk} = \mu + L_i + B(L)_{ij} + \tau_k + e_{ijk}$, where L_i, $B(L)_i$, and τ_k are the location, block within location, and treatment effects, respectively, and the random errors e_{ijk} are assumed i.i.d. $N(0,\sigma^2)$. Its rationale presumes that 1) the reason for having multiple locations is solely to provide a practical way to obtain an adequate sample size and 2) treatment effects are known with certainty not to be location-specific. You can use the following SAS statements to implement the analysis:

```
proc glm data=mloc;
   class loc blk trt;
   model y=trt loc blk(loc);
   means trt;
   lsmeans trt;
```

Output 11.32 shows the results. Normally, you would place TRT last in the model. It is placed before LOC and BLK(LOC) here to illustrate a point. Though not shown here, in practice you would usually add CONTRAST statements or mean comparison options to the MEANS or LSMEANS statements to complete inference about treatment effects.

Output 11.32
An Analysis of
Multi-Location
Data Using the
No LOC×TRT
Model

Source	DF	Sum of Squares	Mean Square	F Value	Pr > F
Model	49	10530.96049	214.91756	4.15	<.0001
Error	94	4869.34944	51.80159		
Corrected Total	143	15400.30993			

Source	DF	Type I SS	Mean Square	F Value	Pr > F
trt	2	1641.777222	820.888611	15.85	<.0001
loc	7	7770.496597	1110.070942	21.43	<.0001
blk(loc)	40	1118.686667	27.967167	0.54	0.9846

Source	DF	Type III SS	Mean Square	F Value	Pr > F
trt	2	1641.777222	820.888611	15.85	<.0001
loc	7	7770.496597	1110.070942	21.43	<.0001
blk(loc)	40	1118.686667	27.967167	0.54	0.9846

Level of trt	N	--------------y-------------- Mean	Std Dev
1	48	41.0562500	6.9066698
2	48	45.2145833	6.8837456
3	48	49.3270833	14.0586876

Least Squares Means

trt	y LSMEAN
1	39.6986111
2	43.8569444
3	47.9694444

You can see that the MEANS and LS means are different. This reflects a different weighting scheme for the L_i effects: The MEANS weight them according to the number of observations per location, whereas the LS means weight them equally. Notice that the differences among pairs of treatment MEANS and LS means, however, are unaffected and the Type I and Type III SS are identical. You get different estimates of means but identical estimates of treatment differences regardless of whether you use MEANS or LS means. This is because differences among the MEANS and LS means eliminate weighting based on the number of observations per location. You can see this by using the E option with the LSMEANS statement (as shown in Chapter 6) to show the weighting scheme. The estimates of the treatment effects are thus disproportionately affected by the locations with the greatest number of observations (in this case locations 7 and 8).

The main risk of using this analysis is that it is very sensitive to the assumption of no location-specific treatment effects. Even minor violations of this assumption can seriously affect the results when you use this model.

11.8.2 A Fixed-Location Analysis with an Interaction

In many, perhaps most, multi-location studies, researchers are not prepared to assume no location×treatment interaction without at least testing the assumption. One approach is to modify the model from Section 11.8.1 by adding an interaction term, yielding the model equation

$$y_{ijk} = \mu + L_i + B(L)_{ij} + \tau_k + (\tau L)_{ik} + e_{ijk},$$ where $(\tau L)_{ik}$ denotes the location×treatment interaction. Use the following SAS statements to implement the analysis:

```
proc glm data=mloc;
    class loc blk trt;
    model y=loc blk(loc) trt loc*trt;
    means trt;
    lsmeans trt loc*trt/slice=loc;
run;
```

As with any factorial arrangement, the appropriate strategy for inference is first to test the location×treatment (LOC*TRT) interaction and then evaluate simple effects of treatment by location (for example, using the SLICE=LOC option) if the interaction is non-negligible, or otherwise, evaluate main effects. The results appear in Output 11.33.

Output 11.33
An Analysis of
Multi-Location
Data Using a
Fixed-Location
Model with an
Interaction

Source	DF	Sum of Squares	Mean Square	F Value	Pr > F
Model	63	13042.45660	207.02312	7.02	<.0001
Error	80	2357.85333	29.47317		
Corrected Total	143	15400.30993			

Source	DF	Type I SS	Mean Square	F Value	Pr > F
loc	7	7770.496597	1110.070942	37.66	<.0001
blk(loc)	40	1118.686667	27.967167	0.95	0.5634
trt	2	1641.777222	820.888611	27.85	<.0001
loc*trt	14	2511.496111	179.392579	6.09	<.0001

Source	DF	Type III SS	Mean Square	F Value	Pr > F
loc	7	7770.496597	1110.070942	37.66	<.0001
blk(loc)	40	1118.686667	27.967167	0.95	0.5634
trt	2	757.254848	378.627424	12.85	<.0001
loc*trt	14	2511.496111	179.392579	6.09	<.0001

Level of trt	N	--------------y-------------- Mean	Std Dev
1	48	41.0562500	6.9066698
2	48	45.2145833	6.8837456
3	48	49.3270833	14.0586876

Output 11.33
(*Continued*)
*An Analysis of
Multi-Location
Data Using a
Fixed-Location
Model with an
Interaction*

```
                        Least Squares Means

              trt          y LSMEAN

               1         40.3229167
               2         44.3520833
               3         46.8500000

           loc*trt Effect Sliced by loc for y

                        Sum of
      loc      DF      Squares     Mean Square   F Value   Pr > F

       1        2      5.508889      2.754444      0.09     0.9109
       2        2    357.762222    178.881111      6.07     0.0035
       3        2    191.775556     95.887778      3.25     0.0438
       4        2   1155.120000    577.560000     19.60    <.0001
       5        2     90.730000     45.365000      1.54     0.2208
       6        2    189.031111     94.515556      3.21     0.0457
       7        2   1055.791667    527.895833     17.91    <.0001
       8        2   1107.553889    553.776944     18.79    <.0001
```

Output 11.33 reveals several points about the data. First, there is very strong evidence of a location×treatment interaction ($F=6.09$, $p<0.0001$). The SLICE output partially reveals the nature of the interaction: Significant treatment effects were observed at locations 2, 3, 4, 6, 7, and 8, but not at locations 1 or 5. You could pursue this by computing the simple effect comparisons among treatments for each comparison using the steps presented in Section 3.7.5 "Simple Effect Comparisons." These comparisons are not shown, but you can inspect the treatment means by location in Output 11.30 to anticipate the results: In locations 2 and 6, the mean of treatment 3 is substantially lower than the means of treatments 1 and 2, whereas in locations 4, 7, and 8, and to a lesser extent locations 3 and 6, the opposite is true.

If you do proceed with inference on main effects, despite the evidence of interaction, then you can see that the MEANS and their associated test using the Type I SS for TRT produce different results than the LS means and their associated test using Type III SS. This mainly results from the fact that the MEANS weight locations 7 and 8 more heavily relative to the other locations, whereas the LS means weight all locations equally. Thus, the large difference between treatment 3 and the others in locations 7 and 8 affects the MEANS to a much greater extent than the LS means.

Recalling the discussion of MEANS and Type I SS versus LS means and Type III SS from Chapter 6, you would want to use the MEANS if the number of observations per location closely reflects the true proportion of populations in the various locations. In other words, if locations 7 and 8, for example, are in communities whose populations are roughly four times the populations of locations 1 through 4, then the proportion of observation is representative. On the other hand, if the number of observations per location is mainly a sampling artifact, and does not reflect the actual size of the populations in each location, then the MEANS may seriously misrepresent the actual treatment effects. Note that if you decide to drop LOC*TRT from the model based on the test for interaction, your subsequent inference is implicitly based on the MEANS.

Keep in mind that the fixed-locations model with interaction makes two critical assumptions about the data. First, recalling the discussion of fixed-effects versus random-effects inference from Chapter 4, the fixed-locations model assumes that the observed locations are the *entire* population. The analysis neither measures nor recovers any information about distribution among locations. Second, fixed-locations analysis assumes that the only relevant source of uncertainty comes from variation among observations within locations, making MS(ERROR) an appropriate error term for testing TRT. If locations are meant to represent a larger population, this assumption is probably untrue and, as you will see in the next section, the tests for treatment shown in this section are incorrect and likely to be misleading. Assuming fixed locations when in fact the location and locationxtreatment effects represent probability distributions can produce severely inflated Type I error rates for the test of treatment effects. Therefore, you should use this model only when the locations in fact are the entire population of inference or when the locations are chosen to represent a second treatment factor associated with known characteristics of the location (for example, soil type or climatic zone in agricultural trials, or socioeconomic group in multi-center clinical trials).

11.8.3 A Random-Location Analysis

In many multi-location studies, locations represent a larger target population. Implicitly, the goal of these studies is to apply inference beyond the observed locations to the entire population. Recalling the criteria for distinguishing fixed from random effects given in the introduction to Chapter 4, location effects are random when the locations actually observed represent a probability distribution of locations that could, in theory, have been sampled. In most multi-location studies, locations are not drawn from a true random sample, but again recalling the discussion in Chapter 4, this is usually a moot point. Location effects are random if the locations plausibly represent the population (and if they don't, you should question either the study design or the use of the data to draw inference beyond the observed locations).

The model equation for random-location analysis is identical to the equation given in Section 11.8.2 for fixed-location analysis with interaction, but the assumptions are different: The location effects, L_i, are assumed i.i.d. $N(0, \sigma_L^2)$, and the locationxtreatment effects are assumed i.i.d. $N(0, \sigma_{LT}^2)$. In addition, the block within treatment effects is assumed random as well. You can obtain the expected mean squares and the overall test for treatment effects using PROC GLM, but, as with other mixed-model examples discussed in previous chapters, the standard errors and tests of various treatment comparisons are wrong or awkward to obtain using PROC GLM. PROC MIXED is a better choice. Use the SAS statements

```
proc mixed method=type3;
   class loc blk  trt;
   model y=trt/ddfm=kr;
   random loc blk(loc) loc*trt;
   lsmeans trt/diff;
```

The METHOD=TYPE3 option is not necessary in practice; it is used here merely to show the expected mean squares, which appear in Output 11.34. The rest of the analysis appears in Output 11.35.

Output 11.34
An Analysis of Variance and Expected Mean Squares for a Random-Locations Analysis of Multi-Location Data

```
                       Type 3 Analysis of Variance

                                    Sum of
              Source      DF        Squares      Mean Square

              trt          2     757.254848      378.627424
              loc          7    7770.496597     1110.070942
              blk(loc)    40    1118.686667       27.967167
              loc*trt     14    2511.496111      179.392579
              Residual    80    2357.853333       29.473167

   Source     Expected Mean Square

   trt        Var(Residual) + 4.3636 Var(loc*trt) + Q(trt)
   loc        Var(Residual) + 5.6786 Var(loc*trt)
              + 3 Var(blk(loc)) + 17.036 Var(loc)
   blk(loc)   Var(Residual) + 3 Var(blk(loc))
   loc*trt    Var(Residual) + 5.6786 Var(loc*trt)
   Residual   Var(Residual)
```

Note the coefficients of the LOC*TRT variance for the TRT main effect and the LOC*TRT interaction. They are different because of the unequal number of observations per location. With unbalanced data and the random location×treatment interaction, the appropriate error term for testing treatment effects is a linear combination of MS(LOC*TRT) and MS(ERROR).

Output 11.35
Random Location Analysis of Multi-Location Data

```
                    Covariance Parameter
                        Estimates

              Cov Parm       Estimate

              loc             54.7194
              blk(loc)        -0.5020
              loc*trt         26.4009
              Residual        29.4732


           Type 3 Tests of Fixed Effects

                     Num     Den
          Effect      DF      DF    F Value   Pr > F

          trt          2     18.1     2.77    0.0893


              Least Squares Means

                                Standard
       Effect   trt   Estimate    Error    DF    t Value   Pr > |t|

       trt       1    40.2770    3.3091   15.8    12.17     <.0001
       trt       2    44.3284    3.3091   15.8    13.40     <.0001
       trt       3    46.9789    3.3091   15.8    14.20     <.0001


            Differences of Least Squares Means

                                   Standard
    Effect   trt   _trt   Estimate   Error    DF    t Value   Pr > |t|

    trt       1     2     -4.0515   2.8690   18.1    -1.41    0.1749
    trt       1     3     -6.7020   2.8690   18.1    -2.34    0.0312
    trt       2     3     -2.6505   2.8690   18.1    -0.92    0.3677
```

You can see from Output 11.35 that the test of treatment effect is considerably more conservative than the corresponding tests in the fixed-locations analyses. This is partly because the MS(LOC*TRT) term is considerably larger than MS(ERROR)—recall the highly significant location×treatment interaction in Output 11.33—and partly because the denominator degrees of freedom depend mainly on the degrees of freedom for LOC*TRT and are thus substantially lower. The BLK(TRT) variance is allowed to remain negative when the METHOD=TYPE3 option is used. The REML default sets the estimate to zero, with some impact on the LOC*TRT variance estimate and some of the test statistics. Section 4.4.2, "Standard Errors for the Two-Way Mixed Model: GLM versus MIXED," discussed the arguments for and against the REML default; this remains an unresolved controversy in mixed-model inference.

Now look at the LS means and the estimates of treatment differences. The estimates are close to the values you would get using the LS means in the fixed location with interaction model in Output 11.33. This means that the random-locations model implicitly weights locations approximately equally. The standard errors and denominator degrees of freedom are considerably different from the fixed-locations analysis because the mixed model uses the LOC*TRT variance.

You can consider location-specific effects with the random-locations model by using best linear unbiased predictors. The following SAS statements obtain the location-specific BLUPs for the differences between treatments 1 and 2 and between 1 and 3, respectively:

```
estimate 't1 vs t2 at loc 1' trt 1 -1 0 | loc*trt 1 -1 0;
estimate 't1 vs t3 at loc 1' trt 1 0 -1 | loc*trt 1 0 -1;
estimate 't1 vs t2 at loc 2' trt 1 -1 0 | loc*trt 0 0 0  1 -1 0;
estimate 't1 vs t3 at loc 2' trt 1 0 -1 | loc*trt 0 0 0  1 0 -1;
estimate 't1 vs t2 at loc 3' trt 1 -1 0
         | loc*trt 0 0 0 0 0  1 -1 0;
estimate 't1 vs t3 at loc 3' trt 1 0 -1
         | loc*trt 0 0 0 0 0  1 0 -1;
estimate 't1 vs t2 at loc 4' trt 1 -1 0
         | loc*trt 0 0 0 0 0 0 0 0 1 -1 0;
estimate 't1 vs t3 at loc 4' trt 1 0 -1
         | loc*trt 0 0 0 0 0 0 0 0  1 0 -1;
estimate 't1 vs t2 at loc 5' trt 1 -1 0
         | loc*trt 0 0 0 0 0 0 0 0 0 0 0 1 -1 0;
estimate 't1 vs t3 at loc 5' trt 1 0 -1
         | loc*trt 0 0 0 0 0 0 0 0 0 0 0 1 0 -1;
estimate 't1 vs t2 at loc 6' trt 1 -1 0
         | loc*trt 0 0 0 0 0 0 0 0 0 0 0 0 0 0  1 -1 0;
estimate 't1 vs t3 at loc 6' trt 1 0 -1
         | loc*trt 0 0 0 0 0 0 0 0 0 0 0 0 0 0  1 0 -1;
estimate 't1 vs t2 at loc 7' trt 1 -1 0
         | loc*trt 0 0 0 0 0 0 0 0 0 0 0 0 0 0 0 0 0  1 -1 0;
estimate 't1 vs t3 at loc 7' trt 1 0 -1
         | loc*trt 0 0 0 0 0 0 0 0 0 0 0 0 0 0 0 0 0  1 0 -1;
estimate 't1 vs t2 at loc 8' trt 1 -1 0
         | loc*trt 0 0 0 0 0 0 0 0 0 0 0 0 0 0 0 0 0 0 0 0  1
-1 0;
estimate 't1 vs t3 at loc 8' trt 1 0 -1
         | loc*trt 0 0 0 0 0 0 0 0 0 0 0 0 0 0 0 0 0 0 0 0  1
0 -1;
```

The results appear in Output 11.36.

*Output 11.36
Location-
Specific Best
Linear
Unbiased
Predictors for
Multi-Location
Data*

```
                              Estimates

                            Standard
        Label              Estimate   Error      DF   t Value   Pr > |t|

        t1 vs t2 at loc 1   -1.4146   3.8811    133    -0.36    0.7161
        t1 vs t3 at loc 1   -0.7973   3.8811    133    -0.21    0.8376
        t1 vs t2 at loc 2   -3.6253   3.8811    133    -0.93    0.3519
        t1 vs t3 at loc 2    6.4178   3.8811    133     1.65    0.1006
        t1 vs t2 at loc 3   -6.0060   3.8811    133    -1.55    0.1241
        t1 vs t3 at loc 3  -10.0044   3.8811    133    -2.58    0.0110
        t1 vs t2 at loc 4   -4.4512   3.8811    133    -1.15    0.2535
        t1 vs t3 at loc 4  -20.7663   3.8811    133    -5.35    <.0001
        t1 vs t2 at loc 5   -4.1345   2.9216    104    -1.42    0.1600
        t1 vs t3 at loc 5   -5.4356   2.9216    104    -1.86    0.0656
        t1 vs t2 at loc 6   -4.1767   2.9216    104    -1.43    0.1558
        t1 vs t3 at loc 6    2.0963   2.9216    104     0.72    0.4747
        t1 vs t2 at loc 7   -5.4148   2.1376   90.7    -2.53    0.0130
        t1 vs t3 at loc 7  -12.6546   2.1376   90.7    -5.92    <.0001
        t1 vs t2 at loc 8   -3.1886   2.1376   90.7    -1.49    0.1392
        t1 vs t3 at loc 8  -12.4716   2.1376   90.7    -5.83    <.0001
```

These estimates are similar to the simple effects you would obtain in the fixed location with interaction analysis in Section 11.8.2, except that the estimates in Output 11.36 are shrinkage estimators to account for the location and location×treatment distributions. Also note that you should use the DDFM=KR option in the MODEL statement of the PROC MIXED program; otherwise, the standard errors of the BLUPs are biased downward.

The main argument for using the random-locations analysis is that it most accurately reflects the inference implicitly intended in most multi-location studies. The primary disadvantage is that in order to get reasonable estimates of LOC and LOC*TRT variance and in order to have sufficient denominator degrees of freedom to test TRT effects, studies must be designed so that there are an adequate number of locations and that the locations plausibly represent the target population.

11.8.4 Further Analysis of a Location×Treatment Interaction Using a Location Index

Closer inspection of the treatment means by location in Output 11.31 and the location-specific BLUPs in Output 11.36 reveals that the treatment 1 minus treatment 3 difference tends to be large favoring treatment 1 in locations with relatively low overall mean responses whereas treatment 3 tends to be favored in locations with relatively high overall mean responses. You formalize this relationship by using an analysis that uses regression on an index characterizing the mean response at each location. This method is closely related to the Tukey test of non-additivity in randomized-complete-blocks designs and has been used by Eberhart and Russell (1966) to characterize genotype-by-environment interactions (which are a special case of multi-location studies). Milliken and Johnson (1989, Chapters 1-3) give an excellent overview of these methods.

The model for this analysis is $y_{ijk} = \mu + L_i + B(L)_{ij} + \tau_k + \beta_k I_i + (\tau L)_{ik} + e_{ijk}$, where all of the terms in the model equation are defined as previously with the addition of a *location index*, I_i, and a regression coefficient, β_k, for the kth treatment. The location index is usually defined as the mean response over all observations on the ith location. You can implement the analysis using the following SAS statements:

```
proc sort data=mloc;
   by loc;
proc means noprint data=mloc;
   by loc; var y;
   output out=env_indx mean=index;
data all;
   merge mloc env_indx;
   by loc;
proc mixed data=all;
   class loc blk trt;
   model y=trt trt*index/noint solution ddfm=satterth;
   random loc blk(loc) loc*trt;
   lsmeans trt/diff;
   contrast 'trt at mean index'
      trt 1 -1 0 trt*index 45.2 -45.2 0,
      trt 1 0 -1 trt*index 45.2 0 -45.2;
```

The PROC SORT and PROC MEANS statements generate a new data set, called ENV_INDX, which contains the means of *Y* by location. The new variable is called INDEX. The ENV_INDX data are then merged with the original data set. You compute the analysis using PROC MIXED. You can see that the MIXED program is similar to the random-locations analysis in Section 11.8.3, except that you add the term TRT*INDEX to the MODEL statement. This term corresponds to $\beta_k I_i$ in the model equation given above. The NOINT and SOLUTION options allow easier interpretation of the output. The CONTRAST statement computes an appropriate test of the equality of treatment effects; because this model is a special case of an unequal slopes analysis-of-covariance model, the test of treatment effects varies with the covariate. The test shown here is for the mean of the INDEX variable over all locations. You could choose different values of INDEX. In fact, in many studies you would want to test treatment effect at several values of the INDEX, say at relatively low and relatively high values, to get an idea of how treatment effects change over locations with different mean responses. Output 11.37 shows the results of the analysis.

Output 11.37
Location Index
Analysis of
Multi-Location
Data

```
                         Covariance Parameter
                             Estimates

                      Cov Parm       Estimate

                       loc                  0
                       blk(loc)             0
                       loc*trt         0.8334
                       Residual       27.9211


                      Solution for Fixed Effects

                              Standard
         Effect       trt     Estimate      Error      DF    t Value   Pr > |t|

         trt           1      12.4035      5.1377    32.7       2.41     0.0215
         trt           2      17.0483      5.1377    32.7       3.32     0.0022
         trt           3     -29.4519      5.1377    32.7      -5.73    <.0001
         index*trt     1       0.6345      0.1128    29.2       5.62    <.0001
         index*trt     2       0.6232      0.1128    29.2       5.52    <.0001
         index*trt     3       1.7423      0.1128    29.2      15.44    <.0001


                      Type 3 Tests of Fixed Effects

                            Num      Den
               Effect        DF       DF     F Value    Pr > F

               trt            3      32.7      16.57    <.0001
               index*trt      3      29.2     100.17    <.0001


                               Contrasts

                              Num      Den
               Label           DF       DF     F Value    Pr > F

         trt at mean index      2      19.2      23.45    <.0001


                         Least Squares Means

                              Standard
         Effect       trt     Estimate      Error      DF    t Value   Pr > |t|

          trt          1      41.0822      0.8483    19.2      48.43    <.0001
          trt          2      45.2182      0.8483    19.2      53.31    <.0001
          trt          3      49.2975      0.8483    19.2      58.12    <.0001


                     Differences of Least Squares Means

                                     Standard
         Effect    trt    _trt    Estimate      Error      DF    t Value   Pr > |t|

          trt       1      2      -4.1360      1.1996    19.2      -3.45     0.0027
          trt       1      3      -8.2153      1.1996    19.2      -6.85    <.0001
          trt       2      3      -4.0793      1.1996    19.2      -3.40     0.0030
```

The "Covariance Parameter Estimates" show that the INDEX accounts for most of the variation among locations. The LOC variance estimate is 0 and the LOC*TRT variance is sharply reduced compared to its estimate in the random-locations analysis in Output 11.35. The "Solution for Fixed Effects" parameters have the following interpretation. The TRT parameter estimate $\mu + \tau_k$ and the INDEX*TRT parameters estimate β_k. Thus, TRT + (INDEX*TRT)×(location index) gives you the expected treatment mean at a given value of location index. For example, at a location whose average response is 45.2, the expected mean of treatment 1 is 12.4035+(0.6345)*(45.2)=41.4, the LS mean shown in the output, aside from round-off error. The INDEX*TRT estimates tell you how the expected treatment mean changes as the location index increases.

The INDEX*TRT estimate is much larger for treatment 3, and the intercept (TRT) is much smaller, which tells you that treatment 3 is expected to have a low mean relative to treatments 1 and 2 in locations with a relatively low mean response. But its mean increases more quickly and thus is expected to have a higher mean relative to treatments 1 and 2 for locations with relatively high mean responses.

The "Type 3 Tests of Fixed Effects" results for TRT and INDEX*TRT test the joint equality of these terms to zero. As such, the test for TRT is usually not of interest. The CONTRAST result for TRT at a given value of INDEX supercedes the TRT test. Use INDEX*TRT to test whether the expected responses vary linearly with location index for the three treatments. You could construct a similar CONTRAST defined on INDEX*TRT to test the equality of the β_k's, which you interpret as a linear location index×treatment interaction.

As discussed earlier, the CONTRAST result tests treatment effects at a location index of 45.2. The "Least Squares Means" and "Differences of Least Squares Means" output are computed for the mean INDEX values determined by the LS means algorithm. You could vary the LS means and differences using the AT option to see what happens in different environments. For example, the following statements compute LS means for INDEX values of 30.9 (roughly the lowest INDEX for any location observed) and 57.9 (roughly the highest INDEX among all locations). For completeness, the LSMEANS statement with the default AT MEANS option is also shown. The SAS statements are

```
lsmeans trt/at index=30.9 diff;
lsmeans trt/at means diff;
lsmeans trt/at index=57.9 diff;
```

The results appear in Output 11.38.

Output 11.38
LS Means and Differences Computed at Various Location Indices

Least Squares Means

Effect	trt	index	Estimate	Standard Error	DF	t Value	Pr > |t|
trt	1	30.90	32.0094	1.7935	36.6	17.85	<.0001
trt	2	30.90	36.3064	1.7935	36.6	20.24	<.0001
trt	3	30.90	24.3842	1.7935	36.6	13.60	<.0001
trt	1	45.20	41.0822	0.8483	19.2	48.43	<.0001
trt	2	45.20	45.2182	0.8483	19.2	53.31	<.0001
trt	3	45.20	49.2975	0.8483	19.2	58.12	<.0001
trt	1	57.90	49.1407	1.6936	19.6	29.02	<.0001
trt	2	57.90	53.1338	1.6936	19.6	31.37	<.0001
trt	3	57.90	71.4255	1.6936	19.6	42.17	<.0001

Differences of Least Squares Means

Effect	trt	_trt	index	Estimate	Standard Error	DF	t Value	Pr > |t|
trt	1	2	30.90	-4.2970	2.5363	36.6	-1.69	0.0987
trt	1	3	30.90	7.6252	2.5363	36.6	3.01	0.0048
trt	2	3	30.90	11.9221	2.5363	36.6	4.70	<.0001
trt	1	2	45.20	-4.1360	1.1996	19.2	-3.45	0.0027
trt	1	3	45.20	-8.2153	1.1996	19.2	-6.85	<.0001
trt	2	3	45.20	-4.0793	1.1996	19.2	-3.40	0.0030
trt	1	2	57.90	-3.9930	2.3951	19.6	-1.67	0.1114
trt	1	3	57.90	-22.2848	2.3951	19.6	-9.30	<.0001
trt	2	3	57.90	-18.2917	2.3951	19.6	-7.64	<.0001

You can see that in locations with the lowest index, or mean response, the expected response of treatment 3 is considerably lower than that for the other two treatments. This is consistent with what was observed in locations 2 and 6 (see Output 11.31). On the other hand, the locations with the highest mean, the expected response of treatment 3 exceeds that of the other two treatments by a considerable margin, as was observed in locations 4, 7, and 8.

As a final note, this analysis suggests that when there are strong location-specific effects, the argument between weighted versus unweighted means, that is, MEANS versus LS means, respectively, from the fixed and random location with interaction models in Sections 11.8.2 and 11.8.3, is probably moot. Evaluating changes in treatment response at different locations and trying to understand why they are occurring is usually more to the point.

11.9 Absorbing Nesting Effects

Nested effects can produce a very large number of dummy variables in models and challenge the capacity of computers. This problem is trivial with modern computers compared with only a few years ago, but it is still an issue. A methodology called absorption greatly reduces the size of the problem by eliminating the need to obtain an explicit solution to the complete set of normal equations. In most applications, nested effects are random, and their estimates might not be required.

Absorption reduces the number of normal equations by eliminating the parameters for one factor from the system before a solution is obtained. This is analogous to the method of solving a set of three equations in three unknowns, x_1, x_2, and x_3. Suppose you combine the first and second equations and eliminate x_3. Next, combine the first and third equations and eliminate x_3. Then, with two equations left involving x_1 and x_2 (the variable x_3 having been absorbed), solve the reduced set for x_1 and x_2.

The use of the ABSORB statement is illustrated with data on 65 steers from Harvey (1975). Several values are recorded for each steer, including line number (LINE), sire number (SIRE), age of dam (AGEDAM), steer age (AGE), initial weight (INTLWT), and the dependent variable, average daily gain (AVDLYGN). Output 11.39 shows the data.

Output 11.39
Data Set Sires

Obs	line	sire	agedam	steerno	age	intlwt	avdlygn
1	1	1	3	1	192	390	2.24
2	1	1	3	2	154	403	2.65
3	1	1	4	3	185	432	2.41
4	1	1	4	4	193	457	2.25
5	1	1	5	5	186	483	2.58
6	1	1	5	6	177	469	2.67
7	1	1	5	7	177	428	2.71
8	1	1	5	8	163	439	2.47
9	1	2	4	9	188	439	2.29
10	1	2	4	10	178	407	2.26
11	1	2	5	11	198	498	1.97
12	1	2	5	12	193	459	2.14
13	1	2	5	13	186	459	2.44
14	1	2	5	14	175	375	2.52
15	1	2	5	15	171	382	1.72
16	1	2	5	16	168	417	2.75
17	1	3	3	17	154	389	2.38
18	1	3	4	18	184	414	2.46
19	1	3	5	19	174	483	2.29
20	1	3	5	20	170	430	2.30
21	1	3	5	21	169	443	2.94
22	2	4	3	22	158	381	2.50
23	2	4	3	23	158	365	2.44
24	2	4	4	24	169	386	2.44
25	2	4	4	25	144	339	2.15
26	2	4	5	26	159	419	2.54
27	2	4	5	27	152	469	2.74
28	2	4	5	28	149	379	2.50

Output 11.39
(Continued)
Data Set Sires

29	2	4	5	29	149	375	2.54
30	2	5	3	30	189	395	2.65
31	2	5	4	31	187	447	2.52
32	2	5	4	32	165	430	2.67
33	2	5	5	33	181	453	2.79
34	2	5	5	34	177	385	2.33
35	2	5	5	35	151	414	2.67
36	2	5	5	36	147	353	2.69
37	3	6	4	37	184	411	3.00
38	3	6	4	38	184	420	2.49
39	3	6	5	39	187	427	2.25
40	3	6	5	40	184	409	2.49
41	3	6	5	41	183	337	2.02
42	3	6	5	42	177	352	2.31
43	3	7	3	43	205	472	2.57
44	3	7	3	44	193	340	2.37
45	3	7	4	45	162	375	2.64
46	3	7	5	46	206	451	2.37
47	3	7	5	47	205	472	2.22
48	3	7	5	48	187	402	1.90
49	3	7	5	49	178	464	2.61
50	3	7	5	50	175	414	2.13
51	3	8	3	51	200	466	2.16
52	3	8	3	52	184	356	2.33
53	3	8	3	53	175	449	2.52
54	3	8	4	54	178	360	2.45
55	3	8	5	55	189	385	1.44
56	3	8	5	56	184	431	1.72
57	3	8	5	57	183	401	2.17
58	3	9	3	58	166	404	2.68
59	3	9	4	59	187	482	2.43
60	3	9	4	60	186	350	2.36
61	3	9	4	61	184	483	2.44
62	3	9	5	62	180	425	2.66
63	3	9	5	63	177	420	2.46
64	3	9	5	64	175	440	2.52
65	3	9	5	65	164	405	2.42

The analysis, as performed by Harvey, can be obtained directly by PROC GLM with the following SAS statements:

```
proc glm;
    class line sire agedam;
    model avdlygn=line sire(line) agedam
        line*agedam age intlwt / solution ss3;
    test h=line e=sire(line);
```

The results appear in Output 11.40.

Output 11.40
Complete Analysis of Variance

The GLM Procedure

Dependent Variable: avdlygn

Source	DF	Sum of Squares	Mean Square	F Value	Pr > F
Model	16	2.52745871	0.15796617	3.14	0.0011
Error	48	2.41191667	0.05024826		
Corrected Total	64	4.93937538			

R-Square	Coeff Var	Root MSE	avdlygn Mean
0.511696	9.295956	0.224161	2.411385

Source	DF	Type III SS	Mean Square	F Value	Pr > F
line	2	0.13620255	0.06810128	1.36	0.2676
sire(line)	6	0.97388905	0.16231484	3.23	0.0095
agedam	2	0.13010623	0.06505311	1.29	0.2834
line*agedam	4	0.45343434	0.11335859	2.26	0.0768
age	1	0.38127612	0.38127612	7.59	0.0083
intlwt	1	0.26970425	0.26970425	5.37	0.0248

Tests of Hypotheses Using the Type III MS for sire(line) as an Error Term

Source	DF	Type III SS	Mean Square	F Value	Pr > F
line	2	0.13620255	0.06810128	0.42	0.6752

Parameter			Estimate	Standard Error	t Value	Pr > |t|
Intercept			2.996269167 B	0.51285394	5.84	<.0001
line	1		0.071824656 B	0.14550628	0.49	0.6238
line	2		0.252468579 B	0.13716655	1.84	0.0719
line	3		0.000000000 B	.	.	.
sire(line)	1	1	0.085729012 B	0.13027803	0.66	0.5137
sire(line)	2	1	-0.121705157 B	0.13622078	-0.89	0.3761
sire(line)	3	1	0.000000000 B	.	.	.
sire(line)	4	2	-0.244601122 B	0.12669287	-1.93	0.0594
sire(line)	5	2	0.000000000 B	.	.	.
sire(line)	6	3	0.105395737 B	0.12908764	0.82	0.4183
sire(line)	7	3	-0.019520926 B	0.12037674	-0.16	0.8719
sire(line)	8	3	-0.330235387 B	0.12566795	-2.63	0.0115
sire(line)	9	3	0.000000000 B	.	.	.
agedam	3		0.370387027 B	0.11455814	3.23	0.0022
agedam	4		0.275459487 B	0.10377628	2.65	0.0107
agedam	5		0.000000000 B	.	.	.
line*agedam	1	3	-0.448936131 B	0.19581259	-2.29	0.0263
line*agedam	1	4	-0.282831924 B	0.16085047	-1.76	0.0851
line*agedam	1	5	0.000000000 B	.	.	.
line*agedam	2	3	-0.260782670 B	0.19528690	-1.34	0.1880
line*agedam	2	4	-0.350258133 B	0.17438656	-2.01	0.0502
line*agedam	2	5	0.000000000 B	.	.	.
line*agedam	3	3	0.000000000 B	.	.	.
line*agedam	3	4	0.000000000 B	.	.	.
line*agedam	3	5	0.000000000 B	.	.	.
age			-0.008530438	0.00309679	-2.75	0.0083
intlwt			0.002026334	0.00087464	2.32	0.0248

NOTE: The X'X matrix has been found to be singular, and a generalized inverse was used to solve the normal equations. Terms whose estimates are followed by the letter 'B' are not uniquely estimable.

The factor AGEDAM is treated as a discrete variable with levels (3, 4, and ≥5). The denominator for the *F*-test for testing LINE is SIRE(LINE).

To introduce the ABSORB statement, Harvey's model has been simplified. All sources of variation except the main effects of SIRE and AGEDAM have been disregarded. For the abbreviated model, the following SAS statements give the desired analysis:

```
proc glm;
    class sire agedam;
    model avdlygn=sire agedam / solution ss1 ss2 ss3;
```

The results appear in Output 11.41.

Output 11.41
Abbreviated
Least-Squares
Analysis of
Variance

```
                                The GLM Procedure

Dependent Variable: avdlygn

                                   Sum of
   Source                DF       Squares     Mean Square   F Value   Pr > F

   Model                 10    1.42537863      0.14253786      2.19   0.0324

   Error                 54    3.51399676      0.06507401

   Corrected Total       64    4.93937538

              R-Square     Coeff Var     Root MSE    avdlygn Mean

              0.288575     10.57882      0.255096     2.411385

   Source                DF     Type I SS     Mean Square   F Value   Pr > F

   sire                   8    1.30643634      0.16330454      2.51   0.0214
   agedam                 2    0.11894229      0.05947115      0.91   0.4071


   Source                DF    Type II SS     Mean Square   F Value   Pr > F

   agedam                 2    0.11894229      0.05947115      0.91   0.4071


   Source                DF   Type III SS     Mean Square   F Value   Pr > F

   agedam                 2    0.11894229      0.05947115      0.91   0.4071


                                         Standard
        Parameter          Estimate         Error    t Value   Pr > |t|

        agedam    3    0.1173825552 B    0.08911680      1.32     0.1933
        agedam    4    0.0482979994 B    0.07715379      0.63     0.5340
        agedam    5    0.0000000000 B           .          .         .

NOTE: The X'X matrix has been found to be singular, and a generalized inverse was used
to solve the normal equations.  Terms whose estimates are followed by the letter 'B'
are not uniquely estimable.
```

If the number of sires were large, then this analysis would be expensive. However, because there is little concern for the actual estimates of the effects of SIRE, considerable expense can be avoided by using the ABSORB statement:

```
proc glm;
    absorb sire;
    class agedam;
    model avdlygn=agedam / solution ss1 ss2 ss3;
```

The results appear in Output 11.42.

Output 11.42
Abbreviated
Least-Squares
Analysis of
Variance
Using the
ABSORB
Statement

The GLM Procedure

Dependent Variable: avdlygn

Source	DF	Sum of Squares	Mean Square	F Value	Pr > F
Model	10	1.42537863	0.14253786	2.19	0.0324
Error	54	3.51399676	0.06507401		
Corrected Total	64	4.93937538			

R-Square	Coeff Var	Root MSE	avdlygn Mean
0.288575	10.57882	0.255096	2.411385

Source	DF	Type I SS	Mean Square	F Value	Pr > F
sire	8	1.30643634	0.16330454	2.51	0.0214
agedam	2	0.11894229	0.05947115	0.91	0.4071

Source	DF	Type II SS	Mean Square	F Value	Pr > F
agedam	2	0.11894229	0.05947115	0.91	0.4071

Source	DF	Type III SS	Mean Square	F Value	Pr > F
agedam	2	0.11894229	0.05947115	0.91	0.4071

Parameter		Estimate	Standard Error	t Value	Pr > |t|
agedam	3	0.1173825552 B	0.08911680	1.32	0.1933
agedam	4	0.0482979994 B	0.07715379	0.63	0.5340
agedam	5	0.0000000000 B	.	.	.

NOTE: The X'X matrix has been found to be singular, and a generalized inverse was used to solve the normal equations. Terms whose estimates are followed by the letter 'B' are not uniquely estimable.

The results in Output 11.41 and Output 11.42 are the same except that the SIRE sums of squares and the SIRE parameter estimates are not printed when SIRE is absorbed. (**Note:** Type I sums of squares for absorbed effects are computed as nested effects.)

Output 11.40 and Output 11.41 include results for the following statements:

```
contrast 'young vs old' agedam .5 .5 -1;
estimate 'young vs old' agedam .5 .5 -1;
```

The output illustrates that the CONTRAST and ESTIMATE statements are legitimate with the ABSORB statement as long as the coefficients of the linear function do not involve absorbed effects—that is, parameter estimates that are not printed (in this case the SIRE parameter estimates). The following ESTIMATE statement would not be legitimate when SIRE is absorbed because it involves the SIRE parameters:

```
estimate 'oldmean' sire .111111 ... .111111 agedam 1;
```

For the same reason, the LSMEANS statement for SIRE is not legitimate with the ABSORB statement.

The ABSORB statement is now applied to the full analysis as given by Harvey (see Output 11.39). If the sums of squares for LINE and SIRE(LINE) are not required, the remaining sums of squares can be obtained with the following statements:

```
proc glm;
    absorb line sire;
    class line agedam;
    model avdlygn=agedam line*agedam age
          intlwt / solution ss3;
```

Output 11.43 contains the output, which is identical to Harvey's original results (see Output 11.40) except that neither the sums of squares nor the parameter estimates for LINE and SIRE(LINE) are computed when LINE and SIRE are absorbed.

Output 11.43
Complete Least-Squares Analysis of Variance Using the ABSORB Statement

```
                              The GLM Procedure
Dependent Variable: avdlygn

                                Sum of
Source                DF        Squares    Mean Square   F Value   Pr > F

Model                 16      2.52745871     0.15796617     3.14    0.0011

Error                 48      2.41191667     0.05024826

Corrected Total       64      4.93937538

              R-Square     Coeff Var     Root MSE     avdlygn Mean

              0.511696     9.295956      0.224161       2.411385

Source                DF     Type III SS   Mean Square   F Value   Pr > F

agedam                 2      0.13010623     0.06505311     1.29    0.2834
line*agedam            4      0.45343434     0.11335859     2.26    0.0768
age                    1      0.38127612     0.38127612     7.59    0.0083
intlwt                 1      0.26970425     0.26970425     5.37    0.0248

Contrast              DF     Contrast SS   Mean Square   F Value   Pr > F

young vs old           1      0.11895160     0.11895160     2.37    0.1305

                                            Standard
Parameter                    Estimate          Error    t Value   Pr > |t|

young vs old               0.09912178     0.06442352       1.54     0.1305

                                            Standard
Parameter                    Estimate          Error    t Value   Pr > |t|

agedam        3          0.3703870271 B    0.11455814       3.23     0.0022
agedam        4          0.2754594872 B    0.10377628       2.65     0.0107
agedam        5          0.0000000000 B         .            .         .
line*agedam 1 3         -.4489361310 B     0.19581259      -2.29     0.0263
line*agedam 1 4         -.2828319237 B     0.16085047      -1.76     0.0851
line*agedam 1 5          0.0000000000 B         .            .         .
line*agedam 2 3         -.2607826701 B     0.19528690      -1.34     0.1880
line*agedam 2 4         -.3502581329 B     0.17438656      -2.01     0.0502
line*agedam 2 5          0.0000000000 B         .            .         .
line*agedam 3 3          0.0000000000 B         .            .         .
line*agedam 3 4          0.0000000000 B         .            .         .
line*agedam 3 5          0.0000000000 B         .            .         .
age                     -.0085304380       0.00309679      -2.75     0.0083
intlwt                   0.0020263340      0.00087464       2.32     0.0248
```

We conclude this section by running a mixed-model analysis using PROC MIXED in which we consider sires to be random. This analysis might be considered preferable to the analysis of variance using PROC GLM because it truly treats sires as random. In some more complicated situations with very large data sets, however, PROC MIXED might overwhelm the computer.

The appropriate statements are

```
proc mixed data=sires;
    class line sire agedam;
    model avdlygn=line agedam line*agedam age intlwt/
    ddfm=satterthwaite;
    random sire(line);
    contrast 'young vs old' agedam .5 .5 -1;
    estimate 'young vs old' agedam .5 .5 -1;
run;
```

Edited results are shown in Output 11.44.

Output 11.44
A Mixed-
Model
Analysis

```
                        Thc Mixed Procedure

                        Model Information

        Data Set                    WORK.SIRES
        Dependent Variable          avdlygn
        Covariance Structure        Variance Components
        Estimation Method           REML
        Residual Variance Method    Profile
        Fixed Effects SE Method     Model-Based
        Degrees of Freedom Method   Satterthwaite

                    Class Level Information

            Class       Levels      Values

            line           3        1 2 3
            sire           9        1 2 3 4 5 6 7 8 9
            agedam         3        3 4 5

                    Covariance Parameter
                         Estimates

                Cov Parm        Estimate

                sire(line)       0.01792
                Residual         0.05028


                Type 3 Tests of Fixed Effects

                        Num     Den
            Effect       DF      DF      F Value     Pr > F

            line          2      7.2       0.43      0.6687
            agedam        2       50       1.21      0.3068
            line*agedam   4      49.5      2.00      0.1095
            age           1      53.8      7.55      0.0082
            intlwt        1      51.6      5.92      0.0185
```

```
                                        Estimates

                                 Standard
      Label            Estimate     Error     DF    t Value    Pr > |t|

      young vs old     0.09581    0.06348    50.6     1.51      0.1374


                                        Contrasts

                               Num    Den
            Label               DF     DF    F Value    Pr > F

            young vs old         1    50.6     2.28      0.1374
```

As you have seen with other examples, the mixed-model analysis provides estimates and tests that use appropriate error terms, at least in principle. The "experimental unit" for LINE is SIRE(LINE), and the table for "Type 3 Tests of Fixed Effects" reflects this, with two numerator DF and 7.2 denominator DF. The other tests have approximately 50 denominator DF, which is essentially the same as the 48 DF for residual error in Output 11.40. Steer in the "experimental unit" for the AGEDAM, LINE*AGEDAM, AGE, and INTLWT. Significance probabilities in Outputs 11.40 and 11.44 agree, for practical purposes, for these effects, but it is worth noting that the ANOVA tests in Output 11.40 are exact F-tests, whereas the mixed-model tests in Output 11.44 are approximate due to estimating the covariance parameters. This illustrates a basic lesson. If you are interested only in the effects AGEDAM, LINE*AGEDAM, AGE, and INTLWT, then there is really no benefit in using mixed-model methodology.

References

Agresti, A. 1990. *Categorical Data Analysis.* New York: Wiley.

Agresti, A. 1996. *An Introduction to Categorical Data Analysis.* New York: Wiley.

Akaike, H. 1974. "A New Look at the Statistical Model Identification." *IEEE Transaction on Automatic Control* AIC-19, 716-723.

Allison, P. D. 1995. *Survival Analysis Using the SAS System: A Practical Guide.* Cary, NC: SAS Institute Inc.

Allison, P. D. 1999. *Logistic Regression Using the SAS System: Theory and Application.* Cary, NC: SAS Institute Inc.

Bancroft, T. A. 1968. *Topics in Intermediate Statistical Methods.* Ames: Iowa State University Press.

Beitler, P. J., and J. R. Landis. 1985. "A Mixed-Effects Model for Categorical Data." *Biometrics* 41: 991-1000.

Belsley, D. A., E. Kuh, and R. E. Welsch. 1980. *Regression Diagnostics.* New York: Wiley.

Box, G. E. P. 1954. "Some Theorems on Quadratic Forms Applied in the Study of Analysis of Variance Problems, II: Effects of Inequality of Variance and of Correlation between Errors in the Two-Way Classification." *Annals of Mathematical Statistics* 25: 484-498.

Breslow, N. E., and D. G. Clayton. 1993. "Approximate Inference in Generalized Linear Mixed Models." *Journal of the American Statistical Association* 88: 9-25.

Burnham, K. P., and D. R. Anderson. 1998. *Model Selection and Inference: A Practical Information-Theoretic Approach.* New York: Springer-Verlag.

Cantor, A. 1997. *Extending SAS Survival Analysis Techniques for Medical Research.* Cary, NC: SAS Institute Inc.

Chew, V. 1976. "Uses and Abuses of Duncan's Multiple Range Test." *Proceedings of Florida State Horticultural Society* 89: 251-253.

Cochran, W. G., and M. G. Cox. 1957. *Experimental Designs.* New York: Wiley.

Dalal, S. R., E. B. Fowlkes, and B. Hoadley. 1989. "Risk Analysis of the Space Shuttle: Pre-*Challenger* Prediction of Failure." *Journal of the American Statistical Association* 84: 945-957.

Diggle, P. J. 1988. "An Approach to the Analysis of Repeated Measures." *Biometrics* 44: 959-971.

Diggle, P. J., K.-Y. Liang, and S. L. Zeger. 1994. *Analysis of Longitudinal Data.* Oxford: Clarendon Press.

Dobson, A. J. 1990. *An Introduction to Generalized Linear Models.* London: Chapman and Hall.

Duncan, D. B. 1955. "Multiple Range and Multiple F-Tests." *Biometrics* 11: 1-42.

Eberhart, S. A., and W. A. Russell. 1966. "Stability Parameters for Comparing Varieties." *Crop Science* 6: 36-40.

Freeman, D. H., Jr. 1987. *Applied Categorical Data Analysis.* New York: Dekker.

Freund, R. J. 1980. "The Case of the Missing Cell." *The American Statistician* 34: 94-98.

Freund, R. J., and R. C. Littell. 2000. *SAS System for Regression.* 3d ed. Cary, NC: SAS Institute Inc.

Freund, R. J., and P. D. Minton. 1979. *Regression Methods.* New York: Dekker.

Graybill, F. A. 1976. *Theory and Application of the Linear Model.* Belmont, CA: Wadsworth.

Greenhouse, S. W., and S. Geisser. 1959. "On Methods in the Analysis of Profile Data." *Psychometrika* 32 (3): 95-112.

Guerin, L., and W. W. Stroup. 2000. "A Simulation Study to Evaluate PROC MIXED Analysis of Repeated Measures Data." *Proceedings of the 12th Annual Conference on Applied Statistics in Agriculture.* Manhattan, KS: Kansas State University.

Hartley, H. O., and S. R. Searle. 1969. "A Discontinuity in Mixed Model Analysis." *Biometrics* 25: 573-576.

Harvey, W. R. 1975. "Least Squares Analysis of Data." Washington, DC: U.S. Department of Agriculture. ARS-H-4, February 1975.

Heck, D. L. 1960. "Charts of Some Upper Percentage Points of the Distribution of the Largest Characteristic Root." *Annals of Mathematical Statistics* XXXI: 625-642.

Henderson, C. R. 1963. *Selection Index and Expected Genetic Advance in Statistical Genetics and Plant Breeding.* Edited by W. D. Hanson and H. F. Robinson, 141-163. Washington, DC: National Academy of Sciences and National Research Council Publication No. 982.

Henderson, C. R. 1975. "Best Linear Unbiased Estimation and Prediction under a Selection Model." *Biometrics* 31: 423-447.

Hicks, C. R., and K. V. Turner, Jr. 2000. *Fundamental Concepts in the Design of Experiments.* 5th ed. New York: Oxford University Press.

Hocking, R. R. 1973. "A Discussion of the Two-Way Mixed Model." *The American Statistician* 27: 148-152.

Hocking, R. R. 1985. *The Analysis of Linear Models.* Monterey, CA: Brooks/Cole.

Hocking, R. R., and F. M. Speed. 1975. "A Full-Rank Analysis of Some Linear Model Problems." *Journal of the American Statistical Association* 70: 706-712.

Hocking, R. R., and F. M. Speed. 1980. "The Cell Means Model for the Analysis of Variance." Twenty-Fourth Annual Technical Conference, Cincinnati, OH, 23-24.

Huynh, H., and L. S. Feldt. 1970. "Conditions under Which Mean Square Ratios in Repeated Measurements Designs Have Exact F-Distributions." *Journal of the American Statistical Association* 65: 1582-1589.

Huynh, H., and L. S. Feldt. 1976. "Estimation of the Box Correction for Degrees of Freedom from Sample Data in the Randomized Block and Split Plot Designs." *Journal of Educational Statistics* 1: 69-82.

Kacker, R. N., and D. A. Harville. 1984. "Approximations for Standard Errors of Estimators of Fixed and Random Effects in Mixed Linear Models." *Journal of the American Statistical Association* 79: 853-862.

Kenward, M. G., and J. H. Roger. 1997. "Small Sample Inference for Fixed Effects from Restricted Maximum Likelihood." *Biometrics* 53: 983-997.

Keselman, H. J., J. Algina, R. K. Kowalchuk, and R. D. Wolfinger. 1998. "A Comparison of Two Approaches for Selecting Covariance Structures in the Analysis of Repeated Measurements." *Communications in Statistics: Simulation and Computation* 27 (3): 591-604.

Keuhl, Robert O. 2000. *Design of Experiments: Statistical Principles of Research Design and Analysis.* Pacific Grove, CA: Duxbury.

Koch, G. C., D. H. Freeman, and J. L. Freeman. 1975. "Strategies in the Multivariate Analysis of Data from Complex Surveys." *International Statistics Review* 43: 59-78.

Lentner, M., J. C. Arnold, and K. Hinkleman. 1989. "The Efficiency of Blocking: How to Use MS(Blocks)/MS(Error) Correctly." *The American Statistician* 43: 106-108.

Leppik, I. E., et al. 1985. "A Double-Blind Crossover Evaluation of Progabide in Partial Seizures." *Neurology* 35: 285.

Liang, K.-Y., and S. L. Zeger. 1986. "Longitudinal Data Analysis for Discrete and Continuous Outcomes Using Generalized Linear Models." *Biometrika* 84: 13-22.

Lindsey, J. K. 1997. *Applying Generalized Linear Models.* New York: Springer-Verlag.

Littell, R. C. 1996. "Analysis of Unbalanced Mixed Model Data: Traditional ANOVA versus Contemporary Methods." *Proceedings of Kansas State University Conference on Applied Statistics in Agriculture.*

Littell, R. C., P. R. Henry, and C. B. Ammerman. 1998. "Statistical Analysis of Repeated Measures Data Using SAS Procedures." *Journal of Animal Science* 76: 1216-1231.

Littell, R. C., P. R. Henry, A. J. Lewis, and C. B. Ammerman. 1997. "Estimation of Relative Bioavailability of Nutrients Using SAS Procedures." *Journal of Animal Science* 75: 2672-2683.

Littell, R. C., and S. B. Linda. 1990. "Computation of Variances of Functions of Parameter Estimates for Mixed Models in GLM." *Proceedings of SAS Users Group, International Conference.*

Littell, R. C., G. A. Milliken, W. W. Stroup, and R. D. Wolfinger. 1996. *SAS System for Mixed Models.* Cary, NC: SAS Institute Inc.

Littell, R. C., J. Pendergast, and R. Natajan. 2000. "Modelling Covariance Structure in the Analysis of Repeated Measures Data." *Statistics in Medicine* 19: 1793-1819.

Little, T. U. 1978. "If Galileo Published in HortiScience." *HortiScience* 13: 504-506.

McCullagh, P. 1983. "Quasi-likelihood Functions." *Annals of Statistics* 11: 59-67.

McCullagh, P., and J. A. Nelder. 1989. *Generalized Linear Models.* 2d ed. London: Chapman and Hall.

McLean, R. A., W. L. Sanders, and W. W. Stroup. 1991. "A Unified Approach to Mixed Linear Models." *The American Statistician* 45: 54-64.

Milliken, G.A. and D. E. Johnson. 1984. *Analysis of Messy Data,* Volume 1: Designed Experiments. New York: Van Nostrand Reinhold.

Milliken, G.A. and D. E. Johnson. 1989. *Analysis of Messy Data,* Volume 2: Nonreplicated Experiments. New York: Van Nostrand Reinhold.

Morrison, D. F. 1976. *Multivariate Statistical Methods.* 2d ed. New York: McGraw-Hill.

Nelder, J. A., and R. W. M. Wedderburn. 1972. "Generalized Linear Models." *Journal of the Royal Statistical Society,* Series A, 135, 370-384.

Petersen, R. G. 1977. "Use and Misuse of Multiple Comparison Procedures." *Agronomy Journal* 69: 205-208.

Pillai, K. C. S. 1960. *Statistical Tables for Tests of Multivariate Hypotheses.* Manila: The Statistical Center, University of the Philippines.

Rao, C. R. 1965. *Linear Statistical Inference and Its Applications.* New York: Wiley.

Samuels, M. L., G. Casella, and G. P. McCabe. 1991. "Interpreting Blocks and Random Factors." *Journal of the American Statistical Association* 86: 798-808.

SAS Institute Inc. 1978. SAS Technical Report R-101: *Tests of Hypotheses in Fixed Effects Linear Models.* Cary, NC: SAS Institute Inc.

SAS Institute Inc. 1990. *SAS/GRAPH Software: Reference, Version 6.* 1st ed. 2 vols. Cary, NC: SAS Institute Inc.

SAS Institute Inc. 1990. *SAS/IML Software: Usage and Reference, Version 6.* 1st ed. Cary, NC: SAS Institute Inc.

SAS Institute Inc. 1990. *SAS Language: Reference, Version 6.* 1st ed. Cary, NC: SAS Institute Inc.

SAS Institute Inc. 1990. *SAS Procedures Guide, Version 6.* 3d ed. Cary, NC: SAS Institute Inc.

SAS Institute Inc. 1990. *SAS/STAT User's Guide, Version 6.* 4th ed. 2 vols. Cary, NC: SAS Institute Inc.

Satterthwaite, F. W. 1946. "An Approximate Distribution of Estimates of Variance Components." *Biometrics Bulletin* 2: 110-114.

Schwarz, G. 1978. "Estimating the Dimension of a Model." *Annals of Statistics* 6: 461-464.

Searle, S. R. 1971. *Linear Models.* New York: Wiley.

Searle, S. R. 1987. *Linear Models for Unbalanced Data.* New York: Wiley.

Snee, R. D. 1973. "Some Aspects of Nonorthogonal Data Analysis." *Journal of Quality Technology* 5: 67-69.

Speed, F. M., R. R. Hocking, and O. P. Hackney. 1978. "Methods of Analysis of Linear Models with Unbalanced Data." *Journal of the American Statistical Association* 73: 105-112.

Steel, R. G. B., and J. H. Torrie. 1980. *Principles and Procedures of Statistics.* 2d ed. New York: McGraw-Hill.

Stokes, M. E., C. S. Davis, and G. G. Koch. 2000. *Categorical Data Analysis Using the SAS System.* 2d ed. Cary, NC: SAS Institute Inc.

Thall, P. F., and S. C. Vail. 1990. "Some Covariance Models for Longitudinal Count Data with Overdispersion." *Biometrics* 46: 657-671.

Waller, R. A., and D. B. Duncan. 1969. "A Bayes Rule for the Symmetric Multiple Comparison Problem." *Journal of the American Statistical Association* 64: 1484-1499.

Wedderburn, R. W. M. 1974. "Quasi-likelihood Functions, Generalized Linear Models, and the Gauss-Newton Method." *Biometrika* 1: 439-447.

Wolfinger, R. D. 1993. "Covariance Structure Selection in General Mixed Models." *Communications in Statistics: Simulation and Computation* 22: 1079-1106.

Young, L. J., N. L. Campbell, and G. A. Capuano. 1999. "Analysis of Overdispersed Count Data from Single-Factor Experiments." *Journal of Agricultural, Biological, and Environmental Statistics* 4: 258-275.

Young, L. J., and G. H. Young. 1998. *Statistical Ecology: A Population Perspective*. Boston: Kluwer Academic.

Zeger, S. L., and K.-Y. Liang. 1986. "Longitudinal Data Analysis for Discrete and Continuous Outcomes." *Biometrics* 42: 121-130.

Zeger, S. L., K.-Y. Liang, and P. S. Albert. 1988. "Models for Longitudinal Data: A Generalized Estimating Equation Approach." *Biometrics* 44: 1049-1060.

Zelterman, D. Book on selected applications for categorical data to be published by Books By Users Press in August 2002.

Index

Books Available from SAS Press

Advanced Log-Linear Models Using SAS®
by **Daniel Zelterman**

Analysis of Clinical Trials Using SAS®: A Practical Guide
by **Alex Dmitrienko, Geert Molenberghs, Walter Offen,** *and*
Christy Chuang-Stein,

Annotate: Simply the Basics
by **Art Carpenter**

*Applied Multivariate Statistics with SAS® Software,
Second Edition*
by **Ravindra Khattree**
and **Dayanand N. Naik**

*Applied Statistics and the SAS® Programming Language,
Fourth Edition*
by **Ronald P. Cody**
and **Jeffrey K. Smith**

An Array of Challenges — Test Your SAS® Skills
by **Robert Virgile**

*Carpenter's Complete Guide to the SAS® Macro Language,
Second Edition*
by **Art Carpenter**

The Cartoon Guide to Statistics
by **Larry Gonick**
and **Woollcott Smith**

*Categorical Data Analysis Using the SAS® System,
Second Edition*
by **Maura E. Stokes, Charles S. Davis,**
and **Gary G. Koch**

Cody's Data Cleaning Techniques Using SAS® Software
by **Ron Cody**

*Common Statistical Methods for Clinical Research with
SAS® Examples, Second Edition*
by **Glenn A. Walker**

*Debugging SAS® Programs: A Handbook of Tools and
Techniques*
by **Michele M. Burlew**

*Efficiency: Improving the Performance of Your SAS®
Applications*
by **Robert Virgile**

The Essential PROC SQL Handbook for SAS® Users
by **Katherine Prairie**

*Fixed Effects Regression Methods for Longitudinal Data
Using SAS®*
by **Paul D. Allison**

Genetic Analysis of Complex Traits Using SAS®
Edited by **Arnold M. Saxton**

A Handbook of Statistical Analyses Using SAS®, Second Edition
by **B.S. Everitt**
and **G. Der**

Health Care Data and the SAS® System
by **Marge Scerbo, Craig Dickstein,**
and **Alan Wilson**

The How-To Book for SAS/GRAPH® Software
by **Thomas Miron**

*In the Know ... SAS® Tips and Techniques From
Around the Globe*
by **Phil Mason**

Instant ODS: Style Templates for the Output Delivery System
by **Bernadette Johnson**

*Integrating Results through Meta-Analytic Review Using
SAS® Software*
by **Morgan C. Wang**
and **Brad J. Bushman**

Learning SAS® in the Computer Lab, Second Edition
by **Rebecca J. Elliott**

The Little SAS® Book: A Primer
by **Lora D. Delwiche**
and **Susan J. Slaughter**

The Little SAS® Book: A Primer, Second Edition
by **Lora D. Delwiche**
and **Susan J. Slaughter**
(updated to include Version 7 features)

The Little SAS® Book: A Primer, Third Edition
by **Lora D. Delwiche**
and **Susan J. Slaughter**
(updated to include SAS 9.1 features)

*Logistic Regression Using the SAS® System:
Theory and Application*
by **Paul D. Allison**

Longitudinal Data and SAS®: A Programmer's Guide
by **Ron Cody**

Maps Made Easy Using SAS®
by **Mike Zdeb**

Models for Discrete Data
by **Daniel Zelterman**

Multiple Comparisons and Multiple Tests Using SAS®
Text and Workbook Set
(books in this set also sold separately)
by **Peter H. Westfall, Randall D. Tobias,**
Dror Rom, Russell D. Wolfinger,
and **Yosef Hochberg**

Multiple-Plot Displays: Simplified with Macros
by **Perry Watts**

Multivariate Data Reduction and Discrimination with
SAS® Software
by **Ravindra Khattree**
and **Dayanand N. Naik**

Output Delivery System: The Basics
by **Lauren E. Haworth**

Painless Windows: A Handbook for SAS® Users, Third Edition
by **Jodie Gilmore**
(updated to include Version 8 and SAS 9.1 features)

PROC TABULATE by Example
by **Lauren E. Haworth**

The Power of PROC FORMAT
by **Jonas V. Bilenas**

Professional SAS® Programming Shortcuts
by **Rick Aster**

Quick Results with SAS/GRAPH® Software
by **Arthur L. Carpenter**
and **Charles E. Shipp**

Quick Results with the Output Delivery System
by **Sunil K. Gupta**

Quick Start to Data Analysis with SAS®
by **Frank C. Dilorio**
and **Kenneth A. Hardy**

Reading External Data Files Using SAS®: Examples Handbook
by **Michele M. Burlew**

Regression and ANOVA: An Integrated Approach Using
SAS® Software
by **Keith E. Muller**
and **Bethel A. Fetterman**

SAS®Applications Programming: A Gentle Introduction
by **Frank C. Dilorio**

SAS® for Forecasting Time Series, Second Edition
by **John C. Brocklebank**
and **David A. Dickey**

SAS® for Linear Models, Fourth Edition
by **Ramon C. Littell, Walter W. Stroup,**
and **Rudolf J. Freund**

SAS® for Monte Carlo Studies: A Guide for Quantitative
Researchers
by **Xitao Fan, Ákos Felsovályi, Stephen A. Sivo,**
and **Sean C. Keenan**

SAS® Functions by Example
by **Ron Cody**

SAS® Macro Programming Made Easy
by **Michele M. Burlew**

SAS® Programming by Example
by **Ron Cody**
and **Ray Pass**

SAS® Programming for Researchers and Social Scientists,
Second Edition
by **Paul E. Spector**

SAS® Survival Analysis Techniques for Medical Research,
Second Edition
by **Alan B. Cantor**

SAS® System for Elementary Statistical Analysis,
Second Edition
by **Sandra D. Schlotzhauer**
and **Ramon C. Littell**

SAS® System for Mixed Models
by **Ramon C. Littell, George A. Milliken, Walter W. Stroup,**
and **Russell D. Wolfinger**

SAS® System for Regression, Third Edition
by **Rudolf J. Freund**
and **Ramon C. Littell**

SAS® System for Statistical Graphics, First Edition
by **Michael Friendly**

The SAS® Workbook and Solutions Set
(books in this set also sold separately)
by **Ron Cody**

Selecting Statistical Techniques for Social Science Data:
A Guide for SAS® Users
by **Frank M. Andrews, Laura Klem, Patrick M. O'Malley,**
Willard L. Rodgers, Kathleen B. Welch,
and **Terrence N. Davidson**

Statistical Quality Control Using the SAS® System
by **Dennis W. King**

A Step-by-Step Approach to Using the SAS® System
for Factor Analysis and Structural Equation Modeling
by **Larry Hatcher**

A Step-by-Step Approach to Using the SAS® System
for Univariate and Multivariate Statistics, Second Edition
by **Norm O'Rourke, Larry Hatcher,**
and **Edward J. Stepanski**

Step-by-Step Basic Statistics Using SAS®: Student Guide
and Exercises
(books in this set also sold separately)
by **Larry Hatcher**

Survival Analysis Using the SAS® System:
A Practical Guide
by **Paul D. Allison**

Tuning SAS® Applications in the OS/390 and z/OS Environments, Second Edition
by **Michael A. Raithel**

Univariate and Multivariate General Linear Models: Theory and Applications Using SAS® Software
by **Neil H. Timm**
and **Tammy A. Mieczkowski**

Using SAS® in Financial Research
by **Ekkehart Boehmer, John Paul Broussard,**
and **Juha-Pekka Kallunki**

Using the SAS® Windowing Environment: A Quick Tutorial
by **Larry Hatcher**

Visualizing Categorical Data
by **Michael Friendly**

Web Development with SAS® by Example
by **Frederick Pratter**

Your Guide to Survey Research Using the SAS® System
by **Archer Gravely**

JMP® Books

JMP® for Basic Univariate and Multivariate Statistics: A Step-by-Step Guide
by **Ann Lehman, Norm O'Rourke, Larry Hatcher,**
and **Edward J. Stepanski**

JMP® Start Statistics, Third Edition
by **John Sall, Ann Lehman,**
and **Lee Creighton**

Regression Using JMP®
by **Rudolf J. Freund, Ramon C. Littell,**
and **Lee Creighton**

WILEY SERIES IN PROBABILITY AND STATISTICS
ESTABLISHED BY WALTER A. SHEWHART AND SAMUEL S. WILKS

Editors: *David J. Balding, Peter Bloomfield, Noel A. C. Cressie, Nicholas I. Fisher, Iain M. Johnstone, J. B. Kadane, Louise M. Ryan, David W. Scott, Adrian F. M. Smith, Jozef L. Teugels*
Editors Emeriti: *Vic Barnett, J. Stuart Hunter, David G. Kendall*

The *Wiley Series in Probability and Statistics* is well established and authoritative. It covers many topics of current research interest in both pure and applied statistics and probability theory. Written by leading statisticians and institutions, the titles span both state-of-the-art developments in the field and classical methods.

Reflecting the wide range of current research in statistics, the series encompasses applied, methodological and theoretical statistics, ranging from applications and new techniques made possible by advances in computerized practice to rigorous treatment of theoretical approaches.

This series provides essential and invaluable reading for all statisticians, whether in academia, industry, government, or research.

ABRAHAM and LEDOLTER · Statistical Methods for Forecasting
AGRESTI · Analysis of Ordinal Categorical Data
AGRESTI · An Introduction to Categorical Data Analysis
AGRESTI · Categorical Data Analysis, *Second Edition*
ANDĚL · Mathematics of Chance
ANDERSON · An Introduction to Multivariate Statistical Analysis, *Second Edition*
*ANDERSON · The Statistical Analysis of Time Series
ANDERSON, AUQUIER, HAUCK, OAKES, VANDAELE, and WEISBERG · Statistical Methods for Comparative Studies
ANDERSON and LOYNES · The Teaching of Practical Statistics
ARMITAGE and DAVID (editors) · Advances in Biometry
ARNOLD, BALAKRISHNAN, and NAGARAJA · Records
*ARTHANARI and DODGE · Mathematical Programming in Statistics
*BAILEY · The Elements of Stochastic Processes with Applications to the Natural Sciences
BALAKRISHNAN and KOUTRAS · Runs and Scans with Applications
BARNETT · Comparative Statistical Inference, *Third Edition*
BARNETT and LEWIS · Outliers in Statistical Data, *Third Edition*
BARTOSZYNSKI and NIEWIADOMSKA-BUGAJ · Probability and Statistical Inference
BASILEVSKY · Statistical Factor Analysis and Related Methods: Theory and Applications
BASU and RIGDON · Statistical Methods for the Reliability of Repairable Systems
BATES and WATTS · Nonlinear Regression Analysis and Its Applications
BECHHOFER, SANTNER, and GOLDSMAN · Design and Analysis of Experiments for Statistical Selection, Screening, and Multiple Comparisons
BELSLEY · Conditioning Diagnostics: Collinearity and Weak Data in Regression
BELSLEY, KUH, and WELSCH · Regression Diagnostics: Identifying Influential Data and Sources of Collinearity
BENDAT and PIERSOL · Random Data: Analysis and Measurement Procedures, *Third Edition*
BERRY, CHALONER, and GEWEKE · Bayesian Analysis in Statistics and Econometrics: Essays in Honor of Arnold Zellner
BERNARDO and SMITH · Bayesian Theory
BHAT and MILLER · Elements of Applied Stochastic Processes, *Third Edition*
BHATTACHARYA and JOHNSON · Statistical Concepts and Methods
BHATTACHARYA and WAYMIRE · Stochastic Processes with Applications
BILLINGSLEY · Convergence of Probability Measures, *Second Edition*
BILLINGSLEY · Probability and Measure, *Third Edition*
BIRKES and DODGE · Alternative Methods of Regression
BLISCHKE AND MURTHY (editors) · Case Studies in Reliability and Maintenance
BLISCHKE AND MURTHY · Reliability: Modeling, Prediction, and Optimization
BLOOMFIELD · Fourier Analysis of Time Series: An Introduction, *Second Edition*
BOLLEN · Structural Equations with Latent Variables
BOROVKOV · Ergodicity and Stability of Stochastic Processes
BOULEAU · Numerical Methods for Stochastic Processes
BOX · Bayesian Inference in Statistical Analysis
BOX · R. A. Fisher, the Life of a Scientist
BOX and DRAPER · Empirical Model-Building and Response Surfaces
*BOX and DRAPER · Evolutionary Operation: A Statistical Method for Process Improvement
BOX, HUNTER, and HUNTER · Statistics for Experimenters: An Introduction to Design, Data Analysis, and Model Building
BOX and LUCEÑO · Statistical Control by Monitoring and Feedback Adjustment
BRANDIMARTE · Numerical Methods in Finance: A MATLAB-Based Introduction
BROWN and HOLLANDER · Statistics: A Biomedical Introduction

*Now available in a lower priced paperback edition in the Wiley Classics Library.

BRUNNER, DOMHOF, and LANGER · Nonparametric Analysis of Longitudinal Data in Factorial Experiments
BUCKLEW · Large Deviation Techniques in Decision, Simulation, and Estimation
CAIROLI and DALANG · Sequential Stochastic Optimization
CHAN · Time Series: Applications to Finance
CHATTERJEE and HADI · Sensitivity Analysis in Linear Regression
CHATTERJEE and PRICE · Regression Analysis by Example, *Third Edition*
CHERNICK · Bootstrap Methods: A Practitioner's Guide
CHERNICK and FRIIS · Introductory Biostatistics for the Health Sciences
CHILÈS and DELFINER · Geostatistics: Modeling Spatial Uncertainty
CHOW and LIU · Design and Analysis of Clinical Trials: Concepts and Methodologies
CLARKE and DISNEY · Probability and Random Processes: A First Course with Applications, *Second Edition*
*COCHRAN and COX · Experimental Designs, *Second Edition*
CONGDON · Bayesian Statistical Modelling
CONOVER · Practical Nonparametric Statistics, *Second Edition*
COOK · Regression Graphics
COOK and WEISBERG · Applied Regression Including Computing and Graphics
COOK and WEISBERG · An Introduction to Regression Graphics
CORNELL · Experiments with Mixtures, Designs, Models, and the Analysis of Mixture Data, *Third Edition*
COVER and THOMAS · Elements of Information Theory
COX · A Handbook of Introductory Statistical Methods
*COX · Planning of Experiments
CRESSIE · Statistics for Spatial Data, *Revised Edition*
CSÖRGŐ and HORVÁTH · Limit Theorems in Change Point Analysis
DANIEL · Applications of Statistics to Industrial Experimentation
DANIEL · Biostatistics: A Foundation for Analysis in the Health Sciences, *Sixth Edition*
*DANIEL · Fitting Equations to Data: Computer Analysis of Multifactor Data, *Second Edition*
DASU and JOHNSON · Exploratory Data Mining and Data Cleaning
DAVID · Order Statistics, *Second Edition*
*DEGROOT, FIENBERG, and KADANE · Statistics and the Law
DEL CASTILLO · Statistical Process Adjustment for Quality Control
DETTE and STUDDEN · The Theory of Canonical Moments with Applications in Statistics, Probability, and Analysis
DEY and MUKERJEE · Fractional Factorial Plans
DILLON and GOLDSTEIN · Multivariate Analysis: Methods and Applications
DODGE · Alternative Methods of Regression
*DODGE and ROMIG · Sampling Inspection Tables, *Second Edition*
*DOOB · Stochastic Processes
DOWDY and WEARDEN · Statistics for Research, *Second Edition*
DRAPER and SMITH · Applied Regression Analysis, *Third Edition*
DRYDEN and MARDIA · Statistical Shape Analysis
DUDEWICZ and MISHRA · Modern Mathematical Statistics
DUNN and CLARK · Applied Statistics: Analysis of Variance and Regression, *Second Edition*
DUNN and CLARK · Basic Statistics: A Primer for the Biomedical Sciences, *Third Edition*
DUPUIS and ELLIS · A Weak Convergence Approach to the Theory of Large Deviations
*ELANDT-JOHNSON and JOHNSON · Survival Models and Data Analysis
ENDERS · Applied Econometric Time Series
ETHIER and KURTZ · Markov Processes: Characterization and Convergence
EVANS, HASTINGS, and PEACOCK · Statistical Distributions, *Third Edition*
FELLER · An Introduction to Probability Theory and Its Applications, Volume I, *Third Edition,* Revised; Volume II, *Second Edition*
FISHER and VAN BELLE · Biostatistics: A Methodology for the Health Sciences
FLEISS · The Design and Analysis of Clinical Experiments
FLEISS · Statistical Methods for Rates and Proportions, *Second Edition*
FLEMING and HARRINGTON · Counting Processes and Survival Analysis
FULLER · Introduction to Statistical Time Series, *Second Edition*
FULLER · Measurement Error Models
GALLANT · Nonlinear Statistical Models
GHOSH, MUKHOPADHYAY, and SEN · Sequential Estimation
GIFI · Nonlinear Multivariate Analysis
GLASSERMAN and YAO · Monotone Structure in Discrete-Event Systems
GNANADESIKAN · Methods for Statistical Data Analysis of Multivariate Observations, *Second Edition*
GOLDSTEIN and LEWIS · Assessment: Problems, Development, and Statistical Issues
GREENWOOD and NIKULIN · A Guide to Chi-Squared Testing
GROSS and HARRIS · Fundamentals of Queueing Theory, *Third Edition*
HAHN and SHAPIRO · Statistical Models in Engineering
HAHN and MEEKER · Statistical Intervals: A Guide for Practitioners
HALD · A History of Probability and Statistics and their Applications Before 1750

*Now available in a lower priced paperback edition in the Wiley Classics Library.

*Now available in a lower priced paperback edition in the Wiley Classics Library.

LINDVALL · Lectures on the Coupling Method

LINHART and ZUCCHINI · Model Selection

LITTLE and RUBIN · Statistical Analysis with Missing Data, *Second Edition*

LLOYD · The Statistical Analysis of Categorical Data

MAGNUS and NEUDECKER · Matrix Differential Calculus with Applications in Statistics and Econometrics, *Revised Edition*

MALLER and ZHOU · Survival Analysis with Long Term Survivors

MALLOWS · Design, Data, and Analysis by Some Friends of Cuthbert Daniel

MANN, SCHAFER, and SINGPURWALLA · Methods for Statistical Analysis of Reliability and Life Data

MANTON, WOODBURY, and TOLLEY · Statistical Applications Using Fuzzy Sets

MARDIA and JUPP · Directional Statistics

MASON, GUNST, and HESS · Statistical Design and Analysis of Experiments with Applications to Engineering and Science,
     Second Edition

McCULLOCH and SEARLE · Generalized, Linear, and Mixed Models

McFADDEN · Management of Data in Clinical Trials

McLACHLAN · Discriminant Analysis and Statistical Pattern Recognition

McLACHLAN and KRISHNAN · The EM Algorithm and Extensions

McLACHLAN and PEEL · Finite Mixture Models

McNEIL · Epidemiological Research Methods

MEEKER and ESCOBAR · Statistical Methods for Reliability Data

MEERSCHAERT and SCHEFFLER · Limit Distributions for Sums of Independent Random Vectors: Heavy Tails in Theory and Practice

*MILLER · Survival Analysis, *Second Edition*

MONTGOMERY, PECK, and VINING · Introduction to Linear Regression Analysis, *Third Edition*

MORGENTHALER and TUKEY · Configural Polysampling: A Route to Practical Robustness

MUIRHEAD · Aspects of Multivariate Statistical Theory

MURRAY · X-STAT 2.0 Statistical Experimentation, Design Data Analysis, and Nonlinear Optimization

MYERS and MONTGOMERY · Response Surface Methodology: Process and Product Optimization Using Designed Experiments,
     Second Edition

MYERS, MONTGOMERY, and VINING · Generalized Linear Models. With Applications in Engineering and the Sciences

NELSON · Accelerated Testing, Statistical Models, Test Plans, and Data Analyses

NELSON · Applied Life Data Analysis

NEWMAN · Biostatistical Methods in Epidemiology

OCHI · Applied Probability and Stochastic Processes in Engineering and Physical Sciences

OKABE, BOOTS, SUGIHARA, and CHIU · Spatial Tesselations: Concepts and Applications of Voronoi Diagrams, *Second Edition*

OLIVER and SMITH · Influence Diagrams, Belief Nets and Decision Analysis

PANKRATZ · Forecasting with Dynamic Regression Models

PANKRATZ · Forecasting with Univariate Box-Jenkins Models: Concepts and Cases

*PARZEN · Modern Probability Theory and Its Applications

PEÑA, TIAO, and TSAY · A Course in Time Series Analysis

PIANTADOSI · Clinical Trials: A Methodologic Perspective

PORT · Theoretical Probability for Applications

POURAHMADI · Foundations of Time Series Analysis and Prediction Theory

PRESS · Bayesian Statistics: Principles, Models, and Applications

PRESS · Subjective and Objective Bayesian Statistics, *Second Edition*

PRESS and TANUR · The Subjectivity of Scientists and the Bayesian Approach

PUKELSHEIM · Optimal Experimental Design

PURI, VILAPLANA, and WERTZ · New Perspectives in Theoretical and Applied Statistics

PUTERMAN · Markov Decision Processes: Discrete Stochastic Dynamic Programming

RAO · Linear Statistical Inference and Its Applications, *Second Edition*

RENCHER · Linear Models in Statistics

RENCHER · Methods of Multivariate Analysis, *Second Edition*

RENCHER · Multivariate Statistical Inference with Applications

RIPLEY · Spatial Statistics

RIPLEY · Stochastic Simulation

ROBINSON · Practical Strategies for Experimenting

ROHATGI and SALEH · An Introduction to Probability and Statistics, *Second Edition*

ROLSKI, SCHMIDLI, SCHMIDT, and TEUGELS · Stochastic Processes for Insurance and Finance

ROSENBERGER and LACHIN · Randomization in Clinical Trials: Theory and Practice

ROSS · Introduction to Probability and Statistics for Engineers and Scientists

ROUSSEEUW and LEROY · Robust Regression and Outlier Detection

RUBIN · Multiple Imputation for Nonresponse in Surveys

RUBINSTEIN · Simulation and the Monte Carlo Method

RUBINSTEIN and MELAMED · Modern Simulation and Modeling

RYAN · Modern Regression Methods

RYAN · Statistical Methods for Quality Improvement, *Second Edition*

SALTELLI, CHAN, and SCOTT (editors) · Sensitivity Analysis

SCHEFFE · The Analysis of Variance

*Now available in a lower priced paperback edition in the Wiley Classics Library.

SCHIMEK · Smoothing and Regression: Approaches, Computation, and Application
SCHOTT · Matrix Analysis for Statistics
SCHUSS · Theory and Applications of Stochastic Differential Equations
SCOTT · Multivariate Density Estimation: Theory, Practice, and Visualization
*SEARLE · Linear Models
SEARLE · Linear Models for Unbalanced Data
SEARLE · Matrix Algebra Useful for Statistics
SEARLE, CASELLA, and McCULLOCH · Variance Components
SEARLE and WILLETT · Matrix Algebra for Applied Economics
SEBER and LEE · Linear Regression Analysis, *Second Edition*
SEBER · Multivariate Observations
SEBER and WILD · Nonlinear Regression
SENNOTT · Stochastic Dynamic Programming and the Control of Queueing Systems
*SERFLING · Approximation Theorems of Mathematical Statistics
SHAFER and VOVK · Probability and Finance: It's Only a Game!
SMALL and McLEISH · Hilbert Space Methods in Probability and Statistical Inference
SRIVASTAVA · Methods of Multivariate Statistics
STAPLETON · Linear Statistical Models
STAUDTE and SHEATHER · Robust Estimation and Testing
STOYAN, KENDALL, and MECKE · Stochastic Geometry and Its Applications, *Second Edition*
STOYAN and STOYAN · Fractals, Random Shapes and Point Fields: Methods of Geometrical Statistics
STYAN · The Collected Papers of T. W. Anderson: 1943–1985
SUTTON, ABRAMS, JONES, SHELDON, and SONG · Methods for Meta-Analysis in Medical Research
TANAKA · Time Series Analysis: Nonstationary and Noninvertible Distribution Theory
THOMPSON · Empirical Model Building
THOMPSON · Sampling, *Second Edition*
THOMPSON · Simulation: A Modeler's Approach
THOMPSON and SEBER · Adaptive Sampling
THOMPSON, WILLIAMS, and FINDLAY · Models for Investors in Real World Markets
TIAO, BISGAARD, HILL, PEÑA, and STIGLER (editors) · Box on Quality and Discovery: with Design, Control, and Robustness
TIERNEY · LISP-STAT: An Object-Oriented Environment for Statistical Computing and Dynamic Graphics
TSAY · Analysis of Financial Time Series
UPTON and FINGLETON · Spatial Data Analysis by Example, Volume II: Categorical and Directional Data
VAN BELLE · Statistical Rules of Thumb
VIDAKOVIC · Statistical Modeling by Wavelets
WEISBERG · Applied Linear Regression, *Second Edition*
WELSH · Aspects of Statistical Inference
WESTFALL and YOUNG · Resampling-Based Multiple Testing: Examples and Methods for p-Value Adjustment
WHITTAKER · Graphical Models in Applied Multivariate Statistics
WINKER · Optimization Heuristics in Economics: Applications of Threshold Accepting
WONNACOTT and WONNACOTT · Econometrics, *Second Edition*
WOODING · Planning Pharmaceutical Clinical Trials: Basic Statistical Principles
WOOLSON and CLARKE · Statistical Methods for the Analysis of Biomedical Data, *Second Edition*
WU and HAMADA · Experiments: Planning, Analysis, and Parameter Design Optimization
YANG · The Construction Theory of Denumerable Markov Processes
*ZELLNER · An Introduction to Bayesian Inference in Econometrics
ZHOU, OBUCHOWSKI, and McCLISH · Statistical Methods in Diagnostic Medicine